DATE			

BAKER & TAYLOR

AGU Reference Shelf 2

Mineral Physics & Crystallography
A Handbook of Physical Constants

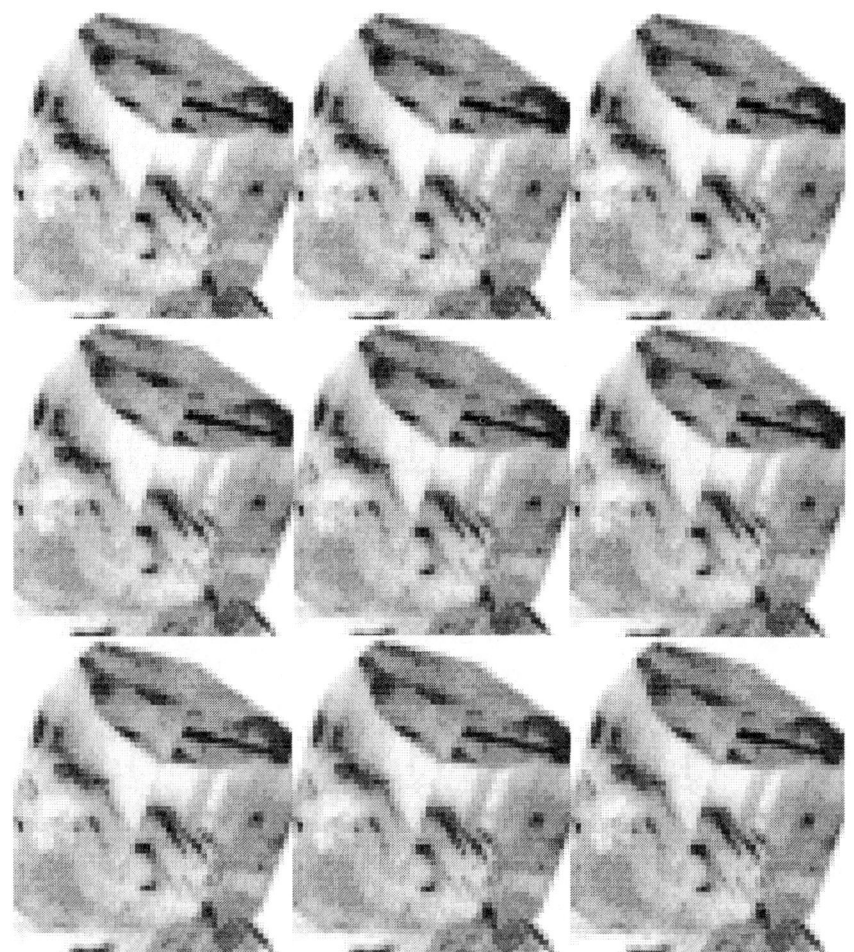

Thomas J. Ahrens, Editor

Published under the aegis of the AGU Books Board

Library of Congress Cataloging-in-Publication Data

Mineral physics and crystallography : a handbook of physical constants/
   Thomas J. Ahrens, editor.
      p. cm. — (AGU reference shelf ISSN 1080-305X; 2)
   Includes bibliographical references and index.
   ISBN 0-87590-852-7 (acid-free)
   1. Mineralogy—Handbooks, manuals, etc. 2. Crystallography—
—Handbooks, manuals, etc.   I. Ahrens, T. J. (Thomas J.), 1936–
II. Series.
  QE366.8.M55   1995
  549'.1—dc20
                                      95-3663
                                       CIP

ISBN 0-87590-852-7
ISSN 1080-305X

This book is printed on acid-free paper. ∞

Published by
American Geophysical Union

Printed in the United States of America.

CONTENTS

CONTENTS

PREFACE

The purpose of this Handbook is to provide, in highly accessible form, selected critical data for professional and student solid Earth and planetary geophysicists. Coverage of topics and authors were carefully chosen to fulfill these objectives.

These volumes represent the third version of the "Handbook of Physical Constants." Several generations of solid Earth scientists have found these handbooks to be the most frequently used item in their personal library. The first version of this Handbook was edited by F. Birch, J. F. Schairer, and H. Cecil Spicer and published in 1942 by the Geological Society of America (GSA) as Special Paper 36. The second edition, edited by Sydney P. Clark, Jr., was also published by GSA as Memoir 92 in 1966. Since 1966, our scientific knowledge of the Earth and planets has grown enormously, spurred by the discovery and verification of plate tectonics and the systematic exploration of the solar system.

The present revision was initiated, in part, by a 1989 chance remark by Alexandra Navrotsky asking what the Mineral Physics (now Mineral and Rock Physics) Committee of the American Geophysical Union could produce that would be a tangible useful product. At the time I responded, "update the Handbook of Physical Constants." As soon as these words were uttered, I realized that I could edit such a revised Handbook. I thank Raymond Jeanloz for his help with initial suggestions of topics, the AGU's Books Board, especially Ian McGregor, for encouragement and enthusiastic support. Ms. Susan Yamada, my assistant, deserves special thanks for her meticulous stewardship of these volumes. I thank the technical reviewers listed below whose efforts, in all cases, improved the manuscripts.

Thomas J. Ahrens, Editor
California Institute of Technology
Pasadena

Carl Agee	Thomas Heaton	William I. Rose, Jr.
Thomas J. Ahrens	Thomas Herring	George Rossman
Orson Anderson	Joel Ita	John Sass
Don Anderson	Andreas K. Kronenberg	Surendra K. Saxena
George H. Brimhall	Robert A. Langel	Ulrich Schmucker
John Brodholt	John Longhi	Ricardo Schwarz
J. Michael Brown	Guenter W. Lugmair	Doug E. Smylie
Bruce Buffett	Stephen Mackwell	Carol Stein
Robert Butler	Gerald M. Mavko	Maureen Steiner
Clement Chase	Walter D. Mooney	Lars Stixrude
Robert Creaser	Herbert Palme	Edward Stolper
Veronique Dehant	Dean Presnall	Stuart Ross Taylor
Alfred G. Duba	Richard H. Rapp	Jeannot Trampert
Larry Finger	Justin Revenaugh	Marius Vassiliou
Michael Gaffey	Rich Reynolds	Richard P. Von Herzen
Carey Gazis	Robert Reynolds	John M. Wahr
Michael Gurnis	Yanick Ricard	Yuk Yung
William W. Hay	Frank Richter	

Crystallographic Data For Minerals

Joseph R. Smyth and Tamsin C. McCormick

With the advent of modern X–ray diffraction instruments and the improving availability of neutron diffraction instrument time, there has been a substantial improvement in the number and quality of structural characterizations of minerals. Also, the past 25 years has seen great advances in high pressure mineral synthesis technology so that many new high pressure silicate and oxide phases of potential geophysical significance have been synthesized in crystals of sufficient size for complete structural characterization by X–ray methods. The object of this work is to compile and present a summary of these data on a selected group of the more abundant, rock–forming minerals in an internally consistent format for use in geophysical and geochemical studies.

Using mostly primary references on crystal structure determinations of these minerals, we have compiled basic crystallographic property information for some 300 minerals. These data are presented in Table 1. The minerals were selected to represent the most abundant minerals composing the crust of the Earth as well as high pressure synthetic phases that are believed to compose the bulk of the solid Earth. The data include mineral name, ideal formula, ideal formula weight, crystal system, space group, structure type, Z (number of formula units per cell), unit cell edges, a, b, and c in Ångstrom units (10^{-10} m) and inter–axial angles α, β, γ in degrees, unit cell volume in $Å^3$, molar volume in cm^3, calculated density in Mg/m^3, and a reference to the complete crystal structure data.

To facilitate geochemical and geophysical modeling, data for pure synthetic end members are presented when available. Otherwise, data are for near end–member natural samples. For many minerals, structure data (or samples) for pure end members are not available, and in these cases, indicated by an asterisk after the mineral name, data for an impure, natural sample are presented together with an approximate ideal formula and formula weight and density calculated from the ideal formula.

In order to conserve space we have omitted the precision given by the original workers in the unit cell parameter determination. However, we have quoted the data such that the stated precision is less than 5 in the last decimal place given. The cell volumes, molar volumes and densities are calculated by us given so that the precision in the last given place is less than 5. The formula weights presented are calculated by us and given to one part in approximately 20,000 for pure phases and one part in 1000 for impure natural samples.

J. R. Smyth, and T. C. McCormick, Department of Geological Sciences, University of Colorado, Boulder, CO 80309-0250

Mineral Physics and Crystallography
A Handbook of Physical Constants
AGU Reference Shelf 2

Table 1. Crystallographic Properties of Minerals.

Mineral	Formula	Crystal System	Space Group	Z	a (Å)	b (Å)	c (Å)	α (°)	β (°)	γ (°)	Unit Cell Vol (Å³)	Molar Vol (cm³)	Density (calc)(Mg/m³)	Ref.
Single Oxides														
Hemi-oxide														
Cuprite	Cu_2O	Cub.	$Pn\bar{3}m$	2	4.2696						77.833	23.439	6.104	25
Monoxides Group														
Periclase	MgO	Cub.	$Fm\bar{3}m$	4	4.211						74.67	11.244	3.585	93
Wustite	FeO	Cub.	$Fm\bar{3}m$	4	4.3108						80.11	12.062	5.956	67
Lime	CaO	Cub.	$Fm\bar{3}m$	4	4.1684						111.32	16.762	3.346	235
Bunsenite	NiO	Cub.	$Fm\bar{3}m$	4	4.446						72.43	10.906	6.850	235
Manganosite	MnO	Cub.	$Fm\bar{3}m$	4	4.8105						87.88	13.223	5.365	195
Tenorite	CuO	Mono.	$C2/c$	4	4.6837	3.4226	5.1288		99.54		81.080	12.209	6.515	11
Montroydite	HgO	Orth.	$Pnma$	4	6.612	5.20	3.531				128.51	19.350	11.193	12
Zincite	ZnO	Hex.	$P6_3mc$	2	3.2427		5.1948				47.306	14.246	5.712	189
Bromellite	BeO	Hex.	$P6_3mc$	2	2.6984		4.2770				26.970	8.122	3.080	189
Sesquioxide Group														
Corundum	Al_2O_3	Trig.	$R\bar{3}c$	6	4.7589		12.9912				254.80	25.577	3.986	157
Hematite	Fe_2O_3	Trig.	$R\bar{3}c$	6	5.038		13.772				302.72	30.388	5.255	23
Eskolaite	Cr_2O_3	Trig.	$R\bar{3}c$	6	4.9607		13.599				289.92	29.093	5.224	157
Karelianite	V_2O_3	Trig.	$R\bar{3}c$	6	4.952		14.002				297.36	29.850	5.021	157
Bixbyite	Mn_2O_3	Cub.	$Ia\bar{3}$	16	9.4146						834.46	31.412	5.027	75
Avicennite	Tl_2O_3	Cub.	$Ia\bar{3}$	16	10.543						1171.9	44.115	10.353	167
Claudetite	As_2O_3	Mono.	$P2_1/n$	4	7.99	4.65	9.12		78.3		331.8	49.961	3.960	176
Arsenolite	As_2O_3	Cub.	$Fd\bar{3}m$	16	11.0744						1358.19	51.127	3.870	177
Senarmontite	Sb_2O_3	Cub.	$Fd\bar{3}m$	16	11.1519						1386.9	52.208	5.583	217
Valentinite	Sb_2O_3	Orth.	$Pccn$	4	4.911	12.464	5.412				331.27	49.881	5.844	216
Dioxide Group														
Brookite	TiO_2	Orth.	$Pbca$	8	9.184	5.447	5.145				257.38	19.377	4.123	17
Anatase	TiO_2	Tetr.	$I4_1/amd$	4	3.7842		9.5146				136.25	20.156	3.895	105
Rutile	TiO_2	Tetr.	$P4_2/mnm$	2	4.5845		2.9533				62.07	18.693	4.2743	204
Cassiterite	SnO_2	Tetr.	$P4_2/mnm$	2	4.737		3.185				71.47	21.523	7.001	15
Stishovite	SiO_2	Tetr.	$P4_2/mnm$	2	4.1790		2.6651				46.54	14.017	4.287	20
Pyrolusite	MnO_2	Tetr.	$P4_2/mnm$	2	4.396		2.871				55.48	86.937	5.203	121
Baddeleyite	ZrO_2	Mono.	$P2_1/c$	4	5.1454	5.2075	5.3107		99.23		140.45	21.149	5.826	208
Uraninite	UO_2	Cub.	$Fm\bar{3}m$	4	5.4682						163.51	24.620	10.968	126
Thorianite	ThO_2	Cub.	$Fm\bar{3}m$	4	5.5997						175.59	26.439	9.987	227
Multiple Oxides														
Chrysoberyl	$BeAl_2O_4$	Orth.	$Pnmb$	4	4.424	9.396	5.471				227.42	34.244	3.708	96
Spinel Group														
Spinel	$MgAl_2O_4$	Cub.	$Fd\bar{3}m$	8	8.0832						528.14	39.762	3.578	61
Hercynite	$FeAl_2O_4$	Cub.	$Fd\bar{3}m$	8	8.1558						542.50	40.843	4.256	99
Magnesioferrite	$MgFe_2O_4$	Cub.	$Fd\bar{3}m$	8	8.360						584.28	43.989	4.547	100
Magnesiochromite	$MgCr_2O_4$	Cub.	$Fd\bar{3}m$	8	8.333						578.63	43.564	4.414	100
Magnetite	$FeFe_2O_4$	Cub.	$Fd\bar{3}m$	8	8.394						591.43	44.528	5.200	100
Jacobsite	$MnFe_2O_4$	Cub.	$Fd\bar{3}m$	8	8.5110						616.51	46.416	4.969	100
Chromite	$FeCr_2O_4$	Cub.	$Fd\bar{3}m$	8	8.3794						588.31	44.293	5.054	100
Ulvoespinel	$TiFe_2O_4$	Cub.	$Fd\bar{3}m$	8	8.536						621.96	46.826	4.775	106

Formula Weights: Cuprite 143.079; Periclase 40.312; Wustite 71.848; Lime 56.079; Bunsenite 74.709; Manganosite 70.937; Tenorite 79.539; Montroydite 216.589; Zincite 81.369; Bromellite 25.012; Corundum 101.961; Hematite 159.692; Eskolaite 151.990; Karelianite 149.882; Bixbyite 157.905; Avicennite 456.738; Claudetite 197.841; Arsenolite 197.841; Senarmontite 291.498; Valentinite 291.498; Brookite 79.890; Anatase 79.890; Rutile 79.890; Cassiterite 150.69; Stishovite 60.086; Pyrolusite 86.94; Baddeleyite 123.22; Uraninite 270.03; Thorianite 264.04; Chrysoberyl 126.97; Spinel 142.27; Hercynite 173.81; Magnesioferrite 200.00; Magnesiochromite 192.30; Magnetite 231.54; Jacobsite 230.63; Chromite 223.84; Ulvoespinel 223.59.

Table 1. Crystallographic Properties of Minerals (continued).

Mineral	Formula	Formula Weight	Crystal System	Space Group	Structure Type	Z	a (Å)	b (Å)	c (Å)	α (°)	β (°)	γ (°)	Unit Cell Vol (Å³)	Molar Vol (cm³)	Density (calc)(Mg/m³)	Ref.
Titanate Group																
Ilmenite	$FeTiO_3$	151.75	Trig.	$R\bar{3}$	Ilmenite	6	5.0884		14.0855				315.84	31.705	4.786	229
Pyrophanite	$MnTiO_3$	150.84	Trig.	$R\bar{3}$	Ilmenite	6	5.137		14.283				326.41	32.766	4.603	235
Perovskite	$CaTiO_3$	135.98	Orth.	$Pbnm$	Perovskite	4	5.3670	5.4439	7.6438				223.33	33.63	4.044	113
Armalcolite	$Mg_{.5}Fe_{.5}Ti_2O_5$	215.88	Orth.	$Bbmm$	Pseudobrookite	4	9.7762	10.0214	3.7485				367.25	55.298	3.904	230
Pseudobrookite	Fe_2TiO_5	239.59	Orth.	$Bbmm$	Pseudobrookite	4	9.767	9.947	3.717				361.12	54.375	4.406	3
Tungstates and Molybdates																
Ferberite	$FeWO_4$	303.70	Mono.	$P2/c$	Ferberite	2	4.730	5.703	4.952		90.0		133.58	40.228	7.549	225
Huebnerite	$MnWO_4$	302.79	Mono.	$P2/c$	Ferberite	2	4.8238	5.7504	4.9901		91.18		138.39	41.676	7.265	231
Scheelite	$CaWO_4$	287.93	Tetr.	$I4_1/a$	Scheelite	4	5.243		11.376				312.72	47.087	6.115	114
Powellite	$CaMoO_4$	200.02	Tetr.	$I4_1/a$	Scheelite	4	5.23		11.44				301.07	45.333	4.412	101
Stolzite	$PbWO_4$	455.04	Tetr.	$I4_1/a$	Scheelite	4	5.46		12.05				359.23	54.091	8.412	101
Wulfenite	$PbMoO_4$	367.12	Tetr.	$I4_1/a$	Scheelite	4	5.435		12.11				357.72	53.864	6.816	101
Hydroxides																
Gibbsite	$Al(OH)_3$	78.00	Mono.	$P2_1/n$	Gibbsite	8	8.684	5.078	9.736		94.54		427.98	32.222	2.421	188
Diaspore	$AlO(OH)$	59.99	Orth.	$Pbnm$	Goethite	4	4.401	9.421	2.845				117.96	17.862	3.377	34
Boehmite	$AlO(OH)$	59.99	Orth.	$Amam$	Boehmite	4	3.693	12.221	2.865				129.30	19.507	3.075	98
Brucite	$Mg(OH)_2$	58.33	Trig.	$P\bar{3}m1$	Brucite	1	3.124		4.766				40.75	24.524	2.377	243
Goethite	$FeO(OH)$	88.85	Orth.	$Pbnm$	Goethite	4	4.587	9.937	3.015				137.43	20.693	4.294	65
Lepidochrosite	$FeO(OH)$	88.85	Orth.	$Cmc2_1$	Boehmite	4	3.08	12.50	3.87				148.99	22.435	3.961	43
Carbonates																
Magnesite	$MgCO_3$	84.32	Trig.	$R\bar{3}c$	Calcite	6	4.6328		15.0129				279.05	28.012	3.010	54
Smithsonite	$ZnCO_3$	125.38	Trig.	$R\bar{3}c$	Calcite	6	4.6526		15.0257				281.68	28.276	4.434	54
Siderite	$FeCO_3$	115.86	Trig.	$R\bar{3}c$	Calcite	6	4.6916		15.3796				293.17	29.429	3.937	54
Rhodochrosite	$MnCO_3$	114.95	Trig.	$R\bar{3}c$	Calcite	6	4.7682		15.6354				307.86	30.904	3.720	54
Otavite	$CdCO_3$	172.41	Trig.	$R\bar{3}c$	Calcite	6	4.923		16.287				341.85	34.316	5.024	26
Calcite	$CaCO_3$	100.09	Trig.	$R\bar{3}c$	Calcite	6	4.9896		17.0610				367.85	36.9257	2.7106	54
Vaterite	$CaCO_3$	100.09	Hex.	$P6_3/mmc$	Vaterite	12	7.151		16.937				750.07	37.647	2.659	146
Dolomite	$CaMg(CO_3)_2$	184.41	Trig.	$R\bar{3}$	Dolomite	3	4.8069		16.0034				320.24	64.293	2.868	182
Ankerite	$CaFe(CO_3)_2$	215.95	Trig.	$R\bar{3}$	Dolomite	3	4.830		16.167				326.63	65.576	3.293	21
Aragonite	$CaCO_3$	100.09	Orth.	$Pmcn$	Aragonite	4	4.9614	7.9671	5.7404				226.91	34.166	2.930	51
Strontianite	$SrCO_3$	147.63	Orth.	$Pmcn$	Aragonite	4	5.090	8.358	5.997				255.13	38.416	3.843	51
Cerussite	$PbCO_3$	267.20	Orth.	$Pmcn$	Aragonite	4	5.180	8.492	6.134				269.83	40.629	6.577	191
Witherite	$BaCO_3$	197.39	Orth.	$Pmcn$	Aragonite	4	5.3126	8.8958	6.4284				303.81	45.745	4.314	51
Azurite	$Cu_3(OH)_2(CO_3)_2$	344.65	Mono.	$P2_1/c$	Azurite	2	5.0109	5.8485	10.345		92.43		302.90	91.219	3.778	245
Malachite	$Cu_2(OH)_2CO_3$	221.10	Mono.	$P2_1/a$	Malachite	4	9.502	11.974	3.240		98.75		364.35	54.862	4.030	244
Nitrates																
Soda Niter	$NaNO_3$	85.00	Trig.	$R\bar{3}c$	Calcite	6	5.0708		16.818				374.51	37.594	2.261	198
Niter	KNO_3	101.11	Orth.	$Pmcn$	Aragonite	4	5.4119	9.1567	6.5189				323.05	48.643	2.079	159
Borates																
Borax	$Na_2B_4O_5(OH)_4 \cdot 8H_2O$	381.37	Mono.	$C2/c$	Borax		11.885	10.654	12.206		106.62		1480.97	223.00	1.710	128

Table 1. Crystallographic Properties of Minerals (continued).

Mineral	Formula	Formula Weight	Crystal System	Space Group	Structure Type	Z	a (Å)	b (Å)	c (Å)	α (°)	β (°)	γ (°)	Unit Cell Vol (Å³)	Mclar Vol (cm³)	Density (calc)(Mg/m³)	Ref.
Kernite	$Na_2B_4O_6(OH)_2 \cdot 3H_2O$	273.28	Mono.	$P2_1/c$	Kernite	4	7.0172	9.1582	15.6774		108.86		953.41	143.560	1.904	48
Colemanite	$CaB_3O_4(OH)_3 \cdot H_2O$	205.55	Mono.	$P2_1/a$	Colemanite	4	8.743	11.264	6.102		110.12		564.30	84.869	2.419	42
Sulfates																
Barite	$BaSO_4$	233.40	Orth.	$Pbnm$	Barite	4	7.157	8.884	5.457				346.97	52.245	4.467	147
Celestite	$SrSO_4$	183.68	Orth.	$Pbnm$	Barite	4	6.870	8.371	5.355				307.96	46.371	3.961	147
Anglesite	$PbSO_4$	303.25	Orth.	$Pbnm$	Barite	4	6.959	8.482	5.398				318.62	47.977	6.321	147
Anhydrite	$CaSO_4$	136.14	Orth.	$Amma$	Anhydrite	4	7.006	6.998	6.245				306.18	46.103	2.953	118
Gypsum	$CaSO_4 \cdot 2H_2O$	172.17	Mono.	$I2/a$	Gypsum	4	5.670	15.201	6.533		118.60		494.37	74.440	2.313	46
Alunite*	$KAl_3(SO_4)_2(OH)_6$	414.21	Trig.	$R\bar{3}m$	Alunite	3	7.020		17.223				735.04	147.572	2.807	145
Jarosite*	$KFe_3(SO_4)_2(OH)_6$	500.81	Trig.	$R\bar{3}m$	Alunite	3	7.304		17.268				797.80	160.172	3.127	112
Antlerite	$Cu_3(SO_4)(OH)_4$	354.71	Orth.	$Pnma$	Antlerite	4	8.244	6.043	11.987				597.19	89.920	2.959	91
Thenardite	Na_2SO_4	142.04	Orth.	$Fddd$	Thenardite	8	9.829	12.302	5.868				709.54	53.419	2.659	90
Arcanite	K_2SO_4	174.27	Orth.	$Pmcn$	Arcanite	4	5.763	10.071	7.476				433.90	65.335	2.667	142
Epsomite	$MgSO_4 \cdot 7H_2O$	246.48	Orth.	$P2_12_12_1$	Epsomite	4	11.846	12.002	6.859				975.18	145.838	1.678	36
Phosphates																
Hydroxyapatite	$Ca_5(PO_4)_3OH$	502.32	Hex.	$P6_3/m$	Apatite	2	9.424		6.879				529.09	159.334	3.153	214
Fluorapatite	$Ca_5(PO_4)_3F$	504.31	Hex.	$P6_3/m$	Apatite	2	9.367		6.884				523.09	157.527	3.201	215
Chlorapatite	$Ca_5(PO_4)_3Cl$	520.77	Hex.	$P6_3/m$	Apatite	2	9.628		6.764				543.01	163.527	3.185	137
Monazite	$CePO_4$	235.09	Mono.	$P2_1/n$	Monazite	4	6.77	7.04	6.46		104.0		298.7	42.98	5.23	76
Xenotime	YPO_4	183.88	Tetr.	$I4_1/amd$	Zircon	4	6.878		6.036				285.54	43.00	4.277	123
Whitlockite	$MgFeCa_{18}H_2(PO_4)_{14}$	2133.	Trig.	$R3c$	Whitlockite	3	10.330		37.103				3428.8	688.386	3.099	38
Triphylite	$LiFePO_4$	157.76	Orth.	$Pmnb$	Olivine	4	10.334	6.010	4.693				291.47	43.888	3.595	237
Lithiophyllite	$LiMnPO_4$	156.85	Orth.	$Pmnb$	Olivine	4	6.05	10.32	4.71				294.07	44.280	3.542	101
Amblygonite*	$LiAl(F,OH)PO_4$	146.9	Tric.	$P\bar{1}$	Amblygonite	2	5.18	7.15	5.04	112.11	97.78	67.88	160.20	48.242	3.045	16
Augelite*	$Al_2(OH)_3PO_4$	199.9	Mono.	$C2/m$	Augelite	4	13.124	7.988	5.066		112.42		490.95	73.924	2.705	101
Berlinite	$AlPO_4$	121.95	Trig.	$P3_121$	Quartz	3	4.943		10.974				232.21	46.620	2.616	206
Orthosilicates																
Garnet Group																
Pyrope	$Mg_3Al_2Si_3O_{12}$	403.15	Cub.	$Ia\bar{3}d$	Garnet	8	11.452						1501.9	113.08	3.565	8
Almandine	$Fe_3Al_2Si_3O_{12}$	497.76	Cub.	$Ia\bar{3}d$	Garnet	8	11.531						1533.2	115.43	4.312	8
Spessartine	$Mn_3Al_2Si_3O_{12}$	495.03	Cub.	$Ia\bar{3}d$	Garnet	8	11.612						1565.7	117.88	4.199	161
Grossular	$Ca_3Al_2Si_3O_{12}$	403.15	Cub.	$Ia\bar{3}d$	Garnet	8	11.845						1661.9	125.12	3.600	161
Andradite	$Ca_3Fe_2Si_3O_{12}$	508.19	Cub.	$Ia\bar{3}d$	Garnet	8	12.058						1753.2	131.99	3.850	161
Uvarovite	$Ca_3Cr_2Si_3O_{12}$	500.48	Cub.	$Ia\bar{3}d$	Garnet	8	11.988						1722.8	129.71	3.859	161
Olivine Group																
Forsterite	Mg_2SiO_4	140.70	Orth.	$Pbnm$	Olivine	4	4.7534	10.1902	5.9783				289.58	43.603	3.227	69
Fayalite	Fe_2SiO_4	203.77	Orth.	$Pbnm$	Olivine	4	4.8195	10.4788	6.0873				307.42	46.290	4.402	69
Tephroite	Mn_2SiO_4	201.96	Orth.	$Pbnm$	Olivine	4	4.9023	10.5964	6.2567				325.02	48.939	4.127	69
Liebenbergite	Ni_2SiO_4	209.50	Orth.	$Pbnm$	Olivine	4	4.726	10.118	5.913				282.75	42.574	4.921	124
Ca-olivine	Ca_2SiO_4	172.24	Orth.	$Pbnm$	Olivine	4	5.078	11.225	6.760				385.32	58.020	2.969	50
Co-olivine	Co_2SiO_4	209.95	Orth.	$Pbnm$	Olivine	4	4.7811	10.2998	6.0004				295.49	44.493	4.719	32
Monticellite	$CaMgSiO_4$	156.48	Orth.	$Pbnm$	Olivine	4	4.822	11.108	6.382				341.84	51.472	3.040	165
Kirschsteinite	$CaFeSiO_4$	188.01	Orth.	$Pbnm$	Olivine	4	4.844	10.577	6.146				314.89	47.415	3.965	32

Table 1. Crystallographic Properties of Minerals (continued).

Mineral	Formula	Crystal System	Space Group	Formula Weight	Z	a (Å)	b (Å)	c (Å)	α (°)	β (°)	γ (°)	Unit Cell Vol (Å³)	Molar Vol (cm³)	Density (calc)(Mg/m³)	Ref.
Zircon Group															
Zircon	$ZrSiO_4$	Tetr.	$I4_1/amd$	183.30	4	6.6042		5.9796				260.80	39.270	4.668	95
Hafnon	$HfSiO_4$	Tetr.	$I4_1/amd$	270.57	4	6.5725		5.9632				257.60	38.787	6.976	212
Thorite*	$ThSiO_4$	Tetr.	$I4_1/amd$	324.1	4	7.1328		6.3188				321.48	48.407	6.696	222
Coffinite*	$USiO_4$	Tetr.	$I4_1/amd$	330.2	4	6.995		6.236				305.13	45.945	7.185	115
Willemite Group															
Phenacite*	Be_2SiO_4	Trig.	$R\bar{3}$	110.10	18	12.472		8.252				1111.6	37.197	2.960	241
Willemite	Zn_2SiO_4	Trig.	$R\bar{3}$	222.82	18	13.971		9.334				1577.8	52.795	4.221	207
Eucryptite*	$LiAlSiO_4$	Trig.	$R\bar{3}$	126.00	18	13.473		9.001				1415.0	47.347	2.661	97
Aluminosilicate Group															
Andalusite	Al_2SiO_5	Orth.	$Pnnm$	162.05	4	7.7980	7.9031	5.5566				342.44	51.564	3.1426	233
Sillimanite	Al_2SiO_5	Orth.	$Pbnm$	162.05	4	7.4883	7.6808	5.7774				332.29	50.035	3.2386	233
Kyanite	Al_2SiO_5	Tric.	$P\bar{1}$	162.05	4	7.1262	7.8520	5.5747	89.99	101.11	106.03	293.72	44.227	3.6640	233
Topaz	$Al_2SiO_4(OH,F)_2$	Orth.	$Pbnm$	182.0	4	4.6651	8.8381	8.3984				346.27	52.140	3.492	242
Humite Group															
Norbergite*	$Mg_3SiO_4F_2$	Orth.	$Pbnm$	203.0	4	4.7104	10.2718	8.7476				423.25	63.73	3.186	73
Chondrodite	$Mg_5(SiO_4)_2F_2$	Mono.	$P2_1/b$	343.7	2	4.7284	10.2539	7.8404		109.06		359.30	108.20	3.158	74
Humite*	$Mg_7(SiO_4)_3F_2$	Orth.	$Pbnm$	484.4	4	4.7408	10.2580	20.8526				1014.09	152.70	3.159	183
Clinohumite*	$Mg_9(SiO_4)_4F_2$	Mono.	$P2_1/b$	624.1	2	4.7441	10.2501	13.6635		100.786		652.68	196.55	3.259	186
Staurolite*	$Fe_4Al_{18}Si_8O_{46}(OH)_2$	Mono.	$C2/m$	1704.	1	7.8713	16.6204	5.6560		90.0		739.94	445.67	3.823	209
Other Orthosilicates															
Titanite	$CaTiSiO_5$	Mono.	$P2_1/a$	196.06	4	7.069	8.722	6.566				370.23	55.748	3.517	213
Datolite	$CaBSiO_4(OH)$	Mono.	$P2_1/c$	159.94	4	4.832	7.608	9.636		90.40		354.23	53.338	2.999	63
Gadolinite*	$RE_2FeB_2Si_2O_{10}$	Mono.	$P2_1/a$	604.5	2	10.000	7.565	4.786		90.31		360.69	108.62	5.565	148
Chloritoid*	$FeAl_2SiO_5(OH)_2$	Tric.	$P\bar{1}$	251.9	2	9.46	5.50	9.15	97.05	101.56	90.10	462.72	69.674	3.616	88
Sapphirine*	$Mg_{3.5}Al_9Si_{1.5}O_{20}$	Mono.	$P2_1/a$	690.0	4	11.266	14.401	9.929		125.46		1312.11	197.57	3.493	149
Prehnite*	$Ca_2Al(Al,Si_3)O_{10}(OH)_2$	Orth.	$Pacm$	412.391	2	4.646	5.483	18.486				470.91	141.82	2.908	170
Pumpelleyite	$Ca_8(Mg_2FeAl)Al_8Si_{12}O_{42}(OH)_{14}$	Mono.	$C2/m$	1915.1	1	8.831	5.894	19.10		97.53		985.6	593.6	3.226	172
Axinite	$HFeCa_2Al_2BSi_4O_{16}$	Tric.	$P\bar{1}$	570.12	2	7.157	9.199	8.959	91.8	98.14	77.30	569.61	171.54	3.324	220
Sorosilicates & Cyclosilicates															
Epidote Group															
Zoisite	$Ca_2Al_3Si_3O_{12}(OH)$	Orth.	$Pnma$	454.36	4	16.212	5.559	10.036				904.47	136.19	3.336	52
Clinozoisite	$Ca_2Al_3Si_3O_{12}(OH)$	Mono.	$P2_1/m$	454.36	2	8.879	5.583	10.155		115.50		454.36	136.83	3.321	52
Hancockite*	$Ca(Pb,Sr)FeAl_2Si_3O_{12}(OH)$	Mono.	$P2_1/m$	590.6	2	8.96	5.67	10.30		114.4		476.5	143.5	4.12	53
Allanite*	$CaRE(Al,Fe)_3Si_3O_{12}(OH)$	Mono.	$P2_1/m$	565.2	2	8.927	5.761	10.150		114.77		473.97	142.74	3.96	53
Epidote*	$Ca_2FeAl_2Si_3O_{12}(OH)$	Mono.	$P2_1/m$	454.4	2	8.8877	5.6275	10.1517		115.383		458.73	138.15	3.465	70
Melilite Group															
Melilite*	$CaNaAlSi_2O_7$	Tetr.	$P\bar{4}2_1m$	258.2	2	7.6344		5.0513				294.41	88.662	2.912	134
Gehlenite*	$Ca_2AlAlSiO_7$	Tetr.	$P\bar{4}2_1m$	274.2	2	7.7173		5.0860				302.91	91.220	3.006	135
Akermanite*	$Ca_2MgSi_2O_7$	Tetr.	$P\bar{4}2_1m$	272.64	2	7.835		5.010				307.55	92.619	2.944	116
Other Sorosilicates and Cyclosilicaes															
Lawsonite	$CaAl_2Si_2O_7(OH)_2H_2O$	Orth.	$Ccmm$	314.24	4	8.795	5.847	13.142				675.82	101.76	3.088	19
Beryl	$Be_3Al_2Si_6O_{18}$	Hex.	$P6/mmc$	537.51	2	9.2086		9.1900				674.89	203.24	2.645	152
Cordierite*	$Mg_2Al_4Si_5O_{18}$	Orth.	$Ccmm$	584.97	4	17.079	9.730	9.356				1554.77	234.11	2.499	45

Table 1. Crystallographic Properties of Minerals (continued).

Mineral	Formula	Formula Weight	Crystal System	Space Group	Structure Type	Z	a (Å)	b (Å)	c (Å)	α (°)	β (°)	γ (°)	Unit Cell Vol (Å³)	Molar Vol (cm³)	Density (calc)(Mg/m³)	Ref.
Tourmaline*	$NaFe_3Al_6B_3Si_6O_{27}(OH)_4$	1043.3	Trig.	$R3m$	Tourmaline	3	15.992		7.190				1592.5	319.7	3.263	66
Vesuvianite*	$Ca_{19}Fe_2MgAl_{10}Si_{18}O_{70}(OH,F)_8$	2935.	Tetr.	$P4/nnc$	Vesuvianite	2	15.533		11.778				2841.8	427.9	3.429	6

Chain Silicates

Enstatite/Ferrosilite Group

Mineral	Formula	Formula Weight	Crystal System	Space Group	Structure Type	Z	a (Å)	b (Å)	c (Å)	α (°)	β (°)	γ (°)	Unit Cell Vol (Å³)	Molar Vol (cm³)	Density (calc)(Mg/m³)	Ref.
Enstatite	$Mg_2Si_2O_6$	200.79	Orth.	$Pbca$	Orthopyroxene	8	18.227	8.819	5.179				832.49	62.676	3.204	197
Ferrosilite	$Fe_2Si_2O_6$	263.86	Orth.	$Pbca$	Orthopyroxene	8	18.427	9.076	5.237				875.85	55.941	4.002	197
Clinoenstatite	$Mg_2Si_2O_6$	200.79	Mono.	$P2_1/c$	Clinoenstatite	4	9.626	8.825	5.188		108.33		418.36	52.994	3.188	150
Clinoferrosilite	$Fe_2Si_2O_6$	263.86	Mono.	$P2_1/c$	Clinoenstatite	4	9.7085	9.0872	5.2284		108.43		437.60	65.892	4.005	33

Clinopyroxene Group

Mineral	Formula	Formula Weight	Crystal System	Space Group	Structure Type	Z	a (Å)	b (Å)	c (Å)	α (°)	β (°)	γ (°)	Unit Cell Vol (Å³)	Molar Vol (cm³)	Density (calc)(Mg/m³)	Ref.
Diopside	$CaMgSi_2O_6$	216.56	Mono.	$C2/c$	Clinopyroxene	4	9.746	8.899	5.251		105.63		438.58	66.039	3.279	39
Hedenbergite	$CaFeSi_2O_6$	248.10	Mono.	$C2/c$	Clinopyroxene	4	9.845	9.024	5.245		104.70		450.72	67.867	3.656	39
Jadeite	$NaAlSi_2O_6$	202.14	Mono.	$C2/c$	Clinopyroxene	4	9.423	8.564	5.223		107.56		401.85	60.508	3.341	39
Acmite	$NaFeSi_2O_6$	231.08	Mono.	$C2/c$	Clinopyroxene	4	9.658	8.795	5.294		107.42		429.06	64.606	3.576	44
Cosmochlor	$NaCrSi_2O_6$	227.15	Mono.	$C2/c$	Clinopyroxene	4	9.579	8.722	5.267		107.37		419.98	63.239	3.592	39
Spodumene	$LiAlSi_2O_6$	186.09	Mono.	$C2/c$	Clinopyroxene	4	9.461	8.395	5.218		110.09		389.15	58.596	3.176	39
Ca-Tschermaks	$CaAlAlSiO_6$	218.20	Mono.	$C2/c$	Clinopyroxene	4	9.609	8.652	5.274		106.06		421.35	63.445	3.438	164

Pyroxenoid Group

Mineral	Formula	Formula Weight	Crystal System	Space Group	Structure Type	Z	a (Å)	b (Å)	c (Å)	α (°)	β (°)	γ (°)	Unit Cell Vol (Å³)	Molar Vol (cm³)	Density (calc)(Mg/m³)	Ref.
Wollastonite	$Ca_3Si_3O_9$	348.49	Tric.	$C\bar{1}$	Wollastonite	4	10.104	11.054	7.305	99.53	100.56	83.44	788.04	118.66	2.937	163
Bustamite*	$(Ca_{2.4}Fe_{.6})Si_3O_9$	358.6	Tric.	$I\bar{1}$	Bustamite	4	9.994	10.946	7.231	99.30	100.56	83.29	764.30	115.09	3.116	163
Rhodonite	$Mn_5Si_5O_{15}$	655.11	Tric.	$P\bar{1}$	Rhodonite	2	7.616	11.851	6.707	92.55	94.35	105.67	579.84	174.62	3.752	155
Pyroxmangite	$Mn_7Si_7O_{21}$	917.16	Tric.	$P\bar{1}$	Pyroxmangite	2	6.721	7.603	17.455	113.18	82.27	94.13	812.31	244.63	3.749	155
Aenigmatite*	$Na_2Fe_5TiSi_6O_{20}$	867.5	Tric.	$P\bar{1}$	Aenigmatite	2	10.406	10.813	8.926	104.93	96.87	125.32	744.52	224.21	3.869	40
Pectolite*	$HNaCa_2Si_3O_9$	332.4	Tric.	$P\bar{1}$	Pectolite	2	7.980	7.023	7.018	90.54	95.14	102.55	382.20	115.10	2.888	163
Petalite	$LiAlSi_4O_{10}$	306.26	Mono.	$P2/a$	Petalite	2	11.737	5.171	7.630		112.54		427.71	128.80	2.378	219

Amphibole Group

Mineral	Formula	Formula Weight	Crystal System	Space Group	Structure Type	Z	a (Å)	b (Å)	c (Å)	α (°)	β (°)	γ (°)	Unit Cell Vol (Å³)	Molar Vol (cm³)	Density (calc)(Mg/m³)	Ref.
Gedrite*	$Na_{.5}(Mg_5Fe_2)Al_2Si_6O_{22}(OH)_2$	853.23	Orth.	$Pnma$	Orthoamphibole	4	18.531	17.741	5.249				1725.65	259.8	3.184	169
Anthophyllite*	$(Mg_5Fe_2)Si_8O_{22}(OH)_2$	843.94	Orth.	$Pnma$	Orthoamphibole	4	18.560	18.013	5.2818				1765.8	265.9	3.111	58
Cummingtonite*	$(Mg_5Fe_2)Si_8O_{22}(OH)_2$	843.94	Mono.	$C2/m$	Amphibole	2	9.51	18.19	5.33		101.92		902.14	271.7	3.14	60
Tremolite*	$Na_{.5}Ca_2Mg_5Si_8O_{22}(OH)_2$	823.90	Mono.	$C2/m$	Amphibole	2	9.863	18.048	5.285		104.79		909.60	273.9	3.01	92
Pargasite*	$NaCa_2FeMg_4Al_2Si_6O_{22}(OH)_2$	864.72	Mono.	$C2/m$	Amphibole	2	9.910	18.022	5.312		105.78		912.96	274.9	3.165	185
Glaucophane*	$Na_2(FeMg_3Al)Si_8O_{22}(OH)_2$	789.44	Mono.	$C2/m$	Amphibole	2	9.541	17.740	5.295		103.67		870.8	262.2	3.135	168

Sheet Silicates

Talc and Pyrophyllite

Mineral	Formula	Formula Weight	Crystal System	Space Group	Structure Type	Z	a (Å)	b (Å)	c (Å)	α (°)	β (°)	γ (°)	Unit Cell Vol (Å³)	Molar Vol (cm³)	Density (calc)(Mg/m³)	Ref.
Talc	$Mg_3Si_4O_{10}(OH)_2$	379.65	Tric.	$C\bar{1}$	Talc	2	5.290	9.173	9.460	90.46	98.68	90.09	453.77	136.654	2.776	175
Pyrophyllite	$Al_2Si_4O_{10}(OH)_2$	360.31	Tric.	$C\bar{1}$	Talc	2	5.160	8.966	9.347	91.18	100.46	89.64	425.16	128.036	2.814	125

Trioctahedral Mica Group

Mineral	Formula	Formula Weight	Crystal System	Space Group	Structure Type	Z	a (Å)	b (Å)	c (Å)	α (°)	β (°)	γ (°)	Unit Cell Vol (Å³)	Molar Vol (cm³)	Density (calc)(Mg/m³)	Ref.
Annite*	$KFe_3(AlSi_3)O_{10}(OH)_2$	511.9	Mono.	$C2/m$	1M	2	5.386	9.324	10.268		100.63		506.82	152.63	3.215	94
Phlogopite*	$KMg_3AlSi_3O_{10}(OH)_2$	417.3	Mono.	$C2/m$	1M	2	5.308	9.190	10.155		100.08		487.69	146.87	2.872	94
Lepidolite*	$KAlLi_2AlSi_3O_{10}(OH)_2$	385.2	Mono.	$C2/c$	2M$_1$	2	5.209	9.053	20.185		99.125		939.82	141.52	2.724	192
Lepidolite*	$KAlLi_2AlSi_3O_{10}(OH)_2$	385.2	Mono.	$C2/c$	2M$_2$	4	9.04	5.22	20.21		99.58		940.38	141.60	2.791	193
Lepidolite*	$KAlLi_2AlSi_3O_{10}(OH)_2$	385.2	Mono.	$C2/m$	1M	2	5.20	9.01	10.09		99.28		466.6	140.5	2.825	194
Zinnwaldite*	$K(AlFeLi)AlSi_3O_{10}(OH)_2$	434.1	Mono.	$C2/m$	1M	2	5.296	9.140	10.096		100.83		480.0	144.55	2.986	82

Table 1. Crystallographic Properties of Minerals (continued).

Mineral	Formula	Formula Weight	Crystal System	Space Group	Structure Type	Z	a (Å)	b (Å)	c (Å)	α (°)	β (°)	γ (°)	Unit Cell Vol (Å³)	Molar Vol (cm³)	Density (calc)(Mg/m³)	Ref.
Dioctahedral Mica Group																
Muscovite*	$KAl_2AlSi_3O_{10}(OH)_2$	398.3	Mono.	$C2/c$	$2M_1$	4	5.1918	9.0153	20.0457		95.74		933.56	140.57	2.834	187
Paragonite*	$NaAl_2AlSi_3O_{10}(OH)_2$	384.3	Mono.	$C2/c$	$2M_1$	4	5.128	8.898	19.287		94.35		877.51	132.13	2.909	129
Margarite*	$CaAl_2AlSi_3O_{10}(OH)_2$	399.3	Mono.	$C2/c$	$2M_2$	4	5.1038	8.8287	19.148		95.46		858.89	129.33	3.061	83
Bityite*	$Ca(LiAl)_2(AlBeSi_2)O_{10}(OH)_2$	387.2	Mono.	$C2/c$	$2M_1$	4	5.058	8.763	19.111		95.39		843.32	126.98	3.049	130
Chlorite Group																
Chlorite*	$(Mg_5Al)(AlSi_3)O_{10}(OH)_2$	555.8	Mono.	$C2/m$	Chlorite–IIb2	2	5.327	9.227	14.327		96.81		699.24	210.57	2.640	109
Chlorite*	$(Mg_5Al)(AlSi_3)O_{10}(OH)_2$	555.8	Tric.	$C\bar{1}$	Chlorite–IIb4	2	5.325	9.234	14.358	90.33	97.38	90.00	700.14	210.85	2.636	108
Clay Group																
Nacrite	$Al_2Si_2O_5(OH)_4$	258.16	Mono.	Cc	Nacrite	4	8.909	5.156	15.697		113.70		658.95	99.221	2.602	24
Dickite	$Al_2Si_2O_5(OH)_4$	258.16	Mono.	Cc	Dickite	4	5.178	8.937	14.738		103.82		662.27	99.721	2.588	22
Kaolinite	$Al_2Si_2O_5(OH)_4$	258.16	Tric.	$P1$	Kaolinite	2	5.1554	8.9448	7.4048	91.700	104.862	89.822	329.89	99.347	2.599	22
Amesite*	$(Mg_2Al)(AlSi)O_5(OH)_4$	278.7	Tric.	$C1$	Amesite	4	5.319	9.208	14.060	90.01	90.27	89.96	688.61	103.69	2.778	86
Lizardite*	$Mg_3Si_2O_5(OH)_4$	277.1	Trig.	$P31m$	Lizardite 1T	1	5.332		7.233				178.09	107.26	2.625	144
Tektosilicates																
Silica Group																
Quartz	SiO_2	60.085	Trig.	$P3_221$	Quartz	3	4.1934		5.4052				113.01	22.688	2.648	127
Coesite	SiO_2	60.085	Mono.	$C2/c$	Coesite	16	7.1464	12.3796	7.1829		120.283		548.76	20.657	2.909	210
Tridymite	SiO_2	60.085	Mono.	Cc	Tridymite	48	18.494	4.991	25.832		117.75		2110.2	26.478	2.269	111
Cristobalite	SiO_2	60.085	Tetr.	$P4_12_12$	Cristobalite	4	4.978		6.948				172.17	25.925	2.318	173
Stishovite	SiO_2	60.085	Tetr.	$P4_2/mnm$	Rutile	2	4.1790		2.6651				46.54	14.017	4.287	20
Feldspar Group																
Sanidine	$KAlSi_3O_8$	278.33	Mono.	$C2/m$	Sanidine	4	8.595	13.028	7.179		115.94		722.48	108.788	2.558	199
Orthoclase	$KAlSi_3O_8$	278.33	Mono.	$C2/m$	Sanidine	4	8.561	12.996	7.192		116.01		719.13	108.283	2.571	47
Microcline	$KAlSi_3O_8$	278.33	Tric.	$C\bar{1}$	Sanidine	4	8.560	12.964	7.215	90.65	115.83	87.70	720.07	108.425	2.567	31
High Albite	$NaAlSi_3O_8$	262.23	Tric.	$C\bar{1}$	Albite	4	8.161	12.875	7.110	93.53	116.46	90.24	667.12	100.452	2.610	234
Low Albite	$NaAlSi_3O_8$	262.23	Tric.	$C\bar{1}$	Albite	4	8.142	12.785	7.159	94.19	116.61	87.68	664.48	100.054	2.621	89
Anorthite	$CaAl_2Si_2O_8$	278.36	Tric.	$P\bar{1}$	Anorthite	8	8.173	12.869	14.165	93.11	115.91	91.261	1336.35	100.610	2.765	228
Celsian	$BaAl_2Si_2O_8$	375.47	Mono.	$I2/c$	Anorthite	8	8.627	13.045	14.408		115.22		1466.90	110.440	3.400	158
Feldspathoid Group																
Leucite	$KAlSi_2O_6$	218.25	Tetr.	$I4_1/a$	Leucite	16	13.09		13.75				2356.	88.69	2.461	139
Kalsilite	$KAlSiO_4$	158.17	Hex.	$P6_3$	Nepheline	2	5.16		8.69				200.4	60.34	2.621	178
Nepheline	$KNa_3Al_4Si_4O_{16}$	584.33	Hex.	$P6_3$	Nepheline	2	9.993		8.374				724.19	218.09	2.679	64
Meionite*	$Ca_4Al_6Si_6O_{24}CO_3$	932.9	Tetr.	$P4_2/n$	Scapolite	2	12.194		7.557				1123.7	338.40	2.757	131
Marialite*	$Na_4Al_6Si_6O_{24}Cl$	863.5	Tetr.	$P4_2/n$	Scapolite	2	12.059		7.587				1103.3	332.26	2.599	132
Zeolite Group																
Analcime*	$Na_{16}Al_{16}Si_{32}O_{96}\cdot16H_2O$	3526.1	Tetr.	$I4_1/acd$	Analcime	1	13.721		13.735				2585.8	1557.4	2.264	138
Chabazite*	$Ca_2Al_4Si_8O_{24}\cdot13H_2O$	1030.9	Trig.	$R\bar{3}m$	Chabazite	1	13.803		15.075				2487.2	499.4	2.065	37
Mordenite*	$K_8Al_8Si_{40}O_{96}\cdot24H_2O$	3620.4	Orth.	$Cmcm$	Mordenite	1	18.167	20.611	7.529				2819.2	1698.0	2.132	153
Clinoptilolite*	$KNa_2CaAl_6Si_{30}O_{72}\cdot24H_2O$	2750.0	Mono.	$C2/m$	Heulandite	1	17.633	17.941	7.400		116.39		2097.1	1263.0	2.177	211
Heulandite*	$Ca_4K_{1.2}Al_{10}Si_{26}O_{72}\cdot26H_2O$	2827.7	Mono.	$C2/m$	Heulandite	1	17.715	17.831	7.430		115.93		2132.2	1284.3	2.221	4
Thomsonite*	$NaCa_2Al_5Si_5O_{20}\cdot6H_2O$	671.8	Orth	$Pncn$	Thomsonite	4	13.089	13.047	13.218				2257.3	339.9	2.373	5

Table 1. Crystallographic Properties of Minerals (continued).

Mineral	Formula	Formula Weight	Crystal System Space Group	Structure Type	Z	a (Å)	b (Å)	c (Å)	α (°)	β (°)	γ (°)	Unit Cell Vol (Å³)	Molar Vol (cm³)	Density (calc)(Mg/m³)	Ref.
Harmotome*	$Ba_2Ca_{.5}Al_5Si_{11}O_{32}\cdot12H_2O$	1466.7	Mono. $P2_1/m$	Phillipsite	1	9.879	14.139	8.693		124.8		996.9	600.5	2.443	184
Phillipsite*	$K_{2.5}Ca_{1.5}Al_5Si_{10}O_{32}\cdot12H_2O$	1291.5	Mono. $P2_1/m$	Phillipsite	1	9.865	14.300	8.668		124.2		1011.3	609.1	2.120	184
Laumontite*	$CaAl_2Si_4O_{12}\cdot4H_2O$	470.44	Mono Am	Laumontite	4	7.549	14.740	13.072	90.	90.	111.9	1349.6	203.2	2.315	202
Natrolite*	$Na_2Al_2Si_3O_{10}\cdot2H_2O$	380.23	Orth. $Fdd2$	Natrolite	8	18.326	18.652	6.601				2256.3	169.87	2.238	174
Sodalite*	$Na_4Al_3Si_3O_{12}Cl$	484.6	Cub. $P\bar{4}3n$	Sodalite	2	8.870						697.86	210.16	2.306	133
Stilbite*	$Na_{.3}Ca_{4.2}Al_9Si_{26}O_{72}\cdot34H_2O$	2968.	Mono. $C2/m$	Stilbite	1	13.64	18.24	11.27		128.0		2210.	1331.	2.23	71
Scolecite*	$CaAl_2Si_3O_{10}\cdot3H_2O$	392.34	Mono Fd	Natrolite	8	18.508	18.981	6.527		90.64		2292.8	172.62	2.273	107
Gonnardite*	$Na_6Ca_2Al_6Si_{11}O_{40}\cdot12H_2O$	1626.04	Tetr. $I\bar{4}2d$	Natrolite	1	13.21		6.622				1155.6	696.00	2.336	141
Edingtonite*	$Ba_2Al_4Si_6O_{20}\cdot8H_2O$	997.22	Tetr. $P\bar{4}2_1m$	Edingtonite	1	9.581		6.526				599.06	350.81	2.764	140
Gismondine*	$Ca_4Al_8Si_8O_{32}\cdot16H_2O$	1401.09	Mono. $P2_1/a$	Gismondine	1	10.024	10.626	9.832		92.40		1024.3	630.21	2.223	226
Garronite*	$NaCa_{2.5}Al_6Si_{10}O_{32}\cdot13H_2O$	1312.12	Tetr. $I\bar{4}m2$	Gismondine	1	9.9266		10.3031				1015.24	611.48	2.146	9
Merlinoite*	$K_5Ca_2Al_9Si_{23}O_{64}\cdot24H_2O$	2620.81	Orth. $Immm$	Merlinoite	1	14.116	14.229	9.946				1982.28	1193.92	2.195	72
Ferrierite*	$Na_2KMgAl_5Si_{31}O_{72}\cdot18H_2O$	2614.2	Mono. $P2_1/a$	Ferrierite	1	18.886	14.182	7.470		90.0		2000.8	1205.1	2.169	79
Ferrierite*	$NaKMgAl_3Si_{29}O_{72}\cdot18H_2O$	2590.3	Orth. $Immm$	Ferrierite	1	19.236	14.162	7.527		90.0		2050.5	1235.0	2.097	80
Faujasite*	$Na_2CaAl_4Si_8O_{24}\cdot16H_2O$	1090.9	Cub. $Fd\bar{3}m$	Sodalite	16	24.74						15142.	570.02	1.914	18
Erionite*	$MgNaK_2Ca_2Al_9Si_{27}O_{72}\cdot18H_2O$	2683.1	Hex. $P6_3/mmc$	Erionite	1	13.252		14.810				2252.4	1356.6	1.978	201
Cancrinite*	$Ca_{1.5}Na_6Al_6Si_6O_{24}\cdot1.6CO_2$	1008.5	Hex. $P6_3$	Cancrinite	1	12.590		5.117				702.4	423.05	2.383	81
Pollucite*	$CsAlSi_2O_6$	312.06	Cub. $Ia3d$	Analcime	16	13.682						2561.2	96.41	3.237	156
Brewsterite*	$SrAl_2Si_6O_{16}\cdot5H_2O$	656.17	Mono. $P2_1/m$	Brewsterite	2	6.767	17.455	7.729		94.40		910.2	274.12	2.394	10

High Pressure Silicates

Phase B Group

Mineral	Formula	Formula Weight	Crystal System Space Group	Structure Type	Z	a (Å)	b (Å)	c (Å)	α (°)	β (°)	γ (°)	Unit Cell Vol (Å³)	Molar Vol (cm³)	Density (calc)(Mg/m³)	Ref.
Phase B	$Mg_{12}Si_4O_{19}(OH)_2$	741.09	Mono. $P2_1/c$	PhsB	4	10.588	14.097	10.073		104.10		1458.4	219.567	3.380	59
Anhydrous B	$Mg_{14}Si_5O_{24}$	864.78	Orth. $Pmcb$	AnhB	2	5.868	14.178	10.048				835.96	252.749	3.435	59
Superhydrous B	$Mg_{10}Si_3O_{14}(OH)_4$	619.40	Orth. $Pnnm$	PhsB	2	5.0894	13.968	8.6956				618.16	186.159	3.327	166

MgSiO₃–Group

MgSiO₃–perovskite	$MgSiO_3$	100.40	Orth. $Pbnm$	Perovskite	4	4.7754	4.9292	6.8969				162.35	24.445	4.107	103
MgSiO₃–ilmenite	$MgSiO_3$	100.40	Trig. $R\bar{3}$	Ilmenite	6	4.7284		13.5591				262.54	26.354	3.810	102
MgSiO₃–garnet	$MgSiO_3$	100.40	Tetr. $I4_1/a$	Garnet	32	11.501		11.480				1518.5	28.581	3.513	7

Wadsleyite Group

| Wadsleyite | Mg_2SiO_4 | 140.71 | Orth. $Imma$ | Wadsleyite | 8 | 5.6983 | 11.4380 | 8.2566 | | | | 538.14 | 40.515 | 3.4729 | 104 |
| β–Co₂SiO₄ | Co_2SiO_4 | 209.95 | Orth. $Imma$ | Wadsleyite | 8 | 5.753 | 11.524 | 8.340 | | | | 552.92 | 41.628 | 5.044 | 151 |

Silicate Spinel Group

γ–Mg₂SiO₄	Mg_2SiO_4	140.71	Cub. $Fd\bar{3}m$	Spinel	8	8.0449						524.56	39.493	3.563	196
γ–Fe₂SiO₄	Fe_2SiO_4	203.78	Cub. $Fd\bar{3}m$	Spinel	8	8.234						558.26	42.030	4.848	236
γ–Ni₂SiO₄	Ni_2SiO_4	209.95	Cub. $Fd\bar{3}m$	Spinel	8	8.138						538.96	40.577	5.174	236
γ–Co₂SiO₄	Co_2SiO_4	209.50	Cub. $Fd\bar{3}m$	Spinel	8	8.044						520.49	39.187	5.346	151

Silica Group

| Coesite | SiO_2 | 60.085 | Mono. $C2/c$ | Coesite | 16 | 7.1464 | 12.3796 | 7.1829 | | 120.283 | | 548.76 | 20.657 | 2.909 | 210 |
| Stishovite | SiO_2 | 60.085 | Tetr. $P4_2/mnm$ | Rutile | 2 | 4.1790 | | 2.6651 | | | | 46.54 | 14.017 | 4.287 | 20 |

Halides

Halite	$NaCl$	58.443	Cub. $Fm\bar{3}m$	Halite	4	5.638						179.22	26.385	2.166	235
Sylvite	KCl	74.555	Cub. $Fm\bar{3}m$	Halite	4	6.291						248.98	37.490	1.989	235
Villiaumite	NaF	41.988	Cub. $Fm\bar{3}m$	Halite	4	4.614						98.23	14.791	2.839	235
Carobbiite	KF	58.100	Cub. $Fm\bar{3}m$	Halite	4	5.34						152.3	22.93	2.53	235

Table 1. Crystallographic Properties of Minerals (continued).

Mineral	Formula	Formula Weight	Crystal System	Space Group	Structure Type	Z	a (Å)	b (Å)	c (Å)	α (°)	β (°)	γ (°)	Unit Cell Vol (Å³)	Molar Vol (cm³)	Density (calc)(Mg/m³)	Ref.
Fluorite	CaF_2	78.077	Cub.	$Fm\bar{3}m$	Fluorite	4	5.460						162.77	24.509	3.186	232
Frankdicksonite	BaF_2	175.34	Cub.	$Fm\bar{3}m$	Fluorite	4	6.1964						237.91	35.824	4.894	180
Sellaite	MgF_2	62.309	Tetr.	$P4_2/mnm$	Rutile	2	4.660		3.078				66.84	20.129	3.096	101
Calomel	Hg_2Cl_2	472.09	Tetr.	$I4/mmm$	Calomel	2	4.45		10.89				215.65	64.94	7.269	101
Cryolite	Na_3AlF_6	209.95	Mono.	$P2_1/n$	Cryolite	2	5.40	5.60	7.78		90.18					101
Neighborite	$NaMgF_3$	104.30	Orth.	$Pcmn$	Perovskite	4	5.363	7.676	5.503				226.54	34.11	3.058	101
Chlorargyrite	$AgCl$	143.32	Cub.	$Fm\bar{3}m$	Halite	4	5.556						171.51	25.83	5.550	101
Iodyrite	AgI	234.77	Hex.	$P6_3mc$	Wurtzite	2	4.58		7.49				136.06	40.98	5.730	101
Nantokite	$CuCl$	98.99	Cub.	$F\bar{4}3m$		4	5.418						159.04	23.95	4.134	101
Sulfides																
Pyrrhotite	Fe_7S_8	647.44	Trig.	$P3_1$	Pyrrhotite	3	6.8673		17.062				696.84	139.90	4.628	62
Pyrite	FeS_2	119.98	Cub.	$Pa3$	Pyrite	4	5.418						159.04	23.95	5.010	29
Cattierite	CoS_2	123.06	Cub.	$Pa\bar{3}$	Pyrite	4	5.5385						169.89	25.582	4.811	162
Vaesite	NiS_2	122.84	Cub.	$Pa\bar{3}$	Pyrite	4	5.6865						183.88	27.688	4.437	162
Marcasite	FeS_2	119.98	Orth.	$Pnnm$	Marcasite	2	4.436	5.414	3.381				81.20	24.45	4.906	30
Troilite	FeS	89.911	Hex.	$P\bar{6}2c$	Troilite	12	5.963		11.754				361.95	18.167	4.839	117
Smythite	$(Fe,Ni)_9S_{11}$	855.3	Trig	$R\bar{3}m$	Smythite	1	3.4651		34.34				357.08	215.07	3.977	221
Chalcopyrite	$CuFeS_2$	183.51	Tetr.	$I\bar{4}2d$	Chalcopyrite	4	5.289		10.423				291.57	43.903	4.180	84
Cubanite	$CuFe_2S_3$	271.43	Orth.	$Pcmn$	Cubanite	4	6.467	11.117	6.231				447.97	67.453	4.024	218
Covellite	CuS	95.60	Hex.	$P6_3/mmc$	Covellite	6	3.7938		16.341				203.68	20.447	4.676	56
Chalcocite	Cu_2S	159.14	Mono.	$P2_1/c$	Chalcocite	48	15.246	11.884	13.494		116.35		2190.9	27.491	5.789	57
Tetrahedrite	$Cu_{12}Fe\,ZnSb_4S_{13}$	1660.5	Cub.	$I\bar{4}3m$	Tetrahedrite	2	10.364						1113.2	335.25	4.953	179
Bornite	Cu_5FeS_4	501.80	Orth.	$Pbca$	Bornite	16	10.950	21.862	10.950				2521.3	98.676	5.085	122
Enargite	Cu_3AsS_4	393.80	Orth.	$Pmn2_1$	Enargite	2	7.407	6.436	6.154				296.63	89.329	4.408	2
Niccolite	$NiAs$	133.63	Hex.	$P6_3/mmc$	NiAs	2	3.619		5.035				57.11	17.199	7.770	240
Cobaltite	$CoAsS$	165.92	Orth.	$Pca2_1$	Cobaltite	4	5.582	5.582	5.582				173.93	26.189	6.335	77
Sphalerite	ZnS	97.434	Cub.	$F\bar{4}3m$	Sphalerite	4	5.4053						157.93	23.780	4.097	239
Wurtzite(2H)	ZnS	97.434	Hex	$P6_3mc$	Wurtzite	2	3.8227		6.2607				79.23	23.860	4.084	119
Greenockite	CdS	144.464	Hex	$P6_3mc$	Wurtzite	2	4.1348		6.7490				99.93	30.093	4.801	235
Pentlandite	$Ni_5Fe_4S_8$	773.5	Cub.	$Fm\bar{3}m$	Halite	4	10.044						1013.26	152.571	5.069	87
Alabandite	MnS	87.02	Cub.	$Fm\bar{3}m$	Halite	4	5.214						141.75	21.344	4.076	224
Galena	PbS	239.25	Cub.	$Fm\bar{3}m$	Halite	4	5.9315						208.69	31.423	7.614	160
Clausthalite	$PbSe$	286.15	Cub.	$Fm\bar{3}m$	Halite	4	6.1213						229.37	34.537	8.285	160
Altaite	$PbTe$	334.79	Cub.	$Fm\bar{3}m$	Halite	4	6.4541						268.85	40.482	8.270	160
Molybdenite(2H)	MoS_2	160.07	Hex	$P6_3/mmc$	Molybdenite	2	3.1602		12.294				106.33	32.021	4.999	28
Tungstenite	WS_2	247.92	Hex	$P6_3/mmc$	Molybdenite-2H2	2	3.1532		12.323				105.77	31.853	7.785	203
Acanthite	Ag_2S	247.80	Mono.	$P2_1/c$	Acanthite	4	4.231	6.930	9.526		125.48		227.45	34.248	7.236	190
Argentite	Ag_2S	247.80	Cub.	$Im3m$	Argentite	2	4.86						114.79	34.569	7.168	41
Proustite	Ag_3AsS_3	494.72	Trig.	$R3c$	Proustite	6	10.82		8.69				881.06	88.44	5.594	55
Pyrargyrite	Ag_3SbS_3	541.55	Trig.	$R3c$	Proustite	6	11.04		8.72				920.42	92.39	5.861	55
Cinnabar	HgS	232.65	Trig.	$P3_221$	Cinnabar	3	4.145		9.496				141.29	28.361	8.202	14
Metacinnabar	HgS	232.65	Cub.	$F\bar{4}3m$	Sphalerite	4	5.8717						202.44	30.482	7.633	13
Coloradoite	$HgTe$	328.19	Cub.	$F\bar{4}3m$	Sphalerite	4	6.440						267.09	40.217	8.161	223
Stibnite	Sb_2S_3	339.69	Orth.	$Pnma$	Stibnite	4	11.302	3.8341	11.222				486.28	73.222	4.639	143
Orpiment	As_2S_3	246.04	Mono.	$P2_1/n$	Orpiment	4	11.475	9.577	4.256		90.68		467.68	70.422	3.494	154

Table 1. Crystallographic Properties of Minerals (continued).

Mineral	Formula	Crystal System	Space Group	Structure Type	Formula Weight	Z	a (Å)	b (Å)	c (Å)	α (°)	β (°)	γ (°)	Unit Cell Vol (Å³)	Molar Vol (cm³)	Density (calc)(Mg/m³)	Ref.
Realgar	AsS	Mono.	$P2_1/n$	Realgar	106.99	16	9.325	13.571	6.587		106.38		799.75	30.107	3.554	154
Bismuthinite	Bi_2S_3	Orth.	$Pmcn$	Stibnite	514.15	4	3.981	11.147	11.305				501.67	75.539	6.806	110
Hazelwoodite	Ni_3S_2	Trig.	$R32$	Hazelwoodite	240.26	1	4.0718			89.459	89.459	89.459	67.50	40.655	5.910	171
Cooperite	PtS	Tetr.	$P4_2/mmc$	Cooperite	227.15	2	3.465		6.104				73.29	22.070	10.292	35
Vysotskite	PdS	Tetr.	$P4_2/m$	Cooperite	138.46	8	6.429		6.611				273.25	20.572	6.731	27
Millerite	NiS	Trig.	$R3m$	Millerite	90.77	9	9.6190		3.1499				252.4	16.891	5.374	181
Linneaite	Co_3S_4	Cub.	$Fd\bar{3}m$	Spinel	305.06	8	9.406						832.2	62.652	4.869	120
Polydymite	Ni_3S_4	Cub.	$Fd\bar{3}m$	Spinel	304.39	8	9.489						854.4	64.326	4.732	49
Violarite	$FeNi_2S_4$	Cub.	$Fd\bar{3}m$	Spinel	301.52	8	9.465						847.93	63.839	4.723	49
Greigite	Fe_3S_4	Cub.	$Fd\bar{3}m$	Spinel	295.80	8	9.875						962.97	72.499	4.080	238
Daubreelite	$FeCr_2S_4$	Cub.	$Fd\bar{3}m$	Spinel	288.10	8	9.995						998.50	75.175	3.832	205
Loellingite	$FeAs_2$	Orth.	$Pnnm$	Loellingite	205.69	2	5.3001	5.9838	2.8821				91.41	27.527	7.472	136
Arsenopyrite	FeAsS	Mono.	$C2_1/d$	Arsenopyrite	162.83	8	6.546	9.451	5.649			89.94	349.48	26.312	6.189	68
Native Elements																
Diamond	C	Cub.	$Fd\bar{3}m$	Diamond	12.011	8	3.56679						45.38	3.4163	3.5158	235
Graphite	C	Hex.	$P6_3/mmc$	Graphite	12.011	4	2.456		6.696				34.98	5.267	2.281	235
Silicon	Si	Cub.	$Fd\bar{3}m$	Diamond	28.086	8	5.43070						160.16	12.058	2.329	235
Sulfur(α)	S	Orth.	$Fddd$	Sulfur	32.064	128	10.467	12.870	24.493				3299.5	15.443	2.076	235
Sulfur(β)	S	Mono.	$P2_1$	Sulfur	32.064	48	10.926	10.885	10.790		95.92		1276.41	16.016	2.002	78
Kamacite	Fe	Cub.	$Im\bar{3}m$	α–Iron	55.847	2	2.8665						23.55	7.093	7.873	235
Taenite	FeNi	Cub.	$Fm\bar{3}m$	Taenite	114.557	32	7.168						368.29	13.864	8.263	235
Nickel	Ni	Cub.	$Fm\bar{3}m$	FCC	58.710	4	3.52387						43.76	6.590	8.910	235
Copper	Cu	Cub.	$Fm\bar{3}m$	FCC	63.540	4	3.61496						47.24	7.113	8.932	235
Arsenic	As	Trig.	$R\bar{3}m$	Arsenic	74.922	18	3.7598		10.5475				129.12	4.321	17.340	200
Tin	Sn	Tetr.	$I4_1/amd$	Tin	118.690	4	5.8197		3.17488				107.54	16.194	7.329	235
Ruthenium	Ru	Hex.	$P6_3/mmc$	HCP	101.070	2	2.7056		4.2803				27.14	8.172	12.368	85
Rhodium	Rh	Cub.	$Fm\bar{3}m$	FCC	102.905	4	3.8031						55.01	8.283	12.424	235
Palladium	Pd	Cub.	$Fm\bar{3}m$	FCC	106.40	4	3.8898						60.16	9.059	11.746	235
Silver	Ag	Cub.	$Fm\bar{3}m$	FCC	107.87	4	4.0862						68.23	10.273	10.500	235
Antimony	Sb	Trig.	$R\bar{3}m$	Arsenic	121.75	6	4.3083		11.2743				180.06	13.075	6.736	235
Tellurium	Te	Trig.	$P3_121$	Selenium	127.60	3	4.456		5.921				101.82	20.441	6.242	1
Iridium	Ir	Cub.	$Fm\bar{3}m$	FCC	192.20	4	3.8394						56.60	8.522	22.553	235
Osmium	Os	Hex.	$P6_3/mmc$	HCP	190.20	2	2.7352		4.3190				27.98	8.427	22.570	235
Platinum	Pt	Cub.	$Fm\bar{3}m$	FCC	195.09	4	3.9231						60.38	9.092	21.458	235
Gold	Au	Cub.	$Fm\bar{3}m$	FCC	196.967	4	4.07825						67.83	10.214	19.285	235
Lead	Pb	Cub.	$Fm\bar{3}m$	FCC	207.190	4	4.9505						121.32	18.268	11.342	235
Bismuth	Bi	Trig.	$R\bar{3}m$	Arsenic	208.980	6	4.54590		11.86225				212.29	21.311	9.806	235

Acknowledgements. The authors thank Stephen J. Guggenheim (University of Illinois) and two anonymous reviewers for constructive criticism of the manuscript. This work was supported by National Science Foundation Grant EAR 91-05391 and U.S. Dept. of Energy Office of Basic Energy Sciences.

REFERENCES

1. Adenis, C., V. Langer, and O. Lindqvist, Reinvestigation of the structure of tellurium, *Acta Cryst., C45*, 941–942, 1989.

2. Adiwidjaja, G. and J. Lohn, Strukturverfeinerung von enargite, Cu_3AsS_4, *Acta Cryst, B26*, 1878–1879, 1970.

3. Akimoto, S., T. Nagata, and T. Katsura, The $TiFe_2O_5 - Ti_2FeO_5$ solid solution series, *Nature, 179*, 37–38, 1957.

4. Alberti, A., and G. Vezzalini, Thermal behavior of heulandites: a structural study of the dehydration of Nadap heulandite, *Tschermaks Mineral. Petrol. Mitteilungen, 31*, 259–270, 1983.

5. Alberti, A., G. Vezzalini, and V. Tazzoli, Thomsonite: a detailed refine–ment with cross checking by crystal energy calculations, *Zeolites, 1*, 91–97, 1981.

6. Allen, F. Chemical and structural variations in vesuvianite. (PhD Thesis) Harvard University, 440p., 1985.

7. Angel, R.J., L.W. Finger, R.M. Hazen, M. Kanzaki, D.J. Weidner, R.C. Liebermann, and D.R. Veblen, Structure and twinning of single–crystal $MgSiO_3$ garnet synthesized at 17GPa and 1800°C, *Am. Mineral., 74*, 509–512, 1989.

8. Armbruster, T., C.A. Geiger, and G.A. Lager, Single–crystal X–ray structure study of synthetic pyrope almandine garnets at 100 and 293K, *Am. Mineral., 77*, 512–521, 1992.

9. Artioli, G., The crystal structure of garronite, *Am. Mineral., 77*, 189–196, 1992.

10. Artioli, G., J.V. Smith, and A. Kvick, Multiple hydrogen positions in the zeolite brewsterite, $(Sr_{0.95}Ba_{0.05})Al_2Si_6O_{16} \cdot 5H_2O$, *Acta Cryst., C41*, 492–497, 1985.

11. Asbrink, S. and L.-J. Norrby, A refinement of the structure of copper (II) oxide with a discussion of some exceptional e.s.d.'s, *Acta Cryst., B26*, 8–15, 1970.

12. Aurivilius, K., The crystal structure of mercury (II) oxide, *Acta Cryst., 9*, 685–686, 1956.

13. Aurivillius, K., An X–ray and neutron diffraction study of metacinnabarite, *Acta Chem., Scand., 18*, 1552–1553, 1964.

14. Auvray, P. and F. Genet, Affinement de la structure cristalline du cinabre α–HgS, *Bull. Soc. Fr. Mineral. Crist., 96*, 218–219, 1973.

15. Baur W. H., Ueber die Verfeinerung der Kristallstrukturbestimmung einiger Vertreter des Rutiltyps: TiO_2 SnO_2, GeO_2 und MnF_2, *Acta Cryst., 9*, 515–520 1956.

16. Baur, W.H. Die Kristallstruktur des Edelamblygonits $LiAlPO_4(OH,F)$, *Acta Cryst., 12*, 988–994, 1959.

17. Baur, W.H., Atomabstaende und Bildungswinkel im Brookite, TiO_2, *Acta Cryst., 14*, 214–216, 1961.

18. Baur, W.H., On the cation and water positions in faujasite, *Am. Mineral., 49*, 697–704, 1964.

19. Baur, W. H., Crystal structure refinement of lawsonite, *Am. Mineral., 63*, 311–315, 1978.

20. Baur, W.H., and A.A. Kahn, Rutile-type compounds IV. SiO_2, GeO_2, and a comparison with other rutile–type compounds, *Acta Cryst. B27*, 2133–2139, 1971.

21. Beran, A. and J. Zemann, Refinement and comparison of the crystal structures of a dolomite and of an Fe–rich ankerite, *Tschermaks Mineral. Petrol. Mitt., 24*, 279–286, 1977.

22. Bish, D.L. and R.B. Von Dreele, Reitveld refinement of non–hydrogen positions in kaolinite. *Clays and Clay Miner., 37*, 289–296, 1989.

23. Blake, R. L., R. E. Hessevick, T. Zoltai, and L. W. Finger, Refinement of the hematite structure, *Am. Mineral., 51*, 123–129, 1966.

24. Blount, A. M., I. M. Threadgold, and S. W. Bailey, Refinement of the crystal structure of nacrite, *Clays and Clay Miner., 17*, 185–194, 1969.

25. Borie, B., Thermally excited forbidden reflections, *Acta Cryst., A30*, 337–341, 1974.

26. Borodin, V. L., V. I. Lyutin, V. V. Ilyukhin, and N. V. Belov, Isomorphous calcite–otavite series, *Dokl. Akad. Nauk SSSR, 245*, 1099–1101, 1979.

27. Brese, N. E., P. J. Squattrito, and J. A. Ibers, Reinvestigation of the structure of PdS, *Acta Cryst., C41*, 1829–1830, 1985.

28. Bronsema, K. D., J. L. de Boer, and F. Jellinek, On the structure of molybdenum diselenide and disulphide, *Z. Anorg. Allg. Chem.*, 540/541, 15–17, 1986.

29. Brostigen, G. and A. Kjekshus, Redetermined crystal structure of FeS_2 (pyrite), *Acta Chem., Scand., 23*, 2186–2188, 1969.

30. Brostigen, G., A. Kjekshus, and C. Romming, Compounds with the marcasite type crystal structure. VIII. Redetermination of the prototype, *Acta Chem. Scand., 27*, 2791–2796 1973.

31. Brown, B. E., and S. W. Bailey, The structure of maximum microcline, *Acta Cryst., 17*, 1391–1400, 1964.

32. Brown G. E., The crystal chemistry of the olivines (PhD Thesis), Virginia Polytechnic Institute and State University Blacksburg, VA 121 p. 1970.

33. Burnham, C. W., Ferrosilite, *Carnegie Inst. Washington, Yb, 65*, 285–290, 1967.

34. Busing, W. R., H. A. Levy, A single crystal neutron diffraction study of diaspore, AlO(OH), *Acta Cryst., 11*, 798–803, 1958.

35. Cabris, L. J., J. H. G. Leflamme, and J. M. Stewart, On cooperite, braggite, and vysotskite, *Am. Mineral., 63*, 832–839, 1978.

36. Calleri, M., A. Gavetti, G. Ivaldi, and M. Rubbo, Synthetic epsomite, $MgSO_4 \cdot 7H_2O$: Absolute configuration and surface features of the complementary {111} forms, *Acta Cryst., B40*, 218–222, 1984.

37. Calligaris, M., G. Nardin, and L.

Randaccio, Cation–site location in antaural chabazite, *Acta Cryst., B38*, 602–605, 1982.

38. Calvo, C., and R. Gopal, The crystal structure of whitlockite from the Palermo quarry, *Am. Mineral. 60*, 120–133, 1975.

39. Cameron, M., S. Sueno, C. T. Prewitt, and J. J. Papike, High–temperature crystal chemistry of acmite, diopside, hedenbergite, jadeite, spodumene, and ureyite, *Am. Mineral., 58*, 594–618, 1973.

40. Cannillo, E., F. Mazzi, J.H. Fang, P.D. Robinson, and Y. Ohya, The crystal structure of aenigmatite, *Am. Mineral., 56*, 427–446. 1971.

41. Cava, R. J., F. Reidinger, and B. J. Weunch, Single crystal neutron diffraction study of the fast–ion conductor β–Ag_2S between 186 and 325°C, *J. Solid St. Chem., 31*, 69–80, 1980.

42. Christ, C.L., J.R. Clark, and H.T. Evans,Jr., Studies of borate minerals (III): The crystal structure of colemanite, $CaB_3O_4(OH)_3 \cdot H_2O$, *Acta Cryst., 11*, 761–770, 1969.

43. Christensen, H., and A.N. Christensen, The crystal structure of lepidochrosite (γ–FeOOH), *Acta Chem. Scand., A32*, 87–88.

44. Clark, J. R., D. E. Appleman, and J. J. Papike, Crystal chemical characterization of clinopyroxenes based on eight new structure refinements, *Mineral. Soc. Am. Spec. Pap., 2*, 31–50, 1969.

45. Cohen, J. P., F. K. Ross, and G. V. Gibbs, An X–ray and neutron diffraction study of hydrous low cordierite, *Am. Mineral., 62*, 67–78, 1977.

46. Cole, W.F., and C.J. Lancucki, A refinement of the crystal structure of gyspum, $CaSO_4 \cdot 2H_2O$, *Acta Cryst., B30*, 921–929. 1974.

47. Colville, A. A., and P. H. Ribbe, The crystal structure of an alularia and a refinement of the crystal structure of orthoclase, *Am. Mineral., 53*, 25–37, 1968.

48. Cooper, W.F., F.K. Larsen, P. Coppens and R.F. Giese, Electron population analysis os accurate diffraction data. V. Structure and one–center charge refinement of the light–atom mineral kernite. $Na_2B_4O_6(OH)_2 \cdot 3H_2O$, *Am.*

Mineral., 58*, 21–31, 1973.

49. Craig, J. R., Violarite stability relations, *Am. Mineral., 56*, 1303–1311, 1971.

50. Czaya R., Refinement of the structure of gamma–Ca_2SiO_4, *Acta Cryst., B27*, 848–849, 1971.

51. De Villiers, J. P. R., Crystal structures of aragonite, strontianite, and witherite, *Am. Mineral., 56*, 758–767, 1971.

52. Dollase W.A., Refinement and comparison of the structures zoisite and clinozoisite, *Am. Mineral., 53*, 1882–1898, 1968.

53. Dollase, W. A., Refinement of the crystal structures of epidote, allanite, and hancockite, *Am. Mineral., 56*, 447–464, 1971.

54. Effenberger, H., K. Mereiter, and J. Zemann, Crystal structure refinements of magnesite, calcite, rhodochrosite, siderite, smithsonite, and dolomite, with discussion of some aspects of the stereochemistry of calcite type carbonates, *Z. Krist., 156*, 233–243, 1981.

55. Engel, P. and W. Nowacki, Die verfeinerung der kristallstruktur von proustit, Ag_3AsS_3, und pyrargyrit, Ag_3SbS_3, *Neues Jb. Miner. Mh., 6*, 181–184, 1966.

56. Evans, H. T., Jr. and J. A. Konnert, Crystal structure refinement of covellite, *Am. Mineral., 61*, 996–1000, 1976.

57. Evans, H. T., Jr., The crystal structures of low chalcocite and djurleite, *Z. Krist., 150*, 299–320, 1979.

58. Finger, L.W., Refinement of the crystal structure of an anthophyllite, *Carnegie Inst. Washington, Yb, 68*, 283–288, 1970.

59. Finger, L.W., R.M. Hazen, and C.T. Prewitt, Crystal structures of $Mg_{12}Si_4O_{19}(OH)_2$ (phase B) and $Mg_{14}Si_5O_{24}$ (phase AnhB), *Am. Mineral., 76*, 1–7, 1991.

60. Fischer, K., A further refinement of the crystal structure of cummingtonite, $(Fe,Mg)_7(Si_4O_{11})_2(OH)_2$, *Am. Mineral., 51*, 814–818, 1966.

61. Fischer P., Neutronenbeugungsuntersuchung der Strukturen von $MgAl_2O_4$– und $ZnAl_2O_4$– spinellen in Abhaengigkeit von der Vorgeschichte, *Z. Krist., 124*, 275–302, 1967.

62. Fleet, M. E., The crystal structure of a

pyrrhotite (Fe_7S_8), *Acta Cryst., B27*, 1864–1867, 1971.

63. Foit, F.F., Jr, M.W. Phillips, and G.V. Gibbs, A refinement of the crystal structure of datolite, *Am. Mineral., 58*, 909–914, 1973.

64. Foreman, N., and D. R. Peacor, Refinement of the nepheline structure at several temperatures, *Z. Krist., 132*, 45–70, 1970.

65. Forsyth, J. B., I. G. Hedley, and C. E. Johnson, The magnetic structure and hyperfine field of goethite (α–FeOOH), *J. of Phys., C1*, 179–188, 1968.

66. Fortier, S., and G. Donnay, Schorl refinement showing composition dependence of the tourmaline structure, *Canad. Mineral., 13*, 173–177, 1975.

67. Foster, P. K., and A. J. E. Welch, Metal oxide solutions: I. Lattice constants and phase relations in ferrous oxide (wustite) and in solid solutions of ferrous oxide and manganous oxide, *Trans. Faraday Soc., 52*, 1626–1634, 1956.

68. Fuess, H., T. Kratz, J. Topel–Schadt, and G. Mieher, Crystal structure refinement and electron microscopy of arsenopyrite, *Z. Krist., 179*, 335–346, 1987.

69. Fugino, K., S. Sasaki, Y Takeuchi, and R. Sadanaga, X–ray determination of electron distributions in forsterite, fayalite, and tephroite, *Acta Cryst., B37*, 513–518, 1981.

70. Gabe, E. J., J. C. Portheine, and S. H. Whitlow, A reinvestigation of the epidote structure: confirmation of the iron location, *Am. Mineral., 58*, 218–223, 1973.

71. Galli, E., Refinement of the crystal structure of stilbite, *Acta Cryst., B27*, 833–841, 1971.

72. Galli, E., G. Gottardi, and D. Pongiluppi, The crystal structure of the zeolite merlinoite, *Neues J. Mineral. Monat.*, 1–9, 1979.

73. Gibbs, G. V, and P. H. Ribbe, The crystal structures of the humite minerals: I. Norbergite. *Am. Mineral., 54*, 376–390, 1969.

74. Gibbs, G. V., P. H. Ribbe, and C. W. Anderson, The crystal structures of the humite minerals II. Chondrodite, *Am.*

Mineral., 55, 1182–1194, 1970.

75. Geller, S., Structures of alpha–Mn_2O_3, $(Mn_{0.983}Fe_{.017})_2O_3$, and $(Mn_{.37}Fe_{.63})_2$ O_3 and relation to magnetic ordering, Acta Cryst., B27, 821–828, 1971.

76. Ghouse K. M., Refinement of the crystal structure of heat–treated monazite crystal, Indian J. Pure Appl. Phys., 6, 265–268, 1968.

77. Giese, R. F., Jr., and P. F. Kerr, The crystal structures of ordered and disordered cobaltite, Am. Mineral., 50, 1002–1014, 1965.

78. Goldsmith, L. M. and C. E. Strouse, Molecular dynamics in the solid state. The order–disorder transition of monoclinic sulfur, J. Am. Chem. Soc., 99, 7580–7589. 1977.

79. Gramlich–Meier, R., V. Gramlich and W.M. Meier, The crystal structure of the monoclinic variety of ferrierite, Am. Mineral., 70, 619–623, 1985.

80. Gramlich–Meier, R., W.M. Meier, and B.K. Smith, On faults in the framework structure of the zeolite ferrierite. Z. Kristal., 168, 233–254.

81. Grundy, H.D. and I. Hassan, The crystal structure of a carbonate–rich cancrinite, Canad. Mineral., 20, 239– 251, 1982.

82. Guggenheim, S., and S. W. Bailey, The refinement of zinnwaldite–1M in subgroup symmetry, Am. Mineral., 62, 1158–1167, 1977.

83. Guggenheim, S., and S. W. Bailey, Refinement of the margarite structure in subgroup symmetry: correction, further refinement, and comments, Am. Mineral., 63, 186–187, 1978.

84. Hall, A. R. , Crystal structures of the chalcopyrite series, Canad. Mineral., 13, 168–172, 1975.

85. Hall, E. O. and J. Crangle, An X–ray investigation of the reported high–temperature allotropy of ruthenium, Acta Cryst., 10, 240–241, 1957.

86. Hall, S. H., and S. W. Bailey, Cation ordering pattern in amesite, Clays and Clay Miner., 27, 241–247, 1979.

87. Hall, S.R. and J.M. Stewart, The crystal structure of argentian pentlandite (Fe, Ni)$_8$AgS$_8$, compared with the refined structure of pentlandite (Fe,Ni)$_9$S$_8$, Canad. Mineral., 12, 169–177, 1973.

88. Hanscom R., The structure of triclinic chloritoid and chloritoid polymorphism, Am. Mineral., 65, 534–539, 1980.

89. Harlow, G. E., and G. E. Brown, Low Albite: an X–ray and neutron diffraction study, Am. Mineral., 65, 986–995, 1980.

90. Hawthorne, F.C. and R.B. Ferguson, Anhydrous sulfates. I: Refinement of the crystal structure of celestite with an appendix on the structure of thenardite, Canad. Mineral., 13, 181–187, 1975.

91. Hawthorne, F.C., L.A. Groat and R.K. Eby, Antlerite, $CuSO_4(OH)_4$, a heteropolyhedral wallpaper structure, Canad. Mineral., 27, 205–209, 1989.

92. Hawthorne, F. C., and H. D. Grundy, The crystal chemistry of the amphiboles: IV. X–ray and neutron refinements of the crystal structure of tremolite, Canad. Mineral., 14, 334–345, 1976.

93. Hazen, R. M., Effects of temperature and pressure on the cell dimension and X–ray temperature factors of periclase, Am. Mineral., 61, 266–271, 1976.

94. Hazen, R. M., and C. W. Burnham, The crystal structures of one–layer phlogopite and annite, Am. Mineral., 58, 889–900, 1973.

95. Hazen, R. M., and L. W. Finger, Crystal structure and compressibility of zircon at high pressure, Am. Mineral., 64, 196–201, 1979.

96. Hazen, R.M. and L.W. Finger, High temperature crystal chemistry of phenacite and chrysoberyl, Phys. Chem. Miner., 14, 426–432, 1987.

97. Hesse, K.–F. Crystal structures of natural and synthetic α–eucryptite, $LiAlSiO_4$, Z. Kristal., 172, 147–151, 1985.

98. Hill, R.J., Hydrogen atoms in boehmite: a single–crystal X–ray diffraction and molecular orbital study, Clays and Clay Miner., 29, 435–445, 1981.

99. Hill R. J., X–ray powder diffraction profile refinement of synthetic hercynite, Am. Mineral., 69, 937–942 1984.

100. Hill, R. J., J. R. Craig G. V. Gibbs, Systematics of the spinel structure type, Phys. Chem. Miner., 4, 317–319 1979.

101. Hoelzel, A.R., Systematics of Minerals. Hoelzel, Mainz, 584pp., 1989.

102. Horiuchi, H., M. Hirano, E. Ito, and Y. Matsui, $MgSiO_3$ (ilmenite–type): Single crystal X–ray diffraction study, Am. Mineral., 67, 788–793, 1982.

103. Horiuchi, H., E. Ito, and D.J. Weidner, Perovskite–type $MgSiO_3$: Single–crystal X–ray diffraction study, Am. Mineral., 72, 357–360, 1987.

104. Horiuchi, H., and H. Sawamoto, β–Mg_2SiO_4: Single–crystal X–ray diffraction study, Am. Mineral., 66, 568–575, 1981.

105. Horn, M., C. F. Schwerdtfeger, and E. P. Meagher , Refinement of the structure of anatase at several temperatures, Z. Krist., 136, 273–281, 1972.

106. Ishikawa, Y., S. Sato, and Y. Syono, Neutron and magnetic studies of a single crystal of Fe_2TiO_4, Tech. rep., Inst. Sol. State Phys., Univ. of Tokyo A 455, 1971.

107. Joswig, W., H. Bartl, and H. Feuss, Structure refinement of scolecite by neutron diffraction, Z. Kristal., 166, 219–223, 1984.

108. Joswig, W., and H. Feuss, Refinement of a one–layer triclinic chlorite. Clays and Clay Miner., 38, 216–218, 1990.

109. Joswig, W., H. Feuss, and S.A. Mason, Neutron diffraction study of a one–layer monoclinic chlorite, Clays and Clay Miner., 37, 511–514, 1989.

110. Kanisceva, A. S., Ju. N. Mikhailov and A. F. Trippel, zv. Akad. Nauk SSSR, Neorg. Mater., 17, 1972–1975, 1981, as cited in Structure Reports, 48A, 31, 1981.

111. Kato, K., and A. Nukui, Die Kristallstruktur des monoklinen tief tridymits, Acta Cryst., B32, 2486–2491, 1976.

112. Kato, T., and Y. Miura, The crystal structures of jarosite and svanbergite, Mineral. J., 8, 419–430, 1977.

113. Kay, H. F., and P. C. Bailey, Structure and properties of $CaTiO_3$, Acta Cryst., 10, 219–226, 1957.

114. Kay, M. I., B. C. Frazer, and I. Almodovar, Neutron diffraction refinement of $CaWO_4$, J. Chem. Phys., 40, 504–506, 1964.

115. Keller, C., Untersuchungen ueber die germanate und silikate des typs ABO_4 der vierwertigen elemente Thorium bis Americium, Nukleonik, 5, 41–48, 1963.

116. Kimata, M., and N. Ii, The crystal structure of synthetic akermanite

$Ca_2MgSi_2O_7$, *Neues Jahrbuch fuer Mineral., Monats.*, 1–10, 1981.

117. King, H. E., Jr., and C. T. Prewitt, High–pressure and high–temperature polymorphism of iron sulfide (FeS), *Acta Cryst., B38*, 1877–1887, 1982.

118. Kirfel, A., and G. Will, Charge density in anhydrite, $CaSO_4$, from X–ray and neutron diffraction measurements, *Acta Cryst., B36*, 2881–2890, 1980.

119. Kisi, E. H. and M. M. Elcombe, U parameters for the wurzite structure of ZnS and ZnO using powder neutron diffraction, *Acta Cryst., C45*, 1867–1870, 1989.

120. Knop, O., K. I. G. Reid, Sutarno, and Y. Nakagawa, Chalcogenides of the transition elements. VI. X–ray, neutron and magnetic investigation of the spinels Co_3O_4, $NiCo_2O_4$, Co_3S_4, and $NiCo_2S_4$, *Canad. J. Chem., 22*, 3463–3476, 1968, as cited in *Structure Reports, 33A*, 290–291, 1968.

121. Kondrasev, J. D., and A. I. Zaslavskij, *Izv. Akad. Nauk SSSR, 15*, 179–186, 1951.

122. Koto, K. and N. Morimoto, Superstructure investigation of bornite, Cu_5FeS_4, by the modified partial patterson function, *Acta Cryst., B31*, 2268, 1975.

123. Krstanovic, I., Redetermination of oxygen parameters in xenotime, YPO_4, *Z. Kristal., 121*, 315–316, 1965.

124. Lager, G. A., and E. P. Meagher, High temperature structural study of six olivines, *Am. Mineral., 63*, 365–377, 1978.

125. Lee, J. H., and S. Guggenheim, Single crystal X–ray refinement of pyrophyllite–1Tc, *Am. Mineral., 66*, 350–357, 1981.

126. Leonova, V. A., Effect of contamination on the lattice parameters of uraninite, *Dokl. Akad. Nauk SSSR, 126*, 1342–1346 1959.

127. LePage, Y., L. D. Calvert, and E. J. Gabe, Parameter variation in low quartz between 94 and 298K, *J. of Phys. and Chem. Solids, 41*, 721–725, 1980.

128. Levy, H.A. and G.C. Lisensky, Crystal structure of sodium sulfate decahydrate (Glauber's salt) and sodium tetraborate decahydrate (borax). Redetermination by neutron diffraction, *Acta Cryst, B34*,

3502–3510, 1978.

129. Lin, C., and S.W. Bailey, The crystal structure of paragonite–2M$_1$, *Am. Mineral., 69*, 122–127, 1984.

130. Lin, J.C., and S. Guggenheim, The crystal structure of a Li,Be–rich brittle mica: a dioctahedral–trioctahedra intermediate. *Am. Mineral. 68*, 130–142, 1983.

131. Lin, S.B., and B.J. Burley, The crystal structure of meionite, *Acta Cryst., B29*, 2024–2026, 1973.

132. Lin, S.B., and B.J. Burley, Crystal structure of a sodium and chlorine–rich scapolite, *Acta Cryst., B29*, 1272–1278, 1973.

133. Löns, J. and H. Schulz, Strukturverfeinerung von sodalith, $Na_8Si_6Al_6O_{24}Cl_2$, *Acta Cryst., 23*, 434–436, 1967.

134. Louisnathan, S.J., The crystal structure of synthetic soda melilite, $CaNaAlSi_2O_7$, *Z. Krist., 131*, 314–321, 1970.

135. Louisnathan, S.J., Refinement of the crystal structure of a natural gehlenite, $Ca_2AlAlSi)_2O_7$, *Canad. Mineral., 10*, 822–837 1970.

136. Lutz, H. D., M. Jung, and G. Waschenbach, Kristallstrukturen des lollingits $FeAs_2$ und des pyrits $RuTe_2$, *Z. Anorg., Chem., 554*, 87–91, 1987, as cited in *Structure Reports, 54A*, 43, 1987.

137. Mackie, P. E., J. C. Elliott and R. A. Young, Monoclinic structure of synthetic $Ca_5(PO_4)_3Cl$, chlorapatite, *Acta Cryst., B28*, 1840–1848, 1972.

138. Mazzi, F., and E. Galli, Is each analcime different? *Am. Mineral., 63*, 448–460, 1978.

139. Mazzi, F., E. Galli, and G. Gottardi, The crystal structure of tetragonal leucite, *Am. Mineral., 61*, 108–115, 1976.

140. Mazzi, F., E. Galli, and G. Gottardi, Crystal structure refinement of two tetragonal edingtonites, *Neues Jahrbuch fur Mineralogie Monat.*, 373–382, 1984.

141. Mazzi, F., A.O. Larsen, G. Gottardi, and E. Galli, Gonnardite has the tetrahedral framework of natrolite: experimental proof with a sample from Norway, *Neues J. Mineral. Monat.*, 219–228, 1986.

142. McGinnety, J.A., Redetermination of

the structures of potassium sulfate and potassium chromate: the effect of electrostatic crystal forces upon observed bond lengths, *Acta Cryst., B28*, 2845–2852, 1972.

143. McKee, D. O. and J. T. McMullan, Comment on the structure of antimony trisulfide, *Z. Krist., 142*, 447–449, 1975.

144. Mellini, M., The crystal structure of lizardite 1T: hydrogen bonds and polytypism, *Am. Mineral., 67*, 587–598, 1982.

145. Menchetti, S. and C. Sabelli, Crystal chemistry of the alunite series: Crystal structure refinement of alunite and synthetic jarosite, *Neues J. Mineral. Monat.*, 406–417, 1976.

146. Meyer, H.J., Struktur und Fehlordnung des Vaterits, *Z. Kristal., 128*, 182–212, 1969.

147. Miyake, M., I. Minato, H. Morikawa, and S. Iwai, Crystal structures and sulfate force constants of barite celestite, and anglesite, *Am. Mineral., 63*, 506–510, 1978.

148. Miyazawa, R., I. Nakai, and K. Magashima, A refinement of the crystal structure of gadolinite, *Am. Mineral., 69*, 948–953, 1984.

149. Moore, P.B., The crystal structure of sapphirine, *Am. Mineral. 54*, 31–49, 1969.

150. Morimoto, N., D. E. Appleman and H. T. Evans, The crystal structures of clinoenstatite and pigeonite, *Z. Krist., 114*, 120–147, 1960.

151. Morimoto, N., M. Tokonami, M. Watanabe, and K. Koto, Crystal structures of three polymorphs of Co_2SiO_4, *Am. Mineral., 59*, 475–485, 1974.

152. Morosin, B., Structure and thermal expansion of beryl, *Acta Cryst., B28*, 1899–1903, 1972.

153. Mortier, L., J. J. Pluth, and J. V. Smith, Positions of cations and molecules in zeolites with the mordenite–type framework IV. Dehydrated and rehydrated K–exchanged "ptilolite", Pergamon, New York, 1978.

154. Mullen, D.J.E. and W. Nowacki, Refinement of the crystal structures of realgar, AsS and orpiment, As_2S_3, *Z. Krist., 136*, 48–65, 1972.

155. Narita, H., K. Koto, and N. Morimoto, The crystal structures of MnSiO$_3$ polymorphs, rhodonite– and pyroxmangite–type, *Mineral. J., 8,* 329–342, 1977.

156. Newnham, R.E., Crystal structure and optical properties of pollucite, *Am. Mineral., 52,* 1515–1518, 1967.

157. Newnham, R.E. and Y. M. deHaan, Refinement of the alpha–Al$_2$O$_3$, Ti$_2$O$_3$, V$_2$O$_3$ and Cr$_2$O$_3$ structures, *Z. Krist., 117,* 235–237, 1962.

158. Newnham, R. E., and H. D. Megaw, The crystal structure of celsian (barium feldspar), *Acta Cryst., 13,* 303–312, 1960.

159. Nimmo, J. K., and B. W. Lucas, A neutron diffraction determination of the crystal structure of alpha–phase potassium nitrate at 25°C and 100°C, *J. of Phys., C6,* 201–211, 1973.

160. Noda, Y., K. Matsumoto, S. Ohba, Y. Saito, K. Toriumi, Y. Iwata, and I. Shibuya, Temperature dependence of atomic thermal paramters of lead chalcogenides, PbS, PbSe and PbTe, *Acta Cryst., C43,* 1443–1445, 1987.

161. Novak, G. A., G. V. Gibbs, The crystal chemistry of the silicate garnets, *Am. Mineral., 56,* 791–825, 1971.

162. Nowack, E., D. Schwarzenbach, W. Gonschorek, and Th. Hahn, Deformationsdichten in CoS$_2$ und NiS$_2$ mit pyritstruktur, *Z. Krist., 186,* 213–215, 1989.

163. Ohashi, Y., and L. W. Finger, The role of octahedral cations in pyroxenoid crystal chemistry I. Bustamite, wollastonite, and the pectolite–schizolite–serandite series, *Am. Mineral., 63,* 274–288, 1978.

164. Okamura, F. P., S. Ghose, and H. Ohashi, Structure and crystal chemistry of calcium Tschermak's pyroxene, CaAlAlSiO$_6$, *Am. Mineral., 59,* 549–557, 1974.

165. Onken H., Verfeinerung der kristallstruktur von monticellite, *Tschermaks Mineral. Petrol. Miit., 10,* 34–44 1965.

166. Pacalo, R.E.G. and J.B. Parise, Crystal structure of superhydrous B, a hydrous magnesium silicate symthesized at 1400°C and 20GPa, *Am. Mineral., 77,* 681–684, 1992.

167. Papamantellos, P. Verfeinerung der Tl$_2$O$_3$–struktur mittels neutronenbeu-

168. Papike, J. J., and J. R. Clark, The crystal structure and cation distribution of glaucophane, *Am. Mineral., 53,* 1156–1173, 1968.

169. Papike, J. J., and M. Ross, Gedrites: crystal structures and intra–crystalline cation distributions, *Am. Mineral., 55,* 1945–1972, 1970.

170. Papike, J.J., and T. Zoltai, Ordering of tetrahedral aluminum in prehnite, *Am. Mineral., 52,* 974–984, 1967.

171. Parise, J. B., Structure of hazelwoodite (Ni$_3$S$_2$), *Acta Cryst., B36,* 1179–1180, 1980.

172. Passaglia, E., and G. Gottardi, Crystal chemistry and nomenclature of pumpellyites and julgoldites, *Canad. Mineral., 12,* 219–223, 1973.

173. Peacor, D. R., High–temperature single–crystal study of cristobalite inversion, *Z. Krist., 138,* 274–298, 1973.

174. Pechar, F., W. Schaefer, and G. Will, A neutron diffraction refinement of the crystal structure of a natural natrolite, Na$_2$Al$_2$Si$_3$O$_{10}$ 2H$_2$O, *Z. Kristal., 164.* 19–24, 1983.

175. Perdikatsis, B., and H. Burzlaff, Strukturverfeinerung am Talk Mg$_3$[(OH)$_2$Si$_4$O$_{10}$], *Z. Kristal., 156,* 177–186, 1981.

176. Pertlik, F., Verfeinerung der kristallstruktur von claudetit (As$_2$O$_3$), *Monats. Chem., 106,* 755–762, 1975.

177. Pertlik F., Strukturverfeinerung von kubischem As$_2$O$_3$ (Arsenolith) mit Einkristalldaten, *Czech. J. Phys., B28,* 170–176, 1978.

178. Perrotta, A. J., and J. V. Smith, The crystal structure of kalsilite, KAlSiO$_4$, *Mineral. Mag., 35,* 588–595, 1967.

179. Peterson, R. C. and I. Miller, Crystal structure and cation distribution in freibergite and tetrahedrite, *Mineral. Mag., 50,* 717–721, 1986.

180. Radke, A. S., and G. E. Brown, Frankdicksonite, BaF$_2$, a new mineral from Nevada, *Am. Mineral., 59,* 885–888, 1974.

181. Rajamani, V. and C. T. Prewitt, The crystal structure of millerite, *Canad. Mineral., 12,* 253–257, 1974.

182. Reeder, R. J., Crystal chemistry of the rhombohedral carbonates, *Rev. Mineral., 11,* 1–47, 1983.

183. Ribbe, P. H., and G. V. Gibbs, Crystal structures of the humite minerals: III.

gung, *Z. Krist., 126,* 143–146, 1968.

Mg/Fe ordering in humite and its relation to other ferromagnesian silicates, *Am. Mineral., 56,* 1155–1169, 1971.

184. Rinaldi, R., J. J. Pluth, and J. V. Smith, Zeolites of the phillipsite family. Refinement of the crystal structures of phillipsite and harmotome, *Acta Cryst., B30,* 2426–2433, 1974.

185. Robinson, K., G. V. Gibbs, P. H. Ribbe, and M. R. Hall, Cation distribution in three hornblendes, *Am. J. Sci., 273A,* 522–535, 1973.

186. Robinson, K., G. V. Gibbs, and P. H. Ribbe, The crystal structures of the humite minerals IV. Clinohumite and titanoclinohumite, *Am. Mineral., 58,* 43–49, 1973.

187. Rothbauer, R., Untersuchung eines 2M$_1$–Muskovits mit Neutronenstrahlen, *Neues J. Mineral. Monat.,* 143–154, 1971.

188. Saalfeld, H., and M. Wedde, Refinement of the crystal structure of gibbsite, Al(OH)$_3$, *Z. Krist., 139,* 129–135, 1974.

189. Sabine, T. M. and S. Hogg, The wurtzite Z parameter for beryllium oxide and zinc oxide, *Acta Cryst., B25,* 2254–2256, 1969.

190. Sadanaga, R. and S. Sueno, X–ray study on the α–β transition of Ag$_2$S, *Mineral. J. Japan, 5,* 124–143, 1967.

191. Sahl, K., Verfeinerung der kristallstruktur von cerussit, PbCO$_3$, *Z. Krist., 139,* 215–222, 1974.

192. Sartori, F., The crystal structure of a 2M$_1$ lepidolite, *Tschermaks Mineral. Petrol. Mitt, 24,* 23–37, 1977.

193. Sartori, F., M. Franzini, and S. Merlino, Crystal structure of a 2M$_2$ lepidolite, *Acta Cryst., B29,* 573–578, 1973.

194. Sartori, F., The crystal structure of a 1M lepidolite, *Tschermaks Mineral. Petrol. Mitt, 23,* 65–75, 1976.

195. Sasaki, S., K. Fujino, Y. Takeuchi, and R. Sadanaga, On the estimation of atomic charges by the X–ray method for some oxides and silicates, *Acta Cryst., A36,* 904–915, 1980.

196. Sasaki, S., C. T. Prewitt, Y. Sato and E. Ito, Single–crystal X–ray study of gamma–Mg$_2$SiO$_4$, *J. of Geophysical Research, 87,* 7829–7832, 1982.

197. Sasaki, S., Y. Takeuchi, K. Fujino, and S. Akimoto, Electron density distributions of three orthopyroxenes,

$Mg_2Si_2O_6$, $Co_2Si_2O_6$, and $Fe_2Si_2O_6$, Z. Krist., 156, 279–297, 1982.

198. Sass, R.L., R. Vidale, and J. Donohue, Interatomic distances and thermal anisotropy in sodium nitrate and calcite, Acta Cryst., 10, 259–265, 1957.

199. Scambos, T. A., J. R. Smyth, and T. C. McCormick, Crystal structure refinement of a natural high sanidine of upper mantle origin, Am. Mineral. 72, 973–978, 1987.

200. Schiferl, D. and C. S. Barrett, The crystal structure of arsenic at 4.2, 78 and 299 °K, J. Appl. Cryst. 2, 30–36, 1969.

201. Schlenker, J.L., J.J. Pluth, and J.V. Smith, Dehydrated natural erionite with stacking faults of the offretite type, Acta Cryst., B33, 3265–3268, 1977.

202. Schramm, V., and K.F. Fischer, Refinement of the crystal structure of laumontite, Molecular Seive Zeolites–I, Advances in Chemistry Series, 101, 259–265, 1971.

203. Schutte, W. J., J. L. de Boer, and F. Jellinek, Crystal structures of tungsten disulphide and diselenide, J. Solid St. Chem., 70, 207–209, 1987.

204. Shintani, H., S. Sato, and Y. Saito, Electron–density distribution in rutile crystals, Acta Cryst., B31, 1981–1982, 1975.

205. Shirane, G. and D. E. Cox, Magnetic structures in $FeCr_2S_4$ and $FeCr_2O_4$, J. Appl. Phys., 35, 954–955, 1964.

206. Shternberg, A.A., G.S. Mironova, and O.V. Zvereva, Berlinite, Kristal., 31, 1206–1211 1986.

207. Simonov, M. A., P. A. Sandomerski, F. K. Egorov–Tesmenko, and N. V. Belov, Crystal structure of willemite, Zn_2SiO_4, Kristal., Dokl., Akad. Nauk USSR, 237, 581–583, 1977.

208. Smith, D. K., and H. W. Newkirk, The crystal structure of baddeleyite (monoclinic ZrO_2) and its relation to the polymorphism of ZrO_2, Acta Cryst., 18, 983–991 1965.

209. Smith, J. V., The crystal structure of staurolite, Am. Mineral., 53, 1139–1155, 1968.

210. Smyth, J. R., G. Artioli, J. V. Smith, and A. Kvick, Crystal structure of coesite, a high–pressure form of SiO_2, at 15 and 298 K from single–crystal neutron and X–ray diffraction data: test of bonding models. J. of Phys. Chem.,

91, 988–992, 1987.

211. Smyth, J.R., A.T. Spaid, and D.L. Bish, Crystal structures of a natural and a Cs–exchanged clinoptilolite. Am. Mineral., 75, 522–528, 1990.

212. Speer, J. A., and B. J. Cooper, Crystal structure of synthetic hafnon, $HfSiO_4$, comparison with zircon and the actinide orthosilicates, Am. Mineral., 67, 804–808, 1982.

213. Speer, J. A., and G. V. Gibbs, The crystal structure of synthetic titanite, $CaTiOSiO_4$, and the domain textures of natural titanites, Am. Mineral., 61, 238–247, 1976.

214. Sudarsanan, K. and R. A. Young, Significant precision in crystal structural details: Holly Springs hydroxyapatite, Acta Cryst., B25, 1534–1543, 1969.

215. Sudarsanan, K., P. E. Mackie, and R. A. Young, Comparison of synthetic and mineral fluorapatite, $Ca_5(PO_4)_3F$, in crystallographic detail, Mat. Res. Bull., 7, 1331–1338, 1972.

216. Svennson, C., The crystal structure of orthorhombic antimony trioxide, Sb_2O_3, Acta Cryst., B30, 458–461, 1974.

217. Svennson, C., Refinement of the crystal structure of cubic antimony trioxide, Sb_2O_3, Acta Cryst., B31, 2016–2018, 1975.

218. Szymanski, J. T., A refinement of the structure of cubanite, $CuFe_2S_3$, Z. Krist., 140, 218–239, 1974.

219. Tagai, Y., H. Ried, W. Joswig, and M. Korekawa, Kristallographische untersuchung eines petalits mittels neutronenbeugung und transmissionselektronmikroskopie, Z. Kristal., 160, 159–170, 1982.

220. Takeuchi, Y., T. Ozawa, T. Ito, T. Araki, T. Zoltai, and J.J. Finney, The $B_2Si_8O_{30}$ groups of tetrahedra in axinite and comments on the deformation of Si tetrahedra in silicates, Z. Kristal., 140, 289–312, 1974.

221. Taylor, L. A., Smythite, $Fe_{3+x}S_4$, and associated minerals from the Silverfields Mine, Cobalt, Ontario, Am. Mineral., 55, 1650–1658, 1970.

222. Taylor, M., and R. C. Ewing, The crystal structures of the $ThSiO_4$ polymorphs: huttonite and thorite, Acta Cryst., B34, 1074–1079, 1978.

223. Thompson, R. M., The telluride minerals and their occurrence in Canada, Am. Mineral., 34, 342–383, 1949.

224. Tornoos, R., Properties of alabandite; alabandite from Finland, Neues Jb., Miner. Abh., 144, 1, 107–123, 1982.

225. Ülkü, D., Untersuchungen zur Kristallstruktur und magnetischen struktur des ferberits $FeWO_4$, Z. Krist., 124, 192– 219 1967.

226. Vezzalini, G. and R. Oberti, The crystal chemistry of gismondines: the non–existence of K–rich gismondines, Bulletin de Mineralogie, 107, 805–812, 1984.

227. Vogel, R. E. and C. P. Kempter, Mathematical technique for the precision determination of lattice constants, U. S. Atomic Energy Commission, LA–2317, 30, 1959.

228. Wainwright, J. E., and J. Starkey, A refinement of the crystal structure of anorthite, Z. Krist., 133, 75–84, 1971.

229. Wechsler, B.A. and C.T. Prewitt, Crystal structure of ilmenite at high temperature and at high pressure, Am. Mineral., 69, 176–185, 1985.

230. Wechsler, B. A., C. T. Prewitt, and J. J. Papike, Chemistry and structure of lunar and synthetic armalcolite, Earth Planet. Sci. Lett., 29, 91–103 1976.

231. Weitzel, H., Kristallstrukturverfeinerung von Wolframiten und Columbiten. Z. Krist., 144, 238–258, 1976.

232. Willis, B.T.M., Anomalous behaviour of the neutron reflexions of fluorite, Acta Cryst., 18, 75–76, 1965.

233. Winter, J. K., and S. Ghose, Thermal expansion and high temperature crystal chemistry of the Al_2SiO_5 polymorphs, Am. Mineral., 64, 573–586, 1979.

234. Winter, J. K., F. P. Okamura, and S. Ghose, A high–temperature structural study of high albite, monalbite, and the analbite–monalbite phase transition, Am. Mineral., 64, 409–423, 1979.

235. Wyckoff, R. W. G. Crystal Structures, John Wiley & Sons, New York, 1963.

236. Yagi, T., F. Marumo, and S. Akimoto, Crystal structures of spinel polymorphs of Fe_2SiO_4 and Ni_2SiO_4, Am. Mineral., 59, 486–490, 1974.

237. Yakubovich, O.V., M.A. Simonov, and N.V. Belov The crystal structure of synthetic triphylite, $LiFe(PO_4)$. Soviet Phys. Dokl., 22, 347–450, 1977.

238. Yamaguchi, S. and H. Wada, Remarques sur une griegite preparee

par processus hydrothermal, *Bull. Soc. Fr. Mineral. Cristallogr., 94*, 549–550, 1971.

239. Yamanaka, T. and M. Tokonami, The anharmonic thermal vibration in ZnX (X=S, Se, Te) and its dependence on the chemical–bond characters, *Acta Cryst., B41*, 298–304, 1985.

240. Yund, R. A., Phase relations in the system Ni–As, *Econ. Geol., 56*, 1273–1296, 1961.

241. Zachariasen, W. H., Refined crystal structure of phenacite, Be_2SiO_4, *Kristallographiya., 16*, 1161–1166, 1971.

242. Zemann, J., E. Zobetz, G. Heger, and H. Voellenkle, Strukturbestimmung eines OH–reichen topases, *Oest. Akad. Wissenschaften, 6*, 145–147, 1979.

243. Zigan, F. and R. Rothbauer, Neutronenbeugungsmessungen am brucit, *Neues J. Mineral. Monat.*, 137–143, 1967.

244. Zigan, F., W. Joswig, H.D. Schuster, and S.A. Mason, Verfeinerung der Struktur von Malachit, $Cu_2(OH)_2CO_3$, durch Neutronenbeugung, *Z. Krist., 145*, 412–426, 1977.

245. Zigan, F., and H.D. Schuster, Verfeinerung der struktur von azurit, $Cu_3(OH)_2(CO_3)_2$, *Z. Krist., 135*, 416–436, 1972.

Thermodynamic Properties of Minerals

Alexandra Navrotsky

1. INTRODUCTION

Thermochemical properties of minerals can be used to calculate the thermodynamic stability of phases as functions of temperature, pressure, component fugacity, and bulk composition. A number of compendia of thermochemical data [4, 5, 7, 9, 10, 13, 15, 16, 18, 19, 31] contain detailed data. The purpose of this summary is to give, in short form, useful data for anhydrous phases of geophysical importance. The values selected are, in the author's opinion, reliable, but no attempt has been made to systematically select values most consistent with a large set of experimental observations. When possible, estimates of uncertainty are given.

2. HEAT CAPACITIES

The isobaric heat capacity, C_p, is the temperature derivative of the enthalpy, $C_p = (dH/\partial T)_p$. For solids, C_p is virtually independent of pressure but a strong function of temperature (see Fig. 1). Contributions to C_p arise from lattice vibrations, and from magnetic, electronic, and positional order-disorder. The relation between heat capacity at constant pressure, C_p, and that at constant volume, $C_v = (\partial E/\partial T)_v$, is given by $C_p - C_v = TV\alpha^2/\beta$, where T = absolute temperature, V = molar volume, α = thermal expansivity = $(1/V) (\partial V/\partial T)_p$ and β = compressibility = inverse bulk modulus = $-(1/V)$

A. Navrotsky, Princeton University, Department of Geological and Geophysical Sciences and Princeton Materials Institute, Guyot Hall, Princeton, NJ 08544

Mineral Physics and Crystallography
A Handbook of Physical Constants
AGU Reference Shelf 2

$(\partial V/\partial P)_T$. For solids, $C_p - C_v$ is on the order of a few percent of C_v, and increases with temperature. The vibrational heat capacity can be calculated using statistical mechanics from the density of states, which in turn can be modeled at various degrees of approximation [20]. The magnetic contributions, important for transition metals, play a major role in iron-bearing minerals [32]. Electronic transitions are usually unimportant in silicates but may become significant in iron oxides and iron silicates at high T and P. Order-disorder is an important complication in framework silicates (Al-Si disorder on tetrahedral sites), in spinels (M^{2+}-M^{3+} disorder over octahedral and tetrahedral sites) and in olivines, pyroxenes, amphiboles, and micas (cation disorder over several inequivalent octahedral sites). These factors must be considered for specific minerals but detailed discussion is beyond the scope of this review.

As T--> 0 K, C_p --> 0 (see Fig. 1). At intermediate temperatures, C_p increases sharply. The Debye temperature is typically 800-1200 K for oxides and silicates. At high temperature, the harmonic contribution to C_v approaches the Dulong and Petit limit of 3nR (R the gas constant, n the number of atoms per formula unit). C_p is then 5-10% larger than 3nR and varies slowly and roughly linearly with temperature (see Fig. 1).

Table 1 lists heat capacities for some common minerals. The values at high temperature may be compared with the 3nR limit as follows: Mg_2SiO_4 (forsterite) 3nR = 175 J/K•mol, C_p at 1500 K = 188 J/K•mol; $MgAl_2O_4$ (spinel) 3nR = 188 J/K•mol, C_p at 1500 K = 191 J/K•mol. Thus the Dulong and Petit limit gives a useful first order estimate of the high temperature heat capacity of a solid, namely 3R per gram atom, irrespective of structural detail.

The entropy,

$$S_T^{\circ} = \int_0^T (C_p / T)dT \qquad (1)$$

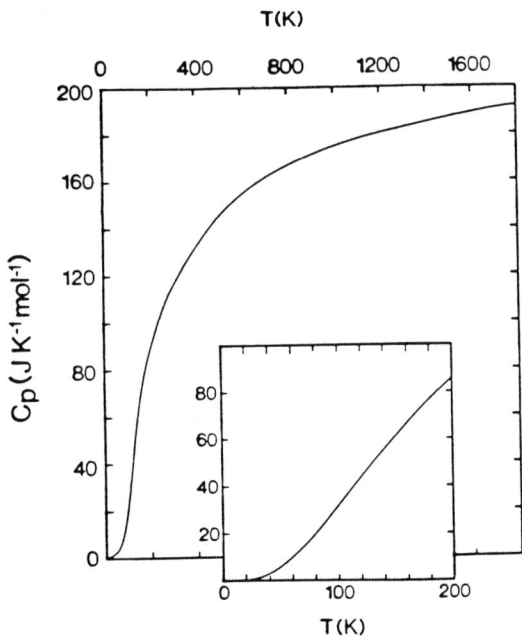

Fig. 1. Heat capacity of Mg_2SiO_4 (forsterite) from 0 to 1800 K, data from [31].

Any "zero point" entropy, arising from "frozen in" configurational disorder, must be added to this calorimetric entropy. Entropies of some common phases are also shown in Table 1.

The sharp dependence of C_p on T at intermediate temperature makes it difficult to fit C_p by algebraic equations which extrapolate properly to high temperature and such empirical equations almost never show proper low temperature behavior. At 298 - 1500 K, an expression of the Maier-Kelley form, [31]

$$C_p = A + BT + CT^{-0.5} + DT^{-2} \qquad (2)$$

gives a reasonable fit but must be extrapolated with care. A form which ensures proper high temperature behavior, recommended by Fei and Saxena [8] is

$$C_p = 3nR[1 + k_1 T^{-1} + k_2 T^{-2} + k_3 T^{-3}] + A + BT + C_p \text{ (disordering)} \qquad (3)$$

Because different authors fit C_p data to a variety of equations and over different temperature ranges, a tabulation of coefficients is not given here but the reader is referred to Robie et al. [31], Holland and Powell [15-

Table 1. Heat Capacities and Entropies of Minerals (J/(K•mol))

	298 K		1000 K		1500 K	
	Cp	S°	Cp	S°	Cp	S°
MgO (periclase)	37.8	26.9	51.2	82.2	53.1	103.5
Al_2O_3 (corundum)	79.0	50.9	124.9	180.2	132.1	232.3
"FeO" (wustite)	48.12	57.6	55.8	121.4	63.6	145.3
Fe_2O_3 (hematite)	103.9	87.4	148.5	252.7	144.6	310.5
Fe_3O_4 (magnetite)	150.8	146.1	206.0	390.2	201.0	471.5
TiO_2 (rutile)	55.1	50.3	73.2	129.2	79.5	160.1
$FeTiO_3$ (ilmenite)	99.5	105.9	133.7	249.3	155.0	307.4
Fe_2TiO_4 (titanomagnetite)	142.3	168.9	197.5	375.1	243.2	463.4
$MgAl_2O_4$ (spinel)	115.9	80.6	178.3	264.5	191.3	339.5
Mg_2SiO_4 (forsterite)	117.9	95.2	175.3	277.2	187.7	350.8
$MgSiO_3$ (enstatite)	82.1	67.9	121.3	192.9	127.6	243.5
$NaAlSi_3O_8$ (low albite)	205.1	207.4	312.3	530.1		
$KAlSi_3O_8$ (microcline)	202.4	214.2	310.3	533.8		
$Mg_3Al_2Si_3O_{12}$ (pyrope)	325.5	222.0	474.0	730.8		
$Ca_3Al_2Si_3O_{12}$ (grossular)	330.1	255.5	491.7	773.0		
$CaSiO_3$ (wollastonite)	85.3	82.0	123.4	213.4		
$CaSiO_3$ (pseudowollastonite)	86.5	87.5	122.3	217.6	132.3	269.1
$CaMgSi_2O_6$ (diopside)	166.5	143.0	248.9	401.7	269.7	506.3
$Mg_2Al_2Si_5O_{18}$ (cordierite)	452.3	407.2	698.3	1126.6	753.6	1420.9
$CaCO_3$ (calcite)	83.5	91.7	124.5	220.2		
$MgCO_3$ (magnesite)	76.1	65.1	131.5	190.5		
$CaMg(CO_3)_2$ (dolomite)	157.5	155.2	253.1	406.0		

Data from [5, 31].

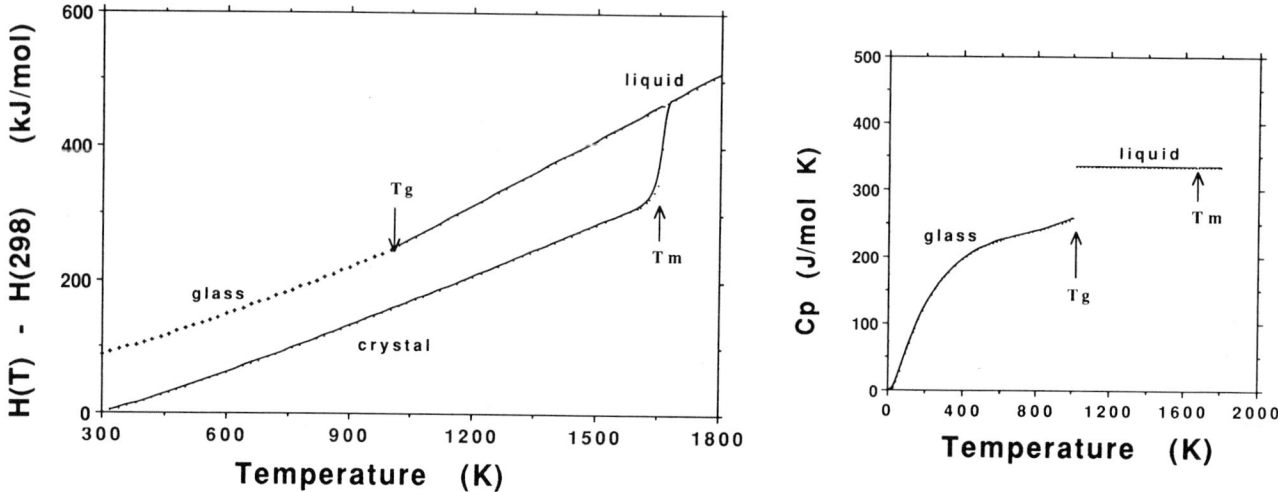

Fig. 2. Enthalpy and heat capacity in CaMgSi$_2$O$_6$, a glass-forming system, data from [21].

Table 2. Heat Capacities of Glasses and Liquids and Glass Transition Temperatures

Composition	Cp glass 298 K J/mol•K	Cp glass (at T$_g$) J/mol•K	T$_g$ (K)	Cp liquid J/mol•K
SiO$_2$	38[31]	74[28,29]	1607[28,29]	81[28,29]
CaMgSi$_2$O$_6$	170[a]	256[28,29]	1005[28,29]	335[28,29]
NaAlSi$_3$O$_8$	210[31]	321[28,29]	1096[28,29]	347[28,29]
KAlSi$_3$O$_8$	209[31]	316[28,29]	1221[28,29]	338[28,29]
CaAl$_2$Si$_2$O$_8$	211[31]	334[28,29]	1160[28,29]	424[28,29]
Mg$_2$SiO$_4$	-----	------	--------	268[11,12]
Na$_2$Si$_2$O$_5$	-----	217[28,29]	703[28,29]	263[28,29]
K$_2$Si$_2$O$_5$	-----	226[28,29]	770[28,29]	259[28,29]
CaSiO$_3$	87[30]	131[28,29,30]	1065	167[28,29]
Mg$_3$Al$_2$Si$_3$O$_{12}$	330[a]	516[28,29]	1020	679[28,29]
Mg$_2$Al$_4$Si$_5$O$_{18}$	460[a]	731[28,29]	1118	928[28,29]

[a]*Estimated from higher temperature data and from comparison with crystalline phases.*

16], Berman [5], JANAF [18], and Fei et al. [9] for such equations.

In glass-forming systems, see Fig. 2, the heat capacity of the glass from room temperature to the glass transition is not very different from that of the crystalline phase. For $CaMgSi_2O_6$ C_p, glass = 170 J/mol·K at 298 K, 256 J/mol·K at 1000 K; C_p, crystal = 167 J/mol·K at 298 K, 249 J/mol·K at 1000 K [21]. At T_g, the viscosity decreases, and the volume and heat capacity increase, reflecting the onset of configurational rearrangements in the liquid [27]. The heat capacity of the liquid is generally larger than that of the glass (see Table 2) and, except for cases with strong structural rearrangements (such as coordination number changes), heat capacities of liquids depend only weakly on temperature.

For multicomponent glasses and liquids with compositions relevant to magmatic processes, heat capacities can, to a useful approximation, be given as a sum of terms depending on the mole fractions of oxide components, i.e., partial molar heat capacities are relatively independent of composition. Then

$$C_p = \sum_i X_i \bar{C}_p \qquad (4)$$

where i is taken over the oxide components of the glass or liquid [22, 33]. The partial molar heat capacities of the oxide components in glasses and melts, $\bar{C}_{p,}$, are given in Table 3.

3. MOLAR VOLUME, ENTROPY, ENTHALPY OF FORMATION

Table 4 lists enthalpies and entropies of formation of selected minerals from the elements and the oxides at several temperatures. These refer to the reaction

$$aA + bB + cC + ^n/_2 O_2 = A_aB_bC_cO_n \qquad (5)$$

and

$$aAO_l + bBO_m + cCO_n = A_aB_bC_cO_n \qquad (6)$$

respectively, where A, B, C are different elements (e.g. Ca, Al, Si), O is oxygen, and reference states are the most stable form of the elements or oxides at the temperature in question. The free energy of formation is then given by

Table 3. Partial Molar Heat Capacities of Oxide Components in Glasses and Melts (J/K·mol)

	Glass[29]		Liquid[22,28,33]	
	298 K	400 K	1000 K	1500 K
SiO_2	44.04	52.39	70.56	82.6
TiO_2	44.92	58.76	84.40	109.2
Al_2O_3	79.22	96.24	124.98	170.3
Fe_2O_3	94.89	115.74	143.65	240.9
FeO	43.23	47.17	70.28	78.8
MgO	35.09	42.89	56.60	94.2
CaO	43.00	45.67	57.66	89.8
Na_2O	74.63	79.09	96.64	97.6
K_2O	75.20	79.43	84.22	98.5
B_2O_3	62.81	77.67	120.96	-----
H_2O	46.45	62.04	78.43	-----

Table 4.   Enthalpies and Entropies of Formation of Selected Compounds from Elements and From Oxides

| Compound | Formation from Elements | | | | Formation from Oxides | | | |
| | 298 K | | 1000 K | | 298 K | | 1000 K | |
	ΔH kJ/mol	ΔS J/mol K	ΔH kJ/mol	ΔS J/mol K	ΔH kJ/mol	ΔS J/mol K	ΔH kJ/mol	ΔS J/mol K
MgO (periclase)	-601.5[5]	-108.4[5]	-608.5[18]	-115.51[18]				
CaO (lime)	-635.1[5]	-106.5[5]	-634.3[18]	-103.6[18]				
Al$_2$O$_3$ (corundum)	-1675.7[5]	-313.8[5]	-1693.4[18]	-332.0[18]				
SiO$_2$ (quartz)	-910.7[5]	-182.6[5]	-905.1[18]	-174.9[18]				
SiO$_2$ (cristobalite)	-907.8[5]	-180.6[5]	-903.2[18]	-173.1[18]				
"FeO" (wustite)	-266.3[5]	-70.9[5]	-263.3[18]	-63.9[18]				
Mg$_2$SiO$_4$ (forsterite)	-2174.4[5]	-400.7[5]	-2182.1[31]	-410.6[31]	-60.7[5]	-1.4[5]	-58.2[31]	-2.9[31]
MgSiO$_3$ (enstatite)	-1545.9[5]	-293.0[5]	-1552.9[31]	-296.5[31]	-33.7[5]	-2.1[5]	-38.2[31]	-4.9[31]
Fe$_2$SiO$_4$ (fayalite)	-1479.4[31]	-335.5[31]	-1472.3[31]	-321.4[31]	-24.6[5]	-12.7[5]	-28.7[31]	-19.9[31]
CaSiO$_3$ (wollastonite)	-1631.5[5]	-286.5[5]	-1630.4[31]	-278.2[31]	-85.7[5]	2.6[5]	-91.1[31]	0.1[31]
CaSiO$_3$ (pseudowollastonite)	-1627.4[5]	-283.0[5]	-1624.7[31]	-274.0[31]	-81.6[5]	6.1[5]	-85.3[31]	4.3[31]
CaMgSi$_2$O$_6$ (diopside)	-3200.5[5]	-585.2[5]	-3209.6[31]	-579.3[31]	-142.6[5]	-5.1[5]	-155.6[31]	-9.4[31]
NaAlSi$_3$O$_8$ (high albite)	-3924.2[31]	-730.6[31]	-3925.8[31]	-735.3[31]	-146.9[31]	39.0[5]	-156.0[31]	24.9[31]
KAlSi$_3$O$_8$ (sanidine)	-3959.6[31]	-737.5[31]	-3962.1[31]	-744.1[31]	-208.0[31]	36.1[5]	222.9[31]	12.3[31]
CaAl$_2$Si$_2$O$_8$ (anorthite)	-4228.7[5]	-756.7[5]	-4239.4[31]	-764.6[31]	-96.5[31]	28.7[5]	-100.9[31]	21.4[31]
Mg$_3$Al$_2$Si$_3$O$_{12}$ (pyrope)	-6286.5[5]	-1176.3[5]	-6317.0[31]	-1211.3[31]	-74.2[31]	-19.9[5]	-79.2[31]	-4.0[31]
Mg$_2$Al$_4$Si$_5$O$_{18}$ (cordierite)	-9158.7[5]	-1702.0[5]	-9200.3[31]	-1750.7[31]	-50.8[31]	55.2[5]	-67.0[31]	23.9[31]
Ca$_3$Al$_2$Si$_3$O$_{12}$ (grossular)	-6632.9[5]	-1214.4[5]	-6649.7[31]	-1214.4[31]	-319.8[31]	-33.3[5]	337.9[31]	-43.0[31]

$\Delta G° = \Delta H° - T\Delta S°$. Fig. 3 shows the equilibrium oxygen fugacity for a series of oxidation reactions

$$A + \frac{n}{2} O_2 = AO_1 \qquad (7)$$

and

$$AO_n + \frac{m}{2} O_2 = AO_{m+n} \qquad (8)$$

as a function of temperature. These curves (see Fig. 3) are the basis for various "buffers" used in geochemistry, e.g. QFM (quartz-fayalite-magnetite), NNO (nickel-nickel oxide) and IW (iron-wurstite).

The free energies of formation from the elements become less negative with increasing temperature, and more reduced species are generally favored as temperature increases. This reflects the large negative entropy of incorporation of oxygen gas in the crystalline phase. Thus the equilibrium oxygen fugacity for a given oxidation-reduction equilibrium increases with increasing temperature. Changes in slope (kinks) in the curves in Fig. 3 reflect phase changes (melting, vaporization, solid-state transitions) in either the reactants (elements) or products (oxides).

The enthalpies of formation of ternary oxides from binary oxides are generally in the range +10 to -250 kJ/mol and become more exothermic with greater difference in "basicity" (or ionic potential = charge/radius) of the components. Thus for Al_2SiO_5, (andalusite) $\Delta H^o_{f, ox, 298} = -1.1$ kJ/mol; for $MgSiO_3$ (enstatite) $\Delta H^o_{f, ox, 298} = -35.6$ kJ/mol, and for $CaSiO_3$ (wollastonite) $\Delta H^o_{f, ox, 298} = -89.4$ kJ/mol. Entropies of formation of ternary oxides from binary components are generally small in magnitude (-10 to +10 J/mol·K) unless major order-disorder occurs.

4. ENTHALPY AND ENTROPY OF PHASE TRANSFORMATION AND MELTING

At constant (atmospheric) pressure, a thermodynamically reversible first order phase transition occurs with increasing temperature if both the enthalpy and entropy of the high temperature polymorph are higher than those of the low temperature polymorph and, at the transformation temperature

$$\Delta G^o_T = 0 = \Delta H^o_T - T\Delta S^o_T \qquad (9)$$

At constant temperature, a thermodynamically reversible phase transition occurs with increasing pressure if the high pressure phase is denser than the low pressure phase

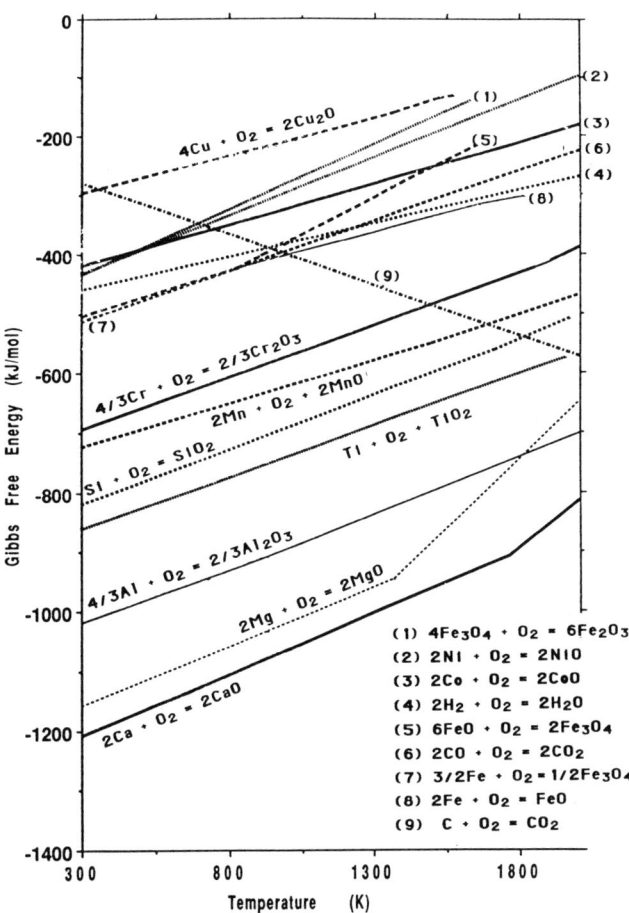

Fig. 3. Gibbs free energy for oxidation-reduction equilibria, per mole of O_2, data from [4, 18, 31].

and the following balance of enthalpy, entropy, and volume terms is reached

$$\Delta G (P, T) = 0 = \Delta H^0_T - T\Delta S^0_T + \int_{1\,atm}^P \Delta V(P, T)\, dP \qquad (10)$$

An equilibrium phase boundary has its slope defined by the Clausius - Clapeyron equation

$$(dP/dT)_{equil} = \Delta S/\Delta V \qquad (11)$$

Thus the phase boundary is a straight line if ΔS and ΔV are independent (or only weakly dependent) of P and T, as is a reasonable first approximation for solid-solid transitions over moderate P-T intervals at high T. A negative P-T slope implies that ΔS and ΔV have opposite signs. Melting curves tend to show decreasing (dT/dP) with increasing pressure because silicate liquids are often

Table 5. Enthalpy, Entropy and Volume Changes for High Pressure Phase Transitions

	$\Delta H°$ (kJ/mol)	$\Delta S°$ (J/mol K)	$\Delta V°$ (cm^3/mol)
Mg_2SiO_4 ($\alpha = \beta$)	30.0 ± 2.8 [a][2]	-7.7 ± 1.9 [a][2]	-3.16 [a][2]
Mg_2SiO_4 ($\alpha = \gamma$)	39.1 ± 2.6 [2]	-15.0 ± 2.4 [2]	-4.14 [2]
Fe_2SiO_4 ($\alpha = \beta$)	9.6 ± 1.3 [2]	-10.9 ± 0.8 [2]	-3.20 [2]
Fe_2SiO_4 ($\alpha = \gamma$)	3.8 ± 2.4 [2]	-14.0 ± 1.9 [2]	-4.24 [2]
$MgSiO_3$ (px = il)	59.1 ± 4.3 [3]	-15.5 ± 2.0 [3]	-4.94 [3]
$MgSiO_3$ (px = gt)	35.7 ± 3.0 [9]	-2.0 ± 0.5 [9]	-2.83 [9]
$MgSiO_3$ (il = pv)	51.1 ± 6.6 [17]	$+5 \pm 4$ [17]	-1.89 [17]
$Mg_2SiO_4(\gamma)= MgSiO_3(pv)+MgO$	96.8 ± 5.8 [17]	$+4 \pm 4$ [17]	-3.79 [17]
SiO_2 (q = co)	2.7 ± 0.5 [1]	-5.0 ± 0.4 [1]	-2.05 [1]
SiO_2 (co = st)	49.0 ± 1.7 [1]	-4.2 ± 1.7 [1]	-6.63 [1]

[a] *ΔH and ΔS are values at 1 atm near 1000 K, ΔV is $\Delta V°_{298}$, for all listings in table, α = olivine, β = spinelloid or wadsleyite, γ = spinel, px = pyroxene, il = ilmenite, gt = garnet, pv = perovskite, q = quartz, co = coesite, st = stishovite*

Table 6. Thermodynamic Parameters for Other Phase Transitions

Transition	$\Delta H°$ (kJ/mol)	$\Delta S°$ (J/K·mol)	$\Delta V°$ (cm^3/mol)
SiO_2 (α-quartz = β-quartz)	0.47 [a,b]	0.35	0.101
SiO_2 (β-quartz = cristobalite)	2.94 [5]	1.93	0.318
GeO_2 (rutile = quartz)	5.6 [23]	4.0	11.51
$CaSiO_3$ (wollastonite = pseudowollastonite)	5.0 [31]	3.6	0.12
Al_2SiO_5 (andalusite = sillimanite)	3.88 [5]	4.50	-0.164
Al_2SiO_5 (sillimanite = kyanite)	-8.13 [5]	-13.5	-0.571
$MgSiO_3$ (ortho = clino)	-0.37 [5]	0.16	-0.002
$MgSiO_3$ (ortho = proto)	1.59 [5]	1.27	0.109
$FeSiO_3$ (ortho = clino)	-0.17 [25]	-0.03	-0.06
$MnSiO_3$ (rhodonite = pyroxmangite)	0.25 [25]	-1.03	-0.39
$MnSiO_3$ (pyroxmangite = pyroxene)	0.88 [25]	-2.66	-0.3
$NaAlSi_3O_8$ (low albite = high albite)	13.5 [31]	14.0	0.40
$KAlSi_3O_9$ (microcline = sanidine)	11.1 [5]	15.0	0.027

[a] *Treated as though all first order, though a strong higher order component.*
[b] *ΔH and ΔS are values near 1000 K, ΔV is $\Delta V°_{298}$ for all listings in table.*

Table 7. Enthalpies of Vitrification and Fusion

Compound	Vitrification ΔH (kJ/mol)	Fusion	
		Melting Point T(K)	$\Delta H(T)$ (kJ/mol)
MgO	------------		
CaO	------------		
Al$_2$O$_3$ -	------------	2323	107.5 ± 5.4[28]
SiO$_2$ (quartz)	------------	1700[b]	9.4 ± 1.0[28]
SiO$_2$ (cristabalite)	------------	1999	8.9 ± 1.0[28]
"FeO" (wustite)	------------	1652	31.3 ± 0.2[28]
Mg$_2$SiO$_4$ (forsterite)	------------	2163	114 ± 20[a]
MgSiO$_3$ (enstatite)	42 ± 1[28]	1834[a]	77 ± 5[28]
Fe$_2$SiO$_4$ (fayatite)	------------	1490	89 ± 10[28]
CaSiO$_3$ (wollastonite)	25.5 ± 0.4[26]	1770[b]	62 ± 4[28]
CaSiO$_3$ (pseudowollastonite)	------------	1817	57 ± 3[28]
CaMgSi$_2$O$_6$ (diopside)	85.8 ± 0.8[24]	1665	138 ± 2[28]
NaAlSi$_3$O$_8$ (high albite)	51.8 ± 08[24]	1373	63 ± 20[28]
KAlSi$_3$O$_8$ (sanidine)		1473[a]	56 ± 4[28]
CaAl$_2$Si$_2$O$_8$ (anorthite)	77.8 ± 0.8[24]	1830	134 ± 4[28]
K$_2$SiO$_3$	9 ± 1[28]	1249	20 ± 4[28]
Mg$_3$Al$_2$Si$_3$O$_{12}$ (pyrope)	------------	1500[a]	243 ± 8[28]
Mg$_2$Al$_4$Si$_5$O$_{18}$ (cordierite)	209 ± 2[6]	1740	346 ± 10[28]

[a] Estimated metastable congruent melting.

[b] Melting of metastable phase.

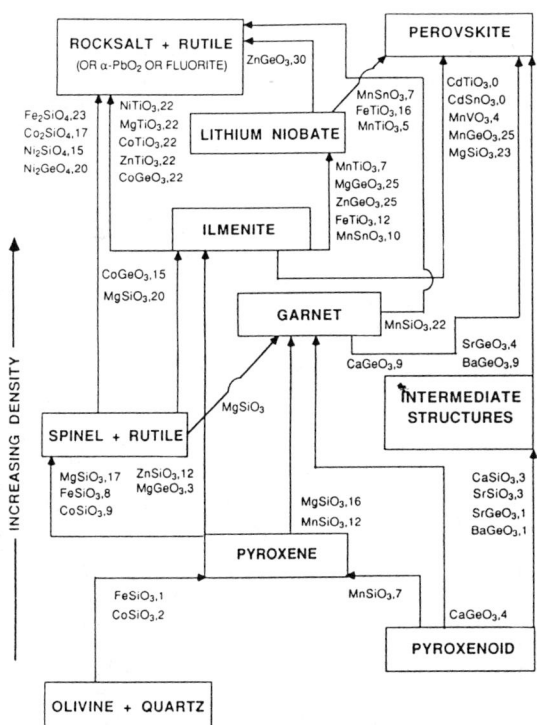

Fig. 4. Schematic diagram showing phase transitions observed in analogue systems of silicates, germanates, and titanates. Numbers refer to pressure in GPa.

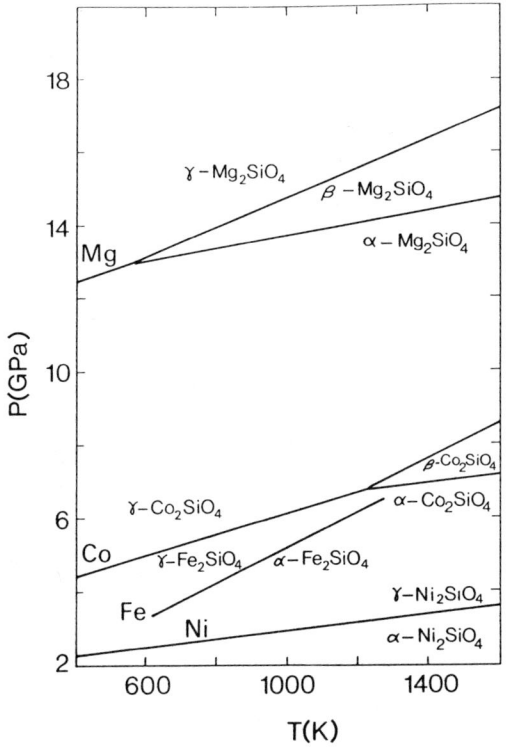

Fig. 5. Phase relations in M$_2$SiO$_4$ systems at high pressure and temperatures [25].

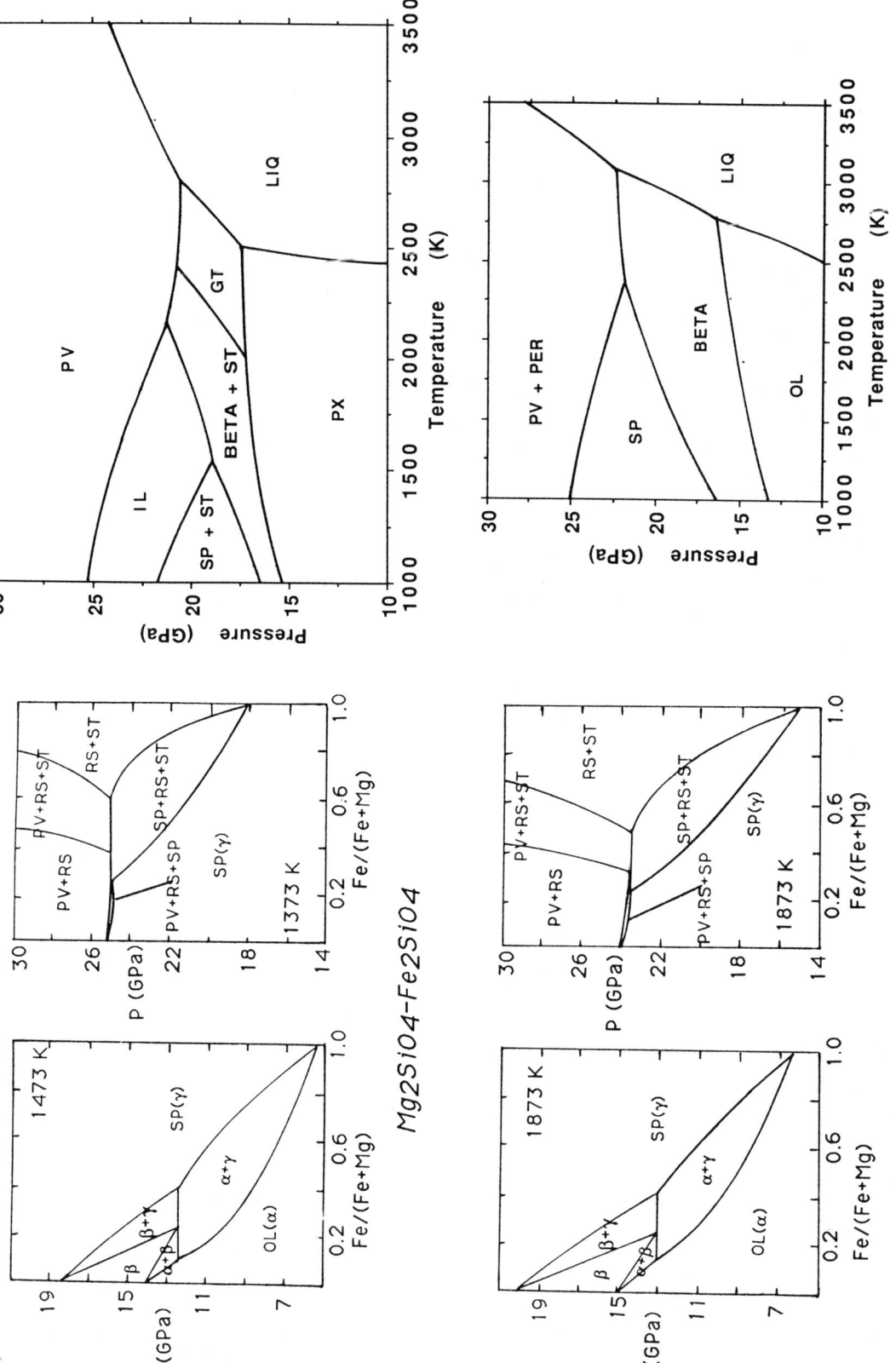

Fig. 6. Phase relations in Mg2SiO4-Fe2SiO4 as a function of pressure [10].

Fig. 7. Phase relations in MgO-SiO2 at high P and T. (a) Mg2SiO4 composition, (b) MgSiO3 composition, [9].

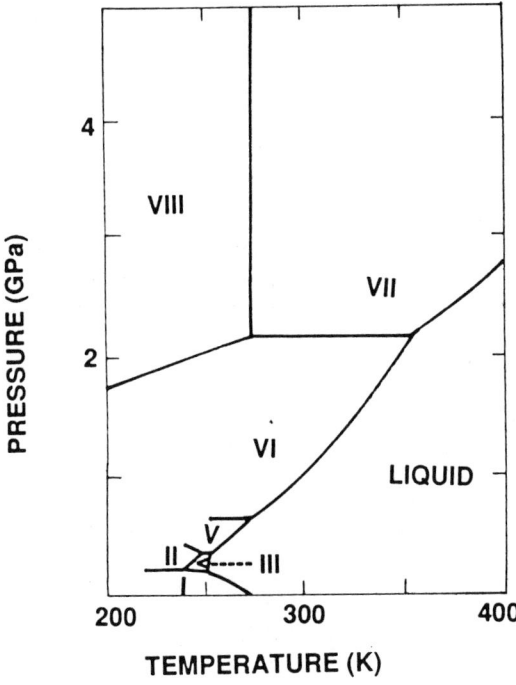

Fig. 8. Equilibrium phase relations in H_2O. Compiled from various sources [14].

substantially more compressible than the corresponding crystals. For reactions involving volatiles (e.g. H_2O and CO_2), phase boundaries are strongly curved in P-T space because the volume of the volatile (gas or fluid) phase depends very strongly on P and T. The section by Presnall gives examples of such behavior.

Table 5 lists entropy, enthalpy, and volume change for high pressure transitions of geophysical significance. Table 6 lists parameters for some other phase transitions. Table 7 presents enthalpies of vitrification (crystal → glass, not an equilibrium process) and enthalpies, entropies, and volumes of fusion at the equilibrium melting point at one atmosphere.

A number of silicates, germanates, and other materials show phase transitions among pyroxene, garnet, ilmenite, perovskite, and related structures, as shown schematically in Fig. 4. Phase relations among olivine, spinel, and beta phase in several silicates are shown in Fig. 5. Relations at high P and T for the system FeO-MgO-SiO_2 at mantle pressures are shown in Figs. 5-7. The wealth of phases in the H_2O phase diagram is shown in Fig. 8.

Acknowledgments. I thank Rebecca Lange and Elena Petrovicova for help with tables and figures.

REFERENCES

1. Akaogi, M. and A. Navrotsky, The quartz-coesite-stishovite transformations: New calorimetric measurements and calculation of phase diagrams, *Phys. Earth Planet. Inter., 36,* 124-134, 1984.
2. Akaogi, M., E. Ito, and A. Navrotsky, Olivine-modified spinel-spinel transitions in the system Mg_2SiO_4-Fe_2SiO_4: Calorimetric measurements, thermochemical calculation, and geophysical application, *J. Geophys. Res., 94,* 15,671-15,686, 1989.
3. Ashida, T., S. Kume, E. Ito, and A. Navrotsky, $MgSiO_3$ ilmenite: heat capacity, thermal expansivity, and enthalpy of transformation, *Phys. Chem. Miner.,* 16, 239-245, 1988.
4. Barin, I. and O. Knacke, Thermochemical properties of inorganic substances, pp. 921, Springer-Verlag, New York, 1973.
5. Berman, R. G., Internally-consistent thermodynamic data for minerals in the system Na_2O-K_2O-CaO-MgO-FeO-Fe_2O_3-Al_2O_3-SiO_2-TiO_2-

H_2O-CO_2, *J. Petrol., 29,* 445-522, 1988.
6. Carpenter, M. A., A. Putnis, A. Navrotsky, and J. Desmond C. McConnell, Enthalpy effects associated with Al/Si ordering in anhydrous Mg-cordierite, *Geochim. Cosmochim. Acta, 47,* 899-906, 1983.
7. Fei, Y. and S. K. Saxena, A thermochemical data base for phase equilibria in the system Fe-Mg-Si-O at high pressure and temperature, *Phys. Chem. Miner., 13,* 311-324, 1986.
8. Fei, Y. and S. K. Saxena, An equation for the heat capacity of solids, *Geochim. Cosmochim. Acta, 51,* 251-254, 1987.
9. Fei, Y., S. K. Saxena, and A. Navrotsky, Internally consistent thermodynamic data and equilibrium phase relations for compounds in the system MgO-SiO_2 at high pressure and high temperature, *J. Geophys. Res., 95,* 6915-6928, 1990.

10. Fei, Y., H.-K. Mao, and B. O. Mysen, Experimental determination of element partitioning and calculation of phase relations in the MgO-FeO-SiO_2 system at high pressure and high temperature, *J. Geophys. Res., 96,* 2157-2169, 1991.
11. Ghiorso, M. S., I. S. E. Carmichael, A regular solution model for met-aluminous silicate liquids: applications to geothermometry, immiscibility, and the source regions of basic magmas, *Contrib. Mineral. Petrol., 71,* 323-342, 1980.
12. Ghiorso, M. S., I. S. E. Carmichael, Modeling magmatic systems: petrologic applications, *Rev. Mineral., 17,* 467-499, 1987.
13. Helgeson, H. C., J. Delany, H. W. Nesbitt, and D. K. Bird, Summary and critique of the thermodynamic properties of rock-forming minerals, *Am. J. Sci., 278A,* 1-229, 1978.
14. Hemley, R. J., L. C. Chen, and H. K. Mao, New transformations between crystalline and amorphous

ice, *Nature, 338,* 638-640, 1989.

15. Holland, T. J. B., R. Powell, An internally consistent thermodynamic dataset with uncertainties and correlations: 2. Data and results, *J. Metamorphic Geol., 3,* 343-370, 1985.

16. Holland, T. J. B., R. Powell, An enlarged and updated internally consistent thermodynamic dataset with uncertainties and correlations: the system K_2O-Na_2O-CaO-MgO-MnO-FeO-Fe_2O_3-Al_2O_3-TiO_2-SiO_2-C-H_2-O_2, *J. Metamorphic Geol., 8,* 89-124, 1990.

17. Ito, E., M. Akaogi, L. Topor, and A. Navrotsky, Negative pressure-temperature slopes for reactions forming $MgSiO_3$ perovskite from calorimetry, *Science, 249,* 1275-1278, 1990.

18. JANAF, Thermochemical Tables, Third Ed., edited by American Chemical Society and American Institute of Physics, 1986.

19. Kelley, K. K., High-temperature heat content, heat capacity, and entropy data for the elements and inorganic compounds, *U.S. Bur. Mines Bull., 584,* 232 pp., 1960.

20. Kieffer, S. W., Heat capacity and entropy: systematic relation to lattice vibrations, *Rev. Mineral., 14,* 65-126, 1985.

21. Lange, R. A., J. J. DeYoreo, and A. Navrotsky, Scanning calorimetric measurement of heat capacity during incongruent melting of diopside, *Amer. Mineral., 76,* 904-912, 1991.

22. Lange, R. A. and A. Navrotsky, Heat capacities of Fe_2O_3-bearing silicate liquids, *Contrib. Mineral. Petrol., 110,* 311-320, 1992.

23. Navrotsky, A., Enthalpies of transformation among the tetragonal, hexagonal, and glassy modifications of GeO_2, *J. Inorg. Nucl. Chem. 33,* 1119-1124, 1971.

24. Navrotsky, A., R. Hon, D. F. Weill, and D. J. Henry, Thermochemistry of glasses and liquids in the systems $CaMgSi_2O_6$-$CaAl_2Si_2O_8$-$NaAlSi_3O_8$, SiO_2-$CaAl_2Si_2O_8$-$NaAlSi_3O_8$ and SiO_2-Al_2O_3-CaO-Na_2O, *Geochim. Cosmochim. Acta, 44,* 1409-1423, 1980.

25. Navrotsky, A., High pressure transitions in silicates, *Prog. Solid St. Chem., 17,* 53-86, 1987.

26. Navrotsky, A., D. Ziegler, R. Oestrike, and P. Maniar, Calorimetry of silicate melts at 1773 K: Measurement of enthalpies of fusion and of mixing in the systems diopside-anorthite-albite and anorthite-forsterite, *Contrib. Mineral. Petrol., 101,* 122-130, 1989.

27. Richet, P., Viscosity and configurational entropy of silicate melts, *Geochim. Cosmochim. Acta., 48,* 471-483, 1984.

28. Richet, P., and Y. Bottinga, Thermochemical properties of silicate glasses and liquids: A review, *Rev. Geophys., 24,* 1-25, 1986.

29. Richet, P., Heat capacity of silicate glasses, *Chem. Geol., 62,* 111-124, 1987.

30. Richet, P., R. A. Robie, and B. S. Hemingway, Thermodynamic properties of wollastonite, pseudowollastonite and $CaSiO_3$ glass and liquid, *Europ. J. Mineral., 3,* 475-485, 1991.

31. Robie, R. A., B. S. Hemingway, and J. R. Fisher, Thermodynamic properties of minerals and related substances at 298.15 K and 1 bar (10^5 pascals) and at higher temperatures, *U. S. Geol. Surv. Bull., 1452,* 456 pp., 1978.

32. Robie, R. A., C. B. Finch, and B. S. Hemingway, Heat capacity and entropy of fayalite (Fe_2SiO_4) between 5.1 and 383 K; comparison of calorimetric and equilibrium values for the QFM buffer reactor, *Amer. Mineral, 67,* 463-469, 1982.

33. Stebbins, J. F., I. S. E. Carmichael, and L. K. Moret, Heat capacities and entropies of silicate liquids and glasses, *Contrib. Mineral. Petrol., 86,* 131-148, 1984.

Thermal Expansion

Yingwei Fei

Since Skinner [75] compiled the thermal expansion data of substances of geological interest, many new measurements on oxides, carbonates, and silicates have been made by x-ray diffraction, dilatometry, and interferometry. With the development of high-temperature x-ray diffraction techniques in the seventies, thermal parameters of many rock-forming minerals were measured [e.g., 14, 22, 28, 45, 68, 77, 97, 99]. Considerable thermal expansion data for important mantle-related minerals such as periclase, stishovite, olivine, wadsleyite, silicate spinel, silicate ilmenite and silicate perovskite were collected by x-ray diffraction methods [e.g., 4, 39, 42, 71] and by dilatometric and interferometric techniques [e.g., 54, 86, 88, 89]. While the data set for 1-bar thermal expansion is expanding, many efforts have recently been made to obtain the pressure effect on thermal expansivity [e.g.,9, 19, 21, 36, 51]. In study of liquid density, a systematic approach is taken to obtain density and its temperature dependence of natural liquids [e.g., 11, 12, 16, 44, 46, 48].

The thermal expansion coefficient α, defined by $\alpha = 1/V(\partial V/\partial T)_P$, is used to express the volume change of a substance due to a temperature change. In a microscopic sense, the thermal expansion is caused by the anharmonic nature of the vibrations in a potential-well model [103]. The Grüneisen theory of thermal expansion leads to a useful relation between volume and temperature [90],

Y. Fei, Carnegie Institution of Washington, Geophysical Laboratory, 5251 Broad Branch Road, NW, Washington, DC 20015-1305

Mineral Physics and Crystallography
A Handbook of Physical Constants
AGU Reference Shelf 2

$$V(T) = \frac{V_0}{2k}\left[1 + 2k - (1 - 4kE/Q_0)^{1/2}\right] \qquad (1)$$

where E is the energy of the lattice vibrations. The constant Q_0 is related to volume (V_0) and bulk modulus (K_0) at zero Kelvin, and the Grüneisen parameter (γ) by $Q_0 = K_0 V_0/\gamma$. The constant k is obtained by fitting to the experimental data. In the Debye model of solids with a characteristic temperature, θ_D, the energy E can be calculated by

$$E = \frac{9nRT}{(\theta_D/T)^3}\int_0^{\theta_D/T} \frac{x^3}{e^x - 1}dx \qquad (2)$$

where n and R are the number of atoms in the chemical formula and the gas constant, respectively.

In this model, four parameters, θ_D, Q_0, k, and V_0, are required to describe the thermal expansion of a solid. When the thermal expansion is accurately measured over a wide temperature range, the four parameters may be uniquely defined by fitting the experimental data to the model. Furthermore, measurements on heat capacity and bulk modulus can provide additional constraints on the model. A simultaneous evaluation of thermal expansion, bulk modulus, and heat capacity through a self-consistent model such as the Debye model [e.g., 81] is, therefore, recommended, especially when extrapolation of data is involved.

In many cases the above model cannot be uniquely defined, either because the accuracy of thermal expansion measurement is not sufficiently high or because the temperature range of measurement is limited. For the purpose of fitting experimental data over a specific temperature range, a polynomial expression for the thermal expansion coefficient may be used

$$\alpha(T) = a_0 + a_1 T + a_2 T^{-2} \qquad (3)$$

where a_0, a_1, and $a_2 (\leq 0)$ are constants determined by fitting the experimental data. The measured volume above room temperature can be well reproduced by

$$V(T_j) = V_{Tr} exp\left[\int_{T_r}^{T} \alpha(T)dT\right] \qquad (4)$$

where V_{Tr} is the volume at reference temperature (T_r), usually room temperature. When the thermal expansion coefficient is independent of temperature over the measured temperature range,

$$V(T) = V_{Tr} exp[\alpha_0(T - T_r)] \qquad (5)$$

The commonly used mean thermal expansion coefficient ($\overline{\alpha}$) can be related to equation (5) by truncating the exponential series of $exp[\alpha_0(T - T_r)]$ at its second order, i.e.,

$$V(T) = V_{Tr} [1 + \overline{\alpha}(T - T_r)] \qquad (6)$$

Table 1 lists thermal expansion coefficients of solids. The coefficients for most substances were obtained by fitting the experimental data to equations (3) and (4). The mean coefficient ($\overline{\alpha}$), listed in the literature, can be converted to α_0, according to equations (5) and (6).

Thermal expansion coefficients of elements and halides (e.g., NaCl, KCl, LiF, and KBr) are not included in this compilation because the data are available in the American Institute of Physics Handbook [41]. Volumes 12 and 13 of Thermophysical Properties of Matter [92,

93] are also recommended as data sources.

The pressure effect on the thermal expansion coefficient may be described by the Anderson-Grüneisen parameter (δ_r),

$$\alpha(P,T)/\alpha(T) = [V(P,T)/V(T)]^{\delta_T} \qquad (7)$$

The thermal expansion coefficient as a function of pressure can be calculated from equation (7) and the third order Birch-Murnaghan equation of state,

$$P = 3f(1 + 2f)^{5/2} K_T \left[1 - \frac{3}{2}(4 - K_T')f\right] \qquad (8)$$

and

$$f = \frac{1}{2}\left[\left(\frac{V(T)}{V(P,T)}\right)^{2/3} - 1\right] \qquad (9)$$

where K_T and K_T' are the bulk modulus and its pressure derivative, respectively. Table 2 lists the values of K_T, K_T', and δ_T for some mantle-related minerals.

The liquid molar volume of a multioxide liquid can be calculated by

$$V_{liq}(T) = \sum X_i \overline{V}_{i,Tr}\left[1 + \overline{\alpha}_i(T - T_r)\right] + V^{ex} \qquad (10)$$

where X_i and $\overline{\alpha}_i$ are the mole fraction and mean thermal expansion coefficient of oxide component i, respectively. $\overline{V}_{i,Tr}$ is the partial molar volume of component i in the liquid at a reference temperature, T_r, and V^{ex} is the excess volume term. Recent measurements on density and thermal expansion coefficient of silicate liquid are summarized in Tables 3a-3d.

TABLE 1. Thermal Expansion Coefficients of Solids

Names		T range	$\alpha_0 (10^{-6})$	$a_0 (10^{-4})$	$a_1 (10^{-8})$	a_2	ref.
Oxides							
αAl_2O_3, corundum	a	293-2298 K	7.3	0.0758	0.1191	-0.0603	[2]
	c	293-2298 K	8.3	0.0773	0.1743	0.0000	[2]
	V	293-2298 K	23.0	0.2276	0.4198	-0.0897	[2]

TABLE 1. (continued)

Names		T range	α_0 (10^{-6})	a_0 (10^{-4})	a_1 (10^{-8})	a_2	ref.
BeAl$_2$O$_4$, chrysoberyl	a	298-963 K	6.6	0.0250	1.3569	0.0000	[30]
	b	298-963 K	8.7	0.0490	1.2777	0.0000	[30]
	c	298-963 K	7.6	0.0540	0.7315	0.0000	[30]
	V	298-963 K	23.8	0.1320	3.5227	0.0000	[30]
BeO	V	292-1272 K	17.8	0.1820	1.3933	-0.4122	[93, cf. 29]
CaO	V	293-2400 K	33.5	0.3032	1.0463	0.0000	[93]
3CaO·Al$_2$O$_3$	V	293-1473 K	19.5	0.2555	0.7564	-0.7490	[75][a]
17CaO·7Al$_2$O$_3$	V	298-1073 K	12.3	0.1230	0.0000	0.0000	[15]
CaO·Al$_2$O$_3$	V	293-1473 K	10.5	0.2232	0.0259	-1.0687	[75]
Co$_3$O$_4$, normal spinel	V	301-995 K	14.8	0.0631	2.8160	0.0000	[49]
Cr$_2$O$_3$, eskolaite	V	293-1473 K	18.6	0.2146	0.1154	-0.2904	[75]
FeAl$_2$O$_4$, hercynite	V	293-1273 K	15.6	0.0977	1.9392	0.0000	[75]
FeCr$_2$O$_4$, chromite	V	293-1273 K	9.9	0.0513	1.5936	0.0000	[75]
FeO, wüstite	V	293-873 K	33.9	0.3203	0.6293	0.0000	[75]
Fe$_2$O$_3$, hematite	a	293-673 K	7.9	0.0350	1.4836	0.0000	[75]
	c	293-673 K	8.0	0.0559	0.7904	0.0000	[75]
	V	293-673 K	23.8	0.1238	3.8014	0.0000	[75]
Fe$_3$O$_4$, magnetite	V	293-843 K	20.6	-0.0353	8.0591	0.0000	[75]
	V	843-1273 K	50.1	0.5013	0.0000	0.0000	[75]
FeTiO$_3$, ilmenit	a	297-1323 K	10.1	0.1006	0.0000	0.0000	[95]
	c	297-1323 K	7.6	0.0638	0.4031	0.0000	[95]
	V	297-1323 K	27.9	0.2689	0.3482	0.0000	[95]
HfO$_2$	V	293-1273 K	15.8	0.1264	1.0368	0.0000	[75]
MgAl$_2$O$_4$, normal spinel	V	293-873 K	24.9	0.2490	0.0000	0.0000	[102]
MgAl$_2$O$_4$, disordered spinel	V	993-1933 K	29.4	0.2940	0.0000	0.0000	[102]
MgCr$_2$O$_4$, picrochromite	V	293-1473 K	16.5	0.1430	1.1191	-0.1063	[75]
MgFe$_2$O$_4$, magnesioferrite	V	293-1473 K	20.5	0.3108	1.2118	-1.2773	[75]
MgGeO$_3$, ilmenite	V	299-1023 K	22.4	0.2244	0.0000	0.0000	[3]
Mg$_2$GeO$_4$, olivine	V	298-1273 K	41.1	0.4110	0.0000	0.0000	[72]
Mg$_2$GeO$_4$, spinel	V	298- 1273 K	32.1	0.3210	0.0000	0.0000	[72]
MgO, periclase	V	303-1273 K	31.6	0.3768	0.7404	-0.7446	[86]
MnO, manganosite	V	293-1123 K	34.5	0.3317	1.2055	-0.2094	[90]
ThO$_2$, thorianite	V	293-1273 K	28.5	0.2853	0.0000	0.0000	[75, cf. 96]
TiO	V	293-1073 K	22.3	0.1832	1.3236	0.0000	[75]
TiO$_2$, rutile	a	298-1883 K	8.9	0.0890	0.0000	0.0000	[85]
	c	298-1883 K	11.1	0.1110	0.0000	0.0000	[85]
	V	298-1883 K	28.9	0.2890	0.0000	0.0000	[85]
UO$_{2.03}$, uraninite	V	293-1273 K	24.5	0.2180	1.2446	-0.0920	[75, cf. 96]
ZrO$_2$, baddeleyite	V	293-1273 K	21.2	0.2042	0.2639	0.0000	[75]
Hydrous minerals							
AlOOH, boehmite	a	100-530 K	9.7	-0.0048	3.4000	0.0000	[7]
	b	100-530 K	25.3	-0.0232	9.2000	0.0000	[7]
	c	100-530 K	0.7	0.0005	0.2000	0.0000	[7]
	V	100-530 K	35.7	-0.0275	12.8000	0.0000	[7]

TABLE 1. (continued)

Names		T range	α_0 (10^{-6})	a_0 (10^{-4})	a_1 (10^{-8})	a_2	ref.
$Ca_2Mg_5Si_8O_{22}(OH)_2$	a	297-973 K	12.0	0.1202	0.0000	0.0000	[83]
tremolite	b	297-973 K	11.7	0.1167	0.0000	0.0000	[83]
	c	297-973 K	5.8	0.0583	0.0000	0.0000	[83]
	β	297-973 K	-2.7	-0.0266	0.0000	0.0000	[83]
	V	297-973 K	31.3	0.3131	0.0000	0.0000	[83]
$KAl_2(AlSi_3O_{10})(OH)_2$	a	293-1073 K	9.9	0.0994	0.0000	0.0000	[25]
muscovite	b	293-1073 K	11.1	0.1110	0.0000	0.0000	[25]
	c	293-1073 K	13.8	0.1379	0.0000	0.0000	[25]
	d001	293-1073 K	13.7	0.1367	0.0000	0.0000	[25]
	V	293-1073 K	35.4	0.3537	0.0000	0.0000	[25]
$Mg(OH)_2$, brucite	a	300-650 K	11.0	0.1100	0.0000	0.0000	[19]
	c	300-650 K	59.0	0.5900	0.0000	0.0000	[19]
	V	300-650 K	80.0	0.8000	0.0000	0.0000	[19]
Carbonates							
$BaCO_3$ (hexagonal)	a	1093-1233 K	-102.0	-1.0200	0.0000	0.0000	[43]
	c	1093-1233 K	297.0	2.9700	0.0000	0.0000	[43]
	V	1093-1233 K	93.0	0.9300	0.0000	0.0000	[43]
$CaCO_3$, aragonite	a	293-673 K	8.3	0.0833	0.0000	0.0000	[75]
	b	293-673 K	18.6	0.1862	0.0000	0.0000	[75]
	c	293-673 K	35.2	0.3520	0.0000	0.0000	[75]
	V	293-673 K	62.2	0.6221	0.0000	0.0000	[75]
$CaCO_3$, calcit	a	297-1173 K	-3.2	-0.0315	0.0000	0.0000	[53]
	c	297-1173 K	13.3	0.1922	2.5183	-1.2140	[53]
	V	297-1173 K	3.8	0.0713	3.3941	-1.2140	[53]
$CdCO_3$, otavite	a	293-593 K	-5.6	-0.0560	0.0000	0.0000	[5]
	c	293-593 K	22.7	0.2270	0.0000	0.0000	[5]
	V	293-593 K	11.5	0.1150	0.0000	0.0000	[5]
$CaMg(CO_3)_2$, dolomite	a	297-973 K	3.2	0.0271	0.6045	-0.1152	[70]
	c	297-973 K	15.6	0.1233	2.2286	-0.3089	[70]
	V	297-973 K	22.8	0.1928	3.1703	-0.5393	[70]
$MgCO_3$, magnesite	a	297-773 K	2.2	0.0775	0.2934	-0.5809	[53]
	c	297-773 K	13.2	0.0037	4.2711	0.0000	[53]
	V	297-773 K	18.2	0.1686	4.7429	-1.1618	[53]
$MnCO_3$, rhodochrosite	a	297-773 K	1.8	0.0180	0.0000	0.0000	[69]
	c	297-773 K	19.2	0.1920	0.0000	0.0000	[69]
	V	297-773 K	22.8	0.2280	0.0000	0.0000	[69]
$FeCO_3$, siderite	a	297-773 K	5.4	0.0540	0.0000	0.0000	[64]
	c	297-773 K	16.1	0.1610	0.0000	0.0000	[64]
	V	297-773 K	26.9	0.2690	0.0000	0.0000	[64]
$SrCO_3$, strontianite	a	293-1073 K	7.1	0.0508	0.6630	0.0000	[75]
	b	293-1073 K	12.1	0.1107	0.3362	0.0000	[75]
	c	293-1073 K	36.5	0.2629	3.4137	0.0000	[75]
	V	293-1073 K	59.2	0.4982	3.1111	0.0000	[75]

TABLE 1. (continued)

Names		T range	α_0 (10^{-6})	a_0 (10^{-4})	a_1 (10^{-8})	a_2	ref.
Sulfides and Sulfates							
FeS$_2$, pyrite	V	293-673 K	25.7	0.1256	4.3873	0.0000	[75]
PbS, galena	V	293-873 K	58.1	0.5027	2.6125	0.0000	[75]
ZnS, sphalerite	V	293-1273 K	17.8	0.2836	0.0000	-0.9537	[75]
ZnS, wurtzite	a	293-1273 K	6.7	0.0763	0.3815	-0.1885	[75]
	c	293-1273 K	6.5	0.0762	0.1134	-0.1274	[75]
	V	293-1273 K	19.0	0.2136	1.0938	-0.5061	[75]
BaSO$_4$, barite	a	298-1158 K	20.7	0.2070	0.0000	0.0000	[73]
	b	298-1158 K	25.5	0.2550	0.0000	0.0000	[73]
	c	298-1158 K	17.2	0.1720	0.0000	0.0000	[73]
	V	298-1158 K	63.7	0.6370	0.0000	0.0000	[73]
K$_2$SO$_4$	a	293-673 K	15.5	-0.1713	10.8705	0.0000	[75]
	b	293-673 K	33.4	0.3337	0.0000	0.0000	[75]
	c	293-673 K	42.6	0.1628	8.7701	0.0000	[75]
	V	293-673 K	91.4	0.3252	19.6406	0.0000	[75]
Silicates							
Akermanite, Ca$_2$MgSi$_2$O$_7$	a	293-693 K	10.7	0.1065	0.0000	0.0000	[31]
	c	293-693 K	5.9	0.0346	0.8280	0.0000	[31]
	V	293-693 K	27.1	0.2453	0.8700	0.0000	[31]
Andalusite, Al$_2$SiO$_5$	a	298-1273 K	12.5	0.1223	0.0963	0.0000	[97]
	b	298-1273 K	8.1	0.0753	0.1918	0.0000	[97]
	c	298-1273 K	2.3	0.0233	0.0000	0.0000	[97]
	V	298-1273 K	22.8	0.2181	0.3261	0.0000	[97]
Beryl, Be$_3$Al$_2$Si$_6$O$_{18}$	a	298-1073 K	2.6	0.0260	0.0000	0.0000	[58]
	c	298-1073 K	-2.9	-0.0290	0.0000	0.0000	[58]
	V	298-1073 K	2.3	0.0230	0.0000	0.0000	[58]
Calcium silicates							
Ca$_3$Si$_2$O$_7$, rankinite	V	293-1473 K	33.1	0.2883	1.4106	0.0000	[75]
β-Ca$_2$SiO$_4$	V	293-1473 K	31.4	0.4601	0.0158	-1.3157	[75]
Ca$_3$SiO$_5$	V	293-1273 K	25.7	0.1852	2.4073	0.0000	[75]
Cancrinite	a	298-673 K	7.0	0.0034	2.2150	0.0000	[75]
	c	298-673 K	16.1	0.0328	4.2629	0.0000	[75]
	V	298-673 K	29.9	0.0364	8.7589	0.0000	[75]
Cordierite							
Mg$_2$Al$_4$Si$_5$O$_{18}$ (hexagonal)	a	298-873 K	2.2	0.0220	0.0000	0.0000	[35]
	c	298-873 K	-1.8	-0.0180	0.0000	0.0000	[35]
	V	298-873 K	2.6	0.0260	0.0000	0.0000	[35, cf. 67][b]
β-Eucryptite, LiAlSiO$_4$	a	296-920 K	8.6	0.0860	0.0000	0.0000	[66]
	c	296-920 K	-18.4	-0.1840	0.0000	0.0000	[66]
	V	296-920 K	-1.2	-0.0120	0.0000	0.0000	[66]
Feldspars							
Celsian, BaAl$_2$Si$_2$O$_8$	V	293-673 K	8.7	0.0605	0.8692	0.0000	[75]
High Albite, NaAlSi$_3$O$_8$	a	297-1378 K	9.6	0.0716	0.8114	0.0000	[68]
	b	297-1378 K	6.6	0.0656	0.0000	0.0000	[68]
	c	297-1378 K	5.2	0.0523	0.0000	0.0000	[68]

TABLE 1. (continued)

Names		T range	α_0 (10^{-6})	a_0 (10^{-4})	a_1 (10^{-8})	a_2	ref.
	α	297-1378 K	-2.1	0.1603	-6.0284	0.0000	[68]
	β	297-1378 K	-2.3	-0.0197	-0.1120	0.0000	[68]
	γ	297-1378 K	-2.6	-0.0252	-0.0252	0.0000	[68]
	V	297-1378 K	26.8	0.2455	0.7621	0.0000	[68, cf. 99]
Low Albite, $NaAlSi_3O_8$	a	298-1243 K	11.7	0.0882	0.9479	0.0000	[98]
	b	298-1243 K	4.7	0.0371	0.3400	0.0000	[98]
	c	298-1243 K	0.3	-0.0113	0.4618	0.0000	[98]
	α	298-1243 K	-2.7	0.0263	-1.7927	0.0000	[98]
	β	298-1243 K	-5.2	-0.0547	0.0987	0.0000	[98]
	γ	298-1243 K	-0.5	0.0061	-0.3641	0.0000	[98]
	V	298-1243 K	22.6	0.1737	1.7276	0.0000	[98]
Adularia, $Or_{88.3}Ab_{9.3}An_{2.4}$	V	293-1273 K	11.2	0.1846	0.5719	-0.8088	[75]
Microcline, $Or_{83.5}Ab_{16.5}$	V	293-1273 K	15.6	0.1297	0.8683	0.0000	[75]
Orthoclase, $Or_{66.6}Ab_{32.8}An_{0.6}$	V	293-1273 K	9.7	-0.0097	3.5490	0.0000	[75]
Plagioclase, $Ab_{99}An_1$	V	293-1273 K	15.4	0.2199	1.0271	-0.8714	[75]
Plagioclase, $Ab_{77}An_{23}$	V	293-1273 K	8.9	0.1612	0.7683	-0.8603	[75]
Plagioclase, $Ab_{56}An_{44}$	V	293-1273 K	10.6	0.1524	0.5038	-0.5550	[75]
Plagioclase, Ab_5An_{95}	V	293-1273 K	14.1	0.1394	0.0597	0.0000	[75, cf. 24[c]]
Garnets							
Almandite, $Fe_3Al_2Si_3O_{12}$	V	294-1044 K	15.8	0.1776	1.2140	-0.5071	[75]
Andradite, $Ca_3Fe_2Si_3O_{12}$	V	294-963 K	20.6	0.2103	0.6839	-0.2245	[75]
Cacium-rich garnet	V	300-1000 K	20.2	0.2647	0.3080	-0.6617	[38]
Grossularite, $Ca_3Al_2Si_3O_{12}$	V	292-980 K	16.4	0.1951	0.8089	-0.4972	[75]
Pyrope, $Mg_3Al_2Si_3O_{12}$	V	283-1031 K	19.9	0.2311	0.5956	-0.4538	[75]
Spessartite, $Mn_3Al_2Si_3O_{12}$	V	292-973 K	17.2	0.2927	0.2726	-1.1560	[75]
Natural garnet (pyrope-rich)	V	298-1000 K	23.6	0.2880	0.2787	-0.5521	[87]
Gehlenite, $Ca_2Al_2SiO_7$	V	293-1473 K	24.0	0.2320	0.2679	0.0000	[75]
Hornblende	V	293-1273 K	23.8	0.2075	1.0270	0.0000	[75]
Kyanite, Al_2SiO_5	a	298-1073 K	7.5	0.0749	0.0000	0.0000	[97]
	b	298-1073 K	6.6	0.0661	0.0000	0.0000	[97]
	c	298-1073 K	10.9	0.1095	0.0000	0.0000	[97]
	V	298-1073 K	25.1	0.2505	0.0000	0.0000	[97]
Merwinite, $Ca_3Mg(SiO_4)_2$	V	293-1473 K	29.8	0.2521	1.5285	0.0000	[75]
Mullite,							
$Al_2O_3(71.2\%)SiO_2(28.6\%)$	a	573-1173 K	3.9	0.0390	0.0000	0.0000	[74]
	b	573-1173 K	7.0	0.0700	0.0000	0.0000	[74]
	c	573-1173 K	5.8	0.0580	0.0000	0.0000	[74]
	V	573-1173 K	16.7	0.1670	0.0000	0.0000	[74]
$Al_2O_3(60.0\%)SiO_2(28.4\%)Cr$		573-1173 K	3.1	0.0310	0.0000	0.0000	[74]
	b	573-1173 K	6.2	0.0620	0.0000	0.0000	[74]
	c	573-1173 K	5.6	0.0560	0.0000	0.0000	[74]
	V	573-1173 K	14.9	0.1490	0.0000	0.0000	[74]
$Al_2O_3(62.1\%)SiO_2(27.4\%)Fe$	a	573-1173 K	3.3	0.0330	0.0000	0.0000	[74]
	b	573-1173 K	7.0	0.0700	0.0000	0.0000	[74]
	c	573-1173 K	5.6	0.0560	0.0000	0.0000	[74]
	V	573-1173 K	15.9	0.1590	0.0000	0.0000	[74]

TABLE 1. (continued)

Names		T range	α_0 (10^{-6})	a_0 (10^{-4})	a_1 (10^{-8})	a_2	ref.
Nephelines							
$(Na_{0.78}K_{0.22})AlSiO_4$	a	293-1073 K	11.1	0.0512	1.9931	0.0000	[75]
	c	293-1073 K	8.3	0.0665	0.5544	0.0000	[75]
	V	293-1073 K	31.3	0.1889	4.1498	0.0000	[75]
$(Na_{0.59}K_{0.41})AlSiO_4$	a	293-1073 K	19.5	0.1952	-0.0211	0.0000	[75]
	c	293-1073 K	19.8	0.2627	-2.1428	0.0000	[75]
	V	293-1073 K	58.5	0.6515	-2.2071	0.0000	[75]
Olivines							
$CaMg_{0.97}Fe_{0.07}SiO_4$	a	298-1068 K	6.4	0.0855	0.1308	-0.2331	[45]
monticellite	b	298-1068 K	7.4	0.0965	0.1806	-0.2575	[45]
	c	298-1068 K	10.3	0.1235	0.4236	-0.2891	[45]
	V	298-1068 K	24.2	0.3114	0.6733	-0.8133	[45]
$CaMn(MgZn)SiO_4$	a	298-1073 K	6.5	0.0976	0.2233	-0.3605	[45]
glaucochroite	b	298-1073 K	6.4	0.0953	0.2091	-0.3536	[45]
	c	298-1073 K	7.2	0.1055	0.2783	-0.3852	[45]
	V	298-1073 K	20.3	0.3007	0.7192	-1.1080	[45]
Mg_2SiO_4, forsterite	a	303-1173 K	6.6	0.0663	0.3898	-0.0918	[86]
	b	303-1173 K	9.9	0.1201	0.2882	-0.2696	[86]
	c	303-1173 K	9.8	0.1172	0.0649	-0.1929	[86]
	V	303-1173 K	26.4	0.3034	0.7422	-0.5381	[86]
Mg_2SiO_4, forsterite	V	296-1293 K	30.6	0.2635	1.4036	0.0000	[28]
Mg_2SiO_4, forsterite	V	298-1273 K	28.2	0.3407	0.8674	-0.7545	[54]
Mg_2SiO_4, forsterite	V	300-1300 K	27.3	0.2854	1.0080	-0.3842	[40]
Mn_2SiO_4, tephroite	a	298-1123 K	5.8	0.0397	0.5249	0.0621	[61]
	b	298-1123 K	8.8	0.1042	0.2744	-0.2188	[61]
	c	298-1123 K	8.0	0.0807	0.3370	-0.0853	[61]
	V	298-1123 K	22.6	0.2307	1.0740	-0.2898	[61]
Ni_2SiO_4, Ni-olivine	a	298-1173 K	9.5	0.1049	0.2093	-0.1409	[45]
	b	298-1173 K	8.9	0.0990	0.1746	-0.1387	[45]
	c	298-1173 K	9.0	0.1004	0.1827	-0.1396	[45]
	V	298-1173 K	27.3	0.3036	0.5598	-0.4204	[45]
Fe_2SiO_4, fayalite	a	298-1123 K	5.5	0.1050	0.0602	-0.4958	[91]
	b	298-1123 K	7.9	0.0819	0.1629	-0.0694	[91]
	c	298-1123 K	9.9	0.1526	-0.1217	-0.4594	[91]
	V	298-1123 K	26.1	0.2386	1.1530	-0.0518	[91, 76, 27]
$(Mg_{0.7}Fe_{0.3})_2SiO_4$	a	297-983 K	6.1	0.0610	0.0000	0.0000	[13]
hortonolite	b	297-983 K	9.6	0.0960	0.0000	0.0000	[13]
	c	297-983 K	9.7	0.0975	0.0000	0.0000	[13]
	V	297-983 K	25.5	0.2557	0.0000	0.0000	[13]
$Mg_{0.75}Fe_{1.10}Mn_{0.15}SiO_4$	a	296-1173 K	9.2	0.0916	0.0000	0.0000	[27]
hortnolite	b	296-1173 K	11.1	0.1109	0.0000	0.0000	[27]
	c	296-1173 K	14.6	0.1456	0.0000	0.0000	[27]
	V	296-1173 K	35.0	0.3504	0.0000	0.0000	[27, cf. 79]
Perovskite							
$MgSiO_3$	a	77-298 K	8.4	0.0840	0.0000	0.0000	[71]
	b	77-298 K	0.0	0.0000	0.0000	0.0000	[71]

TABLE 1. (continued)

Names		T range	α_0 (10^{-6})	a_0 (10^{-4})	a_1 (10^{-8})	a_2	ref.
	c	77-298 K	5.9	0.0590	0.0000	0.0000	[71]
	V	77-298 K	14.5	0.1450	0.0000	0.0000	[71]
	V	298-381 K	22.0	0.2200	0.0000	0.0000	[71]
$(Mg_{0.9}Fe_{0.1})SiO_3$	a	100-250 K	5.8	0.0580	0.0000	0.0000	[62]
	b	100-250 K	5.2	0.0520	0.0000	0.0000	[62]
	c	100-250 K	4.5	0.0450	0.0000	0.0000	[62]
	V	100-250 K	15.5	0.1550	0.0000	0.0000	[62]
	a	250-373 K	8.1	0.0810	0.0000	0.0000	[62]
	b	250-373 K	5.4	0.0540	0.0000	0.0000	[62]
	c	250-373 K	5.4	0.0540	0.0000	0.0000	[62]
	V	250-373 K	18.9	0.1890	0.0000	0.0000	[62]
	V	150-373 K	19.0	0.1900	0.0000	0.0000	[62]
$(Mg_{0.9}Fe_{0.1})SiO_3$	V	298-840 K	30.7	0.3156	0.9421	-0.3271	[42]
Phenakite, Be_2SiO_4	a	298-963 K	5.2	0.0520	0.0000	0.0000	[30]
	c	298-963 K	6.4	0.0640	0.0000	0.0000	[30]
	V	298-963 K	16.8	0.1680	0.0000	0.0000	[30]
Pseudowollastonite, $CaSiO_3$	V	293-1473 K	27.8	0.2474	1.0096	0.0000	[75]
Pyroxenes							
$CaAl_2SiO_6$, CaTs	a	298-1473 K	8.8	0.0882	0.0000	0.0000	[26]
	b	298-1473 K	12.0	0.1204	0.0000	0.0000	[26]
	c	298-1473 K	8.9	0.0888	0.0000	0.0000	[26]
	V	298-1473 K	27.8	0.2780	0.0000	0.0000	[26]
$CaMgSi_2O_6$, diopside	a	297-1273 K	7.8	0.0779	0.0000	0.0000	[14]
	b	297-1273 K	20.5	0.2050	0.0000	0.0000	[14]
	c	297-1273 K	6.5	0.0646	0.0000	0.0000	[14]
	d100	297-1273 K	6.1	0.0606	0.0000	0.0000	[14]
	V	297-1273 K	33.3	0.3330	0.0000	0.0000	[14, cf. 22]
$Ca_{0.015}Mg_{0.305}Fe_{0.68}SiO_3$	a	293-1123 K	13.5	0.1350	0.0000	0.0000	[78]
ferrohypersthene	b	293-1123 K	14.5	0.1450	0.0000	0.0000	[78]
	c	293-1123 K	15.4	0.1540	0.0000	0.0000	[78]
	V	293-1123 K	43.8	0.4380	0.0000	0.0000	[78]
$Ca_{0.015}Mg_{0.305}Fe_{0.68}SiO_3$	a	293-973 K	16.2	0.1620	0.0000	0.0000	[77]
clinohypersthene	b	293-973 K	10.4	0.1040	0.0000	0.0000	[77]
	c	293-973 K	13.8	0.1380	0.0000	0.0000	[77]
	d100	293-973 K	8.3	0.0830	0.0000	0.0000	[77]
	V	293-973 K	32.7	0.3270	0.0000	0.0000	[77]
$CaFeSi_2O_6$, hedenbergite	a	297-1273 K	7.2	0.0724	0.0000	·0.0000	[14]
	b	297-1273 K	17.6	0.1760	0.0000	0.0000	[14]
	c	297-1273 K	6.0	0.0597	0.0000	0.0000	[14]
	d100	297-1273 K	4.8	0.0483	0.0000	0.0000	[14]
	V	297-1273 K	29.8	0.2980	0.0000	0.0000	[14]
$Ca_{0.15}Fe_{0.85}SiO_3$, FsWo	a	297-773 K	18.9	0.1890	0.0000	0.0000	[60]
	b	297-773 K	13.3	0.1330	0.0000	0.0000	[60]
	c	297-773 K	15.2	0.1520	0.0000	0.0000	[60]
	d100	297-773 K	8.9	0.0893	0.0000	0.0000	[60]
	V	297-773 K	37.6	0.3760	0.0000	0.0000	[60]

TABLE 1. (continued)

Names		T range	$\alpha_0\,(10^{-6})$	$a_0\,(10^{-4})$	$a_1\,(10^{-8})$	a_2	ref.
$FeSiO_3$, orthoferrosilite	a	297-1253 K	11.2	0.1120	0.0000	0.0000	[84]
	b	297-1253 K	10.9	0.1090	0.0000	0.0000	[84]
	c	297-1253 K	16.8	0.1680	0.0000	0.0000	[84]
	V	297-1253 K	39.3	0.3930	0.0000	0.0000	[84]
$LiAlSi_2O_6$, spodumene	a	297-1033 K	3.8	0.0380	0.0000	0.0000	[14]
	b	297-1033 K	11.1	0.1110	0.0000	0.0000	[14]
	c	297-1033 K	4.8	0.0475	0.0000	0.0000	[14]
	d100	297-1033 K	6.0	0.0600	0.0000	0.0000	[14]
	V	297-1033 K	22.2	0.2220	0.0000	0.0000	[14]
$Mg_{0.8}Fe_{0.2}SiO_3$, bronzite	a	298-1273 K	16.4	0.1640	0.0000	0.0000	[23]
	b	298-1273 K	14.5	0.1450	0.0000	0.0000	[23]
	c	298-1273 K	16.8	0.1680	0.0000	0.0000	[23]
	V	298-1273 K	47.7	0.4770	0.0000	0.0000	[23]
$MgSiO_3$, enstatite	V	293-1073 K	24.1	0.2947	0.2694	-0.5588	[75]
$MgSiO_3$, protoenstatite	V	1353-1633 K	16.7	0.1670	0.0000	0.0000	[59]
$MnSiO_3$, pyroxmangite	a	297-1073 K	7.6	0.0760	0.0000	0.0000	[65]
	b	297-1073 K	13.8	0.1380	0.0000	0.0000	[65]
	c	297-1073 K	6.7	0.0670	0.0000	0.0000	[65]
	V	297-1073 K	28.1	0.2810	0.0000	0.0000	[65]
$NaAlSi_2O_6$, jadeite	a	297-1073 K	8.5	0.0850	0.0000	0.0000	[14]
	b	297-1073 K	10.0	0.1000	0.0000	0.0000	[14]
	c	297-1073 K	6.3	0.0631	0.0000	0.0000	[14]
	d100	297-1073 K	8.2	0.0817	0.0000	0.0000	[14]
	V	297-1073 K	24.7	0.2470	0.0000	0.0000	[14]
$NaCrSi_2O_6$, ureyite	a	297-873 K	5.9	0.0585	0.0000	0.0000	[14]
	b	297-873 K	9.5	0.0946	0.0000	0.0000	[14]
	c	297-873 K	3.9	0.0390	0.0000	0.0000	[14]
	d100	297-873 K	6.9	0.0691	0.0000	0.0000	[14]
	V	297-873 K	20.4	0.2040	0.0000	0.0000	[14]
$NaFeSi_2O_6$, acmite	a	297-1073 K	7.3	0.0727	0.0000	0.0000	[14]
	b	297-1073 K	12.0	0.1200	0.0000	0.0000	[14]
	c	297-1073 K	4.5	0.0450	0.0000	0.0000	[14]
	d100	297-1073 K	8.0	0.0804	0.0000	0.0000	[14]
	V	297-1073 K	24.7	0.2470	0.0000	0.0000	[14]
Silicate ilmenite, $MgSiO_3$	a	298-876 K	7.1	0.0707	0.0000	0.0000	[4]
	c	298-876 K	10.0	0.0996	0.0000	0.0000	[4]
	V	298-876 K	24.4	0.2440	0.0000	0.0000	[4]
Silicate spinel							
γ-Mg_2SiO_4	V	297-1023 K	18.9	0.2497	0.3639	-0.6531	[88]
γ-Ni_2SiO_4	V	298-973 K	26.8	0.2680	0.0000	0.0000	[101]
γ-Fe_2SiO_4	V	298-673 K	27.0	0.2697	0.0000	0.0000	[101]
γ-Fe_2SiO_4	V	298-673 K	23.0	0.2300	0.0000	0.0000	[52]
Sillimanite, Al_2SiO_5	a	298-1273 K	1.0	0.0231	0.0092	-0.1185	[97]
	b	298-1273 K	7.4	0.0727	0.0470	0.0000	[97]
	c	298-1273 K	4.2	0.0386	0.1051	0.0000	[97]
	V	298-1273 K	13.3	0.1260	0.2314	0.0000	[97]

TABLE 1. (continued)

Names		T range	$\alpha_0 (10^{-6})$	$a_0 (10^{-4})$	$a_1 (10^{-8})$	a_2	ref.
SiO$_2$ group							
Coesite	V	293-1273 K	6.9	0.0597	0.7697	-0.1231	[75]
Cristobalite, low	a	301-491 K	19.5	0.1950	0.0000	0.0000	[63]
	c	301-491 K	52.7	0.5270	0.0000	0.0000	[63]
	V	301-491 K	91.7	0.9170	0.0000	0.0000	[63]
Cristobalite, high	V	673-1473 K	6.0	0.0600	0.0000	0.0000	[75][d]
α-Quartz	V	298-773 K	24.3	0.1417	9.6581	-1.6973	[1][e]
β-Quartz	V	848-1373 K	0.0	0.0000	0.0000	0.0000	[1]
	V	1473-1673 K	-4.4	-0.0440	0.0000	0.0000	[1]
Stishovite	a	291-873 K	7.8	0.0758	0.0656	0.0000	[39]
	c	291-873 K	0.9	0.0060	0.6553	-0.1500	[39]
	V	291-873 K	16.4	0.1574	0.7886	-0.1500	[39]
Stishovite	a	300-693 K	7.5	0.0750	0.0000	0.0000	[18]
	c	300-693 K	3.8	0.0380	0.0000	0.0000	[18]
	V	300-693 K	18.6	0.1860	0.0000	0.0000	[18]
Spodumene, α-LiAlSi$_2$O$_6$	V	293-1073 K	11.0	0.0758	1.1542	0.0000	[75]
Topaz, Al$_2$SiO$_4$(F,OH)$_2$	a	293-1273 K	4.6	0.0316	0.4698	0.0000	[75]
	b	293-1273 K	3.6	0.0245	0.3795	0.0000	[75]
	c	293-1273 K	6.3	0.0485	0.4924	0.0000	[75]
	V	293-1273 K	14.8	0.1098	1.2700	0.0000	[75]
Wadsleyite (β-phase)	a	293-1073 K	6.0	0.0851	0.1388	-0.2662	[89]
Mg$_2$SiO$_4$	b	293-1073 K	5.6	0.0791	0.1165	-0.2487	[89]
	c	293-1073 K	9.3	0.1196	0.3884	-0.3412	[89]
	V	293-1073 K	20.9	0.2893	0.5772	-0.8903	[89]
Zircon, ZrSiO$_4$	a	293-1293 K	3.4	0.0340	0.0000	0.0000	[6]
	c	293-1293 K	5.6	0.0560	0.0000	0.0000	[6]
	V	293-1293 K	12.3	0.1230	0.0000	0.0000	[6, cf. 82]
Perovskites							
BaZrO$_3$, perovskite	V	293-873 K	20.6	0.2060	0.0000	0.0000	[104]
CaGeO$_3$, perovskite	a	295-520 K	13.8	0.1380	0.0000	0.0000	[50]
	b	295-520 K	6.8	0.0680	0.0000	0.0000	[50]
	c	295-520 K	10.5	0.1050	0.0000	0.0000	[50]
	V	295-520 K	31.1	0.3110	0.0000	0.0000	[50]
	a	520-673 K	12.1	0.1210	0.0000	0.0000	[50]
	b	520-673 K	12.1	0.1210	0.0000	0.0000	[50]
	c	520-673 K	10.5	0.1050	0.0000	0.0000	[50]
	V	520-673 K	35.0	0.3500	0.0000	0.0000	[50]
NaMgO$_3$, perovskite	a	298-873 K	40.4	0.4040	0.0000	0.0000	[105]
	b	298-873 K	15.3	0.15300	0.0000	0.0000	[105]
	c	298-873 K	30.6	0.3060	0.0000	0.0000	[105]
	V	288-873 K	88.0	0.8800	0.0000	0.0000	[105]
NaMgO$_3$, cubic	V	1038-1173 K	94.9	0.9490	0.0000	0.0000	[105]
ScAlO$_3$, perovskite	a	293-973 K	10.0	0.1000	0.0000	0.0000	[32]
	b	293-973 K	7.0	0.0700	0.0000	0.0000	[32]
	c	293-973 K	10.0	0.1000	0.0000	0.0000	[32]

TABLE 1. (continued)

Names		T range	$\alpha_0 (10^{-6})$	$a_0 (10^{-4})$	$a_1 (10^{-8})$	a_2	ref.
SrZrO$_3$, perovskite	V	283-1373 K	27.0	0.2700	0.0000	0.0000	[32]
	a	293-973 K	12.4	0.1240	0.0000	0.0000	[104]
	b	293-973 K	7.5	0.0750	0.0000	0.0000	[104]
	c	293-973 K	9.7	0.0970	0.0000	0.0000	[104]
	V	293-973 K	29.8	0.2980	0.0000	0.0000	[104]
	a	973-1123 K	7.6	0.0760	0.0000	0.0000	[104]
	b	973-1123 K	16.1	0.1610	0.0000	0.0000	[104]
	c	973-1123 K	8.2	0.0820	0.0000	0.0000	[104]
	V	973-1123 K	32.4	0.3240	0.0000	0.0000	[104]
	a	1123-1443 K	14.9	0.1490	0.0000	0.0000	[104]
	c	1123-1443 K	6.8	0.0680	0.0000	0.0000	[104]
	V	1123-1443 K	37.5	0.3750	0.0000	0.0000	[104]

[a]For data cited from [75], see [75] for original data sources.

[b]See [56] for orthorhombic cordierite and [33] for hydrous Mg- and Fe-cordierites.

[c]See [24] for plagioclases, An$_{100}$, Ab$_9$An$_{91}$, Ab$_7$An$_{93}$, Ab$_{75}$An$_{22}$Or$_3$, Ab$_{63}$An$_{36}$Or$_1$, Ab$_{41}$An$_{57}$Or$_2$, and Ab$_{23}$An$_{76}$Or$_1$.

[d]Inversion at 491 K. Also see [75] for data on tridymite.

[e]α- and β-quartz transition at 846 K; see [1] for discussion on thermal expansion near the transition.

TABLE 2. Pressure Effect on Thermal Expansion Coefficient of Selected Substances

Phases	K_T, GPa	K_T'	δ_T	references
Fe(bcc)	165.0	5.30	6.5	[34]
Fe(hcp)	212.0	4.00	6.5	[34]
Fe(fcc)	167.0	4.00	6.5	[9]
NaCl	24.0	5.01	5.8	[100]
LiF	65.3	5.10	5.4	[10]
MgO, periclase	160.3	4.13	4.7	[36]
(Mg$_{0.6}$Fe$_{0.4}$)O, magnesiowüstite	157.0	4.00	4.3	[20]
Mg(OH)$_2$	54.3	4.70	4.5	[19]
Mg$_2$SiO$_4$, olivine	129.0	5.37	5.5	[37]
β-(Mg$_{0.84}$Fe$_{0.16}$)$_2$SiO$_4$	174.0	4.00	5.1	[21]
Mg$_2$SiO$_4$, spinel	183.0	4.30	5.8	[55]
(Mg$_{0.9}$Fe$_{0.1}$)SiO$_3$, perovskite	261.0	4.00	6.5	[51]

TABLE 3*a*. Partial Molar Volume and Mean Thermal Expansion Coefficient of Oxide Components [46]

$$V_{liq}(T) = \sum X_i \overline{V}_{i,Tr}\left[1 + \overline{\alpha}_i(T - T_r)\right] + X_{Na_2O}X_{TiO_2}\,\overline{V}_{Na_2O\text{-}TiO_2}$$

	iron-free silicate liquid[a]		64 liquids[b]	
Oxides	$\overline{V}_{i,1773\,K}$	$\overline{\alpha}_i\,(\times 10^5)$	$\overline{V}_{i,1573\,K}$	$\overline{\alpha}_i\,(\times 10^5)$
SiO_2	26.88	-1.2	26.92	0
TiO_2	23.98	36.5	22.43	32.3
Al_2O_3	37.52	2.0	36.80	7.1
Fe_2O_3	-	-	41.44	21.9
FeO	-	-	13.35	21.9
MgO	11.85	0.7	11.24	23.3
CaO	16.84	25.1	16.27	17.9
Na_2O	29.53	26.8	28.02	26.4
K_2O	47.10	72.8	44.61	26.7
Li_2O	17.42	33.4	16.19	32.4
$Na_2O\text{-}TiO_2$	20.10		20.33	

[a] Data were derived from density measurements of melts in iron-free system [8, 46, 80]. Units are in cc/mole and 1/K.
[b] Data were derived from density measurements of 64 melts in the system $Na_2O\text{-}K_2O\text{-}CaO\text{-}MgO\text{-}FeO\text{-}Fe_2O_3\text{-}Al_2O_3\text{-}TiO_2\text{-}SiO_2$ [8, 46, 57, 80].

TABLE 3*b*. Partial Molar Volume and Mean Thermal Expansion Coefficient of Oxide Components in Al-Free Melts [11]

	$\overline{\alpha} = \sum X_i \overline{\alpha}_i$	
Oxides	$\overline{V}_{i,1673}$	$\overline{\alpha}_i\,(\times 10^5)$
SiO_2	26.75	0.1
TiO_2	22.45	37.1
Fe_2O_3	44.40	32.1
FeO	13.94	34.7
MgO	12.32	12.2
CaO	16.59	16.7
Na_2O	29.03	25.9
K_2O	46.30	35.9
Li_2O	17.31	22.0
MnO	14.13	15.1
NiO	12.48	24.9
ZnO	13.64	43.0
SrO	20.45	15.4
BaO	26.20	17.4
PbO	25.71	16.1
Rb_2O	54.22	61.3
Cs_2O	68.33	71.4

[a] Data were derived from density measurements of Al free melts. See [11] for data sources. Units are in cc/mole and 1/K.

TABLE 3*c*. Partial Molar Volume and Mean Thermal Expansion Coefficient of Oxide Components in $CaO\text{-}FeO\text{-}Fe_2O_3\text{-}SiO_2$ Melts [57]

$$V_{liq}(T) = \sum X_i \overline{V}_{i,Tr}\left[1 + \overline{\alpha}_i(T - T_r)\right] + X_{SiO_2}X_{CaO}\,\overline{V}_{SiO_2\text{-}CaO}$$

	> 20 wt% silica[a]		low silica	
Oxides	$\overline{V}_{i,1673\,K}$	$\overline{\alpha}_i\,(\times 10^5)$	$\overline{V}_{i,1673}$	$\overline{\alpha}_i\,(\times 10^5)$
SiO_2	25.727	0	27.801	0
Fe_2O_3	37.501	9.2	35.770	13.3
FeO	14.626	21.0	13.087	19.4
CaO	18.388	12.1	21.460	10.5
$SiO_2\text{-}CaO$	0		-11.042	

[a] Data were derived from density measurements of 30 melts in the system $CaO\text{-}FeO\text{-}Fe_2O_3\text{-}SiO_2$ [16, 57]. Units are in cc/mole and 1/K.

TABLE 3*d*. Partial Molar Volume and Mean Thermal Expansion Coefficient of Oxide Components in $Na_2O\text{-}FeO\text{-}Fe_2O_3\text{-}SiO_2$ Melts [47]

$$V_{liq}(T) = \sum X_i \overline{V}_{i,Tr}\left[1 + \overline{\alpha}_i(T - T_r)\right]$$

Oxides	$\overline{V}_{i,1573}$	$\overline{\alpha}_i\,(\times 10^5)$
SiO_2	26.60	0.3
Fe_2O_3	41.39	12.9
FeO	13.61	18.7
Na_2O	28.48	23.2

[a] Data were derived from density measurements of 12 melts in the system $Na_2O\text{-}FeO\text{-}Fe_2O_3\text{-}SiO_2$ [17] and ferric-ferrous relations [47]. Units are in cc/mole and 1/K.

REFERENCES

1. Ackermann, R. J., and C. A. Sorrell, Thermal expansion and the high-low transformation in quartz. I. High-temperature X-ray studies, *J. Appl. Cryst.*, 7, 461-467, 1974.
2. Aldebert, P., and J. P. Traverse, αAl₂O₃: A high-temperature thermal expansion standard, *High Tempera-ture-High Pressure, 16*, 127-135, 1984.
3. Ashida, T., Y. Miyamoto, and S. Kume, Heat capacity, compressi-bility and thermal expansion coeffi-cient of ilmenite-type MgGeO₃, *Phys. Chem. Minerals, 12*, 129-131, 1985.
4. Ashida, T., S. Kume, E. Ito, and A. Navrotsky, MgSiO₃ ilmenite: heat capacity, thermal expansivity, and enthalpy of tramsformation, *Phys. Chem. Minerals, 16*, 239-245, 1988.
5. Bayer, G., Thermal expansion an-isotropy of dolomite-type borates Me²⁺Me⁴⁺B₂O₆, *Z. Kristallogr., 133*, 85-90, 1971.
6. Bayer, G., Thermal expansion of ABO₄ compounds with zircon and scheelite structures, *J. Less-Common Met., 26*, 255-262, 1972.
7. Berar, J. F., D. Grebille, P. Gregoire, and D. Weigel, Thermal expansion of boehmite, *J. Phys. Chem. Solids, 45*, 147-150, 1984.
8. Bockris, J. O., J. W. Tomlinson, and J. L. White, The structure of liquid silicates: Partial molar volumes and expansivities, *Trans. Faraday Soc., 52*, 299-311, 1956.
9. Boehler, R., N. von Bargen, and A Chopelas, Melting, thermal expan-sion, and phase transitions of iron at high pressures, *J. Geophys. Res., 95*, 21,731-21,736, 1990.
10. Boehler, R., G. C. Kennedy, Thermal expansion of LiF at high pressures, *J. Phys. Chem. Solids, 41*, 1019-1022, 1980.
11. Bottinga, Y., P. Richet, and D. Weill, Calculation of the density and ther-mal expansion coefficient of silicate liquids, *Bull. Mineral., 106*, 129-138, 1983.
12. Bottinga, Y., D. Weill, and P. Richet, Density calculations for silicate liquids. I. Revised method for alumi-nosilicate compositions, *Geochim. Cosmochim. Acta, 46*, 909-919, 1982.
13. Brown, G. E., and C. T. Prewitt, High-temperature crystal structure of hortonolite, *Amer. Mineral., 58*, 577-587, 1973.
14. Cameron, M., S. Sueno, C. T. Prewitt, and J. J. Papike, High-temperature crystal chemistry of acmite, diopside, hedenbergite, jadeite, spodumene, and ureyite, *Amer. Mineral., 58*, 594-618, 1973.
15. Datta, R. K., Thermal expansion of 12CaO·7Al₂O₃, *J. Am. Ceram. Soc., 70*, C-288-C-291, 1987.
16. Dingwell, D. B. and M. Brearley, Melt densities in the CaO-FeO-Fe₂O₃-SiO₂ system and the composi-tional dependence of the partial molar volume of ferric iron in silicate melts, *Geochim. Cosmochim. Acta, 52*, 2815-2825, 1988a.
17. Dingwell, D. B., M. Brearley, and J. E. Dickinson, Jr., Melt densities in the Na₂O-FeO-Fe₂O₃-SiO₂ system and the partial molar volume of tetrahe-drally-coordinated ferric iron in silicate melts, *Geochim. Cosmochim. Acta, 52*, 2467-2475, 1988b.
18. Endo, S., T. Akai, Y. Akahama, M. Wakatsuki, T. Nakamura, Y. Tomii, K. Koto, Y. Ito, and M. Tokonami, High temperature X-ray study of single crystal stishovite synthesized with Li₂WO₄ as flux, *Phys. Chem. Minerals, 13*, 146-151, 1986.
19. Fei, Y., and H. K. Mao, Static com-pression of Mg(OH)₂ to 78 GPa at high temperature and constraints on the equation of state of fluid-H₂O, *J. Geophys. Res., 98*, 11,875-11,884, 1993.
20. Fei, Y., H. K. Mao, J. Shu, J. Hu, *P-V-T* equation of state of magnesio-wüstite (Mg₀.₆Fe₀.₄)O, *Phys. Chem. Miner., 18*, 416-422, 1992a.
21. Fei, Y., H. K. Mao, J. Shu, G. Parthasathy, W. A. Bassett, and J. Ko, Simultaneous high *P-T* x-ray diffraction study of β-(Mg,Fe)₂SiO₄ to 26 GPa and 900 K, *J. Geophys. Res., 97*, 4489-4495, 1992b.
22. Finger, L. W. and Y. Ohashi, The thermal expansion of diopside to 800°C and a refinement of the crystal structure at 700°C, *Amer. Mineral, 61*, 303-310, 1976.
23. Frisillo, A. L., and S. T. Buljan, Linear expansion coefficients of orthopyroxene to 1000 °C, *J. Geophys. Res., 77*, 7115-7117, 1972.
24. Grundy, H. D., and W. L. Brown, A high-temperature X-ray study of low and high plagioclase feldspars, in *The Feldspars, Proceedings of a NATO Advanced Study Institute*, edited by W.S. MacKenzie and J. Zussman, pp. 162-173, University of Manchester Press, 1974.
25. Guggenheim, S., Y-H. Chang, and A. F. K. van Groos, Muscovite dehydro-xylation: High-temperature studies, *Amer. Mineral., 72*, 537-550, 1987.
26. Haselton, Jr., H. T., B. S. Hemingway, and R. A. Robie, Low-temperature heat capacities of CaAl₂SiO₆ glass and pyroxene and thermal expansion of CaAl₂SiO₆ pyroxene, *Amer. Mineral., 69*, 481-489, 1984.
27. Hazen, R. M., Effects of temperature and pressure on the crystal structure of ferromagnesian olivine, *Amer. Mineral., 62*, 286-295, 1977.
28. Hazen, R. M., Effects of temperature and pressure on the crystal structure of forsterite, *Amer. Mineral., 61*, 1280-1293, 1976.
29. Hazen, R. M., and L. W. Finger, High-pressure and high-temperature crystal chemistry of beryllium oxide, *J. Appl. Phys., 59*, 3728-3733, 1986.
30. Hazen, R. M., and L. W. Finger, High-temperature crystal chemistry of phenakite (Be₂SiO₄) and chrysoberyl (BeAl₂O₄), *Phys. Chem. Minerals, 14*, 426-434, 1987.
31. Hemingway, B. S., H. T. Evans, Jr., G. L. Nord, Jr., H. T. Haselton, Jr., R. A. Robie, and J. J. McGee, Akermanite: Phase transitions in heat capacity and thermal expansion, and revised thermodynamic data, *Can.*

Mineral., *24*, 425-434, 1986.

32. Hill, R. J., and I. Jackson, The thermal expansion of $ScAlO_3$ - A silicate perovskite analogue, *Phys. Chem. Minerals*, *17*, 89-96, 1990.

33. Hochella, Jr., M. F., G. E. Brown, Jr., F. K. Ross, and G. V. Gibbs, High-temperature crystal chemistry of hydrous Mg- and Fe-cordierites, *Amer. Mineral.*, *64*, 337-351, 1979.

34. Huang E., W. A. Bassett, and P. Tao, Pressure-temperature-volume relation for hexagonal close packed iron determined by synchrotron radiation, *J. Geophys. Res.*, *92*, 8129-8135, 1987.

35. Ikawa, H., T. Otagiri, O. Imai, M. Suzuki, K. Urabe, and S. Udagawa, Crystal structures and mechanism of thermal expansion of high cordierite and its solid solutions, *J. Am. Ceram. Soc.*, *69*, 492-98, 1986.

36. Isaak, D. G., O. L. Anderson, and T. Goto, Measured elastic moduli of single-crystal MgO up to 1800 K, *Phys. Chem. Minerals*, *16*, 704-713, 1989a.

37. Isaak, D. G., O. L. Anderson, and T. Goto, Elasticity of single-crystal forsterite measured to 1700 K, *J. Geophys. Res.*, *94*, 5895-5906, 1989b.

38. Isaak, D. G., O. L. Anderson, and H. Oda, High-temperature thermal expansion and elasticity of calcium-rich garnet, *Phys. Chem. Minerals*, *19*, 106-120, 1992.

39. Ito, H., K. Kawada, and S. Akimoto, Thermal expansion of stishovite, *Phys. Earth Planet. Inter.*, *8*, 277-281, 1974.

40. Kajiyoshi, K., High temperature equation of state for mantle minerals and their anharmonic properties, M.S. thesis, Okayama Univ., Okayama, Japan, 1986.

41. Kirby, R. K., T. A. Hahn, and B. D. Rothrock, Thermal expansion, in *American Institute of Physics Handbook*, McGraw-Hill, New York, 1972.

42. Knittle, E., R. Jeanloz, and G.L. Smith, Thermal expansion of silicate perovskite and stratification of the Earth's mantle, *Nature*, *319*, 214-216, 1986.

43. Kockel, A., Anisotropie der wärmeausdehnung von hexagonalem bariumcarbonat, *Naturwiss.*, *12*, 646,647, 1972.

44. Kress, V. C. and I. S. E. Carmichael, The lime-iron-silicate system: Redox and volume systematics, *Geochim. Cosmochim. Acta*, *53*, 2883-2892, 1989.

45. Lager, G. A., and E. P. Meagher, High-temperature structural study of six olivines, *Amer. Mineral.*, *63*, 365-377, 1978.

46. Lange, R. A. and I. S. E. Carmichael, Densities of $Na_2O-K_2O-CaO-MgO-Fe_2O_3-Al_2O_3-TiO_2-SiO_2$ liquids: New-measurements and derived partial molar properties, *Geochim. Cosmochim. Acta*, *51*, 2931-2946, 1987.

47. Lange, R. A. and I. S. E. Carmichael, Ferric-ferrous equilibria in $Na_2O-FeO-Fe_2O_3-SiO_2$ melts: Effects of analytical techniques on derived partial molar volumes, *Geochim. Cosmochim. Acta*, *53*, 2195-2204, 1989.

48. Lange, R. L. and I. S. E. Carmichael, Thermodynamic properties of silicate liquids with emphasis on density, thermal expansion and compressibility, in *Modern Methods of Igneous Petrology: Understanding magmatic Processes*, pp. 25-64, The American Mineralogical Society of America, Washington, DC, 1990.

49. Liu, X., and C. T. Prewitt, High-temperature x-ray diffration study of Co_3O_4: Transition from normal to disordered spinel, *Phys. Chem. Minerals*, *17*, 168-172, 1990.

50. Liu, X., Y. Wang, R. C. Liebermann, P. D. Maniar, and A. Navrotsky, Phase transition in $CaGeO_3$ perovskite: Evidence from X-ray powder diffraction, thermal expansion and heat capacity, *Phys. Chem. Minerals*, *18*, 224-230, 1991.

51. Mao, H. K., R. J. Hemley, Y. Fei, J. F. Shu, L. C. Chen, A. P. Jephcoat, Y. Wu, and W. A. Bassett, Effect of pressure, temperature, and composition on lattice parameters and density of $(Fe,Mg)SiO_3$-perovskites to 30 GPa, *J. Geophys. Res.*, *96*, 8069-8079, 1991.

52. Mao, H. K., T. Takahashi, W. A.

Bassett, J. S. Weaver, and S. Akimoto, Effect of pressure and temperature on the molar volumes of wüstite and three $(Mg, Fe)_2SiO_4$ spinel solid solutions, *J. Geophys. Res.*, *74*, 1061-1069, 1969.

53. Markgraf, S. A., and R. J. Reeder, High-temperature structure refinements of calcite and magnesite, *Am. Mineral.*, *70*, 590-600, 1985.

54 Matsui, T., and M. H. Manghnani, Thermal expansion of single-crystal forsterite to 1023 K by Fizeau interferometry, *Phys. Chem. Minerals*, *12*, 201-210, 1985.

55. Meng, Y., D. J. Weidner, G. D. Gwanmesia, R. C. Leibermann, M. T. Vaughan, Y. Wang, K. Leinenweber, R. E. Pacalo, A. Yeganeh-Haeri, and Y. Zhao, In-situ high P-T X-ray diffraction studies on three polymorphs (α, β, γ) of Mg_2SiO_4, *J. Geophys. Res.*, in press, 1993

56. Mirwald, P. W., Thermal expansion of anhydrous Mg-Cordierite between 25 and 950°C, *Phys. Chem. Minerals*, *7*, 268-270, 1981.

57. Mo, X. , I. S. E. Carmichael, M. Rivers, and J. Stebbins, The partial molar volume of Fe_2O_3 in multi-component silicate liquids and the pressure dependence of oxygen fugacity in magmas, *Mineral. Mag.*, *45*, 237-245, 1982.

58. Morosin, B., Structure and thermal expansion of beryl, *Acta Cryst.*, *B28*, 1899-1903, 1972

59. Murakami, T., Y. Takeuchi, and T. Yamanaka, X-ray studies on proto-enstatite, *Z. Kristallogr. 166*, 263-275, 1984.

60. Ohashi, Y., High-temperature structural crystallography of synthetic clinopyroxene $(Ca,Fe)SiO_3$, Ph.D. Thesis, Harvard University, 1973.

61. Okajima, S., I. Suzuki, K. Seya, and Y. Sumino, Thermal expansion of single-crystal tephroite, *Phys. Chem. Minerals*, *3*, 111-115, 1978.

62. Parise, J. B., Y. Wang, A. Yeganeh-Haeri, D. E. Cox, and Y. Fei, Crystal structure and thermal expansion of $(Mg,Fe)SiO_3$ perovskite, *J. Geophys. Lett.*, *17*, 2089-2092, 1990.

63. Peacor, D. R., High-temperature single-crystal study of the cristobalite

inversion, *Z. Kristallogr.*, *138*, S. 274-298, 1973.

64. Pfaff, F., cited in Mellor, J. W., *A Comprehensive Treaties on Inorganic and Theoretical Chemistry*, vol. 14, 359 pp., Longmans, Green & Co., London, 1935.

65. Pinckney, L. R., and C. W. Burnham, High-temperature crystal structure of pyroxmangite, *Amer. Mineral.*, *73*, 809-817, 1988.

66. Pillars, W. W. and D. R. Peacor, The crystal structure of beta eucryptite as a function of temperature, *Amer. Mineral.*, *58*, 681-690, 1973.

67. Predecki, P., J. Haas, J. Faber, Jr., and R. L. Hitterman, Structural aspects of the lattice thermal expansion of hexagonal cordierite, *J. Am. Ceram. Soc.*, *70*, 175-182, 1987.

68. Prewitt, C. T., S. Sueno and J. J. Papike, The crystal structures of high albite and monalbite at high temperatures, *Amer. Mineral.*, *61*, 1213-1225, 1976.

69. Rao, K. V. K., and K. S. Murthy, Thermal expansion of manganese carbonate, *J. Mat. Sci.*, *5*, 82-83, 1970.

70. Reeder, R. J., and S. A. Markgraf, High-temperature crystal chemistry of dolomite, *Am. Mineral.*, *71*, 795-804, 1986.

71. Ross, N. L., and R. M. Hazen, Single crystal X-ray diffraction study of $MgSiO_3$ Perovskite from 77 to 400 K, *Phys. Chem. Mineral*, *16*, 415-420, 1989.

72. Ross, N. L., and A. Navrotsky, The Mg_2GeO_4 olivine-spinel phase transition, *Phys. Chem. Mineral*, *14*, 473-481, 1987.

73. Sawada, H., and Y. Takéuchi, The crystal structure of barite, β-$BaSO_4$, at high temperatures, *Z. Kristallogr.*, *191*, 161-171, 1990.

74. Schneider, H., and E., Eberhard, Thermal expansion of mullite, *J. Am. Ceram. Soc.*, *73*, 2073-76, 1990.

75. Skinner, B. J., Thermal expansion in *Handbook of Physical Constants*, edited by S. P. Clark, Jr., pp. 75-95, Geol. Soc. Am. Mem., 1966.

76. Smyth, J.R., High temperature crystal chemistry of fayalite, *Amer. Mineral.*, *60*, 1092-1097, 1975.

77. Smyth, J. R., The high temperature crystal chemistry of clinohypersthene, *Amer. Mineral.*, *59*, 1069-1082, 1974.

78. Smyth, J. R., An orthopyroxene structure up to 850°C, *Amer. Mineral.*, *58*, 636-648, 1973.

79. Smyth, J. R., R. M. Hazen, The crystal structures of forsterite and hortonolite at several temperatures up to 900°C, *Amer. Mineral.*, *58*, 588-593, 1973.

80. Stein, D. J., J. F. Stebbins, and I. S. E. Carmichael, Density of molten sodium aluminosilicates, *J. Amer. Ceram. Sooc.*, *69*, 396-399, 1986.

81. Stixrude, L., and M. S. T. Bukowinski, Fundamental thermodynamic relations and silicate melting with implications for the constitution of D", *J. Geophys. Res.*, *95*, 19,311-19,325, 1990.

82. Subbarao, E. C., D. K. Agrawal, H. A. McKinstry, C. W. Sallese, and R. Roy, Thermal expansion of compounds of zircon structure, *J. Am. Ceram. Soc.*, *73*, 1246-1252, 1990.

83. Sueno, S., M. Cameron, J. J. Papike, and C.T. Prewitt, The high temperature crystal chemistry of tremolite, *Amer. Mineral.*, *58*, 649-664, 1973.

84. Sueno, S., M. Cameron, and C. T. Prewitt, Orthoferrosilite: High-temperature crystal chemistry, *Amer. Mineral.*, *61*, 38-53, 1976.

85. Sugiyama, K., and Y. Takéuchi, The crystal structure of rutile as a function of temperature up to 1600 °C, *Z. Kristallogr.*, *194*, 305-313, 1991.

86. Suzuki, I., Thermal expansion of periclase and olivine and their anharmonic properties, *J. Phys. Earth*, *23*, 145-159, 1975.

87. Suzuki, I., and O. L. Anderson, Elasticity and thermal expansion of a natural garnet up to 1,000 K, *J. Phys. Earth*, *31*, 125-138, 1983.

88. Suzuki, I., E. Ohtani, and M. Kumazawa, Thermal expansion of γ-Mg_2SiO_4, *J. Phys. Earth*, *27*, 53-61, 1979.

89. Suzuki, I., E. Ohtani, and M. Kumazawa, Thermal expansion of modified spinel, β-Mg_2SiO_4, *J. Phys. Earth*, *28*, 273-280, 1980.

90. Suzuki, I., S. Okajima, and K. Seya, Thermal expansion of single-crystal manganosite, *J. Phys. Earth*, *27*, 63-69, 1979.

91. Suzuki, I., K. Seya, H. Takei, and Y. Sumino, Thermal expansion of fayalite, Fe_2SiO_4, *Phys. Chem. Minerals*, *7*, 60-63, 1981.

92. Touloukian, Y. S., R. K. Kirby, R. E. Taylor, P. D. Desai, Thermal expansion: Metallic elements and alloys, In *Thermophysical Properties of Matter*, vol. 12, edited by Y. S. Touloukian and C. Y. Ho, Plenum, New York, 1975.

93. Touloukian, Y. S., R. K. Kirby, R. E. Taylor, T. Y. R. Lee, Thermal expansion: Nonmetallic solids, In *Thermophysical Properties of Matter*, vol. 13, edited by Y. S. Touloukian and C. Y. Ho, 176 pp., Plenum, New York, 1977.

94. Watanabe, H., Thermochemical properties of synthetic high-pressure compounds relevant to the earths mantle, in *High-Pressure Research in Geophysics*, edited by S. Akimoto and M. H. Manghnani, pp. 411-464, Cent. Acad. Pub. Janpan, Japan, 1982.

95 Wechsler, B. A., and C. T. Prewitt, Crystal structure of ilmenite ($FeTiO_3$) at high temperature and high pressure, *Am. Miner.*, *69*, 176-185, 1984.

96. Winslow, G. H., Thermomechanical properties of real materials: the thermal expansion of UO_2 and ThO_2, *High Temp. Sci.*, *3*, 361-367, 1971.

97. Winter, J. K., and S. Ghose, Thermal expansion and high-temperature crystal chemistry of the Al_2SiO_5 polymorphs, *Amer. Mineral.*, *64*, 573-586, 1979.

98. Winter, J. K., S. Ghose, and F. P. Okamura, A high-temperature study of the thermal expansion and the anisotropy of the sodium atom in low albite, *Amer. Mineral.*, *62*, 921-931, 1977.

99. Winter, J. K., F. P. Okamura, and S. Ghose, A high-temperature structural study of high albite, monalbite, and the analbite-monalbite phase transition, *Amer. Mineral.*, *64*, 409-423,

1979.

100. Yamamoto, S., I. Ohno, and O. L. Anderson, High temperature elasticity of sodium chloride, *J. Phys. Chem. Solids, 48*, 143-151, 1987.

101. Yamanaka, T., Crystal structures of Ni_2SiO_4 and Fe_2SiO_4 as a function of temperature and heating duration, *Phys. Chem. Minerals, 13*, 227-232,

1986.

102. Yamanaka, T., and Y. Takeuchi, Order-disorder transition in $MgAl_2O_4$ spinel at high temperatures up to 1700 °C, *Z. Kristallogr., 165*, 65-78, 1983.

103. Yates, B., *Thermal Expansion,* 121 pp., Plenum Press, New York, 1972.

104. Zhao, Y., and D. J. Weidner, Thermal expansion of $SrZrO_3$ and $BaZrO_3$ perovskites, *Phys. Chem. Minerals, 18*, 294-301, 1991.

105. Zhao, Y., D. J. Weidner, J. B. Parise, and D. E. Cox, Thermal expansion and structure distortion of perovskite: Data for $NaMgO_3$ perovskites, *Phys. Earth Planet. Interiors,* in press, 1993.

Elasticity of Minerals, Glasses, and Melts

Jay D. Bass

INTRODUCTION

In this chapter I present a compilation of the elastic moduli of minerals and related substances which may be of use in geophysical or geochemical calculations. The discipline of elasticity is a mature one. Laboratory measurements of elasticity have been actively investigated for a number of years for a wide variety of materials. Consequently, there are several excellent compilations of elastic moduli available, notably those of Hearmon [46, 47], in the Landolt-Börnstein tables, and Sumino and Anderson [118] (for crystalline materials), and of Bansal and Doremus [6] (for glasses). Here are summarized elastic moduli of most direct geologic importance. Included are many important results published in the last few years which are not available in other summaries.

The main content of the tables consists of elastic moduli, c_{ij}, which are stiffness coefficients in the linear stress-strain relationship [80]:

$$\sigma_{ij} = c_{ijkl}\epsilon_{kl} \qquad (1)$$

where σ_{ij} and ϵ_{kl} are the stress and strain tensors, respectively. We use the standard Voigt notation [80], to represent the moduli as components of a 6×6 matrix c_{ij} where the indices i and j range from 1 to 6. Also listed for each material are the adiabatic bulk modulus and shear modulus for an equivalent isotropic polycrys-

talline aggregate. The isotropic moduli listed are are Hill averages of the Voigt and Reuss bounds [135]. In conjunction with the density, the moduli can be used to calculate acoustic velocities using standard relations [16].

This chapter is not meant to be either historically complete nor encyclopedic in scope. In cases where a material has been the subject of several studies, we have cited the average moduli computed by Hearmon [46, 47], where available. Thus, the results from many older studies are not individually listed, especially where they have been superceded by experiments using more modern techniques. This has made the present summary far more compact than it would otherwise be. However, elastic properties reported after the compilations of Hearmon [46, 47], and by Bansal and Doremus [6], are included as separate entries. Except in a few important cases, only results from single-crystal studies are reported. Results from experiments on polycrystalline samples were uniformly excluded unless no single-crystal data were available.

Since the earlier compilation by Birch [16], the quantity of data related to the equation of state of rocks and minerals has grown considerably. For many materials, complimentary results on the equation of state of minerals from static compression data are found in the chapter by Knittle, with which there is a degree of overlap. Likewise, the chapter by Anderson and Isaak present considerably more detail on the high temperature elasticity of minerals.

The results in this chapter derive from a variety of techniques which have a broad range of precision. We have not made any attempt to assess the relative accuracy of results from different laboratories on a given material.

J. D. Bass, Department of Geology, University of Illinois, 1301 West Green Street, Urbana, IL 61801

Mineral Physics and Crystallography
A Handbook of Physical Constants
AGU Reference Shelf 2

The number of independent elastic constants appropriate to a material depends on the symmetry of that material [80], ranging from two for a noncrystalline substance, to three for an cubic (isometric) crystal, to twenty one for a triclinic crystal. Tables are therefore organized on the basis of crystallographic symmetry, with materials of a similar nature (e.g. elements, garnets, etc.) grouped together.

The notation used throughout the tables is as follows:

Symbol	Units	Description
c_{ij}	GPa	Single-crystal elastic stiffness moduli
K_S	GPa	Adiabatic bulk modulus
$K_{S,0}$	GPa	Adiabatic bulk modulus at zero frequency
G	GPa	Shear modulus
V_P	m/s	Longitudinal wave velocity
T	Kelvins	Temperature

P	GPa	Pressure
ρ	Mg/m^3	Density
		Superscripts
E		Indicates constant electric field
D		Indicates constant electric displacement

Note that for melts, we have cited the zero frequency, or relaxed, bulk modulus where possible. It is not possible within the framework of this review to summarize the frequency dependence of the elastic properties of melts or glasses at high temperature. In cases where the dispersive properties of liquids were investigated, we have listed the results obtained at the lowest frequency.

Most of the entries are for minerals, although some chemically and structurally related compounds of interest are included. In all of the tables, the compositions of solid solutions are given in terms of mole percentages of the end-members, indicated by the subscripts, except where specifically noted.

Table 1. Elastic Moduli of Cubic Crystals at Room P & T

Material	ρ Mg/m^3	11	44	12	K_S GPa	G GPa	References
		\multicolumn{3}{c}{Subscript ij in modulus c_{ij} (GPa)}			

Material	ρ Mg/m^3	Subscript ij in modulus c_{ij} (GPa) 11	44	12	K_S GPa	G GPa	References
\multicolumn{8}{c}{*Elements, Metallic Compounds*}							
Au, Gold	19.283	191	42.4	162	171.7	27.6	47
Ag, Silver	10.500	122	45.5	92	102.0	29.2	47
C, Diamond	3.512	1079	578	124	443.0	535.7	77
Cu, Copper	8.932	169	75.3	122	137.3	46.9	46
Fe, α-Iron	7.874	230	117	135	166.7	81.5	47
$Fe_{0.94}Si_{0.06}$	7.684	221.0	122.3	135.1	163.7	80.40	103
$Fe_{0.94}Si_{0.06}$	7.675	222.3	123	135.5	164.4	81.1	72
$Fe_{0.91}Si_{0.09}$	7.601	216.4	124.6	134	161.4	80.1	72
\multicolumn{8}{c}{*Binary Oxides*}							
BaO	5.992	122	34.4	45	70.7	36.0	47, 126
CaO, Lime	3.346	224	80.6	60	114.7	81.2	46, 111
	3.349	220.5	80.03	57.7	112.0	80.59	81
CoO	6.438	260	82.4	145	183.3	71.3	47
$Fe_{0.92}O$, Wüstite	5.681	245.7	44.7	149.3	181.4	46.1	120
$Fe_{0.943}O$	5.708	218.4	45.5	123.0	154.8	46.4	56
$Fe_{0.95}O$	5.730	217	46	121	153.0	46.8	15
MnO, Manganosite	5.365	227	78	116	153.0	68.1	47
	5.368	223.5	78.1	111.8	149.0	68.3	89
	5.346	226.4	79.0	114.9	152.1	68.7	138
MgO, Periclase	3.584	294	155	93	160.0	130.3	46
	3.584	296.8	155.8	95.3	162.5	130.8	57
	3.584	297.8	155.8	95.1	162.7	131.1	152
NiO, Bunsenite	6.828	344.6	40	141	205	58.8	134

Table 1. (continued)

Material	ρ Mg/m³	Subscript ij in modulus c_{ij} (GPa) 11	44	12	K_S GPa	G GPa	References
SrO	5.009	170	55.6	46	87.3	58.1	46, 116
UO$_2$, Uraninite	10.97	389	59.7	119	209	83	35
Spinel Structured Oxides							
Fe$_3$O$_4$, Magnetite	5.206	275	95.5	104	161.0	91.4	47
		270	98.7	108	162.0	91.2	47
FeCr$_2$O$_4$, Chromite	5.09	322	117	144	203.3	104.9	47
MgAl$_2$O$_4$, Spinel	3.578	282	154	154	196.7	108.3	46, 24
	3.578	282.9	154.8	155.4	197.9	108.5	152
MgO·2.6Al$_2$O$_3$	3.619	298.6	157.6	153.7	202.0	115.3	106
MgO·3.5Al$_2$O$_3$	3.63	300.5	158.6	153.7	202.6	116.4	126
		312	157	168	216.0	114.8	46
		303	156	158	206.3	114.7	46
Mg$_{0.75}$Fe$_{0.36}$Al$_{1.90}$O$_4$, Pleonaste	3.826	269.5	143.5	163.3	198.7	97	130
FeAl$_2$O$_4$, Hercynite	4.280	266.0	133.5	182.5	210.3	84.5	130
γ−Mg$_2$SiO$_4$, Ringwoodite	3.559	327	126	112	184	119	144
Ni$_2$SiO$_4$	5.351	366	106	155	226	106	13
Mg$_2$GeO$_4$	4.389	300	126	118	179	110	140
Sulphides							
FeS$_2$, Pyrite	5.016	361	105.2	33.6	142.7	125.7	108
		402	114	−44	104.7	149.7	47
PbS, Galena	7.597	127	23	24.4	58.6	31.9	47
ZnS, Sphalerite	4.088	102	44.6	64.6	77.1	31.5	47
Binary Halides							
BaF$_2$, Frankdicksonite	4.886	90.7	25.3	41.0	57.8	25.1	46
CaF$_2$, Fluorite	3.181	165	33.9	47	86.3	42.4	46
NaCl, Halite	2.163	49.1	12.8	12.8	24.9	14.7	46
KCl, Sylvite	1.987	40.5	6.27	6.9	18.1	9.4	46
Garnets							
Pyrope (Py), Mg$_3$Al$_2$Si$_3$O$_{12}$	3.567	296.2	91.6	111.1	172.8	92.0	85
	3.563	295	90	117	177	89	67
Grossular,(Gr$_{99}$) Ca$_3$Al$_2$Si$_3$O$_{12}$	3.602	321.7	104.6	91.2	168.4	108.9	11
Uvarovite (Uv) Ca$_3$Cr$_2$Si$_3$O$_{12}$	3.850	304	84	91	162	92	10
Spessartite (Sp$_{95}$) Mn$_3$Al$_2$Si$_3$O$_{12}$	4.195	309.5	95.2	113.5	178.8	96.3	11
Hibschite Ca$_3$Al$_2$(SiO$_4$)$_{1.74}$(H$_4$O$_4$)$_{1.28}$	3.13	187	63.9	57	100	64.3	86
Andradite (An$_{96}$) Ca$_3$Fe$_2^{+3}$Si$_3$O$_{12}$	3.836	289	85	92	157	90	10
An$_{70}$Gr$_{22}$Alm$_4$Py$_3$	3.775	281.2	87.9	80.4	147.3	92.7	5
Gr$_{48}$Py$_{28}$Alm$_{23}$Sp$_1$	3.741	310.2	99.5	100.4	170.4	101.6	84
Alm$_{64}$Py$_{22}$Gr$_1$Sp$_{11}$And$_2$	4.131	306.7	94.9	111.9	176.8	95.9	5
Alm$_{74}$Py$_{20}$Gr$_3$Sp$_3$	4.160	306.2	92.7	112.5	177.0	94.3	111
Py$_{73}$Alm$_{16}$And$_4$Uv$_6$	3.705	296.6	91.6	108.5	171.2	92.6	121
Py$_{62}$Alm$_{36}$Gr$_2$	3.839	301.4	94.3	110.0	173.6	94.9	136

Table 1. (continued)

Material	ρ Mg/m^3	11	44	12	K_S GPa	G GPa	References
Sp$_{54}$Alm$_{46}$	4.249	308.5	94.8	112.3	177.7	96.1	132
Majorite (Mj) – Garnet Solid Solutions							
Mj$_{41}$Py$_{59}$ Mg$_3$(Mg,Si)$_{.82}$Al$_{1.18}$Si$_3$O$_{12}$	3.555				164	89	12
Mj$_{33}$Py$_{67}$	3.545				170	92	150
Mj$_{66}$Py$_{34}$	3.527				172	92	150
Na$_{1.87}$Mg$_{1.18}$Si$_{4.94}$O$_{12}$	3.606	329	114	96	174	115	90

Abbreviations: Py, pyrope; Alm, almandite; Gr, grossular; Uv, uvarovite; An, andradite; Sp, spessartite; Mj, majorite (Si-rich and Al-poor garnet).

Table 2. Elastic Moduli of Hexagonal Crystals (5 Moduli) at Room P & T

Material	ρ Mg/m^3	11	33	44	12	13	K_S GPa	G GPa	References
BeO, Bromellite	3.01	470	494	153	168	119	251	162	14
Beryl	2.724	304.2	277.6	65.3	123.8	114.5	176	78.8	153
Be$_3$Al$_2$Si$_6$O$_{18}$	2.698	308.5	283.4	66.1	128.9	118.5	181	79.2	153
C, Graphite	2.26	1060	36.5	.3	180	15	161.0	109.3	18
Ca$_{10}$(PO$_4$)$_6$(OH)$_2$, Hydroxyapatite	3.146	140	180	36.2	13	69	80.4	45.6	47
Ca$_{10}$(PO$_4$)$_6$F$_2$, Fluorapatite	3.200	141	177	44.3	46	56	212.3	101.8	47
Cancrinite (Na$_2$Ca)$_4$(Al,SiO$_4$)$_6$CO$_3$·nH$_2$O	2.6	79	125	37.2	38	21	48.9	30.7	46
CdS, Greenockite	4.824	86.5	94.4	15.0	54.0	47.3	62.7	16.9	47
c^D		83.8	96.5	15.8	51.1	45.0	60.7	17.5	61
c^E		83.1	94.8	15.3	50.4	46.2	60.7	17.1	61
H$_2$O, Ice-I (257K)		13.5	14.9	3.09	6.5	5.9	8.72	3.48	46
Ice-I (270K)	0.9175	13.70	14.70	2.96	6.97	5.63	8.73	3.40	37
Na$_3$KAl$_4$Si$_4$O$_{16}$, Nepheline	2.571	79	125	37.2	38	21	48.9	30.7	47
β-SiO$_2$ (873K)		117	110	36	16	33	56.4	41.4	47
ZnO, Zincite	5.675	209	218	44.1	120	104	143.5	46.8	46
c^E		207.0	209.5	44.8	117.7	106.1	142.6	46.3	61
c^D		209.6	221.0	46.1	120.4	101.3	142.9	48.2	61
Wurtzite, ZnS	4.084	122	138	28.7	58	42	74.0	33.3	46

Table 3. Elastic Moduli of Trigonal Crystals (6 Moduli) at Room P & T

Mineral	ρ Mg/m³	11	33	44	12	13	14	K_S GPa	G GPa	References
Al₂O₃, Sapphire,	3.999	495	497	146	160	115	−23	251.7	162.5	46
Corundum	3.982	497	501	146.8	162	116	−21.9	253.5	163.2	83
AlPO₄, (c^E)	2.620	64.0	85.8	43.2	7.2	9.6	−12.4	29.3	33.0	25
Berlinite, (c^D)		69.8	87.1	42.2	10.6	14.9	13.4	33.9	32.7	32
CaCO₃, Calcite	2.712	144	84.0	33.5	53.9	51.1	−20.5	73.3	32.0	46
Cr₂O₃, Eskolaite	5.21	374	362	159	148	175	−19	234.0	123.2	1, 46
Fe₂O₃, Hematite	5.254							206.6	91.0	69
MgCO₃, Magnesite	3.009	259	156	54.8	75.6	58.8	−19.0	114.0	68.0	46, 50
NaNO₃, Nitratine	2.260	54.6	34.9	11.3	18.9	19.3	7.5	28.2	12.0	46
Ag₃AsS₃, Proustite	5.59	59.5	39.8	9.97	31.7	29.6	0.18	36.8	11.0	47
SiO₂, α-Quartz	2.648	86.6	106.1	57.8	6.7	12.6	−17.8	37.8	44.3	46
c^E	2.648	86.74	107.2	57.9	6.98	11.9	−17.9	37.8	44.4	46
c^D	2.648	86.47	107.2	58.0	6.25	11.9	−18.1	37.5	44.5	46
Tourmaline, (Na)(Mg,Fe^{+2},Fe^{+3},Al,Li)₃Al₆(BO₃)₃(Si₆O₁₈)(OH,F)₄	3.100	305.0	176.4	64.8	108	51	−6	127.2	81.5	87

Table 4. Elastic Moduli of Trigonal Crystals (7 Moduli) at Room P & T

Material	ρ Mg/m³	11	33	44	12	13	14	15	K_S GPa	G GPa	References
Dolomite, CaMg(CO₃)₂	3.795	205	113	39.8	71.0	57.4	−19.5	13.7	94.9	45.7	46, 50
Phenacite Be₂SiO₄	2.960	341.9	391.0	91.4	148.0	136.0	0.1	3.5	212.8	98.9	148
MgSiO₃ Ilmenite structure	3.795	472	382	106	168	70	−27	24	212	132	141

Table 5. Elastic Moduli of Tetragonal Crystals (6 Moduli) at Room P & T

Material	ρ Mg/m³	11	33	44	66	12	13	K_S GPa	G GPa	References
				Rutile-Structured						
SiO₂, Stishovite	4.290	453	776	252	302	211	203	316	220	143
SiO₂, α-Cristobalite	2.335	59.4	42.4	67.2	25.7	3.8	−4.4	16.4	39.1	151
SnO₂, Cassiterite	6.975	261.7	449.6	103.1	207.4	177.2	155.5	212.3	101.8	22
TeO₂, Paratellurite	6.02	55.7	105.8	26.5	65.9	51.2	21.8	45.0	20.4	93
	5.99	53.2	108.5	24.4	55.2	48.6	21.2	43.7	19.0	122
TiO₂, Rutile	4.260	269	480	124	192	177	146	215.5	112.4	47
GeO₂	6.279	337.2	599.4	161.5	258.4	188.2	187.4	257.6	150.8	131
				Other Minerals						
Ba₂Si₂TiO₈, Fresnoite (c^E)		140	83	33	59	36	24	56.9	42.1	46
		166	100	31.7	69.4	58	44	77.6	43.3	46
Scapolite, (Na,Ca,K)₄Al₃(Al,Si)₃ Si₆O₂₄(Cl,SO₄,CO₃)		99	113	15.6	22.9	35.1	35.4	58.0	23.1	47
		102	140	23.0	30.4	38.9	43.3	65.3	29.1	47
		102	140	23.0	30.4	38.9	43.3	65.3	29.1	47
Vesuvianite Ca₁₀Mg₂Al₄(SiO₄)₅(Si₂O₇)₂(OH)₄		153	166	55.8	54.0	48	44	82.6	55.5	47
ZrSiO₄[a], Zircon	4.675	424.3	489.3	131.1	48.3	69.7	149	227.9	109.0	88
	4.70	256	372	73.5	116	175	214	223.9	66.6	47

[a] nonmetamict.

Table 6. Elastic Moduli of Tetragonal Crystals (7 Moduli) at Room P & T

Material	ρ Mg/m³	11	33	44	66	12	13	16	K_S GPa	G GPa	References
CaMoO₄, Powellite	4.255	144	127	36.8	45.8	65	47	−13.5	81.0	39.9	46
CaWO₄, Scheelite	6.119	141	125	33.7	40.7	61	41	−17	76.5	37.4	46
PbMoO₄, Wulfenite	6.816	109	92	26.7	33.7	68	53	−13.6	72.4	24.5	46
		108	95	26.4	35.4	63	51	−15.8	70.8	25.0	46

Table 7. Elastic Constants of Orthorhombic Crystals at Room P & T

Material	ρ Mg/m³	Subscript ij in modulus c_{ij} (GPa)									K_S GPa	G GPa	References
		11	22	33	44	55	66	12	13	23			
Perovskites													
MgSiO₃	4.108	515	525	435	179	202	175	117	117	139	246.4	184.2	149
NaMgF₃	3.058	125.7	147.3	142.5	46.7	44.8	50.4	49.5	45.1	43.1	75.7	46.7	155
Pyroxenes													
Enstatite (En₁₀₀), MgSiO₃	3.198	224.7	177.9	213.6	77.6	75.9	81.6	72.4	54.1	52.7	107.8	75.7	142
Ferrosilite (Fs₁₀₀), FeSiO₃	4.002	198	136	175	59	58	49	84	72	55	101	52	9
En₉₄Fs₆	3.272	229.3	167.0	193.9	79.7	76.1	77.1	73.6	49.8	46.6	102.3	73.9	31
En₈₄.₅Fs₁₅.₂	3.335	229.9	165.4	205.7	83.1	76.4	78.5	70.1	57.3	49.6	105.0	75.5	64
En₈₀Fs₂₀	3.354	228.6	160.5	210.4	81.8	75.5	77.7	71.0	54.8	46.0	103.5	74.9	31
	3.373	231.0	169.8	215.7	82.8	76.5	78.1	78.9	61.4	49.1	109.4	75.2	137
Protoenstatite, MgSiO₃	3.052	213	152	246	81	44	67	76	59	70	112	63	123
Olivines													
Forsterite (Fo₁₀₀), Mg₂SiO₄	3.221	328	200	235	66.7	81.3	80.9	69	69	73	129.5	81.1	46
Fayalite (Fa₁₀₀), Fe₂SiO₄	4.38	266	168	232	32.3	46.5	57	94	92	92	134	50.7	55
Fo₉₁Fa₉	3.325	320.2	195.9	233.8	63.5	76.9	78.1	67.9	70.5	78.5	129.5	77.6	136
Fo₉₃Fa₇	3.311	323.7	197.6	235.1	64.6	78.1	79.0	66.4	71.6	75.6	129.4	79.1	65
Fo₉₁.₃Fa₈.₁	3.316	324	196	232	63.9	77.9	78.8	71.5	71.5	68.8	128.1	78.7	82
Fo₉₂Fa₈	3.299	319	192	238	63.8	78.3	79.7	59	76	72	126.7	79.0	82
Mn₂SiO₄	4.129	258.4	165.6	206.8	45.3	55.6	57.8	87	95	92	128	54	117
Monticellite, CaMgSiO₄	3.116	216	150	184	50.6	56.5	59.2	59	71	77	106	55.2	92
Ni₂SiO₄	4.933	340	238	253	71	87	78	109	110	113	165	80	13
Co₂SiO₄	4.706	307.8	194.7	234.2	46.7	63.9	64.8	102	105	103	148	62	117
Mg₂GeO₄	4.029	312	187	217	57.2	66.1	71	60	65	66	120	72	140

Table 7. (continued)

Material	ρ Mg/m³	Subscript ij in modulus c_{ij} (GPa)									K_S GPa	G GPa	References
		11	22	33	44	55	66	12	13	23			
Other Silicates													
Wadsleyite, β-Mg₂SiO₄	3.474	360	383	273	112	118	98	75	110	105	174	114	105
Al₂SiO₅													
Andalusite	3.145	233.4	289.0	380.1	99.5	87.8	112.3	97.7	116.2	81.4	162	99.1	126
Sillimanite	3.241	287.3	231.9	388.4	122.4	80.7	89.3	158.6	83.4	94.7	170.8	91.5	126
Sulphates, Sulphides, Carbonates													
Sulphur	2.065	24	20.5	48.3	4.3	8.7	7.6	13.3	17.1	15.9	19.1	6.7	46
		14.2	12.7	18.3	8.27	4.3	4.4	3.0	3.1	8.0	7.2	5.3	46
BaSO₄, Barite	4.473	89.0	81.0	107	12.0	28.1	26.9	47.9	31.7	29.8	55.0	22.8	46
		95.1	83.7	110.6	11.8	29.0	27.7	51.3	33.6	32.8	58.2	23.2	45
CaSO₄, Anhydrite	2.963	93.8	185	112	32.5	26.5	9.3	16.5	15.2	31.7	54.9	29.3	46
SrSO₄, Celestite	3.972	104	106	129	13.5	27.9	26.6	77	60	62	81.8	21.5	46
Na₂SO₄, Thenardite	2.663	80.4	105	67.4	14.8	18.0	23.6	29.8	25.6	16.8	43.4	22.3	46
CaCO₃, Aragonite	2.930	160	87.2	84.8	41.3	25.6	42.7	37.3	1.7	15.7	46.9	38.5	46
Other Minerals													
Chrysoberyl, Al₂BeO₄	3.72	527.7	438.7	465.8	144.4	145.8	151.8	125	111	128	240	160	133
Danburite, CaB₂Si₂O₈	2.99	131	198	211	64.0	59.8	74.9	50	64	34	91.7	64.2	46
Datolite, CaBSi₄O₄OH	3.05	215	155	110	37.1	50.3	78.5	44	50	41	80.4	53.6	46
Staurolite, (Fe,Mg)₂(Al,Fe³⁺)O₆SiO₄(O,OH)₂	3.79	343	185	147	46	70	92	67	61	128	128.2	57.5	46
Topaz, Al₂(F,OH)₂SiO₄	3.563	281	349	294	108	132	131	125	84	88	167.4	114.8	46
Natrolite, Na₂Al₂Si₃O₁₀·2H₂O	2.25	72.2	65.7	138	19.7	24.1	41.1	29.6	25.6	36.9	48.9	27.4	47

Abbreviations: En, enstatite; Fs, ferrosilite; Fo, forsterite; Fa, fayalite.

Table 8. Elastic Constants of Monoclinic Crystals at Room P & T

Material	ρ Mg/m³	Subscript ij in modulus c_{ij} (GPa)													K_S GPa	G GPa	References
		11	22	33	44	55	66	12	13	23	15	25	35	46			
Pyroxenes																	
Acmite, NaFeSi₂O₆	3.50	185.8	181.3	234.4	62.9	51.0	47.4	68.5	70.7	62.6	9.8	9.4	21.4	7.7	112	58.7	3
Augite, (Ca,Na)(Mg,Fe,Al)(Si,Al)₂O₆	3.32	181.6	150.7	217.8	69.7	51.1	55.8	73.4	72.4	33.9	19.9	16.6	24.6	4.3	95	59.0	3
Acmite-Augite	3.42	155.6	151.8	216.1	40.0	46.5	49.2	81.1	66.0	68.4	25.3	26.0	19.2	4.1	102	46.8	3
Diopside, CaMgSi₂O₆	3.31	204	175	238	67.5	58.8	70.5	84.4	88.3	48.2	-19.3	-19.6	-33.6	-11.3	114	64.9	3
CaMgSi₂O₆	3.289	223	171	235	74	67	66	77	81	57	17	7	43	7.3	113	67	68
Diallage	3.30	153.9	149.6	210.8	63.9	62.2	52.3	56.9	37.4	30.5	14.6	14.2	11.9	-8.6	85	61.2	3
Hedenbergite, CaFeSi₂O₆	3.657	222	176	249	55	63	60	69	79	86	12	13	26	-10	120	61	59
Jadeite, NaAlSi₂O₆	3.33	274	253	282	88	65	94	94	71	82	4	14	28	13	143	85	60
Spodumene, LiAlSi₂O₆	3.19	245	199	287	70.1	62.8	70.7	88	64	69	-40	-26.7	-14.2	-7.1	123.5	72.0	46
Feldspars																	
Albite, NaAlSi₃O₈		74	131	128	17.3	29.6	32.0	36.4	39.4	31.0	-6.6	-12.8	-20.0	-2.5	56.9	28.6	46
Anorthite, CaAl₂Si₂O₈		124	205	156	23.5	40.4	41.5	66	50	42	-19	-7	-18	-1	84.2	39.9	47
Hyalophane, (Ba,K)Al₂Si₂O₈		67.4	161	124	13.6	25.3	35.4	42.9	45.1	25.6	-12.8	-7.6	-15.8	-1.7	58.4	26.8	47
Labradorite[a,c]		99.4	158	150	21.7	34.5	37.1	62.8	48.7	26.7	-2.5	-10.7	-12.4	-5.4	74.5	33.7	47
KAlSi₃O₈ Microcline		67.0	169	118	14.3	23.8	36.4	45.3	26.5	20.4	-0.2	-12.3	-15.0	-1.9	55.4	28.1	47
Oligoclase[a]		80.8	163	124	18.7	27.1	35.7	37.9	52.9	32.7	-15.7	-23.7	-6.0	-0.9	62.0	29.3	47
Plagioclase Solid Solutions[a,b]																	
An₉		74.8	137	129	17.4	30.2	31.8	28.9	38.1	21.5	-9.1	-30.7	-19.2	-2.1	50.8	29.3	47
An₂₄		82	145	133	18.1	31.0	33.5	39.8	41.0	33.7	-8.4	-6.3	-18.7	-1.0	62.0	30.6	47

Table 8. (continued)

Subscript ij in modulus c_{ij} (GPa)

Material	ρ Mg/m³	11	22	33	44	55	66	12	13	23	15	25	35	46	K_S GPa	G GPa	References
An29		84.4	151	132	18.9	31.4	34.2	42.1	40.9	32.2	-8.5	-6.5	-18.8	-1.1	63.0	31.4	47
An53		97.1	163	141	20.1	33.1	36.1	51.9	44.0	35.8	-9.4	-9.8	-15.0	-1.4	70.7	33.6	47
An56		98.8	173	141	20.5	34.3	36.8	52.9	43.7	37.2	-10.2	-7.4	-18.0	-1.3	71.9	34.5	47
Na-K Feldspar Solid Solutions[a,b]																	
Or79Ab19	2.56	62.5	172	124	14.3	22.3	37.4	42.8	35.8	24.1	-15.4	-14.3	-11.5	-2.8	53.7	27.2	47
Or75Ab22	2.54	57.2	148	103	13.7	18.0	32.3	32.8	33.3	19.3	-12.4	-6.1	-11.2	-2.5	47.0	23.9	47
Or67Ab29	2.54	58.4	147	99	12.4	18.5	34.3	33.3	34.0	21.6	-10.7	-4.3	-13.0	-3.0	48.0	23.7	47
Or54Ab35	2.58	63.0	152	118	10.1	26.8	35.6	35.9	49.0	36.1	-12.9	-1.8	-18.1	-2.6	57.4	24.2	47
Or65Ab27An4	2.57	59.6	158	105	13.9	20.3	37.0	36.2	36.0	28.5	-11.8	-5.7	-12.9	-2.6	51.4	25.3	47
Or74Ab19	2.57	61.9	158	100	14.1	20.3	36.0	43.4	36.8	21.8	-10.0	-1.8	-12.1	-2.3	53.1	24.9	47
Silicates																	
SiO₂, Coesite	2.911	160.8	230.4	231.6	67.8	73.3	58.8	82.1	102.9	35.6	-36.2	2.6	-39.3	9.9	113.7	61.6	139
Epidote, Ca₂(Al,Fe)₃Si₃O₁₂(OH)	3.40	211.8	238.7	202.0	39.1	43.2	77.5	66.3	45.2	45.6	0.0	-8.2	-14.3	-3.4	106.2	61.2	104
Hornblende, (Ca,Na)₂₋₃(Mg,Fe,Al)₅(Al,Si)₈O₂₂(OH)₂	3.12	116.0	159.7	191.6	57.4	31.8	36.8	44.9	61.4	65.5	4.3	-2.5	10.0	-6.2	87.0	43.0	2, 47
	3.15	130.1	187.7	198.4	61.1	38.7	45.0	61.4	59.2	61.4	9.5	-6.9	-40.6	-0.9	93.3	49.3	2, 47
Muscovite, KAl₃Si₃O₁₀(OH)₂	2.844	184.3	178.4	59.1	16.0	17.6	72.4	48.3	23.8	21.7	-2.0	3.9	1.2	0.5	58.2	35.3	124
Sulphides, Sulphates																	
Gypsum,	2.317	78.6	62.7	72.6	9.1	26.4	10.4	41.0	26.8	24.2	-7.0	3.1	-17.4	-1.6	42.0	15.4	46
CaSO₄		94.5	65.2	50.2	8.6	32.4	10.8	37.9	28.2	32.0	-11.0	6.9	-7.5	-1.1	42.5	15.7	46

Abbreviations: Ab, albite; An, anorthite, Or, orthoclase.

[a] Triclinic, quasi monoclinic.

[b] Subscripts indicate weight percentages of components.

[c] Labradorite is a plagioclase feldspar with composition in the range 50-70% anorthite and 30-50% albite.

Table 9. Elastic Moduli and Velocities in Melts

Composition	T K	ρ kg/m³	$K_{s,\infty}$ GPa	V_P m/s	Frequency $10^6 s^{-1}$	References
Fe	2490	6.54	94.8	3808		48
	3950	5.54	52.4	3075		48
$CaAl_2Si_2O_8$ (An)	1833	2.56	20.6	2850	3.529	100
	1893		20.4		3.0	107
An^a	1923	2.55	17.9			99
$An_{36}Di_{64}$	1677		23.0			107
$An_{36}Di_{64}{}^a$	1673	2.61	24.2			98
$An_{50}Di_{50}$	1673	2.60	21.6	2885	3.635	100
	1573	2.61	22.1	2910	3.922	100
$An_{50}Ab_{50}$	1753	2.44	17.8	2850	3.858	100
$Ab_{50}Di_{50}$	1698	2.45	18.2	2735	3.662	100
	1598	2.46	19.3	2830	3.943	100
$Ab_{75}Di_{25}$	1753	2.39	16.4	2800	3.565	100
	1648	2.40	16.7	3400	3.833	100
$Ab_{33}An_{33}Di_{33}$	1698	2.49	19.5	2805	3.803	100
	1583	2.50	19.8	2880	3.944	100
$BaSi_2O_5$	1793	3.44	19.5	2390	3.906	100
	1693	3.47	20.2	2410	3.652	100
$CaSiO_3$	1836	2.65	27.1	3120	3.484	100
$CaTiSiO_5$	1753	2.96	19.9	2590	4.014	100
	1653	3.01	20.0	2580	4.013	100
$Cs_2Si_2O_5$	1693	3.14	6.4	1450	3.854	100
	1208	3.34	8.8	2345	4.023	100
$CaMgSi_2O_6$ (Di)^a	1773	2.61	22.4			99
Di	1758	2.60	24.2	3040	3.842	100
	1698	2.61	24.1	3020	3.83	100
$Fe_{1.22}Si_{0.89}O_3$	1693	3.48	19.2	2345	3.665	100
	1598	3.51	20.6	2450	3.680	100
Fe_2SiO_4	1653	3.71	21.4	2400	7.65	100
	1503	3.76	22.6	2450	8.67	100
$K_2Si_2O_5$	1693	2.16	10.3	2190	3.955	100
	1408	2.22	11.9	2600	3.951	100
K_2SiO_3	1698	2.10	7.5	1890	4.909	100
	1498	2.17	8.5	1970	5.242	100
$Li_2Si_2O_5$	1693	2.12	15.0	2670	4.100	100
	1411	2.17	16.3	2740	3.852	100
Li_2SiO_3	1543	2.08	20.7	3160	3.712	100
$MgSiO_3$	1913	2.52	20.6	2860	4.040	100
NaCl	1094			1727	8.61	63
	1322			1540	8.61	63
$(Na_2O)_{33}(Al_2O_3)_6(SiO_2)_{61}$	1684		15.8^a	2653	3.707	63
	1599		16.4^a	2695	3.764	63
$(Na_2O)_{32}(Al_2O_3)_{15}(SiO_2)_{52}$	1690		18.6^a	2835	5.558	63
$Na_2Si_2O_5$	1693	2.20	14.0	2525	3.934	100
	1408	2.26	16.2	2680	3.990	100
Na_2SiO_3	1573	2.22	15.7	2663	10.1	100
	1458	2.25	17.0	2752	8.4	100

Table 9. (continued)

Composition	T K	ρ kg/m^3	$K_{s,\infty}$ GPa	V_P m/s	Frequency 10^6s^{-1}	References
Or$_{78}$An$_{22}$	1783	2.33	13.8	4300	3.836	100
	1598	2.35	14.1	5200	3.923	100
Or$_{61}$Di$_{39}$	1768	2.38	16.0	2795	3.656	100
	1578	2.40	16.5	3470	3.673	100
Rb$_2$Si$_2$O$_5$	1693	2.78	7.8	1678	3.945	100
	1408	2.88	9.9	2130	3.974	100
SrSi$_2$O$_5$	1758	3.02	19.6	2550	3.690	100
	1653	3.04	20.1	2570	3.833	100
Tholeitic Basalt	1708	2.65	17.9	2600	3.839	100
	1505	2.68	18.3	2610	3.909	100
Basalt-Andesite	1803	2.55	18.6	2700	3.790	100
	1503	2.59	19.4	2980	3.863	100
Andesite	1783	2.44	16.1	2775	3.827	100
	1553	2.46	16.6	3850	3.889	100
Ryolite	1803	2.29	13.0	4350	3.664	100
	1553	2.31	13.5	5280	3.723	100

Abbreviations: An, CaAl$_2$Si$_2$O$_8$; Di, CaMgSi$_2$O$_6$, Or, KAlSi$_3$O$_8$; Ab, NaAlSi$_3$O$_8$.
[a] From shock wave experiments.

Table 10. Elastic Moduli of Glasses

Composition	ρ kg/m^3	K_s GPa	G GPa	$\delta K/\delta P$	$\delta G/\delta P$	$\delta K/\delta T$ MPa K^{-1}	$\delta G/\delta T$ MPa K^{-1}	References
SiO$_2$	2.204	36.5	31.2	−6	−3.4	16	4	38, 79
MgSiO$_3$	2.761	78.8	41.8					129
CaSiO$_3$	2.880	69.2	36.3					129
CaMgSi$_2$O$_6$	2.863	76.9	39.7					129
	2.847	74.1	38.8					113
CaAl$_2$Si$_2$O$_8$	2.693	69.2	38.7					129
Na$_2$Si$_2$O$_6$	2.494	41.9	24.1					129
(Na$_2$O)$_{35}$(SiO$_2$)$_{65}$[a]	2.495	41.0	23.0	4.6	0.7	−12.2	−10.7	75
NaAlSi$_3$O$_8$	2.369	39.1	29.2					129
Na$_2$Al$_2$Si$_2$O$_8$	2.490	45.1	30.2	2.4	−0.35	−7	−9	44
(Na$_2$O)$_{30}$(TiO$_2$)$_{20}$(SiO$_2$)$_{50}$[a]	2.749	50.0	30.2	4.9	0.5	−8.1	−7.1	74
(K$_2$0)$_{25}$(SiO$_2$)$_{75}$[a]	2.42	30	21	−4 to +4	−1	−2.4	−3.4	75, 36
Obsidian	2.331	37.8	30.1	−1.8	−1.7			79
Andesite	2.571	52.5	33.6	0.6	−0.8			79
Basalt	2.777	62.9	36.5	2.1	−0.3			79

[a] Composition given as mole percentages of oxide components.

Table 11. P and T derivatives of Isotropic Elastic Moduli

Material	$\delta K_S/\delta P$	$\delta G/\delta P$	$\delta K_S/\delta T$ MPa/K	$\delta G/\delta T$ MPa/K	ΔT K	References
Elements, Metallic Compounds						
Ag, Silver	6.09	1.68	−21.5	−12.7	79 − 298	17
Au, Gold	6.13	1.27	−31.0	−8.4	79 − 298	17
C, Diamond	4.0	2.3	−8.7	−5.7	223 − 323	77
α-Fe, (bcc)	5.29	1.82	−31	−27	25 − 300	29, 42
	5.97	1.91	−43	−33	300 − 500	29, 102
	5.13	2.16	−51	−47	500 − 700	29, 128
	4.3	3.4	−43	−43	800 − 900	29, 49
			−18	−14	77 − 300	71
Fe$_{0.94}$Si$_{0.06}$			−19	−17	80 − 298	103
			−33	−30	298 − 900	103
Simple Oxides						
Al$_2$O$_3$, Corundum	4.3		−15	−27	@296	39, 40
			−23	−24	@1000	40
			−19	−24	@1825	40
BaO	5.52	1.12	−23.9	−12.0	281 − 298	26
			−7		195 − 293	127
CaO, Lime	5.23	1.64	−14.3	−13.8	283 − 303	112
	6.0	1.7	−19.2	−15.0	195 − 293	8, 114
			−14.1	−14.7	300 − 1200	81
	4.83	1.78	−12.8	−14.9	281 − 298	26
CoO			20	112	293 − 303	120
Fe$_{0.92}$O, Wüstite			−20	12.4	@298	120
Fe$_{0.943}$O, Wüstite	5.1	0.71				56
Fe$_2$O$_3$, Hematite	4.5	0.73				69
GeO$_2$, (rutile structure)	6.2	1.2	−36	−12	293 − 373	131
MnO, Manganosite	5.28	1.55				138
	4.7	1.2	−20.3	−11	273 − 473	89
			−21	14.6	@298	120
MgO, Periclase	3.85		−15.3		300 − 800	115
	4.5	2.5				4
	4.13	2.5	−14.5	−24	@300K	52, 57
	4.27	2.5	−22.5	−26	@1200K	23, 52
			−21.3	−21	@1800K	52
SrO	5.18	1.61	−17.8	−12.6	281 − 298	26
	6.0		−7.1	−11.9	195 − 293	8, 114
SnO$_2$, Cassiterite	5.50	0.61	−19	−6.7	298 − 373	22
SiO$_2$, Quartz	6.4	0.46	−8.5	−0.8	@293	78, 110, 118
TiO$_2$, Rutile	6.76	0.78	−48.7	−21.0	298 − 583	34, 73, 76
UO$_2$, Uraninite	4.69	1.42				35
Spinel Structured Oxides						
β-Mg$_2$SiO$_4$, Wadsleyite	4.8	1.7				43
MgAl$_2$O$_4$, Spinel	5.66		−15.7	−9.4	293 − 423	70, 152
MgAl$_2$O$_4$	4.89					24
MgO·2.6Al$_2$O$_3$	4.18					106

Table 11. (continued)

Material	$\delta K_S/\delta P$	$\delta G/\delta P$	$\delta K_S/\delta T$ MPa/K	$\delta G/\delta T$ MPa/K	ΔT K	References
$Mg_{0.75}Fe_{0.36}Al_{1.90}O_4$, Pleonaste	4.92	0.29				130
Sulphides						
PbS, Galena	6.28		−39.0		77 − 300	91, 94
ZnS, Wurtzite	4.37	0.00	−9.56	0.00	298 − 373	21
Binary Halides						
BaF_2, Frankdicksonite	5.05		−14.5		195 − 298	145
CaF_2, Fluorite	4.92		−17.5		195 − 298	145
NaCl, Halite	5.27	2.14	−10.8	−9.9	195 − 295	7
	5.256		−11.13		300 − 800	116
			−10.5	−8.2	294 − 338	147
			−15.2	−9.5	745 − 766	147
KCl, Sylvite	5.0	2.0	−7.2	−3.2	300 − 1000	7, 28, 30
			−8.7	−5.6	294 − 865	146
Garnets						
$Py_{73}Alm_{16}And_4Uv_6$			−19.5	−10.2	298 − 1000	121
$Py_{62}Alm_{36}Gr_2$	4.93	1.56				137
$Py_{61}Alm_{36}Gr_2$	4.74		−18.8		298 − 338	20
$Sp_{54}Alm_{46}$	4.95	1.44				132
$Alm_{72}Py_{22}Gr_2And_2Sp_3$	5.43	1.40	−20.1	−10.6	288 − 313	111
$Gr_{98}An_2Py_1$			−14.9	−12.5	300 − 1350	54
$Gr_{76}An_{22}Sp_1$			−14.7	−12.5	300 − 1250	54
Other Minerals						
Forsterite (Fo), Mg_2SiO_4	4.97	1.82	−17.6	−13.6	300 − 700	41
	5.37	1.80	−15.0	−13.0	298 − 306	65
			−16.0	−13.5	293 − 673	119
			−15.7	−13.5	300 − 1700	53
Olivine, Fo_{90}	4.56	1.71				154
$Fo_{93}Fa_7$	5.13	1.79	−15.6	−1.30	298 − 306	65
$Fo_{91}Fa_9$	4.6	1.9				136
$Fo_{90}Fa_{10}$			−18.0	−13.6	300 − 1500	51
$Fo_{92}Fa_8$			−16.9	−13.8	300 − 1500	51
Fayalite (Fa), Fe_2SiO_4			−24	−13	300 − 500	55
Orthopyroxene	10.8	2.06				137
$(Mg_{.8}Fe_{0.2})SiO_3$	9.6	2.38	−26.8	−11.9	298 − 623	33
$AlPO_4$, Berlinite			−7	−2	180 − 298	25
Beryl, $Be_3Al_2Si_6O_{18}$	3.90					153
Calcite[a], $CaCO_3$	4.83					58
Nepheline, $Na_3KAl_4Si_4O_{16}$			−3.7	1.6	298 − 353	19
Zircon, $ZrSiO_4$	6.5	0.78	−21	−9.4	298 − 573	88

Abbreviations: Py, Pyrope $Mg_3Al_2Si_3O_{12}$; Alm, Almandite $Mg_3Al_2Si_3O_{12}$; Gr, Grossular $Ca_3Al_2Si_3O_{12}$; Uv, Uvarovite $Ca_3Cr_2Si_3O_{12}$; And, Andradite $Ca_3Fe_2Si_3O_{12}$; Sp, Spessartite $Mn_3Al_2Si_3O_{12}$; Fo, Forsterite Mg_2SiO_4; Fa, Fayalite Fe_2SiO_4.

[a] Pressure derivative of K_T is given.

Table 12. Higher Order Pressure and Temperature Derivatives

Composition	$\delta^2 K/\delta P^2$ GPa^{-1}	$\delta^2 G/\delta P^2$ GPa^{-1}	$\delta^2 K/\delta T^2$ kPa K^{-2}	$\delta^2 G/\delta T^2$ kPa K^{-2}	References
SiO$_2$ Glass	2.9				38
Grossular Garnet Ca$_3$Al$_2$Si$_3$O$_{12}$			−1.8	−1.1	54
Pyrope Garnet Mg$_3$Al$_2$Si$_3$O$_{12}$	−0.28	−0.08	−1.8	−1.1	136
Forsterite, $T < 760$			−5.2	−2.6	53
Mg$_2$SiO$_4$ $T > 760$			−0.7		53
Olivine,	−0.15	−0.11			136
(Mg,Fe)$_2$SiO$_4$	−0.05	−0.06			154
MgO, Periclase	−0.03				57
Fe$_{0.943}$O, Wüstite	−0.07	−0.10			56
CaO, Lime			−1.4	0.3	81
Orthopyroxene, (Mg,Fe)SiO$_3$	−1.6	−0.12			137
MgAl$_2$O$_4$, Spinel	0.5				24

Acknowledgments: This work was supported in part by the NSF under grant no. EAR-90-18676. The review of O.L. Anderson is appreciated.

REFERENCES

1. Albers, H. L., and J. C. A. Boeyens, The elastic constants and distance dependence of the magnetic interactions of Cr$_2$O$_3$, *J. Magnetism Mag. Mater.*, *2*, 327-333, 1976.

2. Alexandrov, K. S., and T. V. Ryzhova, The elastic properties of rock-forming minerals: pyroxenes and amphiboles, *Izv. USSR Acad. Sci., Geopohys. Ser.*, no. 9, 871-875, 1961.

3. Alexandrov, K. S., T. V. Ryzhova, and Belikov, The elastic properties of pyroxenes, *Sov. Phys. Crystallogr.*, *8*, 589-591, 1964.

4. Anderson, O. L., and P. Andreacht, Jr., Pressure derivatives of elastic constants of single crystal MgO at 23° and -195.8° C, *J. Am. Ceram. Soc.*, *49*, 404-409, 1966.

5. Babuška, V., J. Fiala, M. Kumazawa, I. Ohno, and Y. Sumino, Elastic properties of garnet solid–solutions series, *Phys. Earth Planet. Inter.*, *16*, 157-176, 1978.

6. Bansal, N. P., and R. H. Doremus, *Handbook of Glass Properties*, 680pp., Academic Press, Orlando, FL, 1986.

7. Bartels, R. A., and D. E. Schuele, Pressure derivatives of the elastic constants of NaCl and KCl at 295°K and 195°K, *J. Phys. Chem. Solids*, *33*, 1991-1992, 1965.

8. Bartels, R. A., and V. H. Vetter, The temperature dependence of the elastic constants of CaO and SrO, *J. Phys. Chem. Solids*, *33*, 1991-1992, 1972.

9. Bass, J. D., Elasticity of single-crystal orthoferrosilite, *J. Geophys. Res.*, *89*, 4359–4371, 1984.

10. Bass, J. D., Elasticity of uvarovite and andradite garnets, *J. Geophys. Res.*, *91*, 7505–7516, 1986.

11. Bass, J. D., Elasticity of grossular and spessartite garnets by Brillouin spectroscopy, *J. Geophys. Res.*, *94*, 7621–7628, 1989.

12. Bass, J. D., and M. Kanzaki, Elasticity of a majorite-pyrope solid solution, *Geophys. Res. Lett.*, *17*, 1989–1992, 1990.

13. Bass, J. D., D. J. Weidner, N. Hamaya, M. Ozima, and S. Akimoto, Elasticity of the olivine and spinel polymorphs of Ni$_2$SiO$_4$, *Phys. Chem. Minerals*, *10*, 261–272, 1984.

14. Bentle, G. G., Elastic constants of single-crystal BeO at room temperature, *J. Am. Ceramic Soc.*, *49*, 125-128, 1966.

15. Berger, J., J. Berthon, A. Revcolevschi, and E. Jolles, Elastic constants of Fe$_{1-x}$O single crystals, *J. Am. Ceramic Soc.*, *64*, C153-154, 1981.

16. Birch, F., Compressibility; elastic constants, in *Handbook of Physical Constants*, (revised), pp. 97-173, edited by S. P. Clark, Jr., Geological Soc. Am. Mem. no.

97, New York, 587pp., 1966.

17. Biswas, S. N., P. Van't Klooster, and N. J. Trappeniers, Effect of pressure on the elastic constants of noble metals from -196 to +25°C and up to 2500- bar, *Physica, 103B*, 235-246, 1981.

18. Blakslee, O. L., D. G. Proctor, E. J. Seldin, G. B. Spence, and T. Weng, Elastic constants of compression-annealed pyrolitic graphite, *J. Appl. Phys., 41*, 3373-3382, 1970.

19. Bonczar, L. J. and G. R. Barsch, Elastic and thermoelastic constants of nepheline, *J. Appl. Phys., 46*, 4339-4340, 1975.

20. Bonczar, L. J. and E. K. Graham, The pressure and temperature dependence of the elastic constants of pyrope garnet, *J. Geophys. Res., 82*, 2529-2534, 1977.

21. Chang, E., and G. R. Barsch, Pressure dependence of single crystal elastic constants and anharmonic properties of wurtzite, *J. Phys. Chem. Solids, 34*, 1543-1563, 1973.

22. Chang, E., and E. K. Graham, The elastic constants of cassiterite SnO_2 and their pressure and temperature dependence, *J. Geophys. Res., 80*, 2595-2599, 1975.

23. Chang, Z. P., and G. R. Barsch, Pressure dependence of the elastic constants of single-crystalline magnesium oxide, *J. Geophys. Res., 74*, 3291-3294, 1969.

24. Chang, Z. P., and G. R. Barsch, Pressure dependence of single-crystal elastic constants and anharmonic properties of spinel, *J. Geophys. Res., 78*, 2418-2433, 1973.

25. Chang, Z. P., and G. R. Barsch, Elastic constants and thermal expansion of berlinite, *IEEE Trans. Sonics Ultrasonics, SU-23*, 127-135, 1976.

26. Chang, Z. P., and E. K. Graham, Elastic properties of oxides in the NaCl-structure, *J. Phys. Chem. Solids, 38*, 1355-1362, 1977.

27. Dantl, G., Die elastischen moduln von eis-einkristallen, *Phys. Kondens. Materie., 7*, 390-397, 1968.

28. Demarest, H. H. Jr., R. Ota, and O. L. Anderson, Prediction of high pressure phase transitions by elastic constant data, in: *High Pressure Research, Applications in Geophysics*, edited by M. Manghnani and S. Akimoto, pp. 281-301, Academic Press, New York, 1977.

29. Dever, D. J., Temperature dependence of the elastic constants in α-iron single crystals: *J. Appl. Phys., 43*, 3293-3301, 1972.

30. Dobretsov, A. I., and G. I. Peresada, Dependence of the elastic constants of KCl on pressure, *Sov. Phys. - Solid State, 11*, 1401-1402, 1969.

31. Duffy, T. S., and M. T. Vaughan, Elasticity of enstatite and its relationship to crystal structure, *J. Geophys. Res., 93*, 383-391, 1988.

32. Ecolivet, C., and H. Poignant, Berlinite and quartz $\alpha \leftrightarrow \beta$ phase transition analogy as seen by Brillouin scattering, *Phys. Stat. Sol., 63*, K107-K109, 1981.

33. Frisillo, A. L. and G. R. Barsch, Measurement of single-crystal elastic constants of bronzite as a function of pressure and temperature, *J. Geophys. Res., 77*, 6360-6384, 1972.

34. Fritz, I. J., Pressure and temperature dependences of the elastic properties of rutile (TiO_2), *J. Phys. Chem. Solids, 35*, 817-826, 1974.

35. Fritz, I. J., Elastic properties of UO_2 at high pressure, *J. Appl. Phys., 47*, 4353-4357, 1976.

36. Gamberg, E., D. R. Uhlmann, and D. H. Chung, Pressure dependence of the elastic moduli of glasses in the K_2O-SiO_2 system, *J. Non-Cryst. Solids, 13*, 399-408, 1974.

37. Gammon, P. H., H. Kiefte, and M. J. Clouter, Elastic constants of ice by Brillouin Spectroscopy, *J. Glaciology, 25*, 159-167, 1980.

38. Gerlich, D., and G. C. Kennedy, Second pressure derivatives of the elastic moduli of fused quartz, *J. Phys. Chem. Solids, 39*, 1189-1191, 1978.

39. Gieske, J. H. and G. R. Barsch, Pressure dependence of the elastic constants of single crystalline aluminum oxide, *Phys. Stat. Sol., 29*, 121-131, 1968.

40. Goto, T., O. L. Anderson, I. Ohno, and S. Yamamoto, Elastic constants of corundum up to 1825 K, single crystalline aluminum oxide, *J. Geophys. Res.,* 94, 7588-7602, 1989.

41. Graham, E. K., and G. R. Barsch, Elastic constants of single- crystal forsterite as a function of temperature and pressure, *J. Geophys. Res, 74*, 5949-5960, 1969.

42. Guinan, M. W., and D. N. Beshers, Pressure derivatives of the elastic constants of α-iron to 10 kbs, *J. Phys. Chem. Solids, 29*, 541-549, 1968.

43. Gwanmesia, G. D., S. Rigden, I. Jackson, and R. C. Liebermann, Pressure dependence of elastic wave velocity for β-Mg_2SiO_4 and the composition of the Earth's mantle, *Science, 250*, 794-797, 1990.

44. Halleck, P. M., R. E. Pacalo, and E. K. Graham, The effects of annealing and aluminium substitution on the elastic behavior of alkali silicate glasses, *J. Noncryst. Solids, 86*, 190-203, 1986.

45. Haussühl, S., Elastic and thermoelastic properties of isotypic $KClO_4$, $RbClO_4$, $CsClO_4$, $TlClO_4$, $NHClO_4$, $TlBF_4$, NH_4BF_4 and $BaSO_4$, *Zeitsch. Krist., 192*, 137-145, 1990.

46. Hearmon, R. F. S., The elastic constants of crystals and other anisotropic materials, in *Landolt-Börnstein Tables, III/11*, pp. 1-244, edited by K. H. Hellwege and A. M. Hellwege, Springer-Verlag, Berlin, 854 pp., 1979.

47. Hearmon, R. F. S., The elastic constants of crystals and other anisotropic materials, in *Landolt-Börnstein Tables, III/18*, pp. 1-154, edited by K. H. Hellwege and A. M. Hellwege, Springer-Verlag, Berlin, 559 pp., 1984.

48. Hixson, R. S., M. A. Winkler, and M. L. Hodgdon, Sound speed and thermophysical properties of liquid iron and nickel, *Phys. Rev. B, 42*, 6485-6491, 1990.

49. Hughes, D. S., and C. Maurette, Dynamic elastic moduli of iron, aluminum, and fused quartz, *J. Appl. Phys., 27*, 1184, 1956.

50. Humbert, P., and F. Plicque, Propriétés élastiques de carbonates rhombohédriques monocristallins: calcite, magnésite, dolomie, (in French), *C. R. Acad. Sc. Paris, 275*, series B, 391-394, 1972.

51. Isaak, D. G., High temperature

elasticity of iron-bearing olivines, *J. Geophys. Res., 97*, 1871-1885, 1992.

52. Isaak, D. G., O. L. Anderson, and T. Goto, Measured elastic moduli of single-crystal MgO up to 1800 K, *Phys. Chem. Minerals, 16*, 704-713, 1989a.

53. Isaak, D. G., O. L. Anderson, and T. Goto, Elasticity of single-crystal forsterite measured to 1700 K, *J. Geophys. Res., 94*, 5895-5906, 1989b.

54. Isaak, D. G., O. L. Anderson, and T. Oda, High-temperature thermal expansion and elasticity of calcium-rich garnets, *Phys. Chem. Minerals, 19*, 106-120, 1992.

55. Isaak, D. G., E. K. Graham, J. D. Bass, and H. Wang, The elastic properties of single-crystal fayalite as determined by dynamical measurement techniques, *Pure Appl. Geophys.*, in press, 1993.

56. Jackson, I., and S. K. Khanna, Elasticity, shear-mode softening and high pressure polymorphism of Wüstite (Fe$_{1-x}$O), *J. Geophys. Res., 95*, 21671-21685, 1990.

57. Jackson, I., and H. Niesler, The elasticity of periclase to 3 GPa and some geophysical implications, in *High-Pressure Research in Geophysics, Advances in Earth and Planetary Sciences*, vol. 12, edited by S. Akimoto and M. H. Manghnani, pp. 93-113, Center for Academic Publications, Tokyo, 1982.

58. Kaga, H., Third-order elastic constants of calcite, *Phys. Rev., 172*, 900-919, 1968.

59. Kandelin, J., and D. J. Weidner, Elastic properties of hedenbergite, *J. Geophys. Res., 93*, 1063-1072, 1988a.

60. Kandelin, J., and D. J. Weidner, The single-crystal elastic properties of jadeite, *Phys. Earth Planet. Interiors, 50*, 251-260, 1988b.

61. Kobiakov, I. B., Elastic, Piezoelectric and dielectric properties of ZnO and CdS single crystals in a wide range of temperatures, *Sol. St. Comm., 35*, 305-310, 1980.

62. Koptsik, V. A., and L. A. Ermakova, Electric and elastic parameters of cancrinite as a function of temperature, *Fiz. Tverd. Tela, 2*, 643-646, 1960.

63. Kress, V. C., Q. Williams, and I. S. E. Carmichael, Ultrasonic investigation of melts in the system Na$_2$O-Al$_2$O$_3$-SiO$_2$, *Geochem. Cosmochem. Acta, 52*, 283-293, 1988.

64. Kumazawa, M., The elastic constants of single-crystal orthopyroxene, *J. Geophys. Res., 74*, 5973-5980, 1969.

65. Kumazawa, M., and O. L. Anderson, O. L., Elastic moduli, pressure derivatives, and temperature derivatives of single-crystal olivine and single-crystal forsterite, *J. Geophys. Res., 74*, 5961-5972, 1969.

66. Leese, J., and A. E. Lord, Jr., Elastic stiffness coefficients of single-crystal iron from room temperature to 500°C, *J. Appl. Phys. 39*, 3986-3988, 1968.

67. Leitner, B. J., D. J. Weidner, and R. C. Liebermann, Elasticity of single crystal pyrope and implications for garnet solid solution series, *Phys. Earth Planet. Interiors, 22*, 111-121, 1980.

68. Levien, L., D. J. Weidner, and C. T. Prewitt, Elasticity of diopside, *Phys. Chem. Minerals, 4*, 105-113, 1979.

69. Liebermann, R. C., and E. Schreiber, Elastic constants of polycrystalline hematite as a function of pressure to 3 kilobars, *J. Geophys. Res., 73*, 6585-6590, 1968.

70. Liu, H-P, R. N. Schock, and D. L. Anderson, Temperature dependence of single-crystal spinel (MgAl$_2$O$_4$) elastic constants from 293 to 423°K measured by light-sound scattering in the Raman-Nath region, *Geophys. J. R. astr. Soc., 42*, 217-250, 1975.

71. Lord, A. E. Jr., and D. N. Beshers, Elastic stiffness coefficients of iron from 77° to 673°K, *J. Appl. Phys. 36*, 1620-1623, 1965.

72. Machová, A., and S. Kadečková, Elastic constants of iron-silicon alloy single crystals, *Czech. J. Phys., B27*, 555-563, 1977.

73. Manghnani, M., Elastic constants of single-crystal rutile under pressure to 7.5 kilobars, *J. Geophys. Res., 74*, 4317-4328, 1969.

74. Manghnani, M., Pressure and temperature dependence of the elastic moduli of Na$_2$O-TiO$_2$-SiO$_2$ glasses, *J. Am. Ceram. Soc., 55*, 360-365, 1972.

75. Manghnani, M., and B. K. Singh, Effects of composition, pressure and temperature on the elastic thermal, and ultrasonic attenuation properties of sodium silicate glasses, *Proc. X Int. Cong. on Glass, Kyoto University, Japan*, 11-104–11-114, The Ceramic Society of Japan, 1974.

76. Manghnani, M., E. S. Fisher, and W. S. Brower, Jr., Temperature dependence of the elastic constants of single-crystal rutile between 4° and 583°K, *J. Phys. Chem. Solids, 33*, 2149-2159, 1972.

77. McSkimin, H. J., and W. L. Bond, Elastic moduli of diamond as a function of presure and temperature, *J. Appl. Phys., 43*, 2944-2948, 1972.

78. McSkimin, H. J., P. Andreacht, and R. N. Thurston, Elastic moduli of quartz versus hydrostatic pressure at 25° and -195.8°, *J. Appl. Phys., 36*, 1624-1632, 1965.

79. Meister, R., E. C. Robertson, R. W. Werre, and R. Raspet, Elastic moduli of rock glasses under pressure to 8 kilobars and geophysical implications, *J. Geophys. Res., 85*, 6461-6470, 1980.

80. Nye, J. F., *Physical Properties of Crystals and Their Representation by Tensors and Matrices*, 322 pp., Oxford University Press, Oxford, 1976.

81. Oda, H., O. L. Anderson, D. G. Isaak, and I. Suzuki, Measurement of elastic properties of single-crystal CaO up to 1200 K, *Phys. Chem. Minerals, 19*, 96-105, 1992.

82. Ohno, I., Free vibration of a rectangular parallelepiped crystal and its application to determination of elastic constants of orthorhombic crystals, *J. Phys. Earth, 24*, 355-379, 1976.

83. Ohno, I., S. Yamamoto, and O. L. Anderson, Determination of elastic constants of trigonal crystals by the rectangular parallelepiped resonance method, *J. Phys. Chem. Solids, 47*, 1103-1108, 1986.

84. O'Neill, B., J. D. Bass, J. R. Smyth, and M. T. Vaughan, Elasticity of a grossular-pyrope-almandine garnet, *J. Geophys.*

Res., 94, 17,819-17,824, 1989.

85. O'Neill, B., J. D. Bass, G. R. Rossman, C. A. Geiger, and K. Langer, Elastic properties of pyrope, Phys. Chem. Minerals, 17, 617-621, 1991.

86. O'Neill, B., J. D. Bass, and G. R. Rossman, Elastic properties of hydrogrossular garnet and implications for water in the upper mantle, J. Geophys. Res., 98, 20,031-20,037, 1993.

87. Ozkan, H., Elastic constants of tourmaline, J. Appl. Phys., 50, 6006-6008, 1979.

88. Ozkan, H., and J. C. Jamieson, Pressure dependence of the elastic constants of nonmetamict zircon, Phys. Chem. Minerals, 2, 215-224, 1978.

89. Pacalo, R. E., and E. K. Graham, Pressure and temperature dependence of the elastic properties of synthetic MnO, Phys. Chem. Minerals, 18, 69-80, 1991.

90. Pacalo, R. E. G., D. J. Weidner, and T. Gasparik, Elastic properties of sodium-rich garnet, Geophys. Res. Lett., 19, 1895-1898, 1992.

91. Padaki, V. C., S. T. Lakshmikumar, S. V. Subramanyam, and E. S. R. Gopal, Elastic constants of galena down to liquid helium temperatures, Pramana, 17, 25-32, 1981.

92. Peercy, M. S., and J. D. Bass, Elasticity of monticellite, Phys. Chem. Minerals, 17, 431–437, 1990.

93. Peercy, P. S., I. J. Fritz, and G. A. Samara, Temperature and pressure dependences of the properties and phase transition in paratellurite (TeO_2): Ultrasonic, dielectric and raman and Brillouin scattering results, J. Phys. Chem. Solids, 36, 1105-1122, 1975.

94. Peresada, G. I., E.G. Ponyatovskii, and Zh. D. Sokolovskaya, Pressure dependence of the elastic constants of PbS, Phys. Stat. Sol., A35, K177-K180, 1976.

95. Proctor, T. M., Jr., Low temperature speed of sound in single-crystal ice, J. Acoustic. Soc. Am., 39, 972-977, 1966.

96. Rayne, J. A. and B. S. Chandrasekhar, Elastic constants of iron from 4.2 to 300°K, Phys. Rev., 122, 1714-1716, 1961.

97. Rigden J. A. and B. S. Chandrasekhar, Elastic constants of iron from 4.2 to 300°K, Phys. Rev., 122, 1714-1716, 1961.

98. Rigden S. M., T. J. Ahrens, and E. M. Stolper, Shock compression of molten silicate: results for a model basaltic composition, J. Geophys. Res., 93, 367-382, 1988.

99. Rigden S. M., T. J. Ahrens, and E. M. Stolper, High pressure equation of state of molten anorthite and diopside, J. Geophys. Res., 94, 9508-9522, 1989.

100. Rivers, M. L., and I. S. E. Carmichael, Ultrasonic studies of silicate melts, J. Geophys. Res., 92, 9247-9270, 1987.

101. Robie, R., P. M. Bethke, M. S. Toulmin, and J. L. Edwards, X-ray crystallographic data, densities, and molar volumes of minerals, in Handbook of Physical Constants, (revised), pp. 27-73, edited by S. P. Clark, Jr., Geological Soc. Am. Mem. no. 97, New York, 587pp., 1966.

102. Rotter, C. A. and C. S. Smith, Ultrasonic equation of state of iron I. Low pressure, room temperature, J. Phys. Chem. Solids, 122, 267-276, 1966.

103. Routbort, J. L., C. N. Reid, E. S. Fisher, and D. J. Dever, High-temperature elastic constants and the phase stability of silicon-iron, Acta Metallurgica, 19, 1307-1316, 1971.

104. Ryzhova, T. V., K. S. Aleksandrov, and V. M. Korobkova, The elastic properties of rock-forming inerals V. Additional data on silicates, Izv. Earth Phys., 2, 63-65, 1966.

105. Sawamoto, H., D. J. Weidner, S. Sasaki, and M. Kumazawa, Single crystal elastic properties of the modified spinel (Beta) phase of magnesium orthosilicate, Science, 224, 749-751, 1984.

106. Schreiber, E., Elastic moduli of single-crystal spinel at 25° C and to 2 kbar, J. Appl. Phys., 38, 2508-2511, 1967.

107. Secco, R. A., M. H. Manghnani, and T. C. Liu, The bulk modulus-attenuation-viscosity systematics of diopside-anorthite melts, Geophys. Res. Lett., 18, 93-96, 1991.

108. Simmons, G., and F. Birch, Elastic constants of pyrite, J. Appl. Phys., 34, 2736-2738, 1963.

109. Simmons, G., and H. Wang, Single Crystal Elastic Constants and Calculated Aggregate Properties: A Handbook, 2nd Ed., M.I.T. Press, Cambridge, Mass. 1971. J. Appl. Phys., 34, 2736-2738, 1963.

110. Smagin, A. G., and B. G. Mil'shtein, Elastic constants of α-quartz single crystals, Sov. Phys. Crystallogr., 19, 514-516, 1975.

111. Soga, N., Elastic constants of garnet under pressure and temperature, J. Geophys. Res., 72, 4227-4234, 1967.

112. Soga, N., Elastic constants of CaO under pressure and temperature, J. Geophys. Res., 73, 5385-5390, 1968.

113. Soga, N., H. Yamanaka, C. Hisamoto, and M. Kunugi, Elastic properties and structure of alkaline-earth silicate glasses, J. Non-Cryst. Solids, 22, 67-76, 1976.

114. Son, P. R., and R. A. Bartels, CaO and SrO single crystal elastic constants and their pressure derivatives, J. Phys. Chem. Solids, 33, 819-828, 1972.

115. Spetzler, H., Equation of state of polycrystalline and single-crystal MgO to 8 kilobars and 800°K, J. Geophys. Res., 75, 2073-2087, 1970.

116. Spetzler, H., G. C. Sammis, and R. J. O'Connell, Equation of state of NaCl: Ultrasonic measurements to 8 kilobars and 800°C and static lattice theory, J. Phys. Chem. Solids, 75, 1727-1750, 1972.

117. Sumino, Y., The elastic constants of $Mn_2Fe_2SiO_4$ and Co_2SiO_4, and the elastic properties of the olivine group minerals, J. Phys. Earth, 27, 209-238, 1979.

118. Sumino, Y., and O. L. Anderson, Elastic constants of minerals, in: Handbook of Physical Properties, III, edited by Robert S. Carmichael, CRC Press, Boca Raton, FL, 1984.

119. Sumino, Y., O. Nishizawa, T. Goto, I. Ohno, and M. Ozima, Temperature variation of the elastic constants of single-crystal forsterite between -190° and 400°, J. Phys. Earth, 25, 377-392, 1977.

120. Sumino, Y., M. Kumazawa, O. Nishizawa, and W. Pluschkell, The elastic constants of single-

crystal $Fe_{1-x}O$, MnO and CoO, and the elasticity of stochiometric magnesiowüstite, *J. Phys. Earth, 28*, 475-495, 1980.

121. Suzuki, I., and O. L. Anderson, Elasticity and thermal expansion of a natural garnet up to 1000K, *J. Phys. Earth, 31*, 125-138, 1983.

122. Uchida, N., and Y. Ohmachi, Elastic and photoelastic properties of TeO_2 single crystal, *J. Appl. Phys., 40*, 4692-4695, 1969.

123. Vaughan, M. T., and J. D. Bass, Single crystal elastic properties of protoenstatite: A comparison with orthoenstatite, *Phys. Chem. Minerals, 10*, 62-68, 1983.

124. Vaughan, M. T., and S. Guggenheim, Elasticity of muscovite and its relationship to crystal structure, *J. Geophys. Res., 91*, 4657-4664, 1986.

125. Vaughan, M. T., and D. J. Weidner, The relationship of elasticity and crystal structure in andalusite and sillimanite, *Phys. Chem. Minerals, 3*, 133-144, 1978.

126. Verma, R. K., Elasticity of some high-density crystals, *J. Geophys. Res., 65*, 757-766, 1960.

127. Vetter, V. H., and R. A. Bartels, BaO single crystal elastic constants and their temperature dependence, *J. Phys. Chem. Solids, 34*, 1448-1449, 1973.

128. Voronov, F. F. and L. F. Vereshchagin, The influence of hydrostatic pressure on the elastic properties of metals. I. Experimental data, *Physics Metals Metallogr, 11*, 111, 1961.

129. Wang, Hong, *Elasticity of Silicate Glasses*, M.S. Thesis, University of Illinois at Urbana, 94pp., 1989.

130. Wang, H., and G. Simmons, Elasticity of some mantle crystal structures 1. Pleonaste and hercynite spinel, *J. Geophys. Res., 77*, 4379-4392, 1972.

131. Wang, H., and G. Simmons, Elasticity of some mantle crystal structures 2. Rutile GeO_2, *J. Geophys. Res., 78*, 1262-1273, 1973.

132. Wang, H., and G. Simmons, Elasticity of some mantle crystal structures 3. Spessartite almandine garnet, *J. Geophys. Res., 79*, 2607-2613, 1974.

133. Wang, H., M. C. Gupta, and G. Simmons, Chrysoberyl (Al_2BeO_4): anomaly in velocity-density systematics, *J. Geophys. Res., 80*, 3761-3764, 1975.

134. Wang, J., E. S. Fisher, and M. H. Manghnani, Elastic constants of nickel oxide, *Chinese Phys. Lett., 8*, 153-156, 1991.

135. Watt, J. P., G. F. Davies, and R. J. O'Connell, The elastic properties of composite materials, *Rev. Geophys. Space Phys., 14*, 541-563, 1976.

136. Webb, S. L., The elaticity of the upper mantle orthosilicates olivine and garnet to 3 GPa, *Phys. Chem. Minerals, 16*, 684-692, 1989.

137. Webb, S. L., and I. Jackson, The pressure dependence of the elastic moduli of single-crystal orthopyroxene $(Mg_{0.8}Fe_{0.2})SiO_3$, *Europ. J. Min.*, in press, 1994.

138. Webb, S. L., I. Jackson, and J. D. Fitzgerald, High pressure elasticity, shear-mode softening and polymorphism in MnO, *Phys. Earth Planetary Interiors, 52*, 117-131, 1988.

139. Weidner, D. J., and H. R. Carleton, Elasticity of coesite, *J. Geophys. Res., 82*, 1334-1346, 1977.

140. Weidner, D. J., and N. Hamaya, Elastic properties of the olivine and spinel polymorphs of Mg_2GeO_4, and evaluation of elastic analogues, *Phys. Earth Planetary Interiors, 33*, 275-283, 1983.

141. Weidner, D. J., and E. Ito, Elasticity of $MgSiO_3$ in the ilmenite phase, *Phys. Earth Planetary Interiors, 40*, 65-70, 1985.

142. Weidner, D. J., H. Wang, and J. Ito, Elasticity of orthoenstatite, *Phys. Earth Planetary Interiors, 17*, P7-P13, 1978.

143. Weidner, D. J., J. D. Bass, A. E. Ringwood, and W. Sinclair, The single-crystal elastic moduli of stishovite, *J. Geophys. Res., 87*, 4740-4746, 1982.

144. Weidner, D. J., H. Sawamoto, and S. Sasaki, Single-crystal elastic properties of the spinel phase of Mg_2SiO_4, *J. Geophys. Res., 89*, 7852-7860, 1984.

145. Wong, C., and D. E. Schuele, Pressure and temperature derivatives of the elastic constants of CaF_2 and BaF_2, *J. Phys. Chem. Solids, 29*, 1309-1330, 1968.

146. Yamamoto, S., and O. L. Anderson, Elasticity and anharmonicity of potassium chloride at high temperature, *Phys. Chem. Minerals, 14*, 332-344, 1987.

147. Yamamoto, S., I. Ohno, and O. L. Anderson, High temperature elasticity of sodium chloride, *J. Phys. Chem. Solids, 48*, 143-151, 1987.

148. Yeganeh-Haeri, A., and D. J. Weidner, Elasticity of a beryllium silicate (phenacite: Be_2SiO_4), *Phys. Chem. Minerals, 16*, 360-364, 1989.

149. Yeganeh-Haeri, A., D. J. Weidner, and E. Ito, Single-crystal elastic moduli of magnesium metasilicate perovskite, in *Perovskite: A Structure of Great Interest to Geophysics and Materials Science*, Geophys. Monogr. Ser., vol. 45, pp. 13-25, 1989.

150. Yeganeh-Haeri, A., D. J. Weidner, and E. Ito, Elastic properties of the pyrope-majorite solid solution series, *Geophys. Res. Lett., 17*, 2453-2456, 1990.

151. Yeganeh-Haeri, A., D. J. Weidner, and J. B. Parise, Elasticity of α-cristobalite: A silicon dioxide with a negative Poisson's ratio, *Science, 257*, 650-652f, 1992.

152. Yoneda, A., Pressure derivatives of elastic constants of single crystal MgO and $MgAl_2O_4$, *J. Phys. Earth, 38*, 19-55, 1990.

153. Yoon, H. S., and R. E. Newham, The elastic properties of beryl, *Acta Cryst., A29*, 507-509, 1973.

154. Zaug, J. M., E. H. Abramson, J. M. Brown, and L. J. Slutsky, Sound velocities in olivine at Earth mantle pressures, *Science, 260*, 1487-1489, 1993.

155. Zhao, Y., and D. J. Weidner, The single crystal elastic moduli of neighborite, *Phys. Chem. Minerals, 20*, 419-424, 1993.

Elastic Constants of Mantle Minerals at High Temperature

Orson L. Anderson and Donald G. Isaak

1. ABSTRACT

Data on elastic constants and associated thermoelastic constants at high temperatures for 14 solids of significance to geophysics are presented and discussed. A synopsis of quasiharmonic theory in the high temperature limit shows that anharmonic corrections to the quasiharmonic determination of thermal pressure are not needed in the equation of state throughout conditions of the lower mantle. Equations for extrapolating the bulk and shear moduli to temperatures beyond the limit of experimental measurement are given and evaluated.

2. INTRODUCTION

Though compendiums of elastic constant data for minerals exist [11, 50, 59], they are restricted to temperatures at or near room temperature. Current problems in mantle geophysics and geochemistry often require values of elastic constants at temperatures found in the lower crust and mantle (1000 to 1900 K).

Using the techniques of resonant ultrasound spectroscopy (RUS) [6, 7], elastic constant data have been taken above the Debye temperature of mantle minerals, often as high as 1825 K, which is of the order of $T = 2\Theta$, where Θ is the Debye temperature [13].

O. L. Anderson and D. G. Isaak, Center for Chemistry and Physics of Earth and Planets, Institute of Geophysics and Planetary Physics, UCLA, Los Angeles, CA 90024

Mineral Physics and Crystallography
A Handbook of Physical Constants
AGU Reference Shelf 2

64

The techniques of RUS do not lend themselves to pressure measurement. We note that of the several RUS techniques, the predominant technique used to obtain the data here is the rectangular parallelepiped technique (RPR) pioneered by Ohno [44] and Sumino et al. [60] (see Shankland and Bass [49] for a review of techniques).

We present the elastic constants, C_{ij} versus T, at high T for fourteen solids listed in Tables 1–14; included are silicates, oxides, and two alkali halides. In Tables 15–28, we present isotropic thermoelastic properties, including the adiabatic bulk modulus K_S and the shear modulus G obtained by appropriate averaging schemes (See Section 4). Values of thermal expansivity α and specific heat (at constant P) C_P, coupled with the elasticity data, allow the computation of the Grüneisen ratio γ and then values for the isothermal bulk modulus K_T (computed from K_S) and the specific heat at constant V C_V (computed from C_P). The density ρ is computed from α, which allows the respective isotropic longitudinal and shear sound velocities, v_p and v_s, to be computed from K_S and G.

From the values of properties in Tables 15–28 the temperature derivatives are calculated, thus defining several important dimensionless thermoelastic parameters that are presented in Tables 29–42. We list the Anderson-Grüneisen parameters, δ_S and δ_T; the dimensionless ratio of change of G with T, Γ; and the measure of the rate of change of shear sound velocity with the longitudinal velocity, ν. We also list the Debye temperature, Θ, determined from sound velocities; Poisson's ratio, σ; and αK_T and its integrated value ΔP_{TH}, which is the change of thermal pressure relative to the pressure at 300 K.

Table 1. MgO: Measured single-crystal elastic moduli[†] (GPa) from 300 to 1800 K (max. measured value of T/Θ: 2.22)

T (K)	C_{11}	C_{12}	C_{44}	C_S
300	299.0	96.4	157.1	101.3
	±0.7	±0.6	±0.3	±0.2
400	292.9	97.0	155.8	98.0
500	296.9	97.6	154.3	94.6
600	280.6	98.0	152.8	91.3
700	274.5	98.4	151.3	88.0
800	268.2	98.5	149.7	84.8
900	261.9	98.6	148.1	81.7
1000	255.7	98.7	146.5	78.5
1100	249.5	98.6	144.8	75.5
1200	243.3	98.4	143.1	72.5
1300	237.2	98.1	141.3	69.6
1400	231.0	97.6	139.5	66.7
1500	224.9	97.1	137.9	63.9
1600	219.0	96.4	136.2	61.3
1700	213.4	95.7	134.7	58.9
1800	208.2	95.0	133.1	56.6
	±1.2	±1.2	±0.5	±0.2

$C_S = (1/2)(C_{11} - C_{12})$.
[†]From Isaak et al. [34].

Table 2. CaO: Measured single-crystal elastic moduli[†] (GPa) from 300 to 1200 K (max. measured value of T/Θ: 1.61)

T (K)	C_{11}	C_{12}	C_{44}	C_S
300	220.5	57.67	80.03	81.43
	±0.1	±0.08	±0.02	±0.04
400	215.7	57.96	79.35	78.85
500	210.7	58.23	78.70	76.25
600	205.9	58.44	77.94	73.72
700	201.2	58.66	77.18	71.28
800	196.6	58.81	76.46	68.88
900	192.0	58.98	75.72	66.52
1000	187.2	58.98	74.92	64.13
1100	182.7	58.96	74.17	61.89
1200	178.1	58.99	73.48	59.56
	±0.3	±0.24	±0.09	±0.09

$C_S = (1/2)(C_{11} - C_{12})$.
[†]From Oda et al. [43].

Table 3. Pyrope-rich garnet: Measured single-crystal elastic moduli[†] (GPa) from 300 to 1000 K (max. measured value of T/Θ: 1.62)

T (K)	C_{11}	C_{12}	C_{44}	C_S
300	296.6	108.5	91.6	94.0
	±1.5	±1.4	±0.2	±1.0
350	294.6	107.6	91.2	93.5
400	292.7	106.9	90.8	92.9
450	291.0	106.5	90.4	92.3
500	289.2	105.9	90.0	91.7
550	287.3	105.2	89.6	91.1
600	285.5	104.6	89.1	90.5
650	283.8	104.2	88.7	89.8
700	282.1	103.7	88.3	89.2
750	280.3	103.2	87.8	88.6
800	278.5	102.6	87.4	88.0
850	276.7	102.1	86.9	87.3
900	274.8	101.5	86.5	86.7
950	273.1	101.0	86.0	86.1
1000	271.2	100.3	85.5	85.5
	±2.0	±1.9	±0.2	±1.4

$C_S = (1/2)(C_{11} - C_{12})$.
[†]After Suzuki & Anderson [65].

Table 4. Grossular garnet: Measured single-crystal elastic moduli[†] (GPa) from 300 to 1350 K (max. measured value of T/Θ: 1.89)

T (K)	C_{11}	C_{12}	C_{44}	C_S
300	318.9	92.2	102.9	113.4
	±0.8	±0.7	±0.2	±0.3
400	315.2	91.8	101.4	111.7
500	311.7	91.5	100.4	110.1
600	307.8	91.1	99.8	108.4
700	303.8	90.5	98.7	106.6
800	300.2	90.4	97.6	104.9
900	296.5	90.2	96.5	103.2
1000	292.7	89.9	95.3	101.4
1100	289.1	89.8	94.2	99.7
1200	284.8	89.1	93.0	97.8
1300	280.5	88.6	91.8	96.0
1350	278.8	88.7	91.2	95.0
	±1.4	±1.2	±0.3	±0.4

$C_S = (1/2)(C_{11} - C_{12})$.
[†]After Isaak et al. [36].

Table 5. $MgAl_2O_4$: Measured single-crystal elastic moduli[†] (GPa) from 300 to 1000 K (max. measured value of T/Θ: 1.20)

T (K)	C_{11}	C_{12}	C_{44}	C_S
300	292.2	168.7	156.5	61.8
	±5.2	±5.2	±1.0	±0.3
350	290.1	167.2	155.0	61.5
400	288.6	166.3	155.3	61.1
450	286.2	164.8	154.4	60.7
500	284.4	163.7	153.6	60.3
550	282.8	162.8	152.9	60.0
600	281.1	161.9	152.2	59.6
650	297.1	160.8	151.5	59.1
700	277.2	159.8	150.7	58.7
750	275.3	158.8	149.9	58.2
800	273.3	157.7	149.2	57.8
850	271.1	156.5	148.5	57.3
900	269.2	155.5	147.7	56.9
950	267.3	154.4	146.9	56.4
1000	266.0	154.0	146.1	56.0
	±6.5	±6.5	±1.3	±0.4

$C_S = (1/2)(C_{11} - C_{12})$.

[†] After Cynn [19].

Table 7. KCl: Measured single-crystal elastic moduli[†] (GPa) from 300 to 850 K (max. measured value of T/Θ: 4.42)

T (K)	C_{11}	C_{12}	C_{44}	C_S
300	40.1	6.6	6.35	16.7
	±0.4	±0.5	±0.02	±0.3
350	38.4	6.8	6.28	15.8
400	36.9	7.0	6.21	15.0
450	35.4	7.1	6.15	14.1
500	33.8	7.2	6.11	13.3
550	32.3	7.3	6.05	12.5
600	31.1	7.5	5.96	11.8
650	29.7	7.7	5.87	11.0
700	28.2	7.7	5.79	10.2
750	26.6	7.7	5.69	9.5
800	25.2	7.8	5.57	8.7
850	23.5	7.7	5.57	7.9
	±0.5	±0.5	±0.02	±0.4

$C_S = (1/2)(C_{11} - C_{12})$.

[†] After Yamamoto & Anderson [76].

Table 6. MnO: Measured single-crystal elastic moduli[†] (GPa) from 300 to 500 K (max. measured value of T/Θ: 0.96)

T (K)	C_{11}	C_{12}	C_{44}	C_S
300	223.5	111.8	78.1	55.9
	±4.5	±3.1	±0.9	±2.7
350	220.4	111.8	78.1	54.3
400	217.2	111.8	77.8	52.7
450	214.1	111.7	77.3	51.2
500	210.9	111.7	76.5	49.6
	±4.5	±3.1	±0.9	±2.7

$C_S = (1/2)(C_{11} - C_{12})$.

[†] After Pacalo & Graham [47].

Table 8. NaCl: Measured single-crystal elastic moduli[†] (GPa) from 300 to 750 K (max. measured value of T/Θ: 2.84)

T (K)	C_{11}	C_{12}	C_{44}	C_S
300	49.5	13.2	12.79	18.1
	±0.4	±0.4	±0.02	±0.3
350	47.6	13.3	12.62	17.1
400	45.8	13.4	12.43	16.2
450	44.1	13.5	12.26	15.3
500	42.4	13.6	12.09	14.4
550	40.5	13.5	11.90	13.5
600	38.7	13.2	11.71	12.7
650	37.0	13.1	11.52	11.9
700	35.4	13.1	11.31	11.2
750	33.7	12.9	11.10	10.4
	±0.4	±0.4	±0.02	±0.3

$C_S = (1/2)(C_{11} - C_{12})$.

[†] After Yamamoto et al. [77].

Table 9. Mg_2SiO_4: Measured single-crystal elastic moduli[†] (GPa) from 300 to 1700 K (max. measured value of T/Θ: 2.1)

T (K)	C_{11}	C_{22}	C_{33}	C_{44}	C_{55}	C_{66}	C_{23}	C_{31}	C_{12}
300	330.0	200.0	236.0	67.2	81.5	81.2	72.1	68.0	66.2
	±0.7	±0.4	±0.6	±0.1	±0.2	±0.2	±0.4	±0.5	±0.5
400	326.3	197.2	233.1	65.9	80.1	79.6	71.6	67.0	65.2
500	322.4	194.2	230.1	64.4	78.7	78.0	71.1	66.1	64.0
600	318.6	191.2	226.8	63.0	77.2	76.3	70.4	65.1	62.9
700	314.5	188.0	223.6	61.6	75.8	74.6	69.7	64.3	61.8
800	310.3	184.8	220.3	60.1	74.3	73.0	69.1	63.3	60.7
900	306.3	181.5	216.9	58.8	72.8	71.3	68.3	62.5	59.4
1000	302.0	178.3	213.5	57.4	71.3	69.6	67.8	61.5	58.4
1100	297.4	175.1	209.8	56.1	69.9	67.9	67.2	60.5	57.3
1200	292.8	171.8	206.1	54.7	68.3	66.2	66.6	59.4	56.3
1300	288.3	168.7	202.7	53.3	66.9	64.6	66.0	58.5	55.3
1400	283.8	165.1	199.2	51.9	65.4	62.9	65.2	57.6	54.2
1500	279.1	162.2	195.5	50.6	64.0	61.4	64.6	56.7	53.2
1600	274.4	159.0	192.0	49.3	62.5	59.9	64.0	55.8	52.1
1700	269.8	155.6	188.2	48.0	61.0	58.4	63.3	54.9	51.0
	±1.1	±0.8	±1.0	±0.2	±0.2	±0.3	±0.7	±0.9	±0.8

[†]From Isaak et al. [35].

Table 10. Olivine $Fo_{90}Fa_{10}$: Measured elastic moduli[†] (GPa) from 300 to 1500 K (max. measured value of T/Θ: 2.26)

T (K)	C_{11}	C_{22}	C_{33}	C_{44}	C_{55}	C_{66}	C_{23}	C_{31}	C_{12}
300	320.6	197.1	234.2	63.72	77.6	78.29	74.8	71.2	69.8
	±0.4	±0.3	±0.5	±0.05	±0.1	±0.08	±0.3	±0.4	±0.2
400	316.8	194.1	231.0	62.37	76.3	76.61	74.4	70.3	68.6
500	313.0	190.9	227.6	61.05	74.9	74.97	73.7	69.3	67.4
600	309.0	187.7	224.1	59.73	73.6	73.33	73.0	68.3	66.2
700	305.0	184.6	220.6	58.45	72.3	71.73	72.3	67.2	65.0
800	300.7	181.5	217.2	57.23	71.0	70.17	71.6	66.1	63.6
900	297.0	178.3	214.3	55.91	69.9	68.59	71.2	66.0	62.8
1000	293.1	175.3	210.4	54.68	68.5	67.07	70.3	64.7	61.8
1100	289.0	172.3	206.6	53.47	67.1	65.53	69.4	63.4	60.5
1200	285.1	169.2	202.9	52.28	65.8	64.01	68.6	62.4	59.4
1300	280.9	166.1	199.3	51.06	64.5	62.51	67.8	61.4	58.2
1400	276.6	163.0	195.6	49.83	63.2	61.02	67.0	60.5	57.2
	±0.5	±0.4	±0.7	±0.07	±0.2	±0.11	±0.5	±0.5	±0.4
1500	272.0	159.8	192.1	48.57	62.2	59.52	66.4	59.8	56.2

[†]From Isaak [32].

Table 11. Fe_2SiO_4: Measured single-crystal elastic moduli[†] (GPa) from 300 to 700 K (max. measured value of T/Θ: 1.41)

T (K)	C_{11}	C_{22}	C_{33}	C_{44}	C_{55}	C_{66}	C_{23}	C_{31}	C_{12}
300	266.9	173.5	239.1	32.4	46.7	57.3	97.9	98.7	95.1
	±1.9	±1.1	±1.4	±0.1	±0.1	±0.1	±1.2	±1.6	±1.5
350	264.5	171.8	237.0	31.9	46.2	56.3	97.7	98.2	94.3
400	262.2	170.1	234.7	31.7	46.0	55.3	97.4	97.7	93.4
450	260.7	168.4	232.4	31.4	45.8	54.5	97.2	97.5	92.8
500	258.8	166.6	229.9	31.4	45.8	53.7	96.8	97.0	91.9
550	257.0	164.9	227.5	31.4	45.7	52.9	96.5	96.6	91.0
600	255.0	162.8	225.1	31.5	45.6	52.3	96.0	96.1	90.0
650	252.8	160.9	222.7	31.5	45.6	51.6	95.4	95.5	88.9
700	251.0	159.0	220.5	31.6	45.5	51.0	94.8	94.9	87.7
	±2.2	±1.3	±1.6	±0.1	±0.1	±0.2	±1.3	±1.8	±1.7

[†]After Sumino [58].

Table 12. Mn_2SiO_4: Measured single-crystal elastic moduli[†] (GPa) from 300 to 700 K (max. measured value of T/Θ: 1.28)

T (K)	C_{11}	C_{22}	C_{33}	C_{44}	C_{55}	C_{66}	C_{23}	C_{31}	C_{12}
300	258.3	165.5	206.7	45.3	55.6	57.8	91.7	95.2	87.1
	±1.9	±1.0	±1.3	±0.1	±0.2	±0.2	±1.0	±1.5	±1.3
400	254.8	162.7	203.9	44.4	54.4	56.4	90.6	93.8	85.5
500	251.3	159.8	201.0	89.5	92.3	83.8	43.4	53.2	55.1
600	247.8	157.0	198.2	42.5	52.0	53.7	88.3	90.9	82.2
700	244.3	154.2	195.3	41.5	51.8	52.4	87.2	89.4	80.6
	±2.4	±1.3	±1.7	±0.1	±0.2	±0.2	±1.3	±1.9	±1.6

[†]After Sumino [58].

Table 13. Co_2SiO_4: Measured single-crystal elastic moduli[†] (GPa) from 300 to 700 K (max. measured value of T/Θ: 1.25)

T (K)	C_{11}	C_{22}	C_{33}	C_{44}	C_{55}	C_{66}	C_{23}	C_{31}	C_{12}
300	307.7	194.7	234.1	46.7	63.9	64.8	103.2	105.0	101.6
	±1.2	±0.7	±0.9	±0.1	±0.1	±0.1	±0.7	±1.0	±0.8
400	304.0	192.6	230.7	46.2	62.9	63.8	101.8	103.6	99.8
500	301.1	190.5	227.4	45.7	61.9	62.8	100.5	102.3	97.9
600	297.8	188.8	224.0	45.2	60.8	61.8	99.1	100.9	96.1
700	294.5	186.2	220.6	44.7	59.8	60.8	97.8	99.6	94.3
	±1.7	±1.2	±1.4	±0.1	±0.1	±0.2	±1.1	±1.3	±1.3

[†]After Sumino [58].

Table 14. Al_2O_3: Measured single-crystal elastic moduli[†] (GPa) from 300 to 1800 K (max. measured value of T/Θ: 1.95)

T (K)	C_{11}	C_{33}	C_{44}	C_{12}	C_{13}	C_{14}
300	497.2	500.8	146.7	162.8	116.0	−21.9
	±1.5	±1.8	±0.2	±1.7	±1.0	± 0.2
400	494.7	497.2	144.4	163.8	115.3	−22.5
500	490.6	493.6	141.8	163.7	114.4	−23.0
600	486.0	489.2	139.2	163.1	113.0	−23.3
700	481.5	484.9	136.5	162.9	111.9	−23.4
800	476.8	480.4	133.9	162.4	110.6	−23.7
900	472.3	476.0	131.2	162.4	109.6	−23.9
1000	467.4	471.2	128.6	161.8	108.2	−24.1
1100	462.5	466.4	125.8	161.4	107.1	−24.2
1200	457.3	461.1	123.2	160.7	105.4	−24.3
1300	451.9	456.2	120.4	160.0	104.1	−24.4
1400	446.7	450.8	117.7	159.5	102.4	−24.5
1500	442.2	446.4	115.1	159.4	101.6	−24.5
	±1.9	±2.1	±0.2	±2.2	±2.1	± 0.2
1600	437.2	441.3	112.5	159.0	100.5	−24.6
1700	432.3	436.5	110.0	158.4	99.4	−24.5
1800	427.2	432.5	107.4	158.0	99.1	−24.5

[†]From Goto et al. [26].

Table 15. MgO: Thermal expansivity, specific heat, isotropic elastic constants[‡] and velocities[‡]

T K	ρ g/cm³	α[†] 10^{-5}/K	K_S GPa	G GPa	C_P* J/(gK)	γ	C_V J/(gK)	K_T GPa	v_p km/s	v_s km/s
100	3.602	0.63	165.7	132.0	0.194	1.59	0.194	165.6	9.80	6.13
200	3.597	2.24	164.6	130.3	0.662	1.55	0.658	163.5	9.78	6.10
300	3.585	3.12	163.9	131.8	0.928	1.54	0.915	161.6	9.73	6.06
	±0.005	±0.06	±0.6	±0.5		±0.03		±0.6	±0.01	±0.01
400	3.573	3.57	162.3	129.4	1.061	1.53	1.048	158.9	9.68	6.02
500	3.559	3.84	160.7	126.9	1.130	1.53	1.098	156.1	9.63	5.97
600	3.545	4.02	158.9	124.4	1.173	1.54	1.131	153.2	0.57	5.92
700	3.531	4.14	157.1	121.8	1.204	1.53	1.153	150.4	9.51	5.87
800	3.516	4.26	155.1	119.2	1.227	1.53	1.166	147.4	9.45	5.82
900	3.501	4.38	153.1	116.7	1.246	1.54	1.175	144.3	9.39	5.77
1000	3.486	4.47	151.1	114.1	1.262	1.54	1.181	141.4	9.33	5.72
1100	3.470	4.56	148.9	111.5	1.276	1.53	1.185	138.3	9.26	5.67
1200	3.454	4.65	146.7	109.0	1.289	1.53	1.188	135.1	9.19	5.62
1300	3.438	4.71	144.4	106.4	1.301	1.52	1.190	132.1	9.13	5.56
1400	3.422	4.80	142.0	103.8	1.312	1.52	1.191	128.1	9.05	5.51
1500	3.405	4.89	139.7	101.3	1.323	1.52	1.191	125.7	8.98	5.46
1600	3.388	4.98	137.3	99.0	1.334	1.51	1.191	122.5	8.92	5.41
1700	3.371	5.04	134.9	96.7	1.346	1.50	1.193	119.6	8.85	5.36
1800	3.354	5.13	132.7	94.5	1.358	1.50	1.193	116.6	8.78	5.31
	±0.007	±0.10	±1.1	±1.6		±0.03		±1.1	±0.04	±0.05

[‡]Computed from Table 1; [†]Suzuki [64]; *Garvin et al. [25].

Table 16. Al$_2$O$_3$: Thermal expansivity, specific heat, isotropic moduli[‡] and velocities[‡]

T	ρ	α[†]	K_S	G	C_P^*	γ	C_V	K_T	v_p	v_s
300	3.982	1.62	253.6	163.0	0.779	1.32	0.774	252.0	10.88	6.40
	±0.009	±0.03	±1.7	±2.8		±0.03		±1.7	±0.05	±0.06
400	3.975	1.99	252.6	161.1	0.943	1.34	0.933	249.9	10.84	6.37
500	3.966	2.23	250.9	158.8	1.040	1.36	1.024	247.1	10.80	6.33
600	3.957	2.40	248.6	156.6	1.103	1.37	1.082	243.8	10.75	6.29
700	3.947	2.51	246.6	154.2	1.148	1.36	1.121	240.8	10.70	6.25
800	3.937	2.59	244.4	151.9	1.180	1.36	1.148	237.7	10.65	6.21
900	3.927	2.66	242.4	149.5	1.205	1.36	1.167	234.8	10.61	6.17
1000	3.916	2.73	240.0	147.1	1.223	1.37	1.179	231.4	10.55	6.13
1100	3.906	2.80	237.8	144.6	1.244	1.37	1.194	228.2	10.50	6.09
1200	3.894	2.88	235.2	142.2	1.257	1.38	1.199	224.5	10.44	6.04
1300	3.883	2.96	232.6	139.7	1.267	1.40	1.203	220.8	10.39	6.00
1400	3.872	3.03	230.0	137.2	1.277	1.41	1.205	217.1	10.33	5.95
1500	3.860	3.09	228.1	134.8	1.286	1.42	1.207	214.0	10.28	5.91
1600	3.848	3.15	225.9	133.5	1.296	1.43	1.209	210.7	10.23	5.86
1700	3.835	3.20	224.8	131.2	1.306	1.43	1.212	207.5	10.17	5.82
1800	3.823	3.25	221.8	127.5	1.318	1.43	1.216	204.7	10.12	5.78
	±0.009	±0.06	±2.3	±4.8		±0.03		±2.2	±0.009	±0.11

[‡]Computed from Table 14; [†]White & Roberts [75]; [*]Furukawa et al. [24]; Dimensions as in Table 15.

Table 17. MgAl$_2$O$_4$: Thermal expansivity, specific heat, isotropic moduli[‡] and velocities[‡]

T	ρ	α[†]	K_S	G	C_P^*	γ	C_V	K_T	v_p	v_s
300	3.576	2.11	209.9	108.2	0.819	1.51	0.811	207.9	9.95	5.50
	±0.005	±0.04	±5.2	±2.5		±0.05		±5.2	±0.09	±0.06
350	3.572	2.18	208.2	107.7	0.899	1.41	0.889	205.9	9.92	5.49
400	3.568	2.25	207.1	107.2	0.963	1.36	0.952	204.6	9.91	5.48
450	3.564	2.32	205.3	106.6	1.014	1.32	1.001	202.5	9.87	5.47
500	3.560	2.38	203.9	106.0	1.055	1.30	1.039	200.8	9.85	5.46
550	3.555	2.45	202.8	105.5	1.088	1.28	1.069	199.4	9.83	5.45
600	3.551	2.51	201.6	104.9	1.115	1.28	1.094	197.8	9.81	5.45
650	3.547	2.57	200.3	104.3	1.139	1.28	1.115	196.1	9.78	5.42
700	3.542	2.63	199.0	103.6	1.160	1.27	1.133	194.4	9.76	5.41
750	3.537	2.69	197.7	103.0	1.179	1.27	1.149	192.7	9.73	5.40
800	3.532	2.74	196.2	102.4	1.180	1.27	1.164	190.9	9.71	5.38
850	3.528	2.80	194.7	101.8	1.213	1.27	1.178	189.0	9.68	5.37
900	3.523	2.85	193.4	101.1	1.229	1.27	1.190	187.3	9.65	5.36
950	3.518	2.90	192.0	100.5	1.243	1.27	1.201	185.5	9.63	5.34
1000	3.512	2.94	191.3	99.8	1.253	1.28	1.208	184.4	9.61	5.33
	±0.005	±0.06	±6.5	±2.7		±0.05		±6.3	±0.11	±0.07

[‡]Computed from Table 5; [†]Touloukian et al. [69]; [*]Robie et al. [48]; Dimensions as in Table 15.

Table 18. Mg_2SiO_4: Thermal expansivity, specific heat, isotropic moduli[‡] and velocities[‡]

T	ρ	α[†]	K_S	G	$C_P{}^*$	γ	C_V	K_T	v_p	v_s
300	3.222	2.72	128.6	81.6	0.840	1.29	0.831	127.3	8.58	5.03
	±0.007	±0.05	±0.4	±0.3		±0.02		±0.4	±0.01	±0.01
400	3.213	3.03	127.1	80.3	0.990	1.21	0.976	125.2	8.54	5.00
500	3.203	3.22	125.4	78.9	1.068	1.18	1.048	123.1	8.48	4.96
600	3.192	3.36	123.7	77.4	1.119	1.17	1.093	120.8	8.43	4.93
700	3.181	3.48	121.9	76.0	1.156	1.16	1.124	118.6	8.38	4.89
800	3.170	3.59	120.2	74.5	1.186	1.15	1.148	116.3	8.32	4.85
900	3.159	3.70	118.3	73.1	1.211	1.15	1.167	114.0	8.27	4.81
1000	3.147	3.81	116.6	71.6	1.235	1.14	1.183	111.7	8.21	4.77
1100	3.135	3.92	114.8	70.1	1.256	1.14	1.197	109.4	8.15	4.73
1200	3.122	4.05	112.9	68.6	1.277	1.15	1.210	106.9	8.09	4.69
1300	3.109	4.16	111.1	67.1	1.296	1.15	1.220	104.6	8.03	4.65
1400	3.096	4.27	109.2	65.6	1.315	1.15	1.231	102.2	7.97	4.60
1500	3.083	4.39	107.5	64.1	1.334	1.15	1.240	99.9	7.91	4.56
1600	3.069	4.50	105.6	62.7	1.352	1.15	1.249	97.6	7.85	4.52
1700	3.055	4.62	103.7	61.2	1.370	1.14	1.257	95.2	7.79	4.48
	±0.008	±0.08	±0.5	±0.3		±0.02		±0.5	±0.02	±0.01

[‡]Computed from Table 9; [†]Kajiyoshi [38]; *Barin & Knacke [15]; Dimensions as in Table 15.

Table 19. Olivine $Fo_{90}Fa_{10}$: Thermal expansivity, specific heat, isotropic moduli[‡] and velocities[‡]

T	ρ	α[†]	K_S	G	$C_P{}^*$	γ	C_V	K_T	v_p	v_s
300	3.353	2.66	129.3	78.1	0.816	1.25	0.808	128.0	8.34	4.83
	±0.004	±0.05	±0.3	±0.2		±0.02		±0.3	±0.01	±0.01
400	3.343	2.99	127.7	76.8	0.957	1.19	0.944	125.9	8.29	4.79
500	3.333	3.21	125.9	75.3	1.032	1.17	1.013	123.6	8.24	4.75
600	3.322	3.35	124.1	73.9	1.080	1.16	1.055	121.2	8.19	4.72
700	3.311	3.46	122.2	72.5	1.112	1.14	1.086	118.2	8.13	4.68
800	3.299	3.55	120.3	71.2	1.145	1.13	1.109	116.6	8.07	4.65
900	3.287	3.64	118.9	69.8	1.171	1.12	1.129	114.7	8.03	4.61
1000	3.275	3.71	117.0	68.5	1.194	1.11	1.147	112.1	7.97	4.57
1100	3.263	3.79	115.1	67.1	1.216	1.10	1.163	110.0	7.92	4.54
1200	3.251	3.86	113.2	65.8	1.236	1.09	1.177	107.8	7.86	4.50
1300	3.238	3.93	111.4	64.4	1.256	1.08	1.191	105.6	7.81	4.60
1400	3.225	4.00	109.6	63.1	1.275	1.07	1.203	103.4	7.75	4.22
1500	3.212	4.07	107.8	61.7	1.294	1.06	1.216	101.3	7.69	4.38
	±0.004	±0.08	±0.5	±0.3		±0.02		±0.5	±0.01	±0.01

[‡]Computed from Table 10; [†]Suzuki [64]; *Barin & Knacke [15]; Dimensions as in Table 15.

Table 20. Fe_2SiO_4: Thermal expansivity, specific heat, isotropic moduli[‡] and velocities[‡]

T	ρ	α[†]	K_S	G	C_P[*]	γ	C_V	K_T	v_p	v_s
300	4.400	2.61	138.0	51.0	0.673	1.21	0.667	136.7	6.84	3.40
	±0.009	±0.05	±0.8	±0.5		±0.03		±0.8	±0.02	±0.02
400	4.388	2.74	135.9	49.7	0.746	1.18	0.736	134.1	6.79	3.37
500	4.375	3.00	134.0	48.8	0.793	1.16	0.779	131.7	6.74	3.34
600	4.362	3.12	131.8	48.0	0.830	1.13	0.813	129.0	6.70	3.32
700	4.348	3.22	129.3	47.4	0.863	1.11	0.842	126.1	6.65	3.30
	±0.009	±0.06	±0.9	±0.4		±0.03		±0.8	±0.02	±0.02

[‡]Computed from Table 11; [†]Suzuki et al. [67]; [*]Watanabe [72]; Dimensions as in Table 15.

Table 21. Mn_2SiO_4: Thermal expansivity, specific heat, isotropic moduli[‡] and velocities[‡]

T	ρ	α[†]	K_S	G	C_P[*]	γ	C_V	K_T	v_p	v_s
300	4.129	2.27	128.9	54.5	0.666	1.06	0.661	128.0	6.99	3.634
	±0.005	±0.05	±0.6	±0.3		±0.03		±0.6	±0.01	±0.009
400	4.119	2.57	127.0	53.5	0.736	1.08	0.728	125.6	6.94	3.604
500	4.108	2.77	125.0	52.5	0.781	1.08	0.770	123.1	6.89	3.573
600	4.096	2.91	123.0	51.4	0.818	1.07	0.803	120.8	6.84	3.543
700	4.084	3.03	121.1	50.4	0.850	1.06	0.831	118.4	6.79	3.512
	±0.005	±0.06	±0.8	±0.3		±0.03		±0.8	±0.02	±0.011

[‡]Computed from Table 12; [†]Okajima et al. [46]; [*]Barin & Knacke [15]; Dimensions as in Table 15.

Table 22. Co_2SiO_4: Thermal expansivity, specific heat, isotropic moduli[‡] and velocities[‡]

T	ρ	α[†]	K_S	G	C_P[*]	γ	C_V	K_T	v_p	v_s
300	4.706	2.27	148.2	62.0	0.640	1.12	0.636	147.1	7.00	3.621
	±0.009	±0.05	±0.5	±0.3		±0.03		±0.5	±0.01	±0.009
400	4.695	2.57	146.2	61.4	0.747	1.07	0.739	144.6	6.97	3.611
500	4.682	2.77	144.3	60.7	0.803	1.06	0.791	142.2	6.93	3.594
600	4.669	2.91	142.3	59.9	0.840	1.06	0.825	139.8	6.89	3.575
700	4.655	3.03	140.4	59.1	0.868	1.05	0.849	137.3	6.86	3.557
	±0.009	±0.06	±0.7	±0.3		±0.02		±0.6	±0.01	±0.010

[‡]Computed from Table 13; [†](assume Mn_2SiO_4); [*]Watanabe [72]; Dimensions as in Table 15.

Table 23. MnO: Thermal expansivity, specific heat, isotropic moduli[‡] and velocities[‡]

T	ρ	α[†]	K_S	G	C_P[*]	γ	C_V	K_T	v_p	v_s
300	5.378	3.46	149.0	68.3	0.632	1.51	0.623	146.7	6.68	3.57
	±0.001	±0.07	±2.6	±1.5		±0.04		±2.5	±0.05	±0.04
350	5.369	3.58	148.0	67.6	0.653	1.51	0.641	145.2	6.66	3.55
400	5.359	3.68	146.9	66.7	0.669	1.51	0.655	143.7	6.63	3.53
450	5.349	3.77	145.8	65.6	0.682	1.51	0.665	142.2	6.60	3.50
500	5.339	3.85	144.8	64.4	0.692	1.51	0.673	140.7	6.57	3.47
	±0.001	±0.08	±2.6	±1.6		±0.04		±2.5	±0.05	±0.04

[‡]Computed from Table 6; [†]Suzuki et al. [66]; [*]Barin & Knacke [15]; Dimensions as in Table 15.

Table 24. CaO: Thermal expansivity, specific heat, isotropic moduli[‡] and velocities[‡]

T	ρ	α[†]	K_S	G	C_P*	γ	C_V	K_T	v_p	v_s
300	3.349	3.04	112.0	80.59	0.752	1.35	0.743	110.6	8.094	4.905
	±0.001	±0.06	±0.1	±0.02		±0.03		±0.1	±0.002	±0.001
400	3.338	3.47	110.5	79.15	0.834	1.36	0.819	108.5	8.045	4.869
500	3.327	3.67	109.1	77.71	0.880	1.37	0.858	106.4	7.996	4.834
600	3.314	3.81	107.6	76.22	0.904	1.37	0.877	104.3	7.946	4.796
700	3.301	3.92	106.2	74.76	0.921	1.37	0.888	102.3	7.897	4.759
800	3.288	4.01	104.7	73.33	0.933	1.37	0.894	100.3	7.848	4.723
900	3.275	4.08	103.3	71.90	0.943	1.36	0.898	98.4	7.799	4.686
1000	3.262	4.14	101.7	70.40	0.952	1.36	0.901	96.3	7.745	4.646
1100	3.248	4.20	100.2	68.99	0.959	1.35	0.903	94.3	7.693	4.609
1200	3.234	4.26	98.7	67.56	0.965	1.35	0.903	92.3	7.640	4.571
	±0.002	±0.09	±0.2	±0.08		±0.03		±0.3	±0.006	±0.003

[‡]Computed from Table 2; [†]Oda et al. [43]; *Garvin et al. [25]; Dimensions as in Table 15.

Table 25. Grossular garnet: Thermal expansivity, specific heat, isotropic moduli[‡] and velocities[‡]

T	ρ	α[†]	K_S	G	C_P*	γ	C_V	K_T	v_p	v_s
300	3.597	1.92	167.8	106.9	0.736	1.22	0.730	166.6	9.29	5.453
	±0.006	±0.05	±0.7	±0.2		±0.03		±0.7	±0.01	±0.006
400	3.589	2.28	166.2	105.7	0.865	1.22	0.855	164.4	9.25	5.427
500	3.581	2.49	164.9	104.5	0.945	1.21	0.931	162.5	9.22	5.401
600	3.571	2.61	163.3	103.1	0.995	1.20	0.977	160.3	9.18	5.373
700	3.562	2.71	161.6	101.8	1.028	1.19	1.006	158.1	9.14	5.346
800	3.552	2.78	160.3	100.5	1.052	1.19	1.025	156.2	9.10	5.318
900	3.542	2.83	158.9	99.1	1.072	1.19	1.041	154.3	9.06	5.289
1000	3.532	2.88	157.5	97.7	1.092	1.18	1.056	152.3	9.03	5.259
1100	3.522	2.92	156.2	96.4	1.113	1.16	1.073	150.6	8.99	5.230
1200	3.512	2.97	154.4	94.9	1.139	1.14	1.095	148.3	8.94	5.198
1300	3.501	3.00	152.6	93.4	1.170	1.12	1.121	146.2	8.90	5.165
	±0.006	±0.07	±1.2	±0.2		±0.03		±1.2	±0.02	±0.008

[‡]Computed from Table 4; [†]Isaak et al. [36]; *Krupka et al. [39]; Dimensions as in Table 15.

Table 26. Pyrope-rich garnet: Thermal expansivity, specific heat, isotropic moduli[‡] and velocities[‡]

T	ρ	α[†]	K_S	G	C_P*	γ	C_V	K_T	v_p	v_s
300	3.705	2.36	171.2	92.6	0.726	1.50	0.718	169.4	8.92	5.00
	±0.005	±0.04	±0.8	±0.4		±0.03		±0.8	±0.02	±0.01
400	3.696	2.64	168.9	91.6	0.902	1.34	0.889	166.5	8.87	4.98
500	3.686	2.80	167.0	90.6	0.981	1.29	0.964	164.0	8.84	4.96
600	3.675	2.90	164.9	89.7	1.032	1.26	1.010	161.4	8.80	4.94
700	3.664	2.97	163.2	88.7	1.067	1.24	1.040	159.1	8.76	4.92
800	3.653	3.03	161.3	87.6	1.088	1.23	1.057	156.6	8.72	4.90
900	3.642	3.07	159.3	86.5	1.104	1.22	1.068	154.1	8.68	4.87
1000	3.631	3.11	157.3	85.5	1.116	1.21	1.076	151.6	8.64	4.85
	±0.005	±0.06	±1.1	±0.6		±0.02		±1.1	±0.03	±0.02

[‡]Computed from Table 3; [†]Suzuki & Anderson [65]; *idem; Dimensions as in Table 15.

Table 27. NaCl: Thermal expansivity, specific heat, isotropic elastic moduli[‡] and velocities[‡]

T	ρ	α[†]	K_S	G	C_P^*	γ	C_V	K_T	v_p	v_s
300	2.159	11.8	25.3	14.71	0.868	1.59	0.822	24.0	4.56	2.610
	±0.005	±0.2	±0.3	±0.08		±0.04		±0.3	±0.02	±0.008
350	2.146	12.2	24.8	14.27	0.883	1.60	0.826	23.2	4.52	2.579
400	2.132	12.7	24.2	13.81	0.897	1.61	0.829	22.4	4.47	2.545
450	2.118	13.2	23.7	13.39	0.910	1.62	0.830	21.6	4.43	2.514
500	2.104	13.7	23.2	12.96	0.923	1.64	0.830	20.8	4.39	2.482
550	2.089	14.3	22.5	12.53	0.937	1.64	0.830	19.9	4.33	2.449
600	2.074	14.8	21.7	12.11	0.950	1.63	0.830	19.0	4.27	2.416
650	2.059	15.4	21.1	11.68	0.964	1.63	0.829	18.1	4.22	2.382
700	2.043	16.0	20.5	11.25	0.979	1.63	0.828	17.3	4.17	2.346
750	2.026	16.6	19.8	10.80	0.997	1.63	0.829	16.5	4.11	2.309
	±0.006	±0.3	±0.3	±0.11		±0.04		±0.3	±0.02	±0.012

[‡]Computed from Table 8; [†]Enck & Dommel [22]; *Stull & Prophet [57]; Dimensions as in Table 15.

Table 28. KCl: Thermal expansivity, specific heat, isotropic elastic moduli[‡] and velocities[‡]

T	ρ	α[†]	K_S	G	C_P^*	γ	C_V	K_T	v_p	v_s
300	1.982	11.0	17.8	9.47	0.689	1.44	0.657	17.0	3.92	2.19
	±0.005	±0.2	±0.4	±1.03		±0.04		±0.3	±0.09	±0.12
350	1.971	11.3	17.3	9.18	0.701	1.42	0.664	16.4	3.88	2.16
400	1.959	11.7	17.0	8.91	0.713	1.42	0.669	15.9	3.84	2.13
450	1.948	12.1	16.6	8.64	0.724	1.42	0.672	15.4	3.80	2.11
500	1.935	12.6	16.1	8.39	0.735	1.43	0.674	14.7	3.75	2.08
550	1.923	13.2	15.7	8.13	0.745	1.44	0.675	14.2	3.71	2.06
600	1.910	13.7	15.4	7.85	0.756	1.45	0.676	13.7	3.68	2.03
650	1.897	14.2	15.0	7.57	0.767	1.46	0.676	13.2	3.64	2.00
700	1.883	14.7	14.5	7.29	0.778	1.46	0.677	12.6	3.59	1.97
750	1.869	15.2	14.0	6.98	0.791	1.44	0.679	12.0	3.53	1.93
800	1.855	15.7	13.6	6.67	0.806	1.43	0.683	11.5	3.48	1.90
850	1.840	16.2	12.0	6.41	0.823	1.39	0.691	10.9	3.42	1.87
	±0.005	±0.2	±0.4	±0.13		±0.04		±0.3	±0.03	±0.02

[‡]Computed from Table 7; [†]Enck et al. [23]; *Stull & Prophet [57]; Dimensions as in Table 15.

Table 29. Al$_2$O$_3$: Dimensionless parameters, Debye temperature and thermal pressure

T K	Θ K	σ	δ_S	δ_T	Γ	$\frac{(\delta_T-\delta_S)}{\gamma}$	ν	αK_T MPa/K	ΔP_{TH} GPa
300	1034	0.235	3.30	5.71	5.71	1.82	1.60	4.08	0.00
400	1029	0.237	3.16	5.16	5.16	1.49	1.52	4.98	0.45
500	1022	0.239	3.20	5.03	6.27	1.35	1.46	5.53	0.98
600	1015	0.240	3.31	5.08	6.09	1.29	1.42	5.85	1.55
700	1008	0.241	3.43	5.17	6.05	1.28	1.40	6.03	2.15
800	1001	0.243	3.55	5.29	6.06	1.28	1.38	6.15	2.76
900	994	0.244	3.62	5.37	6.08	1.28	1.36	6.24	1.43
1000	986	0.246	3.66	5.42	6.09	1.29	1.36	6.30	4.01
1100	979	0.247	3.65	5.42	6.07	1.29	1.36	6.40	4.64
1200	971	0.248	3.60	5.39	6.03	1.30	1.37	6.45	5.93
1300	963	0.250	3.51	5.32	5.98	1.29	1.38	6.52	5.93
1400	955	0.251	3.39	5.22	5.93	1.30	1.40	6.57	6.59
1500	947	0.253	3.24	5.08	5.87	1.30	1.43	6.62	7.24
1600	939	0.255	3.06	4.92	5.80	1.30	1.47	6.64	7.91
1700	932	0.257	2.85	4.73	5.74	1.32	1.52	6.64	8.57
1800	924	0.259	2.60	4.50	5.66	1.32	1.58	6.66	9.24

Calculated from Tables 14 & 16.

Table 30. MgO: Dimensionless parameters, Debye temperature and thermal pressure

T	Θ	σ	δ_S	δ_T	Γ	$\frac{(\delta_T-\delta_S)}{\gamma}$	ν	αK_T	ΔP_{TH}
300	945	0.183	2.83	5.26	5.73	1.57	1.40	5.04	0.00
400	937	0.185	2.79	4.83	5.34	1.33	1.40	5.67	0.54
500	928	0.188	2.81	4.69	5.17	1.23	1.38	6.00	1.12
600	920	0.190	2.86	4.67	5.08	1.18	1.37	6.16	1.73
700	911	0.192	2.92	4.70	5.05	1.16	1.35	6.23	2.35
800	902	0.194	2.98	4.74	5.03	1.15	1.34	6.28	2.98
900	894	0.196	3.04	4.78	5.02	1.13	1.32	6.32	3.61
1000	885	0.198	3.12	4.84	5.05	1.12	1.31	6.32	4.24
1100	875	0.200	3.21	4.92	5.08	1.12	1.30	6.31	4.87
1200	806	0.202	3.30	4.99	5.09	1.11	1.28	6.28	5.50
1300	857	0.204	3.41	5.08	5.10	1.10	1.26	6.22	6.12
1400	847	0.206	3.47	5.12	5.04	1.09	1.24	6.19	6.74
1500	838	0.208	3.50	5.13	4.92	1.07	1.22	6.16	7.36
1600	828	0.209	3.46	5.07	4.75	1.07	1.21	6.13	7.97
1700	820	0.211	3.36	4.95	4.56	1.06	1.20	6.03	8.58
1800	811	0.212	3.12	4.66	4.34	1.03	1.23	6.00	9.20

Calculated from Tables 1 & 15. Dimensions as in Table 29.

Table 31. CaO: Dimensionless parameters, Debye temperatures and thermal pressure

T	Θ	σ	δ_S	δ_T	Γ	$\frac{(\delta_T-\delta_S)}{\gamma}$	ν	αK_T	ΔP_{TH}
300	671	0.210	4.15	6.19	6.00	1.51	1.24	3.36	0.00
400	666	0.211	3.75	5.54	5.38	1.31	1.24	3.73	0.36
500	660	0.212	3.60	5.27	5.13	1.22	1.23	3.90	0.74
600	654	0.213	3.54	5.14	5.01	1.17	1.23	3.98	1.13
700	649	0.215	3.52	5.07	4.99	1.13	1.23	4.01	1.53
800	643	0.216	3.53	5.03	4.95	1.10	1.23	4.02	2.34
900	637	0.218	3.55	5.01	4.93	1.07	1.22	4.01	2.34
1000	631	0.219	3.58	5.00	4.94	1.05	1.22	3.99	2.54
1100	625	0.220	3.62	5.01	4.96	1.03	1.22	3.96	3.13
1200	619	0.221	3.65	5.01	4.99	1.01	1.22	3.93	3.53

Calculated from Tables 2 & 24. Dimensions as in Table 29.

Table 32. Grossular garnet: Dimensionless parameters, Debye temperatures and thermal pressure

T	Θ	σ	δ_S	δ_T	Γ	$\frac{(\delta_T-\delta_S)}{\gamma}$	ν	αK_T	ΔP_{TH}
300	824	0.237	4.64	6.30	6.09	1.36	1.22	3.21	0.00
400	820	0.238	3.93	5.36	5.27	1.17	1.23	3.74	0.36
500	816	0.239	3.64	4.98	4.97	1.11	1.24	4.03	0.75
600	811	0.239	3.49	4.80	4.87	1.08	1.25	4.18	1.16
700	806	0.240	3.41	4.70	4.84	1.08	1.26	4.28	1.57
800	801	0.241	3.35	4.64	4.86	1.08	1.27	4.34	1.98
900	796	0.242	3.31	4.60	4.90	1.08	1.28	4.36	2.40
1000	791	0.243	3.29	4.58	4.96	1.09	1.29	4.38	2.83
1100	786	0.244	3.27	4.57	5.03	1.11	1.30	4.41	3.25
1200	780	0.245	3.26	4.57	5.11	1.15	1.31	4.40	3.69
1300	715	0.246	3.25	4.58	5.20	1.18	1.32	4.38	5.40

Calculated from Tables 4 & 25. Dimensions as in Table 29.

Table 33. Pyrope-rich garnet: Dimensionless parameters, Debye temperatures and thermal pressure

T	Θ	σ	δ_S	δ_T	Γ	$\frac{(\delta_T-\delta_S)}{\gamma}$	ν	αK_T	ΔP_{TH}
300	779	0.271	4.81	6.27	4.29	0.97	0.88	4.00	0.00
350	777	0.271	4.52	5.90	4.07	1.00	0.89	4.25	0.21
400	775	0.270	4.36	5.70	3.96	1.00	0.90	4.40	0.42
450	773	0.270	4.24	5.55	3.90	1.00	0.92	4.51	0.65
500	771	0.270	4.16	5.46	3.86	1.00	0.93	4.59	0.87
550	769	0.270	4.11	5.41	3.86	1.02	0.94	4.64	1.10
600	767	0.270	4.07	5.35	3.86	1.04	0.96	4.68	1.34
650	765	0.270	4.05	5.34	3.88	1.04	0.97	4.69	1.57
700	764	0.270	4.01	5.30	3.89	1.04	0.98	4.72	1.81
750	761	0.270	4.00	5.29	3.92	1.05	1.00	4.74	2.04
800	759	0.270	3.98	5.28	3.94	1.06	1.01	4.74	2.28
850	757	0.270	3.98	5.29	3.98	1.07	1.02	4.74	2.52
900	755	0.270	3.98	5.30	4.02	1.09	1.04	4.73	2.75
950	753	0.270	3.97	5.30	4.06	1.10	1.05	4.73	2.99
1000	751	0.270	3.97	5.32	4.10	1.11	1.06	4.71	3.23

Calculated from Tables 3 & 26. Dimensions as in Table 29.

Table 34. Mg_2SiO_4: Dimensionless parameters, Debye temperature and thermal pressure

T	Θ	σ	δ_S	δ_T	Γ	$\frac{(\delta_T - \delta_S)}{\gamma}$	ν	αK_T	ΔP_{TH}
300	763	0.238	4.45	5.94	6.07	1.16	1.20	3.46	0.00
400	757	0.239	4.20	5.58	5.66	1.14	1.21	3.80	0.36
500	751	0.240	4.15	5.49	5.54	1.14	1.20	3.97	0.75
600	744	0.241	4.15	5.48	5.50	1.14	1.20	4.07	1.16
700	738	0.242	4.16	5.49	5.46	1.15	1.20	4.13	1.57
800	731	0.243	4.13	5.47	5.45	1.17	1.18	4.18	1.98
900	724	0.244	4.08	5.46	5.44	1.20	1.20	4.22	2.40
1000	718	0.245	4.05	5.47	5.45	1.25	1.22	4.26	2.83
1100	711	0.246	4.00	5.46	5.43	1.28	1.20	4.31	3.25
1200	704	0.248	4.02	5.49	5.38	1.28	1.21	4.33	3.69
1300	697	0.249	3.97	5.44	5.32	1.28	1.20	4.36	4.13
1400	689	0.250	3.90	5.37	5.24	1.28	1.21	4.37	4.50
1500	682	0.251	3.92	5.38	5.22	1.27	1.23	4.39	5.07
1600	674	0.252	3.93	5.40	5.19	1.28	1.19	4.40	5.43
1700	668	0.254	3.96	5.42	5.16	1.28	1.20	4.39	5.87

Calculated from Tables 9 & 18. Dimensions as in Table 29.

Table 35. Olivine $Fo_{90}Fa_{10}$: Dimensionless parameters, Debye temperatures and thermal pressure

T	Θ	σ	δ_S	δ_T	Γ	$\frac{(\delta_T - \delta_S)}{\gamma}$	ν	αK_T	ΔP_{TH}
300	731	0.249	5.24	6.59	6.56	1.07	1.17	3.37	0.00
400	725	0.250	4.70	5.95	5.92	1.03	1.17	3.76	0.36
500	719	0.251	4.46	5.65	5.63	1.02	1.17	3.97	0.75
600	713	0.252	4.33	5.51	5.50	1.02	1.18	4.05	1.15
700	706	0.252	4.25	5.44	5.42	1.04	1.18	4.11	1.56
800	700	0.253	4.21	5.40	5.38	1.06	1.18	4.14	1.97
900	699	0.255	4.16	5.36	5.36	1.07	1.18	4.18	2.38
1000	688	0.255	4.14	5.36	5.35	1.10	1.18	4.17	2.80
1100	681	0.256	4.13	5.37	5.35	1.13	1.18	4.17	3.22
1200	675	0.257	4.12	5.38	5.36	1.16	1.18	4.16	3.63
1300	669	0.258	4.07	5.35	5.32	1.18	1.18	4.15	4.05
1400	662	0.259	4.10	5.41	5.39	1.23	1.19	4.14	4.46
1500	665	0.260	4.10	5.43	5.41	1.26	1.19	4.13	4.88

Calculated from Tables 10 & 19. Dimensions as in Table 29.

Table 36. Fe_2SiO_4: Dimensionless parameters, Debye temperatures and thermal pressure

T	Θ	σ	δ_S	δ_T	Γ	$\frac{(\delta_T - \delta_S)}{\gamma}$	ν	αK_T	ΔP_{TH}
300	511	0.336	5.99	7.34	9.34	1.12	1.54	3.56	0.00
400	506	0.337	5.56	6.85	7.49	1.09	1.33	3.82	0.37
500	501	0.338	5.35	6.62	6.02	1.09	1.09	3.95	0.76
600	497	0.338	5.24	6.50	4.69	1.11	0.85	4.02	1.16
700	494	0.338	5.18	6.45	3.43	1.14	0.60	4.06	1.56

Calculated from Tables 11 & 20. Dimensions as in Table 29.

Table 37. Mn_2SiO_4: Dimensionless parameters, Debye temperatures and thermal pressure

T	Θ	σ	δ_S	δ_T	Γ	$\frac{(\delta_T - \delta_S)}{\gamma}$	ν	αK_T	ΔP_{TH}
300	535	0.315	6.66	8.19	8.43	1.44	1.19	2.90	0.00
400	530	0.315	5.95	7.35	7.57	1.30	1.19	3.23	0.31
500	525	0.316	5.61	6.96	7.17	1.25	1.19	3.41	0.64
600	520	0.317	5.43	6.76	6.97	1.24	1.19	3.52	0.99
700	515	0.317	5.31	6.63	6.84	1.25	1.20	3.59	1.34

Calculated from Tables 12 & 21. Dimensions as in Table 29.

Table 38. Co_2SiO_4: Dimensionless parameters, Debye temperatures and thermal pressure

T	Θ	σ	δ_S	δ_T	Γ	$\frac{(\delta_T - \delta_S)}{\gamma}$	ν	αK_T	ΔP_{TH}
300	551	0.317	5.81	7.32	5.51	1.35	0.96	3.34	0.00
400	548	0.316	5.19	6.56	4.91	1.27	0.96	3.72	0.35
500	545	0.316	4.88	6.19	4.62	1.23	0.96	3.94	0.74
600	541	0.316	4.71	6.01	4.46	1.22	0.96	4.07	1.14
700	538	0.316	4.60	5.88	4.35	1.21	0.96	4.16	1.55

Calculated from Tables 13 & 22. Dimensions as in Table 29.

Table 39. MnO: Dimensionless parameters, Debye temperatures and thermal pressure

T	Θ	σ	δ_S	δ_T	Γ	$\frac{(\delta_T - \delta_S)}{\gamma}$	ν	αK_T	ΔP_{TH}
300	534	0.301	4.14	5.96	8.33	1.20	1.56	5.07	0.00
350	531	0.302	4.03	5.82	8.14	1.18	1.57	5.20	0.26
400	527	0.303	3.94	5.71	8.01	1.17	1.57	5.29	0.52
450	523	0.305	3.88	5.64	7.95	1.16	1.57	5.36	0.79
500	519	0.307	3.83	5.58	7.94	1.16	1.58	5.41	1.05

Calculated from Tables 6 & 23. Dimensions as in Table 29.

Table 40. $MgAl_2O_4$: Dimensionless parameters, Debye temperature and thermal pressure

T	Θ	σ	δ_S	δ_T	Γ	$\frac{(\delta_T - \delta_S)}{\gamma}$	ν	αK_T	ΔP_{TH}
300	862	0.280	6.03	7.73	5.30	1.12	0.90	4.38	0.00
400	858	0.279	5.72	7.36	5.01	1.20	0.90	4.60	0.45
500	854	0.279	5.49	7.07	4.78	1.22	0.90	4.79	0.92
600	850	0.278	5.27	6.82	4.59	1.21	0.90	4.97	1.41
700	845	0.278	5.10	6.62	4.43	1.20	0.90	5.11	1.91
800	840	0.278	4.96	6.47	4.30	1.19	0.90	5.24	2.43
900	835	0.277	4.85	6.35	4.20	1.18	0.90	5.33	2.96
1000	830	0.278	4.74	6.24	4.11	1.17	0.90	5.43	3.49

Calculated from Tables 5 & 17. Dimensions as in Table 29.

Table 41. NaCl: Dimensionless parameters, Debye temperature and thermal pressure

T	Θ	σ	δ_S	δ_T	Γ	$\frac{(\delta_T - \delta_S)}{\gamma}$	ν	αK_T	ΔP_{TH}
300	304	0.256	3.47	5.56	5.05	1.32	1.29	2.82	0.00
350	300	0.258	3.56	5.62	5.00	1.29	1.26	2.83	0.14
400	296	0.260	3.65	5.69	4.95	1.27	1.23	2.84	0.28
450	291	0.262	3.72	5.74	4.90	1.25	1.20	2.86	0.43
500	287	0.264	3.80	5.82	4.86	1.24	1.18	2.86	0.57
550	283	0.265	3.91	5.95	4.83	1.25	1.16	2.84	0.71
600	278	0.265	4.03	6.10	4.79	1.27	1.13	2.81	0.85
650	274	0.266	4.14	6.24	4.77	1.29	1.11	2.78	0.99
700	269	0.268	4.23	6.37	4.76	1.31	1.10	2.77	1.13
750	264	0.270	4.34	6.53	4.76	1.35	1.08	2.73	1.27

Calculated from Tables 8 & 27. Dimensions as in Table 29.

Table 42. KCl: Dimensionless parameters, Debye temperature and thermal pressure

T	Θ	σ	δ_S	δ_T	Γ	$\frac{(\delta_T - \delta_S)}{\gamma}$	ν	αK_T	ΔP_{TH}
300	230	0.274	3.77	5.84	4.66	1.34	1.17	1.87	0.00
350	227	0.275	3.86	5.88	4.77	1.34	1.17	1.86	0.09
400	224	0.277	3.92	5.88	4.86	1.32	1.17	1.86	0.19
450	221	0.278	3.97	5.88	4.93	1.30	1.17	1.86	0.28
500	218	0.278	4.02	5.88	4.97	1.28	1.17	1.86	0.37
550	214	0.279	4.05	5.87	5.02	1.26	1.17	1.87	0.47
600	211	0.282	4.06	5.84	5.10	1.23	1.18	1.88	0.56
650	208	0.284	4.09	5.83	5.19	1.21	1.18	1.88	0.65
700	204	0.285	4.18	5.90	5.30	1.23	1.18	1.86	0.75
750	200	0.286	4.27	5.98	5.44	1.25	1.18	1.83	0.84
800	196	0.289	4.34	6.04	5.61	1.27	1.19	1.81	0.93
850	192	0.288	4.50	6.19	5.76	1.33	1.19	1.77	1.02

Calculated from Tables 7 & 28. Dimensions as in Table 29.

The appropriate equations used in preparing the tables are presented in Section 2. The various correlations between the thermoelastic constants are presented in Anderson et al. [13] and reviewed in Section 3. The mineral data are set forth in Section 4. Theory appropriate to the high temperature trends of the data is presented in Section 5. Extrapolation equations are reviewed in Section 6.

3. EQUATIONS USED IN TABLED VALUES OF PHYSICAL PROPERTIES

Once the elastic constants have been determined over a wide range in T, the four dimensionless thermoelastic parameters at each T are computed [7] using the following nomenclature:

$$\gamma = \frac{\alpha K_S}{\rho C_P} = \frac{\alpha K_T}{\rho C_V} \tag{1}$$

$$\delta_S = -\left(\frac{1}{\alpha K_S}\right)\left(\frac{\partial K_S}{\partial T}\right)_P = \left(\frac{\partial \ln K_S}{\partial \ln \rho}\right)_P \tag{2}$$

$$\delta_T = -\left(\frac{1}{\alpha K_T}\right)\left(\frac{\partial K_T}{\partial T}\right)_P = \left(\frac{\partial \ln K_T}{\partial \ln \rho}\right)_P \tag{3}$$

$$\Gamma = -\left(\frac{1}{\alpha G}\right)\left(\frac{\partial G}{\partial T}\right)_P = \left(\frac{\partial G}{\partial \ln \rho}\right)_P. \tag{4}$$

The parameter γ is known as the Grüneisen ratio. The parameters δ_T and δ_S are often called the Anderson-Grüneisen parameters [17]. The dimensionless temperatures reached by some of these data (MgO, Al$_2$O$_3$, Mg$_2$SiO$_4$, KCl, NaCl) either exceed or are close to those of the lower mantle, which are in the neighborhood of $T/\Theta = 2.3$ [4]. The measurements must be done at sufficiently high temperature so that one is justified in speaking of observations in the high-temperature region. The definition of Θ used is

$$\Theta = \left(\frac{h}{k}\right)\left(\frac{9\rho N}{4\pi M}\right)^{1/3}\left(v_p^{-3} + 2v_s^{-3}\right)^{-1/3}. \tag{5}$$

Since v_p and v_s decrease with increasing T, Θ also decreases with increasing T.

Since the value of Θ for most mantle minerals is 600–900 K, measurements have to be taken up to the 1300 to 1800 K range so that the high-temperature trends are clearly discernible. Such high T data permit the verification of classical equations. For example, one can test whether C_V is independent of T (as in the Dulong-Petit limit) and whether γ at constant V is independent of T (as required in the derivation of the

Mie-Grüneisen equation of state).

A well-known thermodynamic identity is that the temperature derivative of the pressure at constant V is exactly equal to αK_T by means of calculus definitions, so that

$$\left(\frac{\partial P}{\partial T}\right)_V = \left(\frac{\partial P_{TH}}{\partial T}\right)_V = \alpha K_T, \tag{6}$$

where P_{TH} is the thermal pressure. Thus αK_T is the slope for the P_{TH} versus T curve at constant V.

Using (6) for an isochore,

$$P_{TH} = \int \alpha K_T \, dT, \tag{7}$$

which is equivalent to the statement that

$$P_{TH} = \int \left(\frac{\partial P}{\partial T}\right)_V \, dT.$$

If αK_T is independent of T at constant V and also independent of volume, then αK_T comes out of the integral shown by (7), giving

$$P_{TH}(T) - P_{TH}(\Theta) = \alpha K_T(T - \Theta). \tag{8}$$

As an empirical finding, ΔP_{TH} is linear in T down to much lower temperatures than Θ, and we usually find empirically that the data satisfy

$$\Delta P_{TH} = \alpha K_T(T - 300). \tag{9}$$

The measured C_P data can be used to find C_V from

$$C_V = \frac{C_P}{1 + \alpha\gamma T} \tag{10a}$$

once γ has been determined. Similarly,

$$K_T = \frac{K_S}{1 + \alpha\gamma T}. \tag{10b}$$

In isobaric high-temperature calculations the thermoelastic parameters given by (2) and (3) are useful for many thermodynamic applications relating sound speed or bulk modulus to temperature, the adiabatic case arising from adiabatic elastic constants and the isothermal case arising from isothermal elastic constants.

We list our experimental values for the dimensionless parameter ν defined by

$$\nu = \left(\frac{\partial \ell n \, v_s}{\partial \ell n \, v_p}\right)_P, \tag{11}$$

which is of interest in seismic tomography calculations [1, 20, 37]. We also list values of the adiabatic Poisson's ratio σ given by

$$\sigma = \frac{3K_S - 2G}{6K_S + 2G}. \tag{12}$$

4. SOME CORRELATIONS FOUND FOR PROPERTIES IN THE TABLES

This section deals with relationships between the thermoelastic dimensionless parameters. The equation showing the relationship between $\delta_T - \delta_S$ and γ is [10]:

$$\delta_T - \delta_S = \gamma \frac{\left\{1 + \left[\left(\frac{\partial \ell n \, \alpha}{\partial \ell n \, T}\right)_P + \left(\frac{\partial \ell n \, \gamma}{\partial \ell n \, T}\right)_P\right]\right\}}{(1 + \alpha\gamma T\delta_T)}. \tag{13}$$

If the relative increase in α with T nearly compensates for the relative decrease of γ with T taking into account the change in the denominator of (13), then we may expect that $\delta_T - \delta_S$ is close in value to γ. Since the rate of change of γ with T is seldom the same as the rate of change of α with T, we see that $\delta_T - \delta_S = \gamma$ may be an approximation valid only over a limited range of temperatures, usually near and above Θ.

In some thermodynamics manipulations, the approximation $\delta_T - \delta_S = \gamma$ is useful. In Tables 29–42, we show the variation of $(\delta_T - \delta_S)/\gamma$ with T/Θ for our fourteen minerals. Thus an empirically determined approximation where actual high-temperature data are lacking is

$$\delta_T - \delta_S = \gamma. \tag{14}$$

Equation (14) is seldom a good approximation below $T = \Theta$. We note, however, that even in some high-temperature regions, (14) is in error, especially for Al_2O_3.

The variation of γ with T for all fourteen minerals is listed in Tables 29–42. For some solids, γ decreases with T at high T, but for CaO, Al_2O_3, and Mg_2SiO_4, it appears that $(\partial\gamma/\partial T)_P$ is close to zero at high T.

While the value of $(\partial\gamma/\partial T)_P$ at high T is close to zero for many minerals, the value of $(\partial\gamma/\partial T)_V$ is less and is always a negative number not close to zero (because the correction involves $-\int_{T_0}^{T} \alpha\gamma q \, dT$, and all values in the integrand are positive). This behavior of $(\partial\gamma/\partial T)_V$ has an important effect on the validity of the Mie-Grüneisen formulation of the thermal pressure.

The acoustic version of Θ, Θ_{ac} shows that minerals with higher density have higher values of Θ_{ac}. We know that when the average mass is constant from material to material, the value of Θ_{ac} rises as the 4/3 power of the density [2]. We would therefore expect that Θ_{ac} for perovskite will be high, in the neighborhood of that found for Al_2O_3.

The value of Θ_{ac} for perovskite at ambient conditions calculated by (5) is 1094 K. The values $v_p = 10.94$ km/s, $v_s = 6.69$ km/s, and $\rho_0 = 4.108$ g/cm^3 used in the calculation are given by Yeganeh-Hæri et al. [78]. Θ_{ac} for corundum is 1033 K, and $\rho_0 = 3.981$. This similarity in values of Θ_{ac} and ρ_0 suggests that the measured thermal pressure of corundum could be used as a guide for that of perovskite at mantle temperatures and pressures. The expected variation of Θ_{ac} with T for perovskite is shown in Figure 1, where Θ_{ac} of perovskite is compared to that of forsterite and periclase.

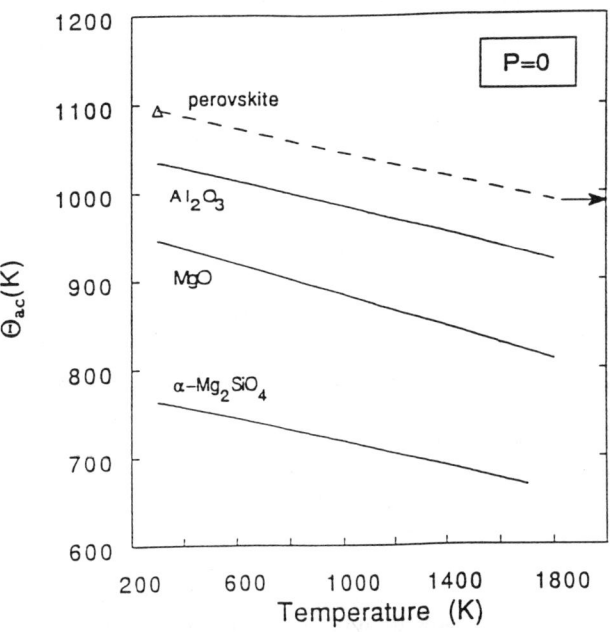

Fig. 1. Plots of Θ (acoustic) versus T. Solid lines show data on Θ from Tables 29, 30, and 34. The Δ is data for Θ (acoustic) for perovskite at room temperature. The dashed line shows the expected variation of Θ (acoustic) with T, yielding a value near 900 at 1900 K.

Another significant correlation found for most minerals is

$$\Gamma \approx \delta_T, \tag{15}$$

which allows one to calculate G for K_T in high temperature ranges where G is not known. It immediately follows from (15) that, because G is less than K_S, $|(\partial G/\partial T)_P|$ must be smaller than $|(\partial K_S/\partial T)_P|$.

5. PRESENTATION OF MINERAL DATA

We tabulate the adiabatic single-crystal elastic moduli (C_{ij}) for 14 minerals in Tables 1 to 14 starting at 300 K and proceeding in intervals of either 50° or 100°K. The errors indicated at selected temperatures in Tables 1–14 are those listed in the original references. The C_{ij} data for thirteen of the fourteen minerals in Tables 1–14 were retrieved by the already defined RUS method (either rectangular parallelepipeds or resonant spheres). The tabulated errors in the C_{ij} from these thirteen minerals are the standard deviations determined from the difference of each measured modal frequency with that modal frequency value calculated from the final set of C_{ij}. Thus the listed uncertainties reflect how consistent the frequency data are in providing C_{ij} values for a particular specimen. The MnO C_{ij} data and their uncertainties are not from RUS measurements, but are those recommended by Pacalo and Graham [47] from a weighted linear regression analyses together with the Pacalo and Graham [47] temperature derivatives.

The isotropic adiabatic bulk K_S and shear G moduli computed from the C_{ij} in Tables 1–14 are included in Tables 15–28. For minerals with cubic crystal symmetry, K_S is given by $(1/3)(C_{11} + 2C_{12})$. We use the Hashin-Shtrikman (HS) [29, 30, 73] scheme to compute G for cubic minerals (MgO, MgAl$_2$O$_4$, MnO, CaO, garnet, and NaCl) and both K_S and G for minerals with orthorhombic symmetry ($[\text{Mg, Fe, Mn, Co}]_2\text{SiO}_4$). Very small variations (typically ≤ 0.1 GPa) between our tabulated values of K_S and G for the orthorhombic minerals and the values found in the original references are due to the differences in K_S (or G) that result when using the HS and Voigt-Reuss-Hill (VRH) schemes. However, the HS scheme usually provides significantly narrower bounds and is preferred. The values in the tables are the averages of the upper and lower bounds found with the HS scheme. We use the VRH scheme to determine the isotropic G for KCl because of difficulties in interpreting the HS upper and lower bounds (see Section g. below). For corundum (Al$_2$O$_3$) we use the K_S and G values tabulated by

Goto et al. [26], which are based on the VRH scheme.

The errors indicated in Tables 15–28 for K_S and G are propagated by standard error techniques in which two sources of error are considered. These errors include both the uncertainty with which the midpoints of the upper and lower bounds of K_S and G are known and the difference between the midpoints and the upper bounds.

We couple thermal expansion α and heat capacity C_P (at constant P) data with K_S and G and calculate the values for several other thermoelastic quantities given in Tables 15–42. We assume α is accurate to 2% (unless specified otherwise in the original reference from which the α data are obtained) and rigorously propagate errors to ρ, C_V, K_T, v_p, and v_s at high and low temperatures in Tables 15–28. The uncertainty in the dimensionless quantities δ_S, δ_T, Γ, and ν should be taken as 5% or more, since the values of these parameters can depend somewhat on the order of polynomial fit used when determining the T derivatives.

5a. MgO

Prior to the work of Isaak et al. [34], high T data on the elastic moduli of single-crystal MgO were available up to 1300 K. Spetzler [55] used an ultrasonic pulse-echo method to obtain ambient P data at 300 and 800 K and provides a table listing T derivatives of elasticity at these two temperatures. Sumino et al. [63], using the rectangular parallelepiped resonance (RPR) method, extended the T range for which data were available up to 1300 K. Isaak et al. [34] (RPR) and Zouboulis and Grimsditch [80] (Brillouin scattering) report C_{ij} data for MgO up to 1800 and 1900 K, respectively.

We list the C_{ij} values as reported by Isaak et al. [34] in Table 1 and note that an alternative source for high T data can be found in Zouboulis and Grimsditch [80]. In their Table 2, Zouboulis and Grimsditch [80] report $C_{11}(T)$, $C_{44}(T)$, and the combined moduli $(C_{11} + C_{12} + 2C_{44})/2$ and $(C_{11} - C_{12})/2$ versus T. However, there is more scatter (especially in the C_{44} modulus) in the data of Zouboulis and Grimsditch [80] than in the data of Isaak et al. [34]. We find excellent agreement in $(\partial C_{ij}/\partial T)_P$ values when comparing the results of Spetzler [55], Sumino et al. [63], and Isaak et al. [34] in their T range of overlap. The high T (i.e., $T > 1400$ K) Zouboulis and Grimsditch [80] data have T dependence nearly identical to that of the Isaak et al. [34] data up to 1800 K.

The values of the dimensionless parameters, δ_S, δ_T, Γ, and ν, in Table 30 are found using sixth order polynomial fits to the K_S, K_T, G, v_p, and v_s results as done by Isaak et al. [34]. We find very little difference in the value of these dimensionless parameters, other than a slightly less rapid decrease when $T > 1500$ K, when a lower order of fit, such as 3, is used.

5b. CaO

The appropriate thermoelastic quantities for CaO [43] are listed in Tables 2, 24, and 31. Oda et al. [43] used the resonant sphere technique (RST) to reach 1200 K. We include the α data of Okajima [45] as tabulated by Oda et al. [43] in our Table 24. The dimensionless parameters in Table 31 are obtained from a second order polynomial fit of the K_S, K_T, G, v_p, and v_s [43].

5c. Pyrope-Rich Garnet

Suzuki and Anderson [65] provide C_{ij} for pyrope-rich garnet over the temperature range 298–993 K at irregular intervals of T. The specimen used by Suzuki and Anderson [65] is a single-crystal natural garnet with composition: pyrope, 72.6%; almandine, 15.7%; uvarovite, 6.1; androdite, 4.3%; spessartine, 0.7%; and grossular, 0.6%. We use the K_S, C_{11}, and C_{44} values given by Suzuki and Anderson [65] (see their Table 1) as the primary data source and interpolate to obtain these moduli given in Table 3 at 100°K increments of T. The C_{44} value at 638 K given by Suzuki and Anderson [65] seems unusually low and is excluded in our fit for interpolating. All errors in Table 3 are either those given by Suzuki and Anderson [65] (i.e., errors for K_S, C_{11}, C_{44}) or are propagated from the errors in K_S, C_{11}, and C_{44} (i.e., errors in C_{12} and C_S). We extrapolate (linear in K_S and C_{11}; quadratic in C_{44}) the Suzuki and Anderson [65] C_{ij} data over seven degrees to include the 1000 K values in Table 3. The dimensionless parameters δ_S and δ_T (Table 33) are found by a linear fit of the K_S and K_T values in Table 26. We find that G and v_s have noticeable curvature, so that second order fits are preferred when calculating Γ and ν. It is worth noting that ν tends to increase gradually from about 0.88 to 1.06 over the 300–1000 K range when second order fits to v_s and v_p versus T are used. If first order fits are assumed, ν is constant with a value of 0.97 over the 300–1000 K range in T. Thermal expansion α data on pyrope-rich garnet are also provided in the Suzuki and Anderson [65] paper.

5d. Grossular Garnet

Isaak et al. [36] provide elasticity data for near end-member grossular (grossular, 96.5%; andradite, 1.6%;

pyrope, 1.2%; less than 0.5% almandine or spessartine) to 1350 K. Table 4 is constructed using the K_S, C_S, and C_{44} of Isaak et al. [36] as primary data. C_{11} and C_{12} and their errors are calculated from these primary data. The C_{ij} values in Table 4 are plotted by Isaak et al. [36], but not explicitly tabulated in their presentation. We follow Isaak et al. [36] in the order of fit used to calculate the T derivatives for the dimensionless parameters. The parameters δ_S and δ_T are found using first order fits to the K_S and K_T data. Second order fits were applied to the G data to obtain Γ and to the v_p and v_s values to obtain ν. Isaak et al. [36] also provide new α data for grossular-rich garnet and point out the possibility that the Skinner [51] thermal expansion values for garnets may be low.

5e. Spinel (MgAl$_2$O$_4$)

Cynn [19] provides C_{ij} for single-crystal MgAl$_2$O$_4$ over the range 298–999 K at irregular intervals of T. There are data [19] up to 1060 K, but a sudden change in the slope of the data near 1000 K is attributed to the effect of cation disordering. We use the K_S, C_S, and C_{44} values from 298–999 K given by Cynn [19] as the primary data source and interpolate to obtain the C_{ij} in Table 5 at regular 50°K increments of T. Errors in Table 5 are either those given by Cynn [19] or are propagated from the errors in K_S, C_S, and C_{44}. The uncertainty in ρ at 300 K (Table 17) is assumed since it is not reported by Cynn [19]. The uncertainties in the C_{ij} for MgAl$_2$O$_4$ are large relative to most other resonance data; thus we present the dimensionless parameters in Table 40 based on linear fits of K_S, K_T, G, v_p, and v_s with T.

5f. MnO

Pacalo and Graham [47] present new data on MnO from 273–473 K. We construct Table 6 using the C_{ij} and $(\partial C_{ij}/\partial T)_P$ values recommended by Pacalo and Graham [47] (see their Table 10) for MnO. For C_{44} we use a second order polynomial to interpolate at intervals of 50°K where the nonzero value of $(\partial^2 C_{ij}/\partial T^2)_P$ is found in Table 6 of Pacalo and Graham [47]. The recommended first temperature derivatives are based on both ultrasonic pulse-echo [47] and resonant [62] measurements. For minerals of cubic symmetry there are only two independent moduli among K_S, C_{11}, and C_{12}. The recommended values for K_S, C_{11}, and C_{12} in Table 10 of Pacalo and Graham [47] are not self-consistent. We take C_{11} and C_{12} from Pacalo and Graham [47] as the primary data and use these to compute K_S for Table 23. Thus a small difference appears between the 300 K value for K_S in our Table 23

(149.0 ± 2.6 GPa) and that given by Pacalo and Graham [47] (150.6±2.5 GPa). The values at 500 K in our Table 6 require an extrapolation of 27°K beyond the maximum T measured, but are below the Néel temperature (522 K for MnO). Temperature derivatives of K_S, K_T, G, v_p, and v_s are done with a linear fit in T when calculating the dimensionless parameters in Table 39.

5g. KCl

Table 7 shows C_{ij} for KCl where we use the Yamamoto and Anderson [76] data to interpolate at even intervals of 100°K. The C_{11}, C_{12}, and C_{44} [76] are taken as the primary data from which C_S and K_S (Table 28) and their errors are derived. In determining the isotropic shear modulus G we used the VRH rather than the HS scheme. When attempting to use the HS scheme [30], we found that the resulting upper bound for G is less than the lower bound for temperatures in the range of 300–450 K. We are unable to resolve this difficulty. This seemingly contradictory result is likely related to the fact that the shear modulus C_S is very near to the value of the bulk modulus K_S at the lower temperatures. In any case we must defer to the VRH scheme for this material and note that there are relatively wide bounds on G (thus also on v_p and v_s), especially so at lower temperatures. Statistical analysis indicates that a second order polynomial fit of K_S, G, v_p, and v_s is warranted when determining the T derivatives, whereas a linear fit of K_T with T is adequate.

5h. NaCl

Yamamoto et al. [77] report elasticity data from 294–766 K for NaCl. Elasticity data up to the melting temperature of 1077 K for NaCl [31] were obtained with the composite oscillator method. To find accurate C_{ij} results with the composite oscillator requires that the resonant frequencies of quartz and silica rods that are coupled to the NaCl specimen be known in order to reduce the NaCl data. We also note that the experimental arrangement of Hunter and Siegel [31] requires a large volume to be heated when increasing temperature. For these reasons we prefer the data of Yamamoto et al. [77], but emphasize that Hunter and Siegel [31] report data 300°K higher in T than do Yamamoto et al. [77]. There are relatively small but measurable differences in the T dependence of the elastic moduli between these two data sets where they overlap. For instance, from 300–700 K the average value of $(\partial K_S/\partial T)_P$ from Hunter and Siegel [31] is −0.015 GPa/K, whereas that found by Yamamoto et

al. [77] is −0.012 GPa/K. The Yamamoto et al. [77] results tend to favor those of Spetzler et al. [56], who obtained elasticity data to 800 K for NaCl. Spetzler et al. [56] find $(\partial K_S/\partial T)_P$ to be −0.011 GPa/K from 300–800 K. The Yamamoto et al. [77] data are out to a slightly lower maximum temperature than Spetzler et al. [56], but contain a much denser set of tabulated C_{ij} values.

We interpolate the C_{11}, C_{12}, and C_{44} found in Table 2 of Yamamoto et al. [77] to regular intervals of 50°K for our primary elasticity data source in Table 8. The uncertainty in the density of the specimen used by Yamamoto et al. [77] is assumed to be 0.005 gm/cm^3 at 300 K (Table 27), since this is not provided by Yamamoto et al. We calculate the dimensionless parameters in Table 41 using a second order fit in T to the K_S, K_T, G, v_p, and v_s values in Table 27, and note that this results in δ_T at 300 K having a value near 5.6 rather than 5.3 as indicated by Yamamoto et al. [77]. We also find that above 600 K, δ_T increases gradually from 6.1 to 6.5, rather than having a value nearer 5.8–5.9 in this temperature range. These differences are due to the difference in method used to calculate the T derivatives. We fit the data over the T range of measurement to a polynomial; Yamamoto et al. [77] find T derivatives by taking a finite difference between two adjacent data points.

5i. Mg₂SiO₄

Following the work of Sumino et al. [61] to 670 K and Suzuki et al. [68] to 1200 K, the Mg₂SiO₄ (forsterite) data of Isaak et al. [35] to 1700 K extend the T limit for which elasticity data are available. Prior to these studies the available data were limited to T near ambient [27, 40]. There is general agreement between the data where they overlap. Sumino et al. [61] provide schemes for extrapolating that are generally confirmed by the higher temperature measurements of Suzuki et al. [68] and Isaak et al. [35] (See Figures 3 and 5). We construct Table 9 from the C_{ij} reported by Isaak et al. [35]. The thermal expansion results of Kajiyoshi [38] are used (see our Table 18) to calculate the thermoelastic properties of Mg₂SiO₄ (see Isaak et al. [35] for a comparison of the Kajiyoshi [38] values of α with other values in the literature). Isaak et al. [35] used the VRH scheme to calculate isotropic K_S and G moduli, whereas Table 18 lists K_S and G obtained from the HS averaging scheme. There is only a small difference (≤ 0.1 GPa) between the two approaches. Isaak et al. [35] show that fits of the K_S, K_T, G, v_p, and v_s data with T imply that third order polynomials are appropriate to describe the data. Thus we use third order fits for all T derivatives required to determine dimensionless parameters in Table 34. There are some small differences in the δ_S, δ_T, and Γ between our Table 34 and Table 5 of Isaak et al. [35] due to different methods used in calculating the T derivatives. Isaak et al. [35] used the interval between the two neighboring points to calculate the derivative at a particular T; here we apply polynomial fits over the whole temperature range from which the derivatives are obtained. The present approach produces smoother variations in δ_S, δ_T, Γ, and ν with T.

5j. Olivine (Fo$_{90}$Fa$_{10}$)

Isaak [32] reports data to 1400 and 1500 K, respectively, on two natural olivine samples with compositions of Fo$_{92.1}$Fa$_{7.7}$ and Fo$_{90.3}$Fa$_{9.5}$. Previously available T data on the elasticity of iron-bearing minerals were limited to temperatures near ambient [40]. We include the C_{ij} for Fo$_{90.3}$Fa$_{9.5}$ (referred to here as olivine Fo$_{90}$Fa$_{10}$) in Table 10. There are some differences between the C_{ij} of the two olivines reported by Isaak [32] that do not appear to be due to fayalite content. The differences are small, but caution should be used when interpolating between specimens with small differences in chemistry. The uncertainties in K_S and G (and in all properties that depend on K_S and G) are somewhat smaller in Table 19 than found in Isaak [32] for these quantities. Isaak [32] estimated the uncertainty by simply adding the uncertainty with which the HS average value is known to the distance between the average value and the HS high value. Here we square both the uncertainty of the HS average and the distance from the HS average to the HS high value, add the squares, and take the root. We use linear fits in T to the K_S, K_T, G, v_p, and v_s data when obtaining the dimensionless parameters δ_S, δ_T, Γ, and ν. We start our olivine tables by interpolating the $T = 296$ and 350 K data of Isaak [32] to 300 K. This shift in starting T causes minor differences in the T derivatives; hence the slight differences between δ_S, δ_T, and Γ in Table 35 and in Table 6b of Isaak [32].

5k. Fe$_2$SiO$_4$, Mn$_2$SiO$_4$, Co$_2$SiO$_4$

Sumino [58] presents T data from 25°–400°C (398–673 K) on the elasticity of single-crystal Fe$_2$SiO$_4$ (fayalite). We interpolate the Sumino [58] fayalite data (see his Table 3, specimen TA) so as to provide C_{ij} from 300–700 K in regular 50°K intervals in Table 11. A small extrapolation over 27°K is required to extend the values represented in Table 3 to 700 K. It should be noted that there are questions regarding the values of some of the C_{ij} and some of the T derivatives of fay-

alite (see Graham et al. [28], Isaak and Anderson [33], and Wang et al. [71]). As an example of the effect of uncertainties in the derivative values, Graham et al. [28] report $(\partial K_S/\partial T)_P = -0.030$ GPa/K, whereas Sumino [58] reports -0.0205 GPa/K. The reasons for these types of discrepancies are under investigation. At present, we defer to the data of Sumino [58] since they involve a significantly wider range in T (up to 673 K) than the maximum temperature of 313 K used by Graham et al. [28]. However, no compelling reasons exist to suggest that the Sumino [58] fayalite data should be preferred over those of Graham et al. [28].

From Figure 2 of Sumino [58] it is clear that K_S is linear in T, whereas G requires a higher order T dependence (at least a fourth order polynomial). In calculating the dimensionless parameters for fayalite in Table 36 we use linear fits in T for K_S and K_T and fourth order fits for G, v_p, and v_s. Sumino [58] attributes the strong nonlinear behavior of the shear modulus to influences of the antiferromagnetic to paramagnetic transition, even though this transition occurs at a much lower temperature (65 K).

Sumino [58] also reports elasticity data on single-crystal Mn$_2$SiO$_4$ and Co$_2$SiO$_4$ at 300 and 673 K. Sumino [58] indicates that all the C_{ij} of Mn$_2$SiO$_4$ and Co$_2$SiO$_4$ are linearly dependent on T with the exception of C_{44} for Co$_2$SiO$_4$. We construct Tables 12 and 13, which show values of C_{ij} at incremental temperatures of 100°K for Mn$_2$SiO$_4$ and Co$_2$SiO$_4$, using the C_{ij} and the $(\partial C_{ij}/\partial T)_P$ given by Sumino [58] in his Tables 4a,b. We assume a linear T dependence of the Co$_2$SiO$_4$ C_{44} modulus in constructing Table 13 since the presence of nonlinearity in this modulus is asserted by Sumino [58] but no quantitative result is provided. As with Fe$_2$SiO$_4$, we must perform a small extrapolation outside of the maximum T measured in order to include the $T = 700$ K values in the tables for Mn$_2$SiO$_4$ and Co$_2$SiO$_4$. When calculating the dimensionless parameters all T derivatives are assumed linear in T (see Figure 2 in Sumino [58]).

5l. Corundum (Al$_2$O$_3$)

The C_{ij} (Table 14), K_S, and G (Table 16) for Al$_2$O$_3$ are those found in Goto et al. [26]. The K_S and G given by Goto et al. [26] are found by the the Voigt-Reuss-Hill averaging scheme. Full account is made of the difference between the Hill (averaged VRH) values for K_S and G and the Voigt (upper bound) values when assigning errors to K_S, G, and other isotropic quantities derived from them. This accounts for larger errors being assigned to the G values in our Table 14 than are found in Table 2 of Goto et al. [26]. Apparently Goto et al. [26] assigned errors in G that include only the uncertainty in the Hill average itself, but do not consider the difference from the Hill average to the upper (or lower) bound. We interpolate the C_{ij}, K_S, G, and ρ provided at 296 and 350 K to start our Tables 14 and 16 at 300 K.

We use α from White and Roberts [75] rather than the α preferred by Goto et al. [26], i.e., those from Wachtman et al. [70].

We use a third order polynomial fit to determine the T derivatives for each of K_S, K_T, G, v_p, and v_s required for the dimensionless parameters shown in Table 29. This order of fit seems appropriate from a statistical consideration, i.e., an F_χ test [18]. Thus, the δ_S, δ_T, and ν found in our Table 29 for Al_2O_3 may differ somewhat at a particular T from those given by Goto et al. [26] in their Tables 4 and 5. The Goto et al. [26] δ_S, δ_T, and ν values are calculated from derivatives found by using a finite interval from the two nearest data points. Thus, we find considerably more scatter in the Goto et al. [26] δ_S and δ_T values than in Table 29.

6. THEORETICAL BASIS FOR OBSERVED PROPERTIES AT HIGH TEMPERATURE

Most, but not quite all, of the high temperature properties of the solids reported here are consistent with the high temperature limit of the quasiharmonic approximation of the statistical formulation of the free energy.

6a. The Helmholtz Free Energy.

The expression for the Helmholtz energy for an insulator is [41]

$$A = E_{ST} + \mathcal{A}_{VIB},$$

where electronic and magnetic effects are ignored. E_{ST} is the potential of a static lattice at absolute zero, and $\mathcal{A}_{VIB}$ is the vibrational energy due to motion of the atoms as each is constrained to vibrate around a lattice point.

The statistical mechanical definition of the vibrational contribution to the Helmholtz energy arising from $3pN$ independent oscillators is

$$\mathcal{A}_{VIB} = \frac{1}{2}\sum_{j=1}^{3pN} \hbar\omega_j + kT \sum_{j=1}^{3pN} \ell n\left(1 - e^{-\hbar\omega_j/kT}\right), \quad (16)$$

where ω_j is the jth modal frequency; N is Avogadro's

number; and p is the number of atoms in each cell (or molecule). The first term on the right is the zero point vibrational energy E_{ZV} given by

$$E_{ZV} = \frac{1}{2}\sum_{j=1}^{3pN} \hbar\omega_j. \quad (17)$$

Note that, unlike the last term in (16), there is no T in E_{ZV}. This term arose by summing the allowed quantum state levels to find the energy of each normal mode, according to the Hamiltonian [42].

We replace ω with y, the dimensionless frequency, so that

$$y_j = \frac{\hbar\omega_j}{kT}. \quad (18)$$

Thus (16) can be given in terms of the thermal energy

$$\mathcal{A}_{VIB} = E_{ZV} + \mathcal{A}_{TH}, \quad (19)$$

where

$$\mathcal{A}_{TH} = kT\sum_{j=1}^{3pN} \ell n\left(1 - e^{-y_j}\right) = kT\sum_{j=1}^{3pN} \mathcal{A}_{TH_j}. \quad (20)$$

$\mathcal{A}_{TH}$ is the energy arising from temperature excitation, called the thermal energy. E_{ZV} is not affected by T, as shown in (17). Thus $\mathcal{A}_{TH} \to 0$ as $T \to 0$. However, $\mathcal{A}_{VIB} \to E_{ZV}$ as $T \to 0$, and E_{ZV} is a nonzero number. E_{ZV} is sufficiently small that for most numerical evaluations it could be dropped, but it is useful to keep this term in $\mathcal{A}_{VIB}$ for algebraic manipulations done later on.

The expression for the Helmholtz energy for an insulator is thus

$$A = E_{ST} + E_{ZV} + \mathcal{A}_{TH}. \quad (21)$$

We need to divide (21) into temperature-dependent and nontemperature-dependent parts, so we use

$$A = E_{T=0} + \mathcal{A}_{TH}, \quad (22)$$

where

$$E_{T=0} = E_{ST} + E_{ZV}. \quad (23)$$

Dividing (21) into its vibrational and nonvibrational parts, we define

$$A = E_{ST} + \mathcal{A}_{VIB}, \quad (24)$$

where $\mathcal{A}_{VIB}$ is given by (16).

6b. The Quasiharmonic Approximation

Before we can find thermodynamic properties such as P and C_V from (20), we must make decisions on the volume and temperature behavior of ω_j. In the quasiharmonic approximation, ω is assumed to be dependent upon V but not upon T. This makes all the thermodynamic properties directly dependent on V. The temperature behavior of the thermodynamic properties is indirect; it comes from the fact that while ω_i is not dependent upon T, the sum $\sum \ell n\,(1 - e^{-y_j})$ depends upon T. When ω_j are different, then y_i are also different, and the sum above becomes T-dependent, especially at low T.

The internal energy $\mathcal{U}$ is found by applying the formula $\mathcal{U} = -T^2\left((\partial \mathcal{A}/T)/\partial T\right)$ [41], yielding

$$\mathcal{U} = E_{T=0} + kT \sum_{j=1} \frac{y_i}{e^{y_i} - 1} = E_{T=0} + kT \sum_{j=1} E_{TH_j}. \tag{25}$$

The pressure P is found by applying $-(\partial/\partial V)_T$ to (21), yielding

$$P = -\left(\frac{\partial E_{ST}}{\partial V}\right)_T + \frac{kT}{2V} \sum_j \gamma_j y_j + \frac{kT}{V} \sum_j \gamma_j \frac{y_i}{e^{y_i} - 1}, \tag{26}$$

where

$$\gamma_j = \frac{\partial \ell n\,\omega_j}{\partial \ell n\,V} \tag{27}$$

and where γ_j are called the mode gammas. Thus

$$P = P_0 + P_{TH}, \tag{28}$$

where the thermal energy (20) effect on the pressure, called the thermal pressure, is

$$P_{TH} = k\frac{T}{V} \sum_j \gamma_j \frac{y_i}{e^{y_i} - 1}. \tag{29}$$

Comparing (29) with (20), we see that

$$P_{TH} = \frac{1}{V} \sum_j \gamma_j \mathcal{A}_{TH_j}. \tag{30}$$

The specific heat is formed by applying $-(\partial/\partial T)_V$ to (25), yielding

$$C_V = k \sum_j \frac{y_j^2 e^{y_j}}{(e^{y_j} - 1)^2} = \sum_j C_{V_j}. \tag{31}$$

Entropy is found by applying the operator $(\partial/\partial T)_V$ to $\mathcal{A}$, obtaining

$$S = k \sum y_i \left(\frac{e^{y_j}}{e^{y_j} - 1}\right) - k \sum \ell n\,\left(1 - e^{-y_i}\right) \tag{32}$$

The expression for αK_T is found by applying $\partial^2/\partial T \partial V$ to (22), yielding

$$\alpha K_T = \frac{\hbar}{V} \sum y_j^2 \left[\gamma_j \frac{e^{y_j}}{\left(e^{y_j} - 1\right)^2}\right]. \tag{33}$$

To get α we divide αK_T (given by (33)) by K_T, defining $K_T = -V(\partial P/\partial V)_T$, and using P given by (26).

Using (30), we find that $\alpha K_T = \gamma C_V/V$, and equating this to (33), we find that [16]

$$\overline{\gamma} = \frac{\sum_j \gamma_i C_{V_j}}{C_V}. \tag{34}$$

The bar over γ indicates that this result is an approximation of $\gamma = \alpha K_T V/C_V$ resulting from invoking the quasiharmonic approximation to the Helmholtz energy.

6c. The High-Temperature Limit of the Quasiharmonic Approximation

At sufficiently high temperatures, the expression for $\mathcal{A}_{TH}$ can be simplified due to the convergence of certain series expansions. The argument in (16) becomes small since $\hbar\omega_j << kT$, that is, y_i is small. At high temperatures, we take advantage of the expansion [79],

$$\ell n\,\left(1 - e^{-y_j}\right) = \ell n\,y_j - \left(\frac{1}{2}y_j^2 + \ldots\right)$$
$$= \ell n\,y_j + \ell n\,\left(1 - \frac{1}{2}y_j\right). \tag{35}$$

By replacing the last logarithmic term above with its argument, which is valid for low values of y_j, we find that

$$\ell n\,\left(1 - e^{-y_j}\right) \simeq \ell n\,\gamma_j - \frac{1}{2}\gamma_j. \tag{36}$$

Thus the high-temperature representation of (20) is

$$A^{ht}_{TH} = kT \sum_{j}^{3pN} \left(\ell n \, y_j - \frac{1}{2} y_j \right). \tag{37}$$

By using (21) and (34), the high T Helmholtz energy for insulators at high T is

$$A^{ht} = E_{ST} + E_{ZV} + kT \left(\sum_{j} \ell n \, y_j - \frac{1}{2} \sum_{j} y_j \right). \tag{38}$$

We see that the last term above just cancels the zero temperature energy given by (17), so that

$$A^{ht} = E_{ST} + kT \sum_{j=1}^{3pN} \left(\ell n \, \hbar\omega_j - \ell n \, kT \right). \tag{39}$$

This can be simplified by defining an average frequency $\overline{\omega}$, given by

$$\ell n \, \overline{\omega} = \frac{1}{3pN} \sum \ell n \, \omega_j, \tag{40}$$

in which case (20) becomes

$$A_{TH} = 3pNT \left(\ell n \, \hbar\overline{\omega} - \ell n \, kT \right). \tag{41}$$

6d. Thermodynamic Properties in the High *T* Limit of the Quasiharmonic Approximation

At high T, $\mathcal{A}$ includes only the static potential and the thermal energy in the quasiharmonic approximation. Using the equation $\mathcal{U} = -1/T^2 \left((\partial\mathcal{A}/T)/\partial T \right)_V$, we find the high temperature internal energy to be

$$\mathcal{U}^{ht} = E_{ST} + 3p\mathbf{R}T, \tag{42}$$

where $\mathbf{R}$ is the gas constant, kN. Equations (41) and (42) are important results for the high temperature limit in the quasiharmonic approximation.

Taking the temperature derivative of $\mathcal{U}$ to find specific heat,

$$C_V^{ht} = 3p\mathbf{R}. \tag{43}$$

Equation (43) is known as the Dulong and Petit limit. Thus the high temperature quasiharmonic approximation leads to the classical limit in specific heats. If at high T the experimentally determined C_V is not parallel to the T axis, we must conclude that the quasiharmonic approximation is not adequate for this property, and anharmonic terms must be added to $\mathcal{A}$.

From Figures 14–26, where the data for specific heat versus T are listed, we see that for several solids of interest to geophysicists $(\partial C_V/\partial T)_P = 0$ at high T. This is true for MgO, CaO, and Al_2O_3 and NaCl, but not for olivine, garnets, spinel, or KCl. We must con-

clude that insofar as specific heat is concerned, the quasiharmonic approximation is not adequate for some solids up to $T \approx 2\Theta$, but is adequate for other solids up to $T \approx 2\Theta$.

The high temperature Grüneisen parameter is found from (34), and considering that as $T \rightarrow \infty$, $C_i \rightarrow k$,

$$\gamma_{ht} = \overline{\gamma}_{(T\rightarrow\infty)} = \frac{1}{3pN} \sum_{j}^{3pN} \gamma_j, \tag{44}$$

which is independent of T. The data on γ is found in Tables 15–28.

We see that there is an approximate trend for γ to be independent of T at high T. For these solids when $d\gamma/dT$ is not zero, it is also true that C_V does not obey the Dulong and Petit limit. Thus departures from quasiharmonic theory in C_V create a departure from the quasiharmonic theory of γ. Above $T = \Theta$, the departure from $(\partial\gamma/\partial T)_P = 0$ is slight, however.

Applying the operator $-(\partial/\partial V)_T$ to (38), we find the high T pressure, P^{ht}, which by using (44) becomes

$$P^{ht}(T,V) = P(0,V) + \frac{3p\mathbf{R}}{V} \gamma_{ht} \, T. \tag{45}$$

Thus the thermal pressure is

$$P_{TH} = \frac{3p\mathbf{R}}{V} \gamma_{ht} \, T.$$

V is a function of T, and it is also equal to M/p where M is the molecular weight, so that the above is

$$P_{TH} = \frac{3\mathbf{R}\gamma_{ht}}{(M/p)} \rho_0 (1 - \alpha T) \, T. \tag{46}$$

We see that the quasiharmonic theory allows for P_{TH} to be virtually linear in T: at the highest measured T, $\alpha T \simeq 0.05$. But γ_{ht} is not, as we shall see, exactly independent of T, though close to it. Thus there are counter effects in T, and the $1 - \alpha T$ term is often obscured. Thus to a very good approximation P_{TH} is proportional to T under the quasiharmonic approximation.

The thermal pressure relative to $T = \Theta$ for $T > \Theta$ is given by

$$\Delta P_{TH} = P_{TH}(T, V_0) - P_{TH}(\Theta, V_0). \tag{47}$$

As shown in Tables 15–28, we see that ΔP_{TH} is indeed linear with T (for $T/\Theta_0 > 1$). (See also Figure 2).

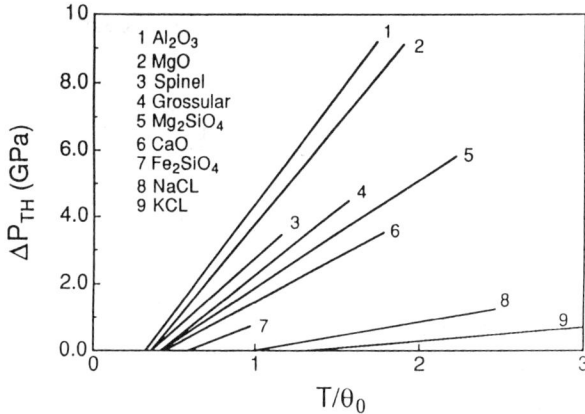

Fig. 2. ΔP_{TH} versus T/Θ_0 for nine minerals showing: 1) ΔP_{TH} is linear in T/Θ_0; and 2) the slope of the $\Delta P_{TH} - T/\Theta_0$ line decreases as $\rho_0/(M/p)$ decreases. Values of $\rho_0/(M/p)$ in ascending order (1–9) are: 0.195, 0.178, 0.176, 0.160, 0.160, 0.151, 0.119, 0.073, and 0.053. Θ_0 is the acoustic Θ at $T = 300$ K.

This means that for the thermal pressure property, the quasiharmonic approximation appears quite adequate up to relative temperatures of $T/\Theta_0 \approx 2.0$. This range covers values of T/Θ_0 for the mantle of the earth. The linearity of P_{TH} with T is not unique to silicates and alkali halides; it is also found for gold (see Figure 10 of Anderson et al. [12]).

The slope of the $\Delta P_{TH} - T$ curve from (47) is empirically a constant. Since $P_{TH} = \int \alpha K_T \, dT$, then we can define an average value of αK_T, $\overline{\alpha K_T}$, good for T above Θ, such that

$$\left(\frac{\partial \Delta P_{TH}}{\partial T}\right)_P = \overline{\alpha K_T} = \frac{3\mathbf{R}\gamma_{ht}}{(M/p)} \rho_0 \left(1 - \alpha T\right) = \text{const.},$$
(48)

where M/p is the mean atomic weight (per atom). For non iron-bearing silicate minerals, M/p varies little from mineral to mineral and is near the value 21 [74]. Also γ_{ht} does not vary greatly from mineral to mineral. Thus from (48) for $M/p = $ constant the slope increases from mineral to mineral as the ambient density increases. The data in Tables 27–30 bear this out, as shown in Figure 2. Since to a good approximation for silicates of constant M/p, $\Theta = \text{const } \rho^{4/3}$ [2], we see that $(\partial P_{TH}/\partial T)_P$ should increase as $\Theta^{3/4}$: In general, the mineral with the highest value of $\rho_0/(M/p)$ will have the highest slope, according to (48), and that is borne out by the experimental results shown in Figure 2.

6e. The Bulk Modulus at High T

Using the operator $-V\left(\partial/\partial V\right)_T$ on (45), we find that the bulk modulus K_T along an isochore is

$$K_T^{ht}(T, V) = K_T(0, V) - \frac{3\mathbf{R}\rho}{(M/p)}\gamma_{ht}\left(1 - q^{ht}\right)T, \quad (49)$$

where $q^{ht} = (\partial \ell n \, \gamma_{ht}/\partial \ell n \, V)$.

Thus at high T, the isochoric bulk modulus is decreasing with T according to the slope

$$\left(\frac{\partial K_T^{ht}}{\partial T}\right)_V = -\frac{3\mathbf{R}\rho}{(M/p)}\gamma_{ht}(1 - q^{ht}). \quad (50)$$

For $q^{ht} = 1$, $(\partial K_T/\partial T)_V$ vanishes, so the determination of (50) ordinarily yields a small quantity. To obtain the isobaric derivative, $(\partial K_T/\partial T)_P$, we use the calculus expression:

$$\left(\frac{\partial K_T}{\partial T}\right)_P = \left(\frac{\partial K_T}{\partial T}\right)_V - \alpha K_T\left(\frac{\partial K_T}{\partial P}\right)_T. \quad (51)$$

Thus at high T, along an isobar,

$$\left(\frac{\partial K_T^{ht}}{\partial T}\right)_P = -\frac{3\mathbf{R}\rho}{(M/p)}\gamma_{ht}(1 - q^{ht}) - \alpha K_T\left(\frac{\partial K_T}{\partial P}\right)_T. \quad (52)$$

The term on the far right of (49) is the thermal bulk modulus, K_{TH}. Using (48) in (52), we have

$$\left(\frac{\partial K_T}{\partial T}\right)_P = \overline{\alpha K_T}\left(q^{th} - 1\right) - \alpha K_T\left(\frac{\partial K_T}{\partial P}\right)_T.$$

Letting $\overline{\alpha K_T} = \alpha K_T$ in the above

$$\left(\frac{\partial K_T}{\partial T}\right)_P = -\overline{\alpha K_T}\left(K' + q^{th} - 1\right). \quad (53)$$

We thus see that the thermal bulk modulus, K_{TH}, contributes a minor amount to $(\partial K/\partial T)_P$.

We evaluate (53) for the case of MgO from the data as an example. The average value of $\overline{\alpha K_T}$ is found from the values of ΔP_{TH} between 1000 K and 1800 K listed in Table 30. In this table $\overline{\alpha K_T}$ is shown to be 0.0064 GPa/K, close to actual measured values of αK_T at high T. The high T value of $K' = 4.5$ [8], and $q^{ht} = 1.4$ [8]. Thus we find from (53) that a predicted value of $(\partial K_T/\partial T)_P = 0.031$ GPa/K. This result agrees well with the measurements found in Table 16 showing $(\partial K_S/\partial T)_P$ to be 0.03 GPa/K in the high T range.

6f. Entropy and αK_T

Applying the operator $(\partial/\partial T)_V$ to (38), the entropy is found

$$\mathcal{S}^{ht} = -3p\mathbf{R}\left(\ell n\, kT - \ell n\, \hbar\overline{\omega} + 1\right). \qquad (54)$$

The entropy increases with $\ell n\, T$ in the quasiharmonic approximation at high T. In contrast, C_V is independent of T in this approximation at high T, as shown by (43).

We note that if P is linear in T, then $(\partial P/\partial T)_V$ is independent of T. From experiments, however, we find that αK_T is independent of T (or approximately so) only at T above Θ. The functional form of αK_T resembles a C_V curve.

Applying the operator $(\partial/\partial T)_V$ to the pressure given by (45), we have for isochores,

$$(\alpha K_T)^{ht} = \frac{3p\mathbf{R}\gamma_{ht}}{V} = \frac{C_V^{ht}\gamma_{ht}}{(M/p)}\rho_0\left(1 - \alpha T\right) \qquad (55)$$

Equation (55) is to be compared to (48). The actual measured αK_T will have more structure than $\overline{\alpha K_T}$ determined from P_{TH}. Nevertheless to a fair approximation αK_T is parallel to the T axis for $T > \Theta$, as Tables 29–42 show.

6g. α Versus T at High T Along Isobars

Using the expression for K_T^{ht} versus T, (49), and the expression for αK_T given by (48), we have

$$\alpha = \frac{\overline{\alpha K_T}}{K_{T_0} + \left(\frac{\partial K_T^{ht}}{\partial T}\right)_P T}, \qquad (56)$$

which is close to, and for some solids exactly equal to,

$$\alpha = \frac{\text{const}}{K_{T_0} - aT} \qquad (57)$$

for the $P = 0$ isobar and where $a = \left(\partial K_T^{ht}/\partial T\right)_P$, given by (53). We note that (57) results from the quasiharmonic approximation.

Examination of (57) shows that at high T we may expect α to increase with T at high T even under the quasiharmonic approximation. This steady increase of α with T at high T does not necessarily require the assumption of additional anharmonic terms in $\mathcal{A}$ beyond the quasiharmonic high T approximation, (41). It arises because the denominator in (57) has a term that decreases with T even in the quasiharmonic approximation. We saw some evidence of anharmonicity

in C_V, but even so, this effect does not appear in properties that are controlled by the volume derivatives of $\mathcal{A}$ as described above. Further, we also showed that there is no evidence of anharmonic effects in P_{TH} because it is linear in T. Thus $(\partial P/\partial T)_V$ should be essentially independent of anharmonicity. If $(\partial P/\partial T)_V$ is independent of T, then α is controlled by (57), a strictly quasiharmonic equation.

6h. δ_T and q From the Quasiharmonic Theory

From the definition of δ_T, (3), and using (51), (52) and (53), we find

$$\delta_T = (q^{ht} - 1) + \left(\frac{\partial K_T}{\partial P}\right)_T. $$

The above equation tells us that δ_T should be virtually independent of T at high T. This is demonstrated in Tables 29–43, which show that roughly speaking, δ_T is parallel to the T axis. We also note that if $(\partial K_T/\partial P)_T = K_0'$ is virtually independent of T, and the experimental evidence is in favor of this, δ_T at high T should be slightly higher than K_0', providing $q > 1$.

We derived a relationship between three important dimensionless thermoelastic constants at high T. The above equation can be expressed as

$$q^{ht} = \delta_T - K_0' + 1. \qquad (58)$$

Equation (58) is appropriate for high T only, arising as it does from the high temperature limit of the quasiharmonic approximation. For the low temperature equation corresponding to (59), we take the logarithmic derivative of γ, as defined by (1), yielding

$$q = \left(\frac{\partial \ell n\, \gamma}{\partial \ell n\, V}\right)_T = \delta_T - K' + 1 - \left(\frac{\partial \ell n\, C_V}{\partial \ell n\, V}\right)_T. \qquad (59)$$

At high T and $P = 0$, (59) reduces to (58) because C_V is independent of V above the Debye temperature.

6i. Summary of the Quasiharmonic Approximation Effects Upon Physical Properties

The data presented here strongly suggest that the quasiharmonic approximation is valid for the following properties at least up to $T = 2\Theta$: P, K_T, αK_T, α, and all the respective volume derivatives of these four properties. They also suggest that the quasiharmonic approximation may or may not be valid for C_V and entropy for temperatures up to $T = 2\Theta$. This means that equations of state should not be corrected for anharmonic effects for pressures and temperatures corresponding to mantle conditions. For an explanation of why P, K_T, αK_T and α do not require anharmonic correction, while at the same time C_V and S may require anharmonic correction, see Anderson et al. [13].

7. HIGH TEMPERATURE EXTRAPOLATION FORMULAS

7a. Introductory Comments

Knowledge of the high temperature thermoelastic properties permits one to generate extrapolation formulas for estimating K_S and G at temperatures higher than measured. We have seen that at high temperature (above Θ), γ and δ_S are virtually independent of T for some minerals, and nearly so for others. By assuming that γ and δ_S are exactly independent of T, good estimates of K_S are found using the definitions of γ and δ_S. Since the values of δ_S and γ are often not steady for $T < \Theta$, it is not safe to use the extrapolation formulas described below when only low T data for δ_S and γ are available. This is true especially for minerals, for in this case ambient values of δ_S and γ are measured at temperatures much lower than Θ.

7b. Extrapolations for K_S Using γ Constant With T

If γ is independent of T, as in the case of MgO, CaO, and Mg_2SiO_4 at high T, and if $\alpha(T)$ and $C_P(T)$ data are available at high T, then the following formula, determined from (1), is useful. Using (1)

$$K_S(T) = \frac{K_{SO}\left[\dfrac{C_P(T)}{C_{PO}}\right]}{\left[\dfrac{\alpha(T)V(T)}{\alpha_0 V_0}\right]},\qquad(60)$$

where K_{S0}, C_{P_0}, α_0, and V_0 all refer to values of these quantities at some reference T beyond which γ is approximately constant. Sumino et al. [61] and Sumino et al. [63], respectively, used (60) to extrapolate K_S from 700 to 2100 K (forsterite) and 1300 to 3100 K (MgO). We emphasize that reliable extrapolations using (60) require that α be known at high T and the assumption that γ is constant.

7c. Extrapolations for K_S Using δ_S Constant With T

Anderson and Nafe [9] pointed out that if δ_S is independent of T along isobars, then

$$\ln K_S = -\delta_S \ln V + \text{constant}\qquad(61)$$

could be used to find K_S vs. T, because V is a function of T. Anderson [3] used (61) in a more convenient form, which follows when δ_S is independent of T. Along isobars,

$$K_S = K_{S_0}\left[\frac{V(T)}{V_0}\right]^{-\delta_S} = K_{S_0}\left[\frac{\rho(T)}{\rho_0}\right]^{\delta_S}.\qquad(62)$$

Using (62) he estimated K_S for MgO up to 2000°C (see Figure 1 and Equation (21) of Anderson [3]). Sumino et al. [61] used (62) to extrapolate values of K_S for forsterite from 700 K to 2000 K.

Duffy and D.L. Anderson [21] present a general formula for extrapolating elastic constants, in which (62) is a special case. Duffy and D.L. Anderson's equation for $G(T)$ in terms of the nomenclature used in this paper is

$$G(T) = G_0\left[\rho(T)/\rho_0\right]^{\Gamma}.\qquad(63)$$

Knowledge of $C_P(T)$ is not required for (62) and (63), but some knowledge of $\alpha(T)$ at high T is required to get $\rho(T)$ at high T. The appropriate equation relating ρ and α is

$$\rho(T) = \rho_0\,\exp\left[-\int_{T_0}^{T}\alpha(T)\,dT\right].\qquad(64)$$

For some solids, $\rho(T)$ at high T may not be known. However there is another method that minimizes the dependence of K_S upon $\alpha(T)$ and $\rho(T)$.

Combining (1) and (2), we have

$$\left(\frac{\partial K_S}{\partial T}\right)_P = -\delta_S\,\alpha K_S = \frac{\delta_S\gamma C_P}{V}.\qquad(65)$$

Integrating (65) with respect to T

$$K_S(T) - K_S(T_0) = -\int\frac{\delta_S\gamma C_P}{V}\,dT.\qquad(66)$$

Assuming now that the product $\delta_S\gamma$ is independent of T at high T, and taking $V = V_0(1+\alpha T)$, we find to a very good approximation [54]

$$K_S(T) - K_S(T_0) = -\frac{\gamma\delta_S}{V_0}\big[H(T) - H(T_0)\big]$$
$$\left[1 - \frac{\alpha(T - T_0)}{2}\right],\qquad(67)$$

where $H(T) = \int C_P \, dT$ is the enthalpy.

Equation (67) shows that $K_S(T)$ should be linear in $H(T)$. This was proven experimentally for three minerals for $300 < T < 1200$ K by Soga et al. [54], and for the same three minerals for $300 < T < 1800$ K by Anderson [5]. The term $\alpha(T - T_0)/2$ is of little consequence in the calculation and can be ignored. Anderson [5] used (67) to extrapolate the velocities of sound from Θ to the melting point. Equation (67) is very useful because $\alpha(T)$ is not needed to find $K_S(T)$, and there is much enthalpy data published listing H to temperatures about or greater than 2500 K.

7d. Extrapolations of K_S by Sumino et al. [61]

Sumino et al. [61] measured the elastic constants of forsterite between $-190°$C and $400°$C. They then extrapolated the values of K_S up to 2200 K using (60), (62), and (67) assuming γ and δ_S constant in the high temperature regime. The extrapolation to 2000 K is shown in Figure 3. Isaak et al. [35] reported measured values of K_S up to 1700 K, and their values are also plotted in Figure 3. Equations (62) and (63) gave good results for extrapolations, since the actual measured values are quite close to the extrapolated ones. However, Figure 3 illustrates the point that (60) is not a good extrapolation formula, at least for forsterite.

Sumino et al. [63] measured the elastic constants of MgO from 80 to 1300 K. They used equation (60) to extrapolate values of K_S above the maximum measurement (1300 K) up to 3000 K. This is shown in Figure 4, in which the measured values of K_S [35] up to 1800 K are also plotted. It is clear that the extrapolation formula given by (60) is reasonably successful in the case of MgO.

7e. Extrapolation of G to High T

Soga and Anderson [52] measured K and G for MgO and Al_2O_3 up to 1200 K and found that G was linear in T above 800 K. They [53] later measured G and K_S up to 1100 K for Mg_2SiO_4 and $MgSiO_3$ and also found linearity for G and K_S. They used the empirical formula

$$G(T) = G_0 - \left(\frac{\partial G}{\partial T}\right)_P \bigg|_{T_0} (T - T_0) \tag{68}$$

to estimate shear velocities up to 2500 K.

Sumino et al. [61] measured G for forsterite up to $400°$C and reported extrapolated values from 700 K to 2200 K. Isaak et al. [35] measured G up to 1700 K. These results are plotted in Figure 5. Equation (68) is

used for one extrapolation and an equation involving linearity in Poisson's ratio is also used

$$\sigma(T) = \sigma_0 + \left(\frac{\partial \sigma}{\partial T}\right)_P \bigg|_{T_0} (T - T_0). \tag{69}$$

Both extrapolation equations give reasonably good re-

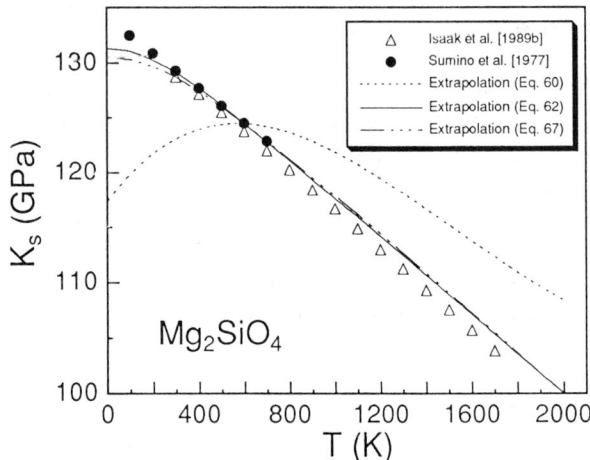

Fig. 3. Observed and extrapolated values of K_S versus T for forsterite. Experimental (plotted symbols): Isaak et al. [35]; Sumino et al. [61]. Extrapolations by Sumino et al. [61]: (a) using Equation (60) where 600 K is the reference T; (b) using Equation (62); and (c) using Equation (67). Two extrapolations made by Sumino et al. [61] are confirmed to within 2% out to 1700 K by the measurements of Isaak et al. [35].

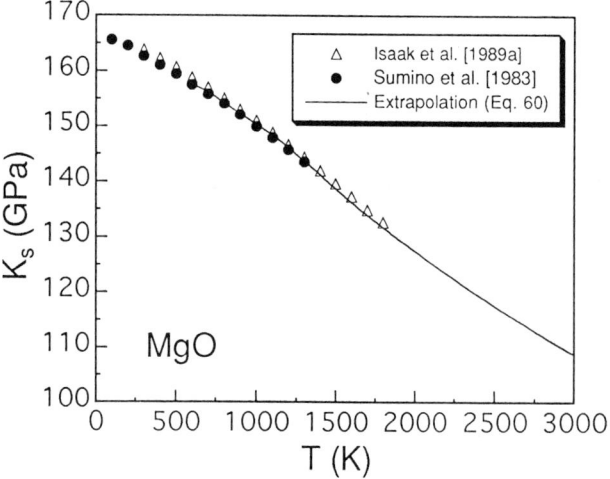

Fig. 4. Observed and extrapolated values of K_S versus T for MgO. Experimental (plotted symbols): Isaak et al. [34]; Sumino et al. [63]. Extrapolation by Sumino et al. [63] using Equation (60).

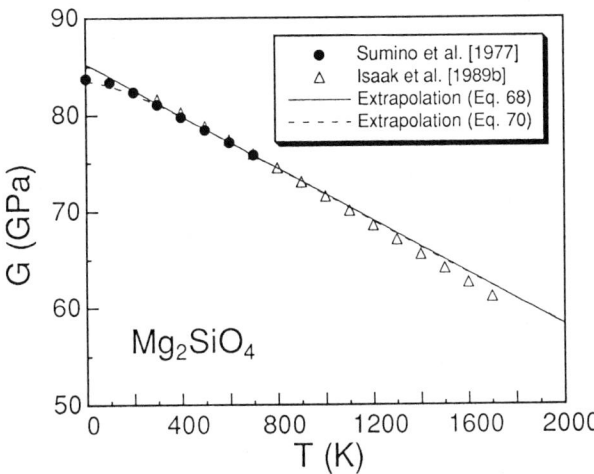

Fig. 5. Observed and extrapolated values of G versus T for forsterite. Experimental (plotted symbols): Isaak et al. [35]; Sumino et al. [61]. Extrapolations by Sumino et al. [61]: (a) using Equation (68); (b) using Equation (70). The extrapolations made by Sumino et al. [61] are confirmed to within 1.5% out to 1700 K by the measurements of Isaak et al. [35].

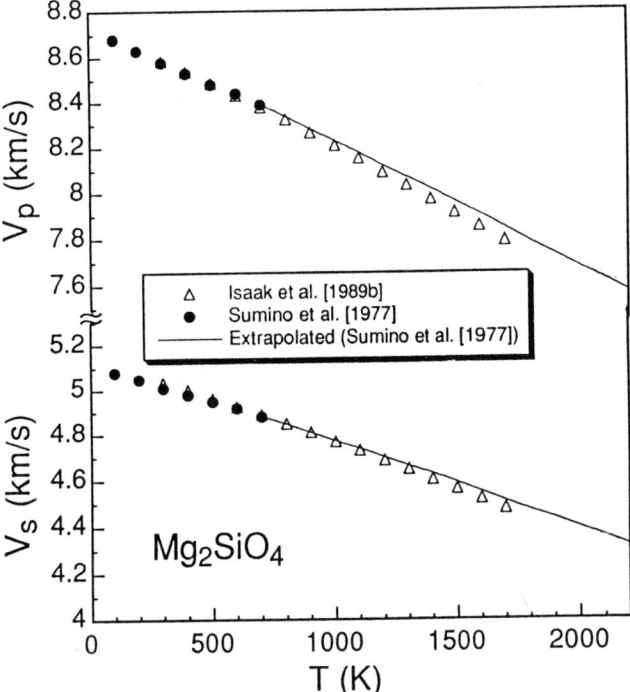

Fig. 6. Observed and extrapolated values of v_p and v_s versus T for forsterite. Experimental (plotted symbols): Sumino et al. [61] up to 700 K; Isaak et al. [35] up to 1700 K. Extrapolations by Sumino et al. [61] are from 700 to 2200 K and are confirmed to within 1.5% at 1700 K by the measurements of Isaak et al. [35].

sults when compared with the later measurements of Isaak et al. [35]

The equation of G is

$$G(T) = \frac{3}{2} \frac{(1 - 2\sigma(T))}{(1 + \sigma(T))} K_S(T). \qquad (70)$$

7f. Extrapolations for v_s and v_p to High T

If K and G are successfully extrapolated, then v_s and v_p are extrapolated. Sumino et al. [61] gave values for the extrapolated values of v_s and v_p for forsterite up to 2200 K using their 100 K to 200 K measurements. These are shown in Figure 6. Comparison with the actual measurements of v_s and v_p by Isaak et al. [35] and presented in Table 15 are also plotted. It is clear that the extrapolations by Sumino et al. [61] were quite successful.

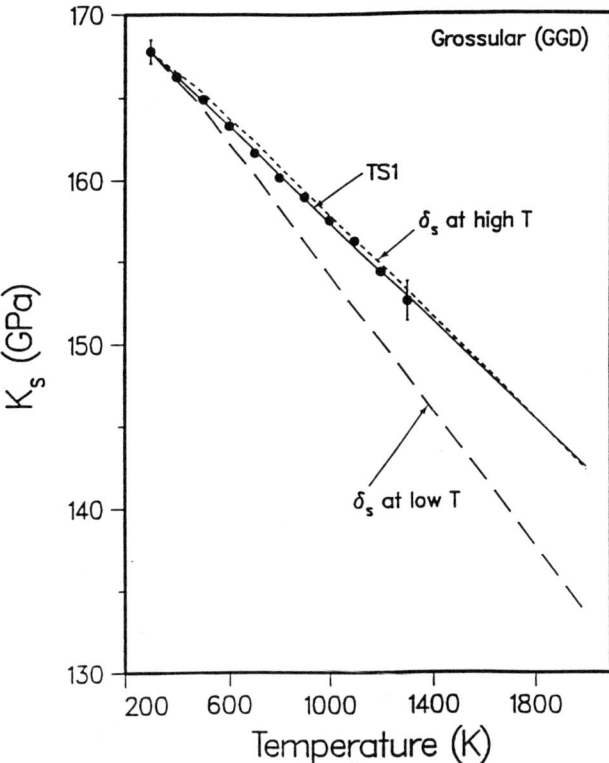

Fig. 7. Observed and extrapolated values of K_S vs. T for grossular garnet. Experimental points shown with symbols [36]. Extrapolations: TS1 refers to first order Taylor expansion using lower temperature data; long dashed line using Equation (62) with low temperature value of δ_S; short dashed line using Equation (62) with high temperature values of δ_S.

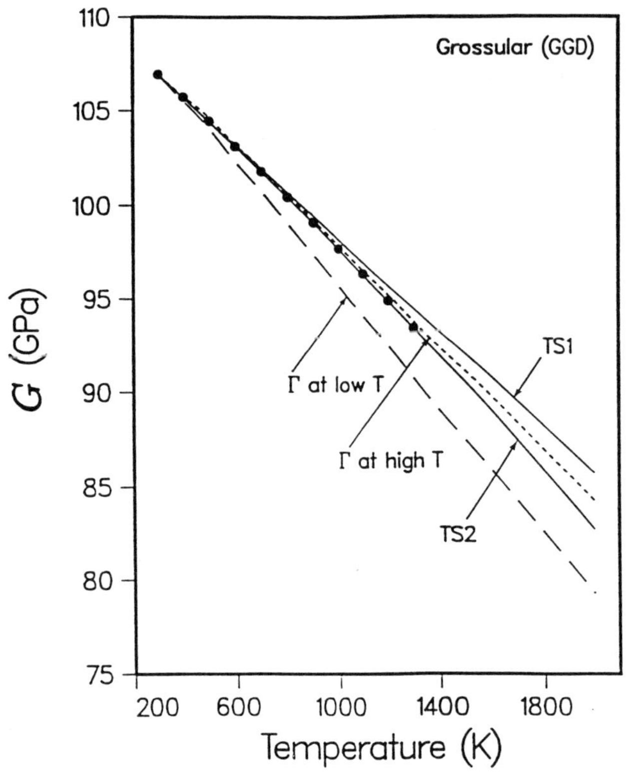

Fig. 8. Observed and extrapolated values of G (GPa) versus T for grossular garnet. Experimental points shown with symbols [36]. Extrapolations: TS1 and TS2 refer to first and second order Taylor expansions using low T data; long dashed lines using Equation (63) with low T values of Γ; short dashed lines using Equation (63) with high temperature value of Γ.

7g. The Importance of Using High T Values of δ_S and Γ

Isaak et al. [36] reported K_S and G for grossular garnet up to 1350 K, measuring δ_S and Γ over a wide T range. Figures 7 and 8 show the extrapolations presented by them up to 2000 K, with 4 methods. They tried both a first order polynomial and a second order polynomial. The fits were marginally successful, as shown in Figures 7 and 8. They also found Equations (62) and (63) to be a much better representation of $G(T)$ provided that the exponents, δ_S and Γ, are the average high temperature values, and not the low temperature values of Γ ($T < \Theta$). Therefore we recommend Equation (62) for extrapolations of K_S and Equation (63) for extrapolations of G, provided that values of δ_S and Γ measured above Θ are used in the equations. If enthalpy data are available, we recommend (67) for extrapolations of K_S.

Acknowledgements. The authors thank former members of the mineral physics laboratory who contributed to the data bank published here, including M. Kumazawa, I. Ohno, Y. Sumino, I. Suzuki, S. Yamamoto, T. Goto, H. Oda, S. Ota, and H. Cynn. We acknowledge with thanks the works of E. K. Graham, N. Soga, R. Liebermann, H. Spetzler, D. L. Weidner, J. Bass, I. Jackson, and M. Manghnani that affected our research programs. Sung Kim of Azusa Pacific University provided much help in the statistical analysis required to arrive at values for the various dimensionless parameters. The salary of one of us was provided by grants from the LLNL-IGPP branch under the auspices of the Department of Energy under contract W-7405-ENG 48. Support for Mr. Kim was provided by the Office of Naval Research through Grant no. N00014-93-10544. IGPP contribution no. 3849.

REFERENCES

1. Anderson, D. L., A seismic equation of state. II. Shear properties and thermodynamics of the lower mantle, *Phys. Earth Planet. Inter.*, *45*, 307–327, 1987.

2. Anderson, O. L., Determination and some uses of isotropic elastic constants of polycrystalline aggregates using single-crystal data, in *Physical Acoustics*, vol. 3, edited by W. P. Mason, pp. 43–95, Academic Press, New York, 1965.

3. Anderson, O. L., A proposed law of corresponding state for oxide compounds, *J. Geophys. Res.*, *71*, 4963–4971, 1966.

4. Anderson, O. L., Evidence supporting the approximation $\gamma\rho$ for the Grüneisen constant of the earth's lower mantle, *J. Geophys. Res.*, *84*, 3537–3542, 1979.

5. Anderson, O. L., The relationship between adiabatic bulk modulus and enthalpy for mantle minerals, *Phys. Chem. Miner.*, *16*, 559–562, 1989.

6. Anderson, O. L., Rectangular parallelepiped resonance–a technique of resonant ultrasound and its applications to the determination of elasticity at high pressures, *J. Acoust. Soc.*, *91*, 2245–2253, 1992.

7. Anderson, O. L., and T. Goto, Measurement of elastic constants of mantle-related minerals at temperatures up to 1800 K, *Phys. Earth Planet. Inter.*, *55*, 241–253, 1989.

8. Anderson, O. L., and D. G. Isaak, The dependence of the Anderson-Grüneisen parameter δ_T upon compression at extreme conditions, *J. Phys. Chem. Solids*, *54*, 221–227, 1993.

9. Anderson, O. L., and J. F. Nafe, The bulk modulus-volume relationship for oxide compounds and related geophysical problems, *J. Geophys. Res.*, *70*, 3951–3963, 1965.

10. Anderson, O. L., and S. Yamamoto, The interrelationship of thermodynamic properties obtained by the piston-cylinder high pressure experiments and RPR high temperature experiments for NaCl, in *High Pressure Research in Mineral Physics, Geophys. Monogr. Ser.*, vol. 39, edited by M. H. Manghnani and Y. Syono, pp. 289–298, AGU, Washington, D.C., 1987, 486 pp.

11. Anderson, O. L., E. Schreiber, R. Liebermann, and N. Soga, Some elastic constant data on minerals relevant to geophysics, *Rev. Geophys.*, *6*, 491–524, 1968.

12. Anderson, O. L., D. G. Isaak, and S. Yamamoto, Anharmonicity and the equation of state for gold, *J. Appl. Phys.*, *65*, 1534–1543, 1989.

13. Anderson, O. L., D. Isaak, and H. Oda, High temperature elastic constant data on minerals relevant to geophysics, *Rev. Geophys.*, *30*, 57–90, 1992.

14. Anderson, O. L., H. Oda, A. Chopelas, and D. Isaak, A thermodynamic theory of the Grüneisen ratio at extreme conditions: MgO as an example, *Phys. Chem. Minerals*, *19*, 369–380, 1993.

15. Barin, I., and O. Knacke, *Thermo-chemical Properties of Inorganic Substances*, 921 pp., Springer-Verlag, New York, 1973.

16. Barron, T. H. K., The Grüneisen parameter for the equation of state of solids, *Ann. Phys.*, *1*, 77–89, 1957.

17. Barron, T. H. K., A note on the Anderson-Grüneisen functions, *J. Phys. C.*, *12*, L155–L159, 1979.

18. Bevington, P. R., and D. K. Robinson, *Data Reduction and Error Analysis for the Physical Sciences*, 2nd Edition, 328 pp., McGraw-Hill, New York, 1992.

19. Cynn, H., Effects of cation disordering in $MgAl_2O_4$ spinel on the rectangular parallelepiped resonance and Raman measurements of vibrational spectra, Ph.D. thesis, 169 pp., UCLA, Los Angeles, November 1992.

20. Duffy, T. S., and T. J. Ahrens, Lateral variations in lower mantle seismic velocity, in *High Pressure Research: Application to Earth and Planetary Sciences*, edited by Y. Syono and M. H. Manghnani, Terra, Washington, D.C., in press, 1993.

21. Duffy, T. S., and D. L. Anderson, Sound velocities in mantle minerals and the mineralogy of the upper mantle, *J. Geophys. Res.*, *94*, 1895–1912, 1989.

22. Enck, F. D., and J. D. Dommel, Behavior of the thermal expansion of NaCl at elevated temperature. Part 1., *J. Appl. Phys.*, *36*, 839–844, 1965.

23. Enck, F. D., D. G. Engle, and K. I. Marks, Thermal expansion of KCl at elevated temperatures, *J. Appl. Phys.*, *36*, 2070–2072, 1962.

24. Furukawa, G. T., T. B. Douglas, R. E. McCoskey, and D. C. Ginning, Thermal properties of aluminum oxide from 0° to 1200 K, *J. Res. Natl. Bur. Stand.*, *57*, 121–131, 1968.

25. Garvin, D., V. B. Parker, and H. J. White, Jr. (Eds.), *Codata Thermodynamic Tables: Selections for Some Compounds of Calcium and Related Mixtures: A Prototype Set of Tables*, 356 pp., Hemisphere Publishing Corporation, Washington, D.C., 1987.

26. Goto, T., S. Yamamoto, I. Ohno, and O. L. Anderson, Elastic constants of corundum up to 1825 K, *J. Geophys. Res.*, *94*, 7588–7602, 1989.

27. Graham, E. K., Jr., and G. R. Barsch, Elastic constants of single-crystal forsterite as a function of temperature and pressure, *J. Geophys. Res.*, *74*, 5949–5960, 1969.

28. Graham, E. K., J. A. Schwab, S. M. Sopkin, and H. Takei, The pressure and temperature dependence of the elastic properties of single-crystal fayalite Fe_2SiO_4, *Phys. Chem. Miner.*, *16*, 186–198, 1988.

29. Hashin, Z., and S. Shtrikman, On some variational principles in anisotropic and nonhomogeneous elasticity, *J. Mech. Phys. Solids*, *10*, 335–342, 1962a.

30. Hashin, Z., and S. Shtrikman, A variational approach to the theory of the elastic behavior of polycrystals, *J. Mech. Phys. Solids*, *10*, 343–352, 1962b.

31. Hunter, L., and S. Siegel, The variation with temperature of the principal elastic moduli of NaCl near the melting point, *Phys. Rev.*, *61*, 1942.

32. Isaak, D. G, High temperature elasticity of iron-bearing olivines, *J. Geophys. Res.*, *97*, 1871–1885, 1992.

33. Isaak, D. G., and O. L. Anderson, Elasticity of single-crystal fayalite (abstract), *Eos Trans. AGU*, *70*, 1418, 1989.

34. Isaak, D. G., O. L. Anderson, and T. Goto, Measured elastic modulus of single-crystal MgO up to

1800 K, *Phys. Chem. Miner.*, *16*, 703–704, 1989*a*.

35. Isaak, D. G., O. L. Anderson, and T. Goto, Elasticity of single-crystal forsterite measured to 1700 K, *J. Geophys. Res.*, *94*, 5895–5906, 1989*b*.

36. Isaak, D. G., O. L. Anderson, and H. Oda, Thermal expansion and high temperature elasticity of calcium-rich garnets, *Phys. Chem. Miner.*, *19*, 106–120, 1992*a*.

37. Isaak, D. G., O. L. Anderson, and R. Cohen, The relationship between shear and compressional velocities at high pressure: Reconciliation of seismic tomography and mineral physics, *Geophys. Res. Lett.*, *19*, 741–744, 1992*b*.

38. Kajiyoshi, K., High temperature equation of state for mantle minerals and their anharmonic properties, M. S. thesis, 30 pp., Okayama Univ., Okayama, Jpn., 1986.

39. Krupka, K. M., R. A. Robie, and B. S. Hemingway, High-temperature heat capacities of corundum, periclase, anorthite, $CaAl_2Si_2O_8$ glass, muscovite, pyrophyllite, $KAlSi_3O_8$ glass, grossular, and $NaAlSi_3O_8$ glass, *Am. Mineral.*, *64*, 86–101, 1979.

40. Kumazawa, M., and O. L. Anderson, Elastic moduli, pressure derivatives, and temperature derivatives of single-crystal olivine and single-crystal forsterite, *J. Geophys. Res.*, *74*, 5961–5972, 1969.

41. Landau, L. D., and E. M. Lifshitz, *Statistical Physics*, 483 pp., (trans. from Russian), Addison-Wesley Publishing Co., Inc., Reading, MA, 1958.

42. Morse, P. A., *Thermal Physics*, 410 pp., W. A. Benjamin, Inc., New York, 1964.

43. Oda, H., O. L. Anderson, and D. G. Isaak, Measurement of elastic properties of single crystal CaO up to 1200 K, *Phys. Chem. Miner.*, *19*, 96–105, 1992.

44. Ohno, I., Free vibration of rectangular parallelepiped crystal and its application to determination of elastic constants of orthorhombic crystals, *J. Phys. Earth*, *24*, 355–379, 1976.

45. Okajima, S., Study of thermal properties of rock-forming minerals, M.S. thesis, 53 pp., Okayama Univ., Okayama, Jpn., 1978.

46. Okajima, S., I. Suzuki, K. Seya, and Y. Sumino, Thermal expansion of single-crystal tephroite, *Phys. Chem. Minerals*, *3*, 111–115, 1978.

47. Pacalo, R. E., and E. K. Graham, Pressure and temperature dependence of the elastic properties of synthetic MnO, *Phys. Chem. Miner.*, *18*, 69–80, 1991.

48. Robie, R. A., B. S. Hemingway, and J. R. Fisher, Thermodynamic properties of minerals and related substances at 298.15 K and 1 bar (10^5 Pascals) pressure and at higher temperatures, *Geol. Surv. Bull.*, *1452*, 1–456, 1978.

49. Shankland, T. J., and J. D. Bass, *Elastic Properties and Equations of State*, 568 pp., Mineral Physics Reprint, AGU, Washington, DC, 1988.

50. Simmons, G., and H. Wang, *Single Crystal Elastic Constants and Calculated Aggregate Properties*, 310 pp., MIT Press, Cambridge, MA, 1971.

51. Skinner, B. J., Physical properties of end-members of the garnet group, *Am. Mineral*, *41*, 428–436, 1956.

52. Soga, N., and O. L. Anderson, High temperature elastic properties of polycrystalline MgO and Al_2O_3, *J. Am. Ceram. Soc.*, *49*, 355–359, 1966.

53. Soga, N., and O. L. Anderson, High temperature elasticity and expansivity of forsterite and steatite, *Am. Ceramic Soc.*, *50*, 239–242, 1967.

54. Soga, N., E. Schreiber, and O. L.

Anderson, Estimation of bulk modulus and sound velocities at very high temperatures, *J. Geophys. Res.*, *71*, 5315–5320, 1966.

55. Spetzler, H., Equation of state of polycrystalline and single-crystal MgO to 8 kilobars and 800° K, *J. Geophys. Res.*, *75*, 2073–2087, 1970.

56. Spetzler, H., C. G. Sammis, and R. J. O'Connell, Equation of state of NaCl: Ultrasonic measurements to 8 kbar and 800°C and static lattice theory, *J. Phys. Chem. Solids*, *33*, 1727–1750, 1972.

57. Stull, D. R., and H. Prophet (Eds.), *JANAF Thermochemical Tables*, (2nd ed.)., 1141 pp., U. S. Department of Commerce, National Bureau of Standards, Washington, DC, 1971.

58. Sumino, Y., The elastic constants of Mn_2SiO_4, Fe_2SiO_4, and CO_2SiO_4 and the elastic properties of olivine group minerals at high temperatures, *J. Phys. Earth*, *27*, 209–238, 1979.

59. Sumino, Y., and O. L. Anderson, Elastic constants of minerals, in *Handbook of Physical Properties of Rocks*, VIII, edited by R. S. Carmichael, pp. 39–137, CRC, Boca Raton, Fl., 1984.

60. Sumino, Y., I. Ohno, T. Goto, M. Kumazawa, Measurement of elastic constants and internal friction on single-crystal MgO by rectangular parallelepiped resonance, *J. Phys. Earth*, *24*, 263–273, 1976.

61. Sumino, Y., O. Nishizawa, T. Goto, I. Ohno, and M. Ozima, Temperature variation of elastic constants of single-crystal forsterite, between −190°C and 900°C, *J. Phys. Earth*, *25*, 377–392, 1977.

62. Sumino, Y., M. Kumazawa, O. Nishizawa, and W. Pluschkell, The elastic constants of single crystal Fe_{1-x}, MnO and CaO and the elasticity of stoichiometric

magnesiowüstite, *J. Phys. Earth*, *28*, 475–495, 1980.

63. Sumino, Y., O. L. Anderson, and Y. Suzuki, Temperature coefficients of single crystal MgO between 80 and 1300 K, *Phys. Chem. Miner.*, *9*, 38–47, 1983.

64. Suzuki, I., Thermal expansion of periclase and olivine, and their anharmonic properties, *J. Phys. Earth*, *23*, 145–159, 1975.

65. Suzuki, I., and O. L. Anderson, Elasticity and thermal expansion of a natural garnet up to 1000 K, *J. Phys. Earth*, *31*, 125–138, 1983.

66. Suzuki, I., S. Okajima, and K. Seya, Thermal expansion of single-crystal manganosite, *J. Phys. Earth*, *27*, 63–69, 1979.

67. Suzuki, I., K. Seya, H. Takai, and Y. Sumino, Thermal expansion of fayalite, *Phys. Chem. Miner.*, *1*, 60–63, 1981.

68. Suzuki, I., O. L. Anderson, and Y. Sumino, Elastic properties of a single-crystal forsterite Mg_2SiO_4 up to 1,200 K, *Phys. Chem. Miner.*, *10*, 38–46, 1983.

69. Touloukian, Y. S., R. K. Kirby, R. E. Taylor, and T. Y. R. Lee, *Thermal Expansion, Nonmetallic Solids: Thermophysical Properties of Matter*, vol. 13, 1658 pp.,

Plenum, New York-Washington, 1977.

70. Wachtman, J. B., Jr., T. G. Scuderi, and G. W. Cleek, Linear thermal expansion of aluminum oxide and thorium oxide from 100° to 1,100°K, *J. Am. Ceram. Soc.*, *45*, 319–323, 1962.

71. Wang, H., J. D. Bass, and G. R. Rossman, Elastic properties of Fe-bearing pyroxenes and olivines (abstract), *Eos Trans. AGU*, *70*, 474, 1989.

72. Watanabe, H., Thermochemical properties of synthetic high-pressure compounds relevant to earth's mantle, in *High Pressure Research in Geophysics*, vol. 12, *Advances in Earth and Planetary Sciences*, edited by S. Akimoto and M. H. Manghnani, pp. 441–464, Center for Academic Publications, Tokyo, 1982.

73. Watt, J. P., Hashin-Shtrikman bounds on the effective elastic moduli of polycrystals with orthorhombic symmetry, *J. Appl. Phys.*, *50*, 6290–6295, 1979.

74. Watt, J. P., T. J. Shankland, and N. Mao, Uniformity of mantle composition, *Geology*, *3*, 91–99, 1975.

75. White, G. K., and R. B. Roberts,

Thermal expansion of reference materials: Tungsten and α-Al_2O_3, *High Temp. High Pressures*, *15*, 321–328, 1983.

76. Yamamoto, S., and O. L. Anderson, Elasticity and anharmonicity of potassium chloride at high temperature, *Phys. Chem. Minerals*, *14*, 332–340, 1987.

77. Yamamoto, S., I. Ohno, and O. L. Anderson, High temperature elasticity of sodium chloride, *Phys. Chem. Solids*, *48*, 143–151, 1987.

78. Yeganeh-Haeri, A., D. J. Weidner, and E. Ito, Single crystal elastic moduli of magnesium metasilicate perovskite, in *Perovskite, Geophys. Monogr. Ser.*, vol. 45, edited by A. Navrotsky and D. J. Weidner, pp. 13–26, AGU, Washington, D. C., 1989.

79. Zharkov, V. N., and V. A. Kalinin, Equations of state for solids at high pressures and temperatures, 257 pp., translated from Russian by A. Tybulewicz, Consultants Bureau, New York, 1971.

80. Zouboulis, E. S., and M. Grimsditch, Refractive index and elastic properties of MgO up to 1900 K, *J. Geophys. Res.*, *96*, 4167–4170, 1991.

Static Compression Measurements of Equations of State

Elise Knittle

1. INTRODUCTION

Generation of high pressures and temperatures in the laboratory is essential in exploring the behavior of solids and liquids at the pressure and temperature conditions of the Earth's interior. Indeed, the primary insights into magma genesis, phase transitions and the equations of state of deep Earth constituents, as well as the geochemical behavior of materials at depth in the planet have been derived from experiments involving static compression. In this chapter, the emphasis is on static-compression measurements of the equations of state of minerals, elements and related materials. To date, the vast majority of static compression experiments have focused on measuring isothermal equations of state (P-V measurements at constant temperature), in order to determine the isothermal bulk modulus ($K_{0T} = -1/V(dP/dV)_T$) and its pressure derivative ($dK_{0T}/dP = K_{0T}'$) as well as characterizing the pressure conditions of structural phase transitions. With the bulk modulus and its pressure derivative, it is possible to extrapolate the density of a material to any pressure condition. Here, I tabulate bulk moduli data for a variety of minerals and related materials.

2. EXPERIMENTAL TECHNIQUES

There are a number of different experimental techniques and apparatuses for achieving high pressures statically, or

pressures maintainable for long periods of time. The bulk moduli data have been collected using primarily either x-ray diffraction or piston displacement in conjunction with a high-pressure apparatus to measure the change in volume of a material as a function of pressure. In general, high pressure apparatuses are optimized for either: 1) allowing optical access to the sample while held at high pressures (diamond-anvil cells); or 2) maintaining large sample volumes (large-volume presses).

2.1. Diamond-Anvil Cell

The most widely used high-pressure apparatus for measuring static equations of state is the diamond-anvil cell coupled with x-ray diffraction. There are several excellent recent reviews on the use of the diamond cell [88, 97, 215], and thus no details of its operation are presented here. For equation-of-state measurements, two basic diamond cell designs are used (see Figure 1). The Merrill-Bassett-type diamond cells are commonly used for single-crystal x-ray diffraction measurements, as their small size makes them convenient for mounting on a goniometer to obtain single-crystal x-ray patterns. In general, these diamond cells are used for pressures below about 10 GPa. However, for single-crystal measurements, it is necessary to maintain the sample in a hydrostatic environment. Since liquid pressure media (such as 4:1 methanol:ethanol mixtures) are only hydrostatic to ~12 GPa, the pressure limitation of the diamond cell is not a limiting factor for the experiment.

The second type of diamond cell, a Mao-Bell or "megabar"-type diamond cell, is optimal for obtaining x-ray diffraction data to pressures up to 300 GPa. Usually, in this type of diamond cell, the sample is a powder and the x-ray diffraction data collected is directly analogous to a Debye-Scherrer powder pattern. A typical sample configuration in the diamond-cell is shown in Figure 2.

E. Knittle, University of California, Santa Cruz, Institute of Tectonics, Mineral Physics Laboratory, Santa Cruz, CA 95064

Mineral Physics and Crystallography
A Handbook of Physical Constants
AGU Reference Shelf 2

Copyright 1995 by the American Geophysical Union. 98

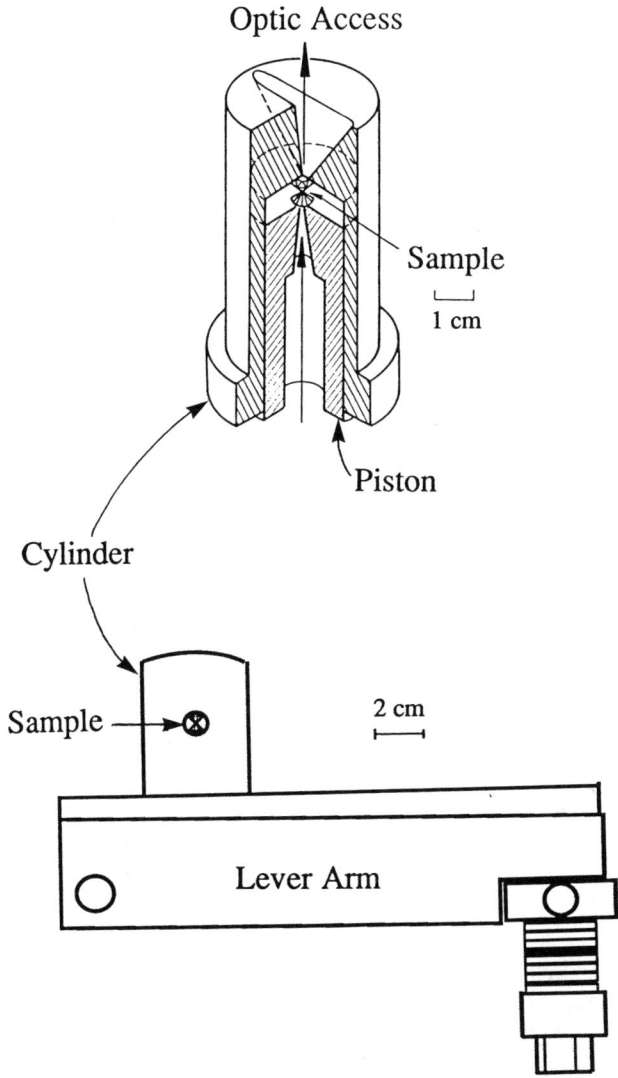

Fig. 1*a*. The Mao-Bell-type diamond cell capable of generating pressures up to 550 GPa. The diamonds are glued onto half-cylinders of tungsten carbide which are mounted in a piston and cylinder made of hardened tool steel (shown here in a cutaway view at the top). The diamonds can be translationally aligned by means of set screws and tilted with respect to each other by rocking the WC half-cylinders. The piston and cylinder are placed in a lever arm assembly (bottom) designed to advance the piston within the cylinder, apply force uniformly to the sample held between the diamonds, and exert a minimum amount of torque on the piston.

A schematic diagram of the geometry of an x-ray diffraction experiment in shown in Figure 3. Usually, MoK_α x-rays are passed through the diamonds and sample along the axis of force of the diamond cell. X-ray diffraction from a crystal occurs when the Bragg equation is satisfied:

$$\lambda = 2d\sin\theta \qquad (1)$$

where λ is the wavelength of the incoming x-ray, 2θ is the angle between the diffracted x-ray beam and the transmitted beam, and d is the d-spacing or the distance between the diffracting set of atomic planes in the crystal. The diffracted x-rays, either Laue spots for single crystal samples or diffracted cones of radiation for powdered samples, are detected by a solid state detector or on film. The set of d-spacings for a crystal is used to determine the lattice parameters and thus the volume of the crystallographic unit cell as a function of pressure. The relationship between d-spacings and lattice parameters for the various lattice symmetries are given in most standard texts on x-ray diffraction [e.g., 36].

Pressure in the diamond cell is usually measured using the ruby fluorescence technique [20, 128, 129]. Here, the R_1 and R_2 fluorescence bands of ruby, at zero-pressure wavelengths of 694.24 nm and 692.92 nm, are excited with blue or green laser light (typically He:Cd or Ar-ion lasers are used). These fluorescence bands have a strong wavelength shift as a function of pressure. This shift has been calibrated using simultaneous x-ray diffraction measurements on metals whose equations of state (pressure-volume relation) are independently known from shock-wave measurements (see Chapter on Shock-wave Measurements). Therefore, the accuracy of the pressure measured in the diamond cell is ultimately determined by the accuracy of measurements of shock-wave equations of state. In addition to average pressure measurements, ruby can be finely dispersed in the sample (fluorescence measurements can be obtained for ruby grains less than 1 μm in diameter), thus enabling pressure gradients to be accurately determined within diamond cell samples.

Because diamonds are transparent and allow optical access to the sample, other methods for measuring equations of state are possible. For example, the changes in the volume of the sample chamber as a function of pressure can be directly measured [cf., 139], or changes in the sample dimensions with pressure may be directly measured [cf., 143]. Additionally, accurate adiabatic moduli and their derivatives may be obtained using Brillouin spectroscopy in conjunction with the diamond cell (see the chapter on Elasticity).

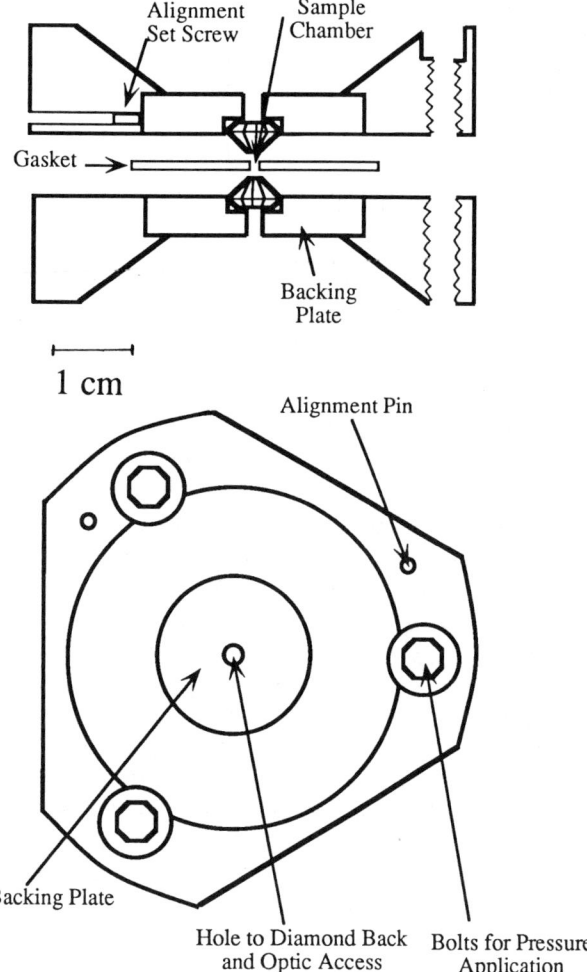

Fig. 1*b*. Merrill-Bassett-type diamond-anvil cell shown in schematic from both side (top) and top views (bottom). This device can be used to generate pressures of ~ 30 GPa. The single-crystal diamonds are glued on flat, circular backing plates typically made of hardened steel, tungsten carbide or copper-beryllium alloy. Holes or slits in the backing plates provide optical access to the sample, which is contained between the diamonds in a metal gasket. The backing plates are held in place in recesses in larger steel or inconel plates by means of set screws, and also allow the diamonds to be translationally aligned relative to one another. Alignment pins ensure that the the top and bottom halves of the cell are correctly oriented. Pressure is applied by sequentially turning the three bolts which compress the top and bottom plates together.

2.2. Large-Volume Presses

Equation of state measurements have also been carried out using a variety of large-volume, high-pressure apparatuses: piston-cylinder devices, Bridgman-anvil presses, Drickamer presses, cubic and tetrahedral-anvil presses and multianvil presses (Figure 4). For recent reviews of the design and use of large-volume presses see Graham [59] and Holloway and Wood [85].

One of the most widely used high-pressure, large-volume devices for equation of state measurements is the piston-cylinder (Figure 4a), where the pressure is applied uniaxially by a piston on a sample embedded in a pressure medium and contained by the cylinder. As the sample is compressed its volume decreases, and by careful measurement of the resulting piston displacement as a function of pressure, the P-V equation of state can be determined. The routine pressure limit of the piston-cylinder at room temperature is between about 3 and 5 GPa.

Other types of high-pressure, large-volume devices utilizing opposed anvils for equation of state measurements are Bridgman anvils and the Drickamer press. In these devices, two tungsten carbide anvils are driven together to uniaxially compress a sample contained within a large retaining ring which acts as a gasket. If all or a portion of the retaining ring is made of a material transparent to x-rays (such as B or Be), diffraction measurements can be made on a sample held at high pressures. The pressure limit of the Bridgman-anvil press is about 10-12 GPa at room temperature, and the limit of the Drickamer press is about 30-35 GPa.

A second family of large-volume apparatuses is the multi-anvil press. Rather than uniaxially compressing the sample, these devices compress simultaneously from several directions. Such instruments are more difficult to use than a piston-cylinder, as the anvils must be carefully aligned and synchronized to compress the sample uniformly and thus prevent failure of the anvils; however, much higher pressures can be attained in multi-anvil devices. A number of designs exist for multi-anvil presses of which the ones most commonly used for equation of state measurements are the tetrahedral-anvil and cubic-anvil presses (Figure 4b), and multiple-stage systems.

In tetrahedral and cubic-anvil presses, as their names imply, the samples are typically contained within either a tetrahedron or a cube composed of oxides, with the sample occupying a cylindrical cavity inside the cube or tetrahedron. The sample assembly, which varies in complexity depending on the experiment, is compressed by four hydraulic rams in the case of the tetrahedral press, or six hydraulic rams for the cubic press. As illustrated in Figure 4c for a cubic-anvil press, in-situ x-ray diffraction measurements can be made through these presses if the sample is contained in an x-ray-transparent material. Equation of state measurements have typically been made to pressures of about 12 GPa in the cubic-anvil press. To achieve higher pressures, the sample assembly in a cubic-anvil press can be replaced by a second, pressure-intensifying stage. Such a multi-anvil design can reach typical pressures of about 25 GPa.

The pressure inside a large-volume device is determined either by including an internal stress monitor within the sample, or by calibration of the external load on the sample. In the latter case, which represents the most common method for estimating the pressure in all large-volume devices, the oil pressure of the press is calibrated against a series of materials or fixed points, with well-characterized phase transformations, such as those occuring in Bi (the I-II transition at 2.5 GPa and V-VI transition at 7.7 GPa). Specific disadvantages of x-ray diffraction measurements in large volume presses include: 1) indirect pressure calibration; 2) an inability to accurately characterize pressure gradients; and 3) the small volume of the sample relative to the containing material, which necessitates both careful experimental design and a high intensity x-ray source. Furthermore, the diffraction angles available through the apparatus may be limited, producing few redundancies in the measurement of lattice parameters.

3.   ANALYSIS OF THE DATA

To determine the values of K_{0T} and K_{0T}' from pressure-volume data, an equation of state formalism must be used. Here, I summarize the most common methods for data reduction. A linear fit to pressure-volume data is often used for static compression data limited to low pressures (small compressions). The bulk modulus is calculated from the thermodynamic definition: $K_{0T} = -1/V(dP/dV)_T$, where K_{0T} is the isothermal bulk modulus, V is the volume and P is the pressure. This analysis implies that $K_{0T}' = 0$, and thus is at best approximate, and is most appropriate for rigid materials over narrow pressure ranges (preferably less than 5 GPa). In general, because K_{0T}' is usually positive, this analysis method will tend to overestimate K_{0T}.

An equation of state formalism used extensively prior to 1980 and sporadically since is the Murnaghan equation [153], in which:

$$P = K_{0T}/K_{0T}'\{(V_0/V)^{K_{0T}'} -1\} \qquad (2)$$

The Murnaghan equation is derived from finite strain theory with the assumption that the bulk modulus is a linear function of pressure. At modest compressions, this

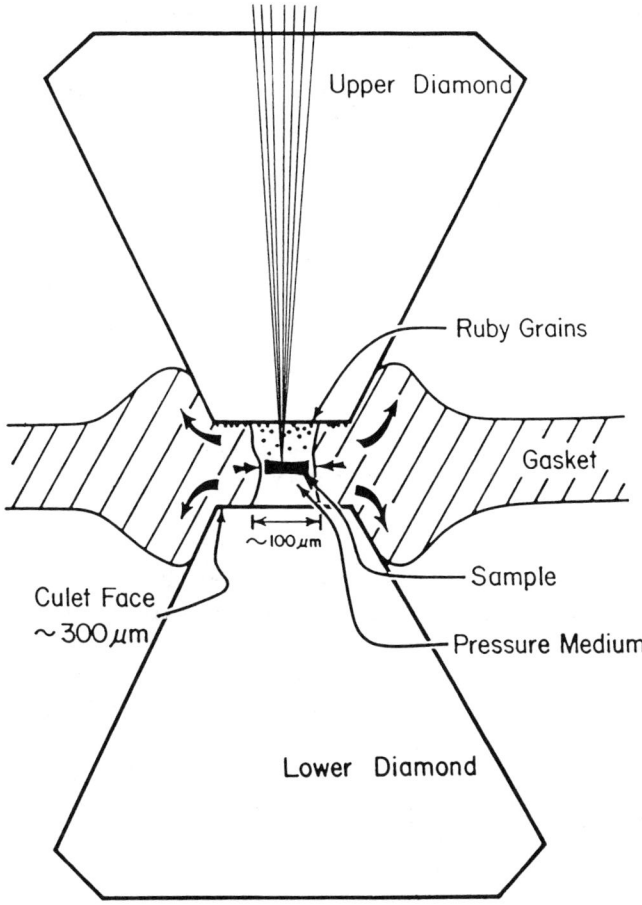

Fig. 2. Schematic view of the sample configuration in the diamond cell. A hole is drilled in a metal gasket (here a typical diameter of 0.10 mm is shown), and the gasket is centered between the diamond anvils (shown with a typical culet diameter of 0.30 mm). The sample is placed in the hole in the gasket along with a pressure medium (generally either an alcohol mixture, or a noble gas such as argon) and small ruby chips for pressure calibration. The transparent diamonds allow for a wide range of different in situ experiments to be performed on the samples. For equation of state measurements, x-ray diffraction patterns can be obtained at high pressure along the axis of force of the diamond cell for either single-crystals or powders. (From [215]).

assumption is sufficiently accurate that the Murnaghan equation accurately fits static compression data.

One of the most widely used equations of state in reducing static compression data is the Eulerian finite-strain formalism proposed by Birch, and referred to as the Birch-Murnaghan equation [25]. This equation has been empirically shown to describe the behavior of a wide variety of materials over a large range of compressions. Pressure is expressed as a Taylor series in strain:

$$P = 3f(1+2f)^{5/2}K_{0T}(1 + x_1 f + x_2 f^2 + ...) \qquad (3)$$

where f is the Eulerian strain variable defined as

$$f = 1/2[(V/V_0)^{2/3} -1]$$

and the coefficients in (3) are:

$$x_1 = 3/2(K_{0T}' - 4)$$

$$x_2 = 3/2 [K_{0T}K_{0T}'' + K_{0T}'(K_{0T}'-7) + 143/9].$$

Note that $K_{0T}' = 4$ reduces equation (3) to a second order equation in strain. Indeed, K_{0T}' is close to 4 for many materials, and in the reduction of pressure-volume data, is often assumed to equal 4. In addition, minerals are seldom compressible enough for K_{0T}'' to be significant; therefore, the x_2 term is usually dropped from the equation.

Recently, another equation of state formalism has been proposed by Vinet et al. [205] and is known as the Universal equation of state. Here:

$$P = 3 K_{0T} (V_0/V)^{2/3}[1-(V/V_0)^{1/3} \\ \exp\{3/2(K_{0T}'-1) [1-(V/V_0)^{1/3}]\}. \qquad (4)$$

This equation is successful in describing the equation of state of a material at extremely high compressions and is being used with increasing frequency in static compression studies.

4. EXPLANATION OF THE TABLE

The Table lists the isothermal bulk moduli (K_{0T}) and pressure derivatives ($dK_{0T}/dP=K_{0T}'$) for minerals, mineral analogues and chemical elements studied using static compression techniques (i.e. x-ray diffraction, piston displacement measurements or direct volume measurements at high pressures). All the data reported is from the literature from 1966 to the present: static compression data from before 1966 can be found in the previous Handbook of Physical Constants. Not every entry in the Table is a

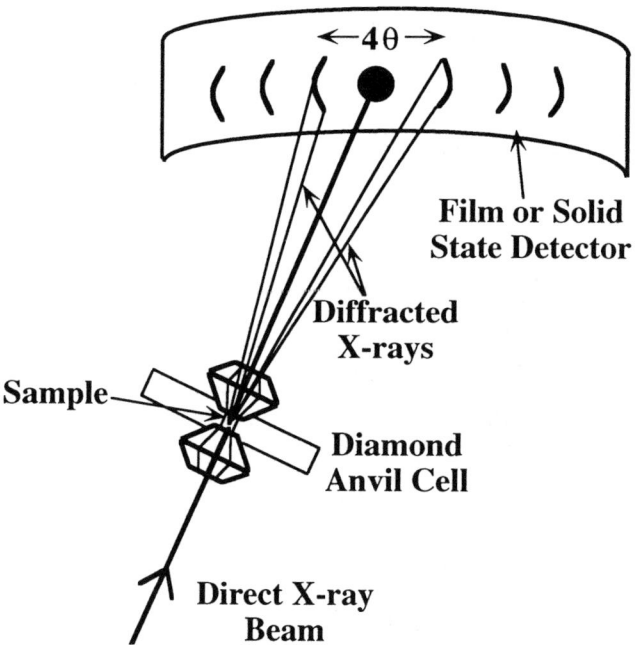

Fig. 3. Schematic of an x-ray diffraction experiment through the diamond cell. $Mo_{K\alpha}$ x-rays are passed through the sample and the diffracted x-rays are recorded either on film (shown here) or by a solid-state detector. Here the schematic x-ray pattern is that of a powder diffraction experiment, where the film intersects diffracted cones of radiation. The maximum distance between the corresponding x-ray line on either side of the transmitted x-ray beam is 4θ, which is used in the Bragg equation to determine the set of d-spacings of the crystal at each pressure.

mineral or native element. Oxides, sulfides, halides and hydrides were included which are useful as mineral analogues, either due to structural similarities to minerals, or to provide the systematics of the behavior of a given structural type. Specifically not included are data for non-mineral III-V compounds and other non-naturally occurring semiconductors. All the K_{0T} and K_{0T}' data listed are for the structure of the common room-temperature, atmospheric-pressure phase, except where noted. In particular, the structures of the high pressure phases are explicitly given, as are the specific structure for solids with more than one polymorph at ambient conditions. It should be noted that entries with no errors simply reflect that error bars were not reported in the original references.

The silicate minerals are categorized following Deer, Howie and Zussman's *An Introduction to the Rock-*

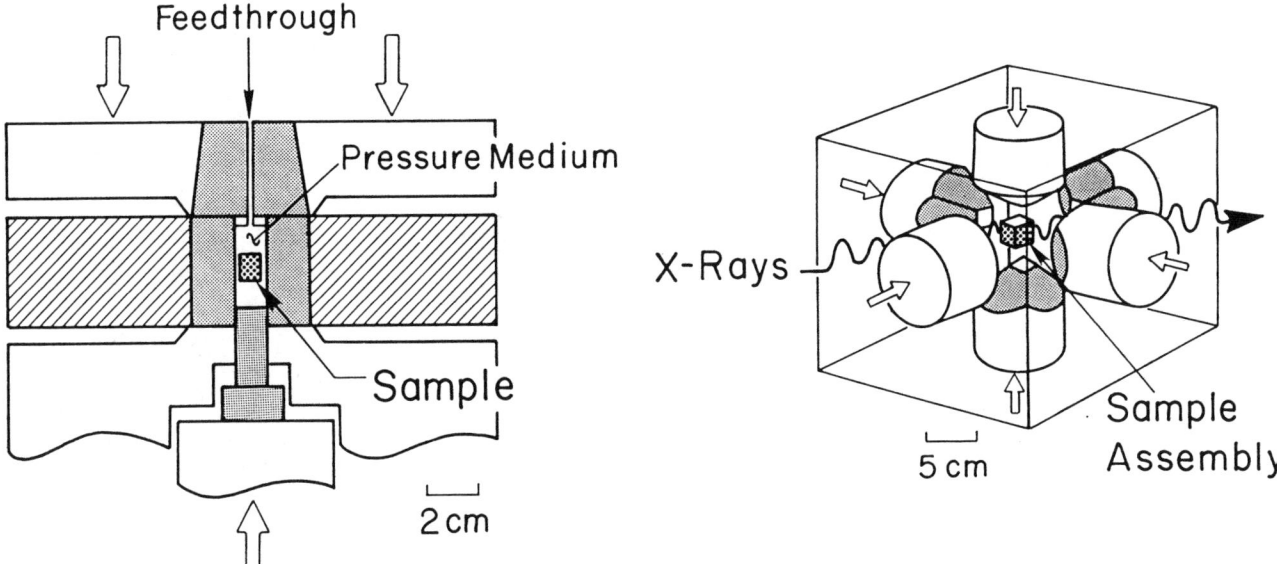

Fig. 4a. A piston-cylinder apparatus shown in cross-section. Here, a uniaxial force is used to advance a piston into a cylinder, compressing a pressure medium and thus applying hydrostatic or quasi-hydrostatic pressure to a sample contained within the medium. The piston and cylinder are usually made of tungsten carbide and the maximum routine pressure limit of this device is 5 GPa at room temperature. The feedthrough into the cylinder gives a way of monitoring the sample environment. For example, electrical leads can be introduced into the sample to measure temperature or as a pressure gauge. Another use of the feedthrough is to provide current to a resistance heater surrounding the sample (not shown). Equation of state measurements can be made in this device by determining the piston displacement as a function of pressure. (From [215]).

Fig. 4b A cubic-anvil press in which six, synchronized rams are used to compress a cubic sample. The rams are usually made of steel with tungsten carbide or sintered diamond tips. The sample geometries can vary from relatively simple, with the sample assembly embedded in a cube of pressure medium which is directly compressed by the rams, to a sample contained in a second, cubic, pressure-intensifying set of anvils (a multi-anvil press). Electrical leads can be brought in between the rams of the press to heat the sample, monitor temperature and pressure, and measure pressure-induced changes in electrical conductivity. In addition, by using a low-atomic weight pressure medium such as B or Be, x-ray diffraction measurements can be obtained on samples held at high pressures. (From [215]).

Forming Minerals [38]. The data for high-pressure phases are given in the same category as for the low-pressure phase. For example, the bulk moduli for perovskite-structured silicates are given with the data for the pyroxenes, which have the same stoichiometry, under the "Chain Silicates" category.

The data for the iron polymorphs and iron alloys which are important as possible core constituents are listed in a separate category ("Iron and Iron Alloys"). Therefore, iron sulfides, $Fe_{1-x}O$, iron hydride, and iron-silicon, iron-nickel and iron-cobalt alloys are found in this portion of the Table, while more oxidized iron minerals such as hematite (Fe_2O_3) and magnetite (Fe_3O_4) are listed with the oxides.

Also, measurements are not given which clearly seemed to be incorrect, based on comparison with other static compression measurements, ultrasonic data or Brillouin spectroscopic results. Finally, I note that a range of materials have been statically compressed with the primary motivation being to characterize the pressure at which phase transitions occur. Such studies often do not report either bulk moduli or the numerical static compression data, and are accordingly not included in the Table (an example of this type of study are several static compression results for Ge).

Table 1. Bulk Moduli from Static Compression Data

Chemical Formula[1]	Density (Mg/m^3)	Isothermal Bulk Modulus (GPa)[2]	dK_{0T}/dP[3]	Technique and Notes[4]	Ref.
Framework Silicates:					
SiO_2 *Quartz*	2.65	37.1 ± 0.2 (fixed from ultrasonic data)	6.2 ± 1	DAC, scXRD, P ≤ 5 GPa, B-M EOS	115
SiO_2 *Quartz*	2.65	36.4 ± 0.5	6.3 ± 0.4	PC, ND, P ≤ 2.5 GPa, M EOS	98
SiO_2 *Quartz*	2.65	38.0	5.4	BA, XRD, P ≤ 12 GPa, O-H method	157
SiO_2 *Coesite*	2.92	96 ± 3	8.4 ± 1.9	DAC, scXRD, P ≤ 5.2 GPa, B-M EOS	113
SiO_2 *Stishovite*	4.29	313 ± 4	1.7 ± 0.6	DAC, scXRD, P ≤ 16 GPa, B-M EOS	176
		306 (fixed from Brillouin data)	2.8 ± 0.2		
SiO_2 *Stishovite*	4.29	300 ± 30	--	TAP, XRD, P ≤ 8.5 GPa, linPVfit	21
SiO_2 *Stishovite*	4.29	288 ± 13	6	BA, XRD, P ≤ 12 GPa, O-H method	158
SiO_2 *Stishovite*	4.29	281	5	CAP, XRD , P ≤ 12 GPa, B-M EOS	179
$NaAlSi_3O_8$ *Low Albite*	2.62	70	--	DAC, scXRD, P ≤ 5 GPa, linPVfit	14
$KAlSi_3O_8$ *High Sanidine*	2.56	67	--	DAC, scXRD, P ≤ 5 GPa, linPVfit	14
$CaAl_2Si_2O_8$ *Anorthite* (low-pressure phase)	2.76	94	--	DAC, scXRD, P ≤ 5 GPa, linPVfit	14
$CaAl_2Si_2O_8$ *Anorthite* (high-pressure phase)	2.76	106	--	DAC, scXRD, P ≤ 5 GPa, linPVfit	14
$BeAlSiO_4OH$ *Euclase*	3.12	159 ± 3	4 (assumed)	DAC, scXRD, P ≤ 5 GPa, B-M EOS	77
$Be_4Si_2O_7(OH)_2$ *Betrandite*	2.61	70 ± 3	5.3 ± 1.5	DAC, scXRD, P ≤ 5 GPa, B-M EOS	63

Table 1. Bulk Moduli from Static Compression Data (continued)

Chemical Formula[1]	Density (Mg/m^3)	Isothermal Bulk Modulus (GPa)[2]	dK$_{0T}$/dP[3]	Technique and Notes[4]	Ref.
12NaAlSiO$_4$ ·27H$_2$O *Zeolite 4A*	2.44	21.7 (in glycerol) 142.8 (in water)	4 (assumed)	DAC, scXRD, P ≤ 4 GPa, B-M EOS	73
NaAlSi$_2$O$_6$·H$_2$O *Analcite*	2.24	40 ± 1	---	DAC, scXRD, P < 3 GPa, linPVfit	69
Na$_4$Al$_3$Si$_3$O$_{12}$Cl *Sodalite*	2.30	52 ± 8	4 (assumed)	DAC, scXRD, P ≤ 5 GPa, B-M EOS	76
0.17(Ca$_4$Al$_6$Si$_6$O$_{24}$ CO$_3$) · 0.83(Na$_4$Al$_3$ Si$_9$O$_{24}$Cl) *Scapolite*	2.54	60	---	DAC, scXRD, P < 3 GPa, linPVfit	33
0.68(Ca$_4$Al$_6$Si$_6$O$_{24}$ CO$_3$) · 0.32(Na$_4$Al$_3$ Si$_9$O$_{24}$Cl) *Scapolite*	2.66	86	---	DAC, scXRD, P ≤ 4 GPa, linPVfit	33
0.88(Ca$_4$Al$_6$Si$_6$O$_{24}$ CO$_3$) · 0.12(Na$_4$Al$_3$ Si$_9$O$_{24}$Cl) *Scapolite*	2.71	90 ± 12	4 (assumed)	DAC, scXRD, P ≤ 5 GPa, B-M EOS	76

Chain silicates:

CaMgSi$_2$O$_6$ *Diopside*	3.22	114 ± 4	4.5 ± 1.8	DAC, scXRD, P ≤ 5.3 GPa, B-M EOS	112
CaMgSi$_2$O$_6$ *Diopside*	3.22	113 ± 2	4.8 ± 0.7	DAC, scXRD, P ≤ 5.3 GPa, B-M EOS	116
CaMgSi$_2$O$_6$ *Diopside*	3.22	122 ± 2	4 (assumed)	DAC, scXRD, P ≤ 6 GPa, B-M EOS	142
CaFeSi$_2$O$_6$ *Hedenbergite*	3.56	119 ± 2	4 (assumed)	DAC, pXRD, P ≤ 10 GPa, B-M EOS	228
Ca(Mg$_{0.4}$Fe$_{0.6}$) Si$_2$O$_6$ *Hedenbergite*	3.42	82.7 ± 1	4 (assumed)	DAC, pXRD, P ≤ 10 GPa, B-M EOS	228
(Ca,Na)(Mg,Fe,Al) Si$_2$O$_6$ *Omphacite* (vacancy-rich)	3.26	129 ± 3	4 (assumed)	DAC, scXRD, P ≤ 6 GPa, B-M EOS	142

Table 1. Bulk Moduli from Static Compression Data (continued)

Chemical Formula[1]	Density (Mg/m^3)	Isothermal Bulk Modulus $(GPa)[2]$	$dK_{0T}/dP[3]$	Technique and Notes[4]	Ref.
(Ca,Na)(Mg,Fe,Al) Si_2O_6 *Omphacite* (vacancy-poor)	3.29	139 ± 4	4 (assumed)	DAC, scXRD, P ≤ 6 GPa, B-M EOS	142
$MgSiO_3$ *Enstatite*	3.22	125	5 (assumed)	BA, XRD, P ≤ 6.2 GPa, O-H method	159
$MgSiO_3$ (high-pressure tetragonal-garnet structure)	3.52	161	4 (assumed)	DAC, pXRD, P ≤ ?, B-M EOS	223
$(Mg_4Si_4O_{12})_{0.6}$ $(Mg_3Al_2Si_3O_{12})_{0.4}$ *Majorite*	3.55	159.8 ± 0.6	4 (assumed)	CAP, XRD, P ≤ 6 GPa, B-M EOS	224
$(Fe_4Si_4O_{12})_{0.2}$ $(Fe_3Al_2Si_3O_{12})_{0.8}$ *Majorite*	4.41	164.6 ± 1.1	4 (assumed)	CAP, XRD, P ≤ 6 GPa, B-M EOS	224
$Mg_{0.79}Fe_{0.21}SiO_3$ *Majorite*	3.74	221 ± 15	4.4 ± 4.8	DAC, XRD, P ≤ 8 GPa, B-M EOS	90
$MgSiO_3$ (high-pressure perovskite structure)	4.10	247	4 (assumed)	DAC, scXRD, P ≤ 10 GPa, B-M EOS	106
$MgSiO_3$ (high-pressure perovskite structure)	4.10	254 ± 4	4 (assumed)	DAC, scXRD, P ≤ 13 GPa, B-M EOS	173
$MgSiO_3$ (high-pressure perovskite structure)	4.10	258 ± 20	4 (assumed)	DAC, pXRD, P ≤ 7 GPa, B-M EOS	220
$Mg_{0.88}Fe_{0.12}SiO_3$ (high-pressure perovskite structure)	4.26	266 ± 6	3.9 ± 0.4	DAC, pXRD, P ≤ 112 GPa, B-M EOS	102
$Mg_{1.0-0.8}Fe_{0-0.2}$ SiO_3 (high-pressure perovskite structure)	4.10 - 4.33	261 ± 4	4 (assumed)	DAC, pXRD, P ≤ 30 GPa, B-M EOS dK_{0T}/dT = -6.3 (±0.5) x 10^{-2} GPa/K	138
$CaSiO_3$ (high-pressure perovskite structure)	4.25	281 ± 4	4 (assumed)	DAC, pXRD, P ≤ 134 GPa, B-M EOS	133

Table 1. Bulk Moduli from Static Compression Data (continued)

Chemical Formula[1]	Density (Mg/m^3)	Isothermal Bulk Modulus (GPa)[2]	dK$_{0T}$/dP[3]	Technique and Notes[4]	Ref.
CaSiO$_3$ (high-pressure perovskite structure)	4.25	325 ± 10	4 (assumed)	DAC, pXRD, P ≤ 31.5 GPa, B-M EOS	199
CaSiO$_3$ (high-pressure perovskite structure)	4.25	275 ± 15	4 (assumed)	DAC, pXRD, P ≤ 85 GPa, B-M EOS	200
ZnSiO$_3$ (ilmenite structure)	2.96	216 ± 2	4 (assumed)	CAP, XRD, P ≤ 10 GPa, B-M EOS	184
Ca$_2$Mg$_5$Si$_8$O$_{22}$(OH)$_2$ *Tremolite*	2.99	85	---	DAC, scXRD, P ≤ 4 GPa, linPVfit	35
Na$_2$Mg$_3$Al$_2$Si$_8$O$_{22}$(OH)$_2$ *Glaucophane*	3.10	96	---	DAC, scXRD, P ≤ 4 GPa, linPVfit	35
NaCa$_2$Mg$_4$AlSi$_6$Al$_2$O$_{22}$(OH)$_2$ *Pargasite*	3.10	97	---	DAC, scXRD, P ≤ 4 GPa, linPVfit	35
(Na,K,Fe,Mg,Al)$_7$(Si,Al)$_8$O$_{22}$(OH,F,Cl)$_2$ *Grunerite*	3.50	50 ± 1	13 ± 1	DAC, scXRD, P ≤ 5.1 GPa, B-M EOS	229

Ortho- and Ring Silicates:

Chemical Formula[1]	Density (Mg/m^3)	Isothermal Bulk Modulus (GPa)[2]	dK$_{0T}$/dP[3]	Technique and Notes[4]	Ref.
Mg$_2$SiO$_4$ *Forsterite*	3.22	135.7 ± 1.0	3.98 ± 0.1	DAC, pXRD, P ≤ 30 GPa, M EOS	214
Mg$_2$SiO$_4$ *Forsterite*	3.22	120	5.6	BA, XRD, P ≤ 10 GPa, O-H method	159
Mg$_2$SiO$_4$ *Forsterite*	3.22	122.6	4.3	DAC, scXRD, P < 15 GPa, B-M EOS	105
Fe$_2$SiO$_4$ *Fayalite*	4.39	123.9 ± 4.6	5.0 ± 0.8	DAC, pXRD, P < 38 GPa, B-M EOS	216
Fe$_2$SiO$_4$ *Fayalite*	4.39	119 ± 10	7 ± 4	CAP, XRD, P < 7 GPa, B-M EOS	222
		124 ± 2	5 (assumed)		

Table 1. Bulk Moduli from Static Compression Data (continued)

Chemical Formula[1]	Density (Mg/m^3)	Isothermal Bulk Modulus (GPa)[2]	dK_{0T}/dP[3]	Technique and Notes[4]	Ref.
Fe$_2$SiO$_4$ *Fayalite*	4.39	124 ± 2	4 (assumed)	DAC, pXRD, P ≤ 30 GPa, B-M EOS	131
Fe$_2$SiO$_4$ *Fayalite*	4.39	134	4 (assumed)	DAC, scXRD, P < 14 GPa, B-M EOS	104
Fe$_2$SiO$_4$ *Fayalite*	4.39	103.8 (at 400 °C)	7.1 (at 400 °C)	DAC, pXRD, P < 8.0 GPa at 400 °C, dK/dT= -5.4 x 10^{-2} GPa/K	168
CaMgSiO$_4$ *Monticellite*	3.05	113 ± 3	4 (assumed)	DAC, scXRD, P ≤ 6.2 GPa, B-M EOS	186
ZrSiO$_4$ *Zircon*	4.65	227 ± 2	--	DAC, scXRD, P ≤ 5 GPa, linPVfit	67
(Mg$_{0.84}$Fe$_{0.16}$)$_2$ SiO$_4$ *Wadsleyite or β-Phase*	3.53	174 ± 3 168 ± 4	4 (assumed) 4.7 (fixed from ultrasonic data)	DAC, pXRD, P ≤ 27 GPa, B-M EOS, dK_{0T}/dT = -2.7 ± 0.3 x 10^{-2} GPa/K	52
(Mg$_{0.84}$Fe$_{0.16}$)$_2$ SiO$_4$ *Wadsleyite or β-Phase*	3.53	164 ± 2	4 (assumed)	DAC, scXRD, P ≤ 5 GPa, B-M EOS	78
(MgFe)$_2$SiO$_4$ 1.00≥Mg/(Mg+Fe)≥ 0.75 *Wadsleyite or β-Phase*	3.47 - 3.78	171 ± 0.6	4 (assumed)	Re-analysis of [78]	92
Mg$_2$SiO$_4$ *Wadsleyite or β-Phase*	3.47	166 ± 40	--	MAP, XRD, P ≤ 10 GPa, linPVfit	151
Mg$_2$SiO$_4$ *Ringwoodite or γ-Phase* (spinel structure)	3.55	213 ± 10	--	MAP, XRD, P ≤ 10 GPa, linPVfit	151
(Mg$_{0.6}$Fe$_{0.4}$)$_2$SiO$_4$ *Ringwoodite or γ-Phase* (spinel structure)	4.09	183 ± 2	5.38 ± 0.24	DAC, pXRD, P ≤ 50 GPa, B-M EOS	227
Fe$_2$SiO$_4$ (spinel structure)	4.85	196 ± 6	--	DAC, scXRD, P ≤ 4 GPa, linPVfit	56

Table 1. Bulk Moduli from Static Compression Data (continued)

Chemical Formula[1]	Density (Mg/m^3)	Isothermal Bulk Modulus (GPa)[2]	dK$_{0T}$/dP[3]	Technique and Notes[4]	Ref.
Fe$_2$SiO$_4$ (spinel structure)	4.85	197 ± 2	4 (assumed)	CAP, XRD, P ≤ 8 GPa, B-M EOS	178
Fe$_2$SiO$_4$ (spinel structure)	4.85	212	4 (assumed)	DAC, pXRD, P ≤ 26 GPa, B-M EOS	131
Ni$_2$SiO$_4$ (spinel structure)	5.35	227 ± 4	--	DAC, scXRD, P ≤ 4 GPa, linPVfit	56
Ni$_2$SiO$_4$ (spinel structure)	5.35	227 ± 4	4 (assumed)	CAP, XRD, P ≤ 8 GPa, B-M EOS	178
Ni$_2$SiO$_4$ (spinel structure)	5.35	214	4 (assumed)	DAC, pXRD, P ≤ 30 GPa, B-M EOS	127
Co$_2$SiO$_4$ (spinel structure)	5.17	210 ± 6	4.0 ± 0.6	DAC, pXRD, P ≤ 30 GPa, B-M EOS	123
Co$_2$SiO$_4$ (spinel structure)	5.17	206 ± 2	4 (assumed)	CAP, XRD, P ≤ 8 GPa, B-M EOS	178
Ca$_3$Al$_2$(O$_4$H$_4$)$_3$ *Katoite or Hydrogrossular*	2.52	66 ± 4	4.1 ± 0.5	DAC, pXRD, P ≤ 43 GPa, B-M EOS	156
Ca$_3$Al$_2$Si$_3$O$_{12}$ *Grossular*	3.59	168 ± 25	6.2 ± 4	DAC, pXRD, P ≤ 19 GPa, B-M EOS	156
		168.4 (fixed from Brillouin data)	6.1 ± 1.5		
Ca$_3$Al$_2$Si$_3$O$_{12}$ *Grossular*	3.59	139 ± 5	--	DAC, scXRD, P ≤ 6 GPa, linPVfit	66
Mn$_3$Al$_2$Si$_3$O$_{12}$ *Spessartine*	4.19	171.8	7.4 ± 1	DAC, pXRD, P ≤ 25 GPa, B-M EOS	108
		174.2 (fixed from ultrasonic data)	7.0 ± 1.0		
Mg$_3$Al$_2$Si$_3$O$_{12}$ *Pyrope*	3.58	175 ± 1	4.5 ± 0.5	DAC, scXRD, P ≤ 5 GPa, B-M EOS	114
Mg$_3$Al$_2$Si$_3$O$_{12}$ *Pyrope*	3.58	171 ± 3	1.8 ± 0.7	CAP, XRD, P ≤ 8 GPa, B-M EOS	183
Mg$_3$Al$_2$Si$_3$O$_{12}$ *Pyrope*	3.58	212 ± 8	3.5 ± 0.6	DAC, XRD, P ≤ 30 GPa, B-M EOS	195

Table 1. Bulk Moduli from Static Compression Data (continued)

Chemical Formula[1]	Density (Mg/m^3)	Isothermal Bulk Modulus (GPa)[2]	dK_{0T}/dP[3]	Technique and Notes[4]	Ref.
$Mg_3Al_2Si_3O_{12}$ *Pyrope*	3.58	179 ± 3	4 (assumed)	DAC, scXRD, $P \leq 6$ GPa, B-M EOS	75
$Mg_3Al_2Si_3O_{12}$ *Pyrope*	3.58	175 ± 0.3	3.3 ± 1	DAC, pXRD, $P \leq 25$ GPa, B-M EOS	108
		172.8 (fixed from Brillouin data)	3.8 ± 1		
$Fe_3Al_2Si_3O_{12}$ *Almandine*	4.32	175 ± 7	1.5 ± 1.6	CAP, XRD, $P \leq 8$ GPa, B-M EOS	183
$Fe_3Al_2Si_3O_{12}$ *Almandine*	4.32	190 ± 5	3 ± 0.5	DAC, pXRD, $P \leq 30$ GPa, B-M EOS	195
$Ca_3Cr_2Si_3O_{12}$ *Uvarovite*	3.83	162	4.7 ± 0.7	DAC, pXRD, $P \leq 25$ GPa, B-M EOS	108
$Ca_3Fe_2Si_3O_{12}$ *Andradite*	3.86	159 ± 2	4 (assumed)	DAC, scXRD, $P \leq 5$ GPa, B-M EOS	75
Al_2SiO_5 *Andalusite*	3.15	135 ± 10	4 (assumed)	DAC, scXRD, $P \leq 3.7$ GPa, B-M EOS	169
$Mg_{1.3}Fe_{0.7}Al_2Si_2O_{10}(OH)_4$ *Magnesiochloritoid*	3.68	148 ± 5 GPa	---	DAC, scXRD, $P \leq 4.2$ GPa, B-M EOS	34
$Be_3Al_2Si_6O_{18}$ *Beryl*	2.66	170 ± 5	4 (assumed)	DAC, scXRD, $P \leq 5$ GPa, B-M EOS	77
$(Mg,Fe)_2Al_4Si_5O_{18}$ $\cdot n(H_2O, CO_2; Na^+, K^+)$ *Cordierite*	2.63	110	--	DAC and PC, XRD and PD, $P \leq 3$ GPa, linPVfit	150
Be_2SiO_4 *Phenakite*	2.96	201 ± 8	2 ± 4	DAC, scXRD, $P \leq 5$ GPa, B-M EOS	63

Sheet silicates:

$KMg_3AlSi_3O_{10}(F,OH)_2$ *Phlogopite*	2.75	58.5 ± 2	--	DAC, scXRD, $P \leq 4.7$ GPa, linPVfit	65

Table 1. Bulk Moduli from Static Compression Data (continued)

Chemical Formula[1]	Density (Mg/m^3)	Isothermal Bulk Modulus (GPa)[2]	dK_{0T}/dP[3]	Technique and Notes[4]	Ref.
$(Mg,Fe,Al)_6(Si,Al)_4$ $O_{10}(F,OH)_2$ *Chlorite*	2.65	55.0 ± 10	--	DAC, scXRD, P ≤ 4.7 GPa, linPVfit	65
$KAl_3Si_3O_{10}(OH)_2$ *Muscovite*	2.80	61.4 ± 4.0	6.9 ± 1.4	DAC, pXRD, P < 18 GPa, B-M EOS	49
$BaFeSi_4O_{10}$ *Gillespite I* (low-pressure tetragonal structure)	3.72	62 ± 3	4 (assumed)	DAC, scXRD, 0 < P<1.9 GPa, B-M EOS	71
$BaFeSi_4O_{10}$ *Gillespite II* (high-pressure orthorhombic structure)	3.76	66 ± 3	4 (assumed)	DAC, scXRD, 1.9<P<4.6 GPa, B-M EOS	71

Oxides and Hydroxides:

BeO *Bromellite*	3.01	212 ± 3	4 (assumed)	DAC, scXRD, P ≤ 5 GPa, B-M EOS	74
MgO *Periclase*	3.56	156 ± 9	4.7 ± 2	DAC, pXRD, P ≤ 95 GPa, M EOS	126
MgO *Periclase*	3.56	178	4.0	DP, XRD, P ≤ 30 GPa, M EOS	41
$(Mg_{0.6}Fe_{0.4})O$ *Magnesiowüstite*	4.54	157	4 (assumed)	DAC, pXRD, P ≤ 30 GPa, B-M EOS, T ≤ 800 K, $(dK_T/dT)_P = -2.7 \pm 3 \times 10^{-2}$ GPa/K	51
CaO *Lime*	3.38	111 ± 1	4.2 ± 0.2	DAC, pXRD, P ≤ 55 GPa, B-M EOS	170
CaO *Lime*	3.38	112	3.9	DP, XRD, P ≤ 30 GPa, M EOS	41
CaO (high-pressure B2 structure)	3.79	130 ± 20	3.5 ± 0.5	DAC, pXRD, 55<P<135 GPa, B-M EOS	170
SrO	4.70	90.6 ± 2.4	4.4 ± 0.3	DAC, pXRD, P ≤ 34 GPa, B-M EOS	120

Table 1. Bulk Moduli from Static Compression Data (continued)

Chemical Formula[1]	Density (Mg/m^3)	Isothermal Bulk Modulus (GPa)[2]	dK_{0T}/dP[3]	Technique and Notes[4]	Ref.
SrO (high-pressure B2 structure)	6.14	160 ± 19	4 (assumed)	DAC, pXRD, 36<P<59 GPa, B-M EOS	182
BaO	5.72	66.2 ± 0.8	5.7 (fixed from ultrasonics)	DAC, pXRD, P ≤ 10 GPa, B-M EOS	211
BaO (high-pressure PH4I structure)	6.09	33.2 ± 1.9	6.02 ± 0.3	DAC, pXRD, 18<P<60.5 GPa, B-M EOS	211
MnO *Manganosite*	5.46	162 ± 17	4.8 ± 1.1	DAC, pXRD, P ≤ 60 GPa, B-M EOS	93
MnO *Manganosite*	5.46	144	3.3	DP, XRD, P ≤ 30 GPa, M EOS	41
NiO *Bunsenite*	6.67	199	4.1	DP, XRD, P ≤ 30 GPa, M EOS	32
		190 (fixed from ultrasonics)	5.0		
CoO	6.45	190.5	3.9	DP, XRD, P ≤ 30 GPa, M EOS	41
CdO *Monteponite*	8.15	108.0	9.0	DP, XRD, P ≤ 30 GPa, M EOS	41
PbO *Massicot*	9.63	22.7 ± 6.2	17.8 ± 1.6	DAC, scXRD, P ≤ 5 GPa, fit to Hooke's Law	2
EuO	8.25	97.0	4 (assumed)	DAC, pXRD, P ≤ 13 GPa, B-M EOS	231
Cu2O *Cuprite*	5.91	131	5.7	DAC, pXRD, P ≤ 10 GPa, M EOS	208
GeO2 (quartz structure)	4.23	63.9 ± 0.7	4 (assumed)	DAC, pXRD, P ≤ 5.7 GPa, B-M EOS	225
GeO2 (quartz structure)	4.23	39.1 ± 0.4	2.2 ± 0.5	PC, ND, P ≤ 2.5 GPa, M EOS	98
GeO2 (quartz structure)	4.23	26.5	16.8	BCAC, EXAFS, P ≤ 6 GPa, M EOS	86
		34.3	--		

Table 1. Bulk Moduli from Static Compression Data (continued)

Chemical Formula[1]	Density (Mg/m^3)	Isothermal Bulk Modulus (GPa)[2]	dK_{0T}/dP[3]	Technique and Notes[4]	Ref.
GeO_2 *Argutite* (rutile structure)	6.24	394.9 ± 0.2	4 (assumed)	DAC, pXRD, P ≤ 9.6 GPa, B-M EOS	225
GeO_2 *Argutite* (rutile structure)	6.24	258 ± 5	7 (assumed)	DAC, scXRD, P ≤ 5 GPa, B-M EOS	70
		265 ± 5	4 (assumed)		
TiO_2 *Rutile*	4.24	216 ± 5	7 (assumed)	DAC, scXRD, P ≤ 5 GPa, B-M EOS	70
		222 ± 5	4 (assumed)		
TiO_2 *Rutile*	4.24	203	4 (assumed)	CAP, XRD, P ≤ 12 GPa, B-M EOS	178
		197	6.8 (fixed from ultrasonics)		
TiO_2 *Rutile*	4.24	187	4 (assumed)	DAC, scXRD, P < 9 GPa, B-M EOS	104
SnO_2 *Cassiterite*	7.00	218 ± 2	7 (assumed)	DAC, scXRD, P ≤ 5 GPa, B-M EOS	70
		224 ± 2	4 (assumed)		
TeO_2 *Paratellurite* (tetragonal structure)	6.26	44.4 ± 1.6	5.8 ± 0.5	DAC, pXRD, P ≤ 25 GPa, B-M EOS	119
UO_2 *Uraninite*	10.96	207	7.2	DAC, XRD, P ≤ ? GPa, B-M EOS	24
CeO_2 *Cerianite*	7.13	230 ± 10	4 (assumed)	DAC, pXRD, P ≤ 31 GPa, B-M EOS	45
CeO_2 (high-pressure α-$PbCl_2$-type structure)	7.84	304 ± 25 GPa	4 (assumed)	DAC, pXRD, 31<P<70 GPa, B-M EOS	45
RuO_2	6.97	270 ± 6	4 (assumed)	DAC, scXRD, P ≤ 5 GPa, B-M EOS	70
ZrO_2 *Baddeleyite*	5.81	95 ± 8	4-5 (assumed)	DAC, pXRD, 0<P<10 GPa, B-M EOS	110
ZrO_2 (high-pressure orthorhombic-I phase)	6.17	220	5 (assumed)	DAC, pXRD,10<P<25 GPa, B-M EOS	110

Table 1. Bulk Moduli from Static Compression Data (continued)

Chemical Formula[1]	Density (Mg/m^3)	Isothermal Bulk Modulus (GPa)[2]	dK_{0T}/dP[3]	Technique and Notes[4]	Ref.
HfO_2	9.68	145	5 (assumed)	DAC, pXRD,0<P<10 GPa, B-M EOS	109
HfO_2 (high-pressure orthorhombic-II phase)	10.14	210	5 (assumed)	DAC, pXRD,10<P<26 GPa, B-M EOS	109
HfO_2 (high-pressure orthorhombic-III phase)	10.98	475	5 (assumed)	DAC, pXRD,26<P<42 GPa, B-M EOS	109
HfO_2 (high-pressure tetragonal phase)	11.88	550	5 (assumed)	DAC, pXRD,42<P<50 GPa, B-M EOS	109
V_3O_5	4.73	269 ± 3	--	DAC, scXRD,0<P< 5.5 GPa, linPVfit	16
V_3O_5 (high pressure phase)	5.39	175 ± 11	--	DAC, scXRD,6.3<P< 7.5 GPa, linPVfit	16
$AlPO_4$ *Berlinite*	2.57	36	4 (assumed)	DAC, scXRD, P ≤ 8.5 GPa, B-M EOS	190
$MgAl_2O_4$ *Spinel*	3.55	194 ± 6	4 (assumed)	DAC, scXRD, P ≤ 4 GPa, B-M EOS	55
Fe_3O_4 *Magnetite*	5.20	186 ± 3	4 (assumed)	DAC, scXRD, P ≤ 4 GPa, B-M EOS	55
		183 ± 5	5.6 (from ultrasonics)		
Fe_3O_4 *Magnetite*	5.20	181 ± 2	5.5 ± 15	DAC, scXRD, P ≤ 4.5 GPa, B-M EOS	154
Fe_3O_4 *Magnetite*	5.20	155 ± 12	4 (assumed)	DAC, pXRD, P ≤ ? GPa, B-M EOS	212
Fe_3O_4 *Magnetite*	5.20	183 ± 10	4 (assumed)	DAC, pXRD, P ≤ 32 GPa, B-M EOS	130
Mn_3O_4 *Hausmannite*	4.83	137.0 ± 3.8	4 (assumed)	DAC, pXRD, 0<P<10 GPa, B-M EOS	163
Mn_3O_4 (high-pressure marokite-type structure)	5.33	166.6 ± 2.7	4 (assumed)	DAC, pXRD, 10<P <39 GPa, B-M EOS	163

Table 1. Bulk Moduli from Static Compression Data (continued)

Chemical Formula[1]	Density (Mg/m³)	Isothermal Bulk Modulus (GPa)[2]	dK_{0T}/dP[3]	Technique and Notes[4]	Ref.
Al_2BeO_4 *Chrysoberyl*	3.70	242 ± 5	4 (assumed)	DAC, scXRD, P ≤ 6.3 GPa, B-M EOS	62
ReO_3	6.9	200 ± 4	--	see ref., P < 0.5 GPa, linPVfit	23
$LiNbO_3$	4.63	134 ± 3	2.9 ± 0.5	DAC, XRD, P ≤ 13 GPa, B-M EOS	37
$CaTiO_3$ *Perovskite*	3.98	210 ± 7	5.6 (assumed)	DAC, pXRD, P ≤ 10.4 GPa, B-M EOS	218
$SrTiO_3$ *Tausonite*	5.12	174.2 (fixed from ultrasonic value)	5.3	DAC, pXRD, P ≤ 20 GPa, B-M EOS	57
$SrTiO_3$ *Tausonite*	5.12	176	4.4	DAC, pXRD, P ≤ 20 GPa, M EOS	47
$FeTiO_3$ *Ilmenite*	4.72	170 ± 7	8 ± 4	DAC, scXRD, P ≤ 5 GPa, B-M EOS	207
		177 ± 3	4 (assumed)		
$(Fe,Mg)TiO_3$ *Ilmenite*	4.44	168 ± 13	5 ± 1	DAC, pXRD, P < 28 GPa, B-M EOS	123
$MnTiO_3$-I *Pyrophanite*	4.54	70 ± 9	4 (assumed)	DAC, scXRD, P ≤ 5 GPa, B-M EOS	175
$MnTiO_3$-II ($LiNbO_3$ structure)	4.68	158 ± 9	4 (assumed)	DAC, scXRD, P ≤ 3 GPa, B-M EOS	175
$MnTiO_3$-III (perovskite structure)	4.88	227 ± 4	4 (assumed)	DAC, scXRD, 2.4 <P<5 GPa, B-M EOS	175
$MnSnO_3$ (perovskite structure)	6.12	196 ± 20	4 (assumed)	DAC, pXRD, 7<P<20 GPa, B-M EOS	111
$MgGeO_3$ (ilmenite structure)	4.97	187 ± 2	4 (assumed)	CAP, XRD, P ≤ 8 GPa, B-M EOS	184
$MgGeO_3$ (ilmenite structure)	4.97	195	3.6	DAC, pXRD, P ≤ 22.5 GPa, B-M EOS	17
$CuGeO_3$	5.11	67.8	--	DAC, pXRD, P ≤ 7.3 GPa, linPVfit	1

Table 1. Bulk Moduli from Static Compression Data (continued)

Chemical Formula[1]	Density (Mg/m^3)	Isothermal Bulk Modulus (GPa)[2]	dK_{0T}/dP[3]	Technique and Notes[4]	Ref.
$Na_{0.55}WO_3$ (perovskite structure)	7.23	105	--	DAC, scXRD, P < 5.3 GPa, linPVfit	72
$Na_{0.62}WO_3$ (perovskite structure)	7.44	119	--	DAC, scXRD, P < 5.3 GPa, linPVfit	72
$Na_{0.70}WO_3$ (perovskite structure)	7.61	91	--	DAC, scXRD, P < 5.3 GPa, linPVfit	72
Fe_2O_3 *Hematite*	5.25	225	4 (assumed)	DAC, scXRD, P ≤ 5 GPa, B-M EOS	54
Fe_2O_3 *Hematite*	5.25	199 ± 6	4 (assumed)	DAC, pXRD, P ≤ 30 GPa, B-M EOS	212
Fe_2O_3 *Hematite*	5.25	228 ± 15	4 (assumed)	DAC, XRD, P ≤ 30 GPa, B-M EOS	22
Fe_2O_3 *Hematite*	5.25	178 ± 4	4 (assumed)	CAP, XRD, P ≤ 12 GPa, B-M EOS	181
Al_2O_3 *Ruby*	3.98	253 ± 1	5.0 ± 0.4	DAC, pXRD, P ≤ 65 GPa, B-M EOS	171
Al_2O_3 *Corundum*	3.98	226 ± 2 / 239 ± 4	4 (assumed) / 0.9 ± 0.8	CAP, XRD, P ≤ 12 GPa, B-M EOS	181
Al_2O_3 *Ruby*	3.98	257 ± 6	4 (assumed)	DAC, scXRD, P ≤ 5 GPa, B-M EOS	53
Al_2O_3 *Corundum*	3.98	254.4 ± 2.0	4.275 ± 0.006	DAC, scXRD, P ≤ 10 GPa, B-M EOS	40
V_2O_3 *Karelianite*	4.87	171 ± 1 / 175 ± 3	4 (assumed) / 3.1 ± 0.7	CAP, XRD, P ≤ 12 GPa, B-M EOS	181
V_2O_3 *Karelianite*	4.87	195 ± 6	4 (assumed)	DAC, scXRD, P ≤ 5 GPa, B-M EOS	54
Cr_2O_3 *Eskolaite*	5.21	222 ± 2 / 231 ± 5	4 (assumed) / 2.0 ± 1.1	CAP, XRD, P ≤ 12 GPa, B-M EOS	181
Cr_2O_3 *Eskolaite*	5.21	238 ± 4	4 (assumed)	DAC, scXRD, P ≤ 5 GPa, B-M EOS	54

Table 1. Bulk Moduli from Static Compression Data (continued)

Chemical Formula[1]	Density (Mg/m^3)	Isothermal Bulk Modulus (GPa)[2]	dK$_{0T}$/dP[3]	Technique and Notes[4]	Ref.
KVO$_3$	?	84.7	---	DAC, scXRD, P ≤ 5 GPa, linPVfit	3
RbVO$_3$	?	42.9	---	DAC, scXRD, P ≤ 5 GPa, linPVfit	3
CsVO$_3$	?	62.9	---	DAC, scXRD, P ≤ 5 GPa, linPVfit	3
NaNO$_2$	2.17	21.9 ± 0.2	4.3 ± 0.8	DAC, scXRD, P ≤ 2.6 GPa, B-M EOS	68
NaNO$_3$ *Nitratine*	2.26	25.8 ± 0.6	6.6 ± 1.5	DAC, scXRD, P ≤ 2.6 GPa, B-M EOS	68
Mg(OH)$_2$ *Brucite*	2.39	54.3 ± 1.5	4.7 ± 0.2	DAC, pXRD, P < 35 GPa, B-M EOS dK$_T$/dP= -0.018±0.003 GPa/K	50
Ca(OH)$_2$ *Portlandite*	2.24	37.8 ± 1.8	5.2 ± 0.7	DAC, pXRD, P ≤ 11 GPa, B-M EOS	144

Carbon-Bearing Minerals:

SiC *Moissonite*	3.22	227 ± 3	4.1 ± 0.1	DAC, XRD, P ≤ 43 GPa	5
CaMg(CO$_3$)$_2$ *Dolomite*	2.86	94	4 (assumed)	DAC, scXRD, P ≤ 4.7 GPa, B-M EOS	174
CaMg$_{0.3}$Fe$_{0.7}$(CO$_3$)$_2$ *Ankerite*	3.05	91	4 (assumed)	DAC, scXRD, P ≤ 4.0 GPa, B-M EOS	174
BaCO$_3$ *Witherite*	4.30	50 ± 3	4 (assumed)	DAC, DVM, 1<P 6.5 GPa, B-M EOS	139
SrCO$_3$ *Strontianite*	3.73	58 ± 5	4 (assumed)	DAC, DVM, 1<P 6.5 GPa, B-M EOS	139
MnCO$_3$ *Rhodocrosite*	3.67	95 ± 9	4 (assumed)	DAC, DVM, 2.5<P 4.5 GPa, B-M EOS	139
CaCO$_3$ *Calcite*	2.71	71.1	4.15	PC, PD, P ≤ 1.5 GPa, polyPVfit	189

Table 1. Bulk Moduli from Static Compression Data (continued)

Chemical Formula[1]	Density (Mg/m^3)	Isothermal Bulk Modulus (GPa)[2]	dK_{0T}/dP[3]	Technique and Notes[4]	Ref.
$CaCO_3$ *Calcite II*	2.71	32.7 (at 1.4 GPa)	4.4 (at 1.4 GPa)	PC, PD, 1.4<P< 1.7 GPa, polyPVfit	189
$CaCO_3$ *Calcite III*	2.85	75.2 (at 1.7 GPa)	?	PC, PD, 1.7<P<4 GPa, polyPVfit	189
$CaCO_3$ *Calcite III*	2.85	84.0 ± 8	4 (assumed)	DAC, DVM, 3<P< 5 GPa, B-M EOS	139

Sulfides and Tellurides:

Chemical Formula[1]	Density (Mg/m^3)	Isothermal Bulk Modulus (GPa)[2]	dK_{0T}/dP[3]	Technique and Notes[4]	Ref.
HgS *Cinnabar*	8.13	19.4 ± 0.5	11.1	DAC, pXRD, P ≤ 24 GPa, B-M EOS	210
HgTe *Coloradoite*	8.09	16.0 ± 0.5	7.3	DAC, pXRD, P ≤ 24 GPa, B-M EOS	210
CaS *Oldhamite*	2.50	56.7	4.9	DP, XRD, P ≤ 30 GPa, M EOS	41
CdS *Greenockite*	4.82	86.7	4.36	DAC, pXRD, P ≤ 55 GPa, B-M EOS	192
CdS *Greenockite*	4.82	94.0	7.6	DP, XRD, P ≤ 30 GPa, M EOS	41
MnS *Alabandite*	3.99	81.0	3.3	DP, XRD, P ≤ 30 GPa, M EOS	41
MnS *Alabandite*	3.99	72 ± 2	4.2 ± 1.3	DAC, pXRD, P ≤ 21 GPa, B-M EOS	140
NiS (NiAs-type structure)	5.50	156 ± 10	4.4 ± 0.1	DAC, pXRD, P < 45 GPa, B-M EOS	27
BaS	4.25	55.1 ± 1.4	5.5 (assumed)	DAC, pXRD, P ≤ 6.5 GPa, B-M EOS	211
BaS (high pressure B2-structure)	4.67	21.4 ± 0.3	7.8 ± 0.1	DAC, pXRD, 6.5<P<89 GPa, B-M EOS	211
ZnS *Sphalerite*	4.02	76.5	4.49	----	see 230
ZnS (high-pressure phase)	4.72	85.0 ± 3.8	4 (assumed)	DAC, pXRD,11<P<45 GPa, B-M EOS	230

Table 1. Bulk Moduli from Static Compression Data (continued)

Chemical Formula[1]	Density (Mg/m^3)	Isothermal Bulk Modulus (GPa)[2]	dK_{0T}/dP[3]	Technique and Notes[4]	Ref.
YbS	7.38	60 ± 3	4 (assumed)	DAC, pXRD, P ≤ 8 GPa, B-M EOS	194
CeS	5.93	82	2.2	DAC, pXRD, P ≤ 25 GPa, B-M EOS	204
ErS	8.38	101.5 ± 1.0	5.4 ± 0.2	see ref., P < 3 GPa, lsqPV fit	89
ThS	9.56	145 ± 6	5.4 ± 0.1	DAC, pXRD, P ≤ 45 GPa, M EOS	160
US	10.87	92 ± 9	9.1 ± 0.2	DAC, pXRD, P ≤ 45 GPa, M EOS	160
NiS_2 *Vaesite*	4.45	109 ± 6	—	BA, XRD, P ≤ 4 GPa, linPVfit	48
NiS_2 (high-pressure metallic phase)	4.91	141 ± 11	—	BA, XRD, 4<P<11 GPa, linPVfit	48
MnS_2 *Hauerite*	3.46	76.0	5.4	DAC, pXRD, 0<P<10 GPa, B-M EOS	30
MnS_2 (high-pressure orthorhombic marcasite-type structure)	4.02	213.8	5.0	DAC, pXRD, 10 <P<35 GPa, B-M EOS	30
CoS_2 *Cattierite*	4.27	118.3	—	DAC, scXRD, P≤ 3.6 GPa, linPVfit	58
SnS_2 *Berndtite*	4.50	29	—	DAC, scXRD, P ≤ 5 GPa, linPVfit	64
As_4S_3 *Dimorphite*	3.58	17.0	5.5	BA, XRD, P ≤ 12 GPa, M EOS	31
$CuFe_2S_3$ *Cubanite*	4.11	55.3 ± 1.7	4 (assumed)	DAC, scXRD, P ≤ 3.7 GPa, B-M eqn	141
$CuGaS_2$ *Gallite*	4.36	96 ± 10	6.5	DAC, XRD, P < 16 GPa, M EOS	209
$AgGaS_2$	4.70	60 ± 8	6	DAC, XRD, P < 5 GPa, M EOS	209

Table 1. Bulk Moduli from Static Compression Data (continued)

Chemical Formula[1]	Density (Mg/m^3)	Isothermal Bulk Modulus (GPa)[2]	dK_{0T}/dP[3]	Technique and Notes[4]	Ref.
La$_6$CoSi$_2$S$_{14}$	4.37	79.2 ± 0.4	--	DAC, scXRD, P ≤ 5 GPa, linPVfit	145
La$_6$NiSi$_2$S$_{14}$	4.37	75.5 ± 0.5	--	DAC, scXRD, P ≤ 5 GPa, linPVfit	145
Halides:					
LiF	2.64	66.5	3.5	CAP, XRD, P ≤ 9 GPa, B-M EOS	219
LiF	2.64	62.7	6.8	PC, PD, P ≤ 4.5 GPa, M EOS	202
LiF	2.64	65.0	4.7	DP, XRD, P ≤ 30 GPa, M EOS	41
LiCl	2.07	31.9	3.4	PC, PD, P ≤ 4.5 GPa, M EOS	202
LiBr	3.46	24.3	3.5	PC, PD, P ≤ 4.5 GPa, M EOS	202
LiI	4.08	16.8	4.3	PC, PD, P ≤ 4.5 GPa, M EOS	202
NaF *Villiaumite*	2.56	45.9	4.4	CAP, XRD, P ≤ 9 GPa, B-M EOS	219
NaF *Villiaumite*	2.56	46.4 ± 6.2	4.9 ± 1.2	DAC, pXRD, P ≤ 23 GPa, B-M EOS	180
NaF *Villiaumite*	2.56	46.7	5.2	PC, PD, P ≤ 4.5 GPa, M EOS	202
NaF *Villiaumite*	2.56	45.6	5.7	DP, XRD, P ≤ 30 GPa, M EOS	41
NaF (high-pressure, B2 structure)	3.16	103 ± 19	4 (assumed)	DAC, pXRD, 23<P<60 GPa, B-M EOS	180
NaCl *Halite*	2.17	26.4	3.9	DP, XRD, P ≤ 30 GPa, M EOS	41
NaCl *Halite*	2.17	23.2	4.9	PC, PD, P ≤ 4.5 GPa, M EOS	202
NaCl *Halite*	2.17	23.8 ± 7.5	4.0 ± 3.9	DAC, pXRD, P ≤ 29 GPa, B-M EOS	180

Table 1. Bulk Moduli from Static Compression Data (continued)

Chemical Formula[1]	Density (Mg/m^3)	Isothermal Bulk Modulus (GPa)[2]	dK$_{0T}$/dP[3]	Technique and Notes[4]	Ref.
NaCl (high-pressure B2 structure)	2.34	36.2 ± 4.2	4 (assumed)	DAC, pXRD, 25<P<70 GPa, B-M EOS	80
NaBr	3.20	20.3	4.2	PC, PD, P ≤ 4.5 GPa, M EOS	202
NaBr	3.20	18.5 ± 3.4	5.8 ± 1.1	DAC, pXRD, P ≤ 39 GPa, B-M EOS	180
NaI	3.67	15.0	4.1	DP, XRD, P ≤ 30 GPa, M EOS	41
NaI	3.67	14.7 ± 1.1	5.7 ± 0.5	DAC, pXRD, P ≤ 36 GPa, B-M EOS	180
NaI	3.67	15.1	4.2	PC, PD, P ≤ 4.5 GPa, M EOS	202
KF *Carobbiite*	2.48	29.3	5.4	CAP, XRD, P ≤ 9 GPa, B-M EOS	219
KF (high-pressure, B2 structure)	2.87	37.0	5.4	CAP, XRD, P ≤ 9 GPa, B-M EOS	219
KCl *Sylvite*	1.98	37.0	5.0	DP, XRD, P ≤ 30 GPa, M EOS	41
KCl (high-pressure B2 structure)	2.34	28.7 ± 0.6	4 (assumed)	DAC, pXRD, P ≤ 56 GPa, B-M EOS	26
CsCl	3.99	18.0	4.8	DP, XRD, P ≤ 30 GPa, M EOS	41
CsCl	3.99	18.2	5.1	CAP, XRD, P ≤ 9 GPa, B-M EOS	219
CsCl	3.99	17.1	5.1	PC, PD, P ≤ 4.5 GPa, M EOS	202
CsBr	4.44	14.4	5.3	PC, PD, P ≤ 4.5 GPa, M EOS	202
CsBr	4.44	19.1 ± 0.9	5.0 ± 0.1	DAC, pXRD, P < 53 GPa, B-M EOS	103
CsI	4.51	13.3 ± 2.3	5.9 ± 0.9	DAC, pXRD, P < 61 GPa, B-M EOS	101

Table 1. Bulk Moduli from Static Compression Data (continued)

Chemical Formula[1]	Density (Mg/m^3)	Isothermal Bulk Modulus (GPa)[2]	dK_{0T}/dP[3]	Technique and Notes[4]	Ref.
CsI	4.51	12.5	4.5	PC, PD, P ≤ 4.5 GPa, M EOS	202
CsI	4.51	13.5 ± 0.2	5.45 ± 0.06	DAC, pXRD, P ≤ 100 GPa, V EOS	137
AgCl *Cerargyrite*	5.56	41.5	6.0	PC, PD, P ≤ 4.5 GPa, M EOS	202
AgBr *Bromyrite*	6.47	38.2	5.9	PC, PD, P ≤ 4.5 GPa, M EOS	202
CuBr	4.98	36.2	2.9	PC, PD, P ≤ 4.5 GPa, M EOS	202
MnF_2	3.98	94 ± 3	4 (assumed)	DAC, scXRD, P ≤ 5 GPa, B-M EOS	70
CaF_2 *Fluorite*	3.18	81.0 ± 1.2	5.22 ± 0.35	DAC, scXRD, P ≤ 9 GPa, B-M EOS	13
SrF_2	4.24	69.1	5.2	PC, PD, P ≤ 4 GPa, M EOS	149
$BaCl_2$	3.87	69.8 ± 0.5	1.40 ± 0.05	DAC, XRD, 0<P<10 GPa, B-M EOS	107
$BaCl_2$ (high pressure hexagonal phase)	4.73	69.3 ± 2	8.6 ± 0.1	DAC, XRD, 10<P<50 GPa, B-M EOS	107

Group I Elements:

H_2	0.088	0.362 ± 0.003 (4 K)	4.71 ± 0.03 (4 K)	DAC, scXRD, 5.4<P<26.5 GPa, B-M EOS	83 136
		0.172 ± 0.004 (4 K)	7.19 ± 0.04 (4 K)	V EOS	
H_2	0.088 (4 K)	0.166 (4 K)	7.3 (4 K)	PC, PD, P ≤ 2 GPa, lsqPVfit	8
D_2	0.20	0.46 ± 0.05	5.2 ± 0.2	DAC, scXRD, 6.5<P<14.2 GPa, B-M EOS	83

Table 1. Bulk Moduli from Static Compression Data (continued)

Chemical Formula[1]	Density (Mg/m³)	Isothermal Bulk Modulus (GPa)[2]	dK$_{0T}$/dP[3]	Technique and Notes[4]	Ref.
		0.35 ± 0.03	6.6 ± 0.2	V EOS	
Li	0.53	11.556 ± 0.033	3.51 ± 0.06	PC, PD, P ≤ 2 GPa, lsqPVfit	11
Na	0.97	6.06 ± 0.02	4.13 ± 0.04	PC, PD, P ≤ 2 GPa, lsqPVfit	10
K	0.86	2.99 ± 0.02	4.15 ± 0.10	DAC, XRD, P ≤ 12 GPa, B-M EOS	118
K	0.86	2.963 ± 0.001	4.208 ± 0.003	PC, PD, P ≤ 2 GPa, lsqPVfit	10
K	0.86	3.10 ± 0.01	3.91 ± 0.01	see ref., P < 0.7 GPa, M EOS	99
Rb	1.53	2.301 ± 0.003	4.15 ± 0.1	PC, PD, P ≤ 2 GPa, lsqPVfit	10
Rb	1.53	2.61	3.62	DAC, pXRD, P ≤ 11 GPa, B-M EOS	197
Cs	1.90	1.698 ± 0.006	3.79 ± 0.02	PC, PD, P ≤ 2 GPa, lsqPVfit	11

Group II Elements:

Chemical Formula[1]	Density (Mg/m³)	Isothermal Bulk Modulus (GPa)[2]	dK$_{0T}$/dP[3]	Technique and Notes[4]	Ref.
Mg	1.74	33.6	4.8	PC, PD, P ≤ 4.5 GPa, M EOS	203
Ca	1.54	18.7	2.5	PC, PD, P ≤ 4.5 GPa, M EOS	201
Ca	1.54	17.4 ± 0.1	3.7 ± 0.1	PC, PD, P ≤ 2 GPa, lsqPVfit	12
Sr	2.60	12.1	2.5	PC, PD, P ≤ 4.5 GPa, M EOS	203
Sr	2.60	11.83 ± 0.07	2.47 ± 0.07	PC, PD, P ≤ 2 GPa, lsqPVfit	12
Ba	3.50	9.4	2.1	PC, PD, P ≤ 4.5 GPa, M EOS	201
Ba	3.50	8.93 ± 0.6	2.76 ± 0.05	PC, PD, P ≤ 2 GPa, lsqPVfit	12

Table 1. Bulk Moduli from Static Compression Data (continued)

Chemical Formula[1]	Density (Mg/m^3)	Isothermal Bulk Modulus (GPa)[2]	dK_{0T}/dP[3]	Technique and Notes[4]	Ref.
Transition Metals:					
Ti	4.51	109.4	3.4	PC, PD, P ≤ 4.5 GPa, M EOS	203
V	6.10	154 ± 5	4.27 (fixed from ultrasonics)	DAC, XRD, P ≤ 10 GPa, B-M EOS	147
V	6.10	176 .4 ± 3.0	4	DAC, XRD, P ≤ 60 GPa, B-M EOS	19
V	6.10	139.4	18.2	PC, PD, P ≤ 4.5 GPa, M EOS	203
Cr	7.19	253.0 ± 11.0	8.9	DAC, XRD, P ≤ 40 GPa, B-M EOS	19
Cr	7.19	193 ± 6	4.89 (fixed from ultrasonics)	DAC, XRD, P ≤ 10 GPa, B-M EOS	147
Mn	7.43	131 ± 6	6.6 ± 7	DAC, XRD, P ≤ 42 GPa, B-M EOS	198
Co	8.90	167.1	17.3	PC, PD, P ≤ 4.5 GPa, M EOS	203
Ni	8.92	190.5	4.0	PC, PD, P ≤ 4.5 GPa, M EOS	201
Cu	8.96	162.5	4.24	pPC, PD, P ≤ 4.5 GPa, M EOS	201
Cu	8.96	137.4	5.52	DAC, XRD, P ≤ 10 GPa, B-M EOS	124
Zn	7.14	59.8	4.4	PC, PD, P ≤ 4.5 GPa, M EOS	201
Y	4.46	44.9	2.2	PC, PD, P ≤ 4.5 GPa, M EOS	203
Zr	4.46	104	2.05	DAC, XRD, P ≤ 30 GPa, B-M EOS	217
Zr	6.49	102.8	3.1	PC, PD, P ≤ 4.5 GPa, M EOS	203

Table 1. Bulk Moduli from Static Compression Data (continued)

Chemical Formula[1]	Density (Mg/m³)	Isothermal Bulk Modulus (GPa)[2]	dK_{0T}/dP[3]	Technique and Notes[4]	Ref.
Nb	8.41	144.2	14.5	PC, PD, P ≤ 4.5 GPa, M EOS	203
Nb	8.40	171 ± 7	4.03	DAC, XRD, P ≤ 10 GPa, B-M EOS	147
Nb	8.40	175.7 ± 2.7	4	DAC, XRD, P ≤ 60 GPa, B-M EOS	19
Mo	10.20	266.0	3.5	PC, PD, P ≤ 4.5 GPa, M EOS	201
Mo	10.20	267 ± 11	4.46 (fixed from ultrasonics)	DAC, XRD, P ≤ 10 GPa, B-M EOS	147
Pd	12.00	128.0 ± 5.0	5	DAC, XRD, P ≤ 60 GPa, B-M EOS	19
Ag	10.50	103 ± 5	5.6 ± 0.8	DAC, XRD, P ≤ 25 GPa, B-M EOS	60
Ag	10.50	116.7 ± 0.7	3.4	DAC, XRD, P ≤ 10 GPa, B-M EOS	124
Ag	10.50	120.9	5.2	PC, PD, P ≤ 4.5 GPa, M EOS	201
Ag	10.50	106.1	4.7	TCOA, XRD, P ≤ 12 GPa, B-M EOS	193
Cd	8.65	44.8	4.9	PC, PD, P ≤ 4.5 GPa, M EOS	201
La	6.17	24.5	1.6	PC, PD, P ≤ 4.5 GPa, M EOS	201
Ta	16.60	205.7	3.7	PC, PD, P ≤ 4.5 GPa, M EOS	201
Ta	16.60	194 ± 7	3.80 (fixed from ultrasonics)	DAC, XRD, P ≤ 10 GPa, B-M EOS	147
W	19.30	300.1	19.1	PC, PD, P ≤ 4.5 GPa, M EOS	203
W	19.30	307 ± 11	4.32 (fixed from ultrasonics)	DAC, XRD, P ≤ 10 GPa, B-M EOS	147

Table 1. Bulk Moduli from Static Compression Data (continued)

Chemical Formula[1]	Density (Mg/m³)	Isothermal Bulk Modulus (GPa)[2]	dK$_{0T}$/dP[3]	Technique and Notes[4]	Ref.
Au	19.30	166.4 ± 2.6	7.3	DAC, XRD, P ≤ 10 GPa, B-M EOS	124
Au	19.30	163.5 ± 8.3	4.42-5.16 (fixed from ultrasonics)	DAC, XRD, P ≤ 20 GPa, B-M EOS	148
Au	19.30	166.6 ± 10.8	5.5 ± 0.8	DAC, XRD, P ≤ 70 GPa, B-M EOS	81

Group III Elements:

Al	2.70	71.7 ± 3.6	5.31-6.43 (fixed from ultrasonics)	DAC, XRD, P ≤ 20 GPa, B-M EOS	148
Al	2.70	77.9	4.6	PC, PD, P ≤ 4.5 GPa, M EOS	201
Al	2.70	72.7	4.3	TCOA, XRD, P ≤ 12 GPa, B-M EOS	193
In	7.31	39.1	5.2	C, PD, P ≤ 4.5 GPa, M EOS	201
In	7.31	38 ± 2	5.5 ± 0.3	DAC, XRD, P < 67 GPa, B-M EOS	185
Tl	11.85	36.6	3.0	PC, PD, P ≤ 4.5 GPa, M EOS	201

Group IV Elements:

C *Diamond*	3.51	444 ± 3	1.9 ± 0.3	DAC, XRD, P ≤ 42 GPa, M EOS	6
C *Graphite*	2.25	33.8 ± 3	8.9 ± 1.0	DAC, XRD, P ≤ 14 GPa, M EOS	61
C *Graphite*	2.25	30.8	4 (assumed)	DAC, XRD, P ≤ 11 GPa, V EOS	226
C$_{60}$ *Fullerite*	1.67	18.1 ± 1.8	5.7 ± 0.6	DAC, scXRD, P ≤ 20 GPa, B-M EOS	44

Table 1. Bulk Moduli from Static Compression Data (continued)

Chemical Formula[1]	Density (Mg/m³)	Isothermal Bulk Modulus (GPa)[2]	dK_{0T}/dP[3]	Technique and Notes[4]	Ref.
Si	2.33	100.8	4.7	PC, PD, P ≤ 4.5 GPa, M EOS	203
Si	2.33	97.9	4.16	DAC, XRD, P ≤ 15 GPa. B-M EOS	191
Si (high-pressure hcp phase)	3.09	72 ± 2	3.91 ± 0.07	DAC, XRD, 42<P<79 GPa, B-M EOS	43
Si (high-pressure fcc phase)	3.25	82 ± 2	4.22 ± 0.05	DAC, XRD, 79<P<248 GPa, B-M EOS	43
Sn	7.30	54.9	4.8	PC, PD, P ≤ 4.5 GPa, M EOS	201
Sn	7.30	50.2 ± 0.5	4.9	DAC, XRD, P ≤ 10.3 GPa, B-M EOS	121
Sn	7.30	56.82 ± 2.19	2.3 ± 0.8	DAC, XRD, P ≤ 9.5 GPa, B-M EOS dK/dT = -1.38 (± 0.13) x 10⁻² GPa/K	28
Sn (high-pressure, body-centered tetragonal structure)	7.49	82.0 ± 1.2 (at 9.5 GPa)	5.5 (at 9.5 GPa)	DAC, XRD, 9.5<P< 53.4 GPa, B-M EOS	121
Sn (high-pressure, body-centered tetragonal structure)	7.49	56.65 ± 9.04	4.53 ± 0.81	DAC, XRD, 9<P<17 GPa, B-M EOS dK/dT = -4.63 (± 1.24) x 10⁻² GPa/K	167
Sn (high-pressure, body-centered cubic structure)	7.49	76.4	4.04	DAC, XRD, 45<P<120 GPa, B-M EOS	39
Pb	11.40	40.0	5.8	PC, PD, P ≤ 4.5 GPa, M EOS	201
Pb	11.40	43.2	4.87	DAC, XRD, P ≤ 10 GPa, B-M EOS	206

Table 1. Bulk Moduli from Static Compression Data (continued)

Chemical Formula[1]	Density (Mg/m^3)	Isothermal Bulk Modulus (GPa)[2]	dK_{0T}/dP[3]	Technique and Notes[4]	Ref.
Pb (combined hcp and bcc high pressure phases)	11.62	39.9± 0.2	6.13 ± 0.10	DAC, XRD, 15<P< 238 GPa, V EOS	134
Pb (high-pressure hcp phase)	11.57	46.63	5.23	DAC, XRD, 10<P< 100 GPa, B-M EOS	206
Pb (high-pressure bcc phase)	11.66	29.02	7.16	DAC, XRD, 100<P< 272 GPa, B-M EOS	206

Group V Elements:

N_2 (cubic δ-phase)	1.03	2.69	3.93	DAC, XRD, 5<P<16 GPa, O-H method	155
N_2 (hexagonal ε-phase)	1.21	2.98	3.78	DAC, XRD, 16<P<44 GPa, O-H method	155
P (orthorhombic)	2.70	36 ± 2	4.5 ± 0.5	DAC, pXRD, 0<P<5.5 GPa, M EOS	98a
P (high-pressure rhombohedral structure)	2.97	46 ± 4	3.0 ± 0.6	DAC, pXRD, 5.5 <P<10 GPa, M EOS	98a
P (high-pressure simple cubic structure)	3.08	95 ± 5	2.1 ± 0.8	DAC, pXRD, 10 <P<32 GPa, M EOS	98a
P (high-pressure simple cubic structure)	3.08	114	--	DAC, XRD, 11<P<60 GPa, B-M EOS	187
Sb	6.62	40.4	4.3	PC, PD, P ≤ 4.5 GPa, M EOS	203
Bi	9.80	29.7	2.4	PC, PD, P ≤ 4.5 GPa, M EOS	201

Group VI Elements:

O_2 (high-pressure ε-phase indexed as monoclinic)	1.32	37.5	3.31	DAC, XRD, 10<P<62 GPa, B-M EOS	177

Table 1. Bulk Moduli from Static Compression Data (continued)

Chemical Formula[1]	Density (Mg/m^3)	Isothermal Bulk Modulus (GPa)[2]	dK$_{0T}$/dP[3]	Technique and Notes[4]	Ref.
O$_2$ (high-pressure ε-phase indexed as orthorhombic)	1.32	16.7	4.09	DAC, XRD, 10<P<62 GPa, B-M EOS	177
S	2.07	8.8	6.5	PC, PD, P ≤ 4.5 GPa, M EOS	203
S	2.07	14.5	7	DAC, XRD, 0<P<5 GPa, B-M EOS	125
S (high-pressure phase)	2.72	17.3	5	DAC, XRD, 5<P<24 GPa, B-M EOS	125
Se	4.79	48.1 ± 0.2 (at 7.7 GPa)	4.33 ± 0.04 (at 7.7 GPa)	DAC, XRD, 0<P<12 GPa, M EOS	165
Se	4.79	7.9	5.8	PC, PD, P ≤ 4.5 GPa, M EOS	203
Te	6.24	18.2	8.4	PC, PD, P ≤ 4.5 GPa, M EOS	203
Te	6.24	24 ± 2 (at 2 GPa)	2.3 ± 0.2 (at 2 GPa)	DAC, XRD, 0<P<4 GPa, M EOS	164

Group VII Elements:

Chemical Formula[1]	Density (Mg/m^3)	Isothermal Bulk Modulus (GPa)[2]	dK$_{0T}$/dP[3]	Technique and Notes[4]	Ref.
Cl$_2$	2.09	11.7 ± 0.9	5.2 (assumed)	DAC, XRD, P ≤ 45 GPa. B-M EOS	46
Br$_2$	4.10	13.3 ± 0.7	5.2 (assumed)	DAC, XRD, P ≤ 45 GPa. B-M EOS	46
I$_2$	4.94	8.4	6.0	PC, PD, P ≤ 4.5 GPa, M EOS	203
I$_2$	4.94	13.6 ± 0.2	5.2 (assumed)	DAC, XRD, P ≤ 45 GPa. B-M EOS	46

Noble Gases:

Chemical Formula[1]	Density (Mg/m^3)	Isothermal Bulk Modulus (GPa)[2]	dK$_{0T}$/dP[3]	Technique and Notes[4]	Ref.
He (solid)	0-0.43 GPa: 0.23	0.085	--	see reference	42
	>0.18 GPa: 0.21	0.082			

Table 1. Bulk Moduli from Static Compression Data (continued)

Chemical Formula[1]	Density (Mg/m[3])	Isothermal Bulk Modulus (GPa)[2]	dK_{0T}/dP[3]	Technique and Notes[4]	Ref.
He (solid)	0.23 (0 K)	P-V data tabulated only		DAC, XRD, P < 24 GPa	135
Ne (solid)	1. 51 (4 K)	1.097	9.23 ± 0.03	DAC,XRD,4.7<P< 110 GPa, B-M EOS	84
Ar (solid)	1.73 (4 K)	2.86 (at 4 K)	7.2 (at 4 K)	PC, PD, P ≤ 2 GPa, lsqPVfit	9
	1.65 (77 K)	1.41 (at 77 K)	8.4 (at 77 K)		
Ar (solid)	1.73 (0 K)	P-V data tabulated only		DAC, XRD, P < 80 GPa	172
Kr (solid)	2.317	1.41	4.3 ± 0.1	DAC, XRD, P ≤ 32 GPa, M EOS	166
Kr (solid)	3.09 (4 K)	3.34 (at 4 K)	7.2 (at 4 K)	PC, PD, P ≤ 2 GPa, lsqPVfit	9
	2.82 (110 K)	1.58 (at 110 K)	7.6 (at 110 K)		
Xe (solid)	3.78 (0 K)	5.18 ± 0.32 GPa (0 K)	5.48 ± 0.24 (0 K)	DAC, XRD, P ≤ 32 GPa, B-M EOS	15
Xe (solid)	3.78 (4 K)	3.63 (at 4 K)	7.2 (at 4 K)	PC, PD, P ≤ 2 GPa, lsqPVfit	9
	3.41 (159 K)	1.48 (at 159 K)	8.8 (at 159 K)		

Lanthanides:

Pr	6.77	30.2	1.6	PC, PD, P ≤ 4.5 GPa, M EOS	203
Nd	7.00	32.6	3.0	PC, PD, P ≤ 4.5 GPa, M EOS	203
Gd	8.23	35.5	4.8	PC, PD, P ≤ 4.5 GPa, M EOS	203
Gd	8.23	22.71 ± 2.07	4.31 ± 0.29	DAC, XRD, P ≤ 106 GPa, B-M EOS	7
Dy	8.78	40.3	5.1	PC, PD, P ≤ 4.5 GPa, M EOS	203

Table 1. Bulk Moduli from Static Compression Data (continued)

Chemical Formula[1]	Density (Mg/m^3)	Isothermal Bulk Modulus (GPa)[2]	dK$_{0T}$/dP[3]	Technique and Notes[4]	Ref.
Er	9.37	44.9	3.5	PC, PD, P ≤ 4.5 GPa, M EOS	203
Actinides:					
Th	11.70	54.0	4.9	PC, PD, P ≤ 4.5 GPa, M EOS	203
Pa	15.40	157	1.5	DAC, XRD, P ≤ 38 GPa, B-M EOS	24
Iron and Iron Alloys:					
Fe α (bcc) phase	7.86	164 ± 7	4 (assumed)	DAC, XRD, P ≤ 16 GPa, B-M EOS	212
Fe α (bcc) phase	7.86	175.8	3.7	PC, PD, P ≤ 4.5 GPa, M EOS	201
Fe α (bcc) phase	7.86	162.5 ± 5	5.5 ± 0.8	DAC, XRD, P ≤ 30 GPa, B-M EOS	196
Fe-5.2 wt. % Ni α (bcc) phase	7.88	156 ±7	4.2 ± 0.8	DAC, XRD, P ≤ 30 GPa, B-M EOS	196
Fe-10.3 wt. % Ni α (bcc) phase	7.88	153 ± 7	5.7 ± 0.8	DAC, XRD, P ≤ 30 GPa, B-M EOS	196
Fe-7.2 wt. % Si α (bcc) phase	7.39	175 ± 8	4.3 ± 1.0	DAC, XRD, P ≤ 23 GPa, B-M EOS	188
Fe-8 wt. % Si α (bcc) phase	7.35	174.0	4.6	DP, XRD, P ≤ 30 GPa, M EOS	41
Fe-25 wt. % Si α (bcc) phase	6.77	214 ± 9	3.5 ± 0.8	DAC, XRD, P ≤ 23 GPa, B-M EOS	188
Fe$_3$Si *Suessite*	6.58	250.0	-2.0	DP, XRD, P ≤ 30 GPa, M EOS	41
Fe ε (hcp) phase	8.30	208 ± 10	4 (assumed)	DAC, XRD, P ≤ 30 GPa, B-M EOS	196
Fe ε (hcp) phase	8.30	192.7 ± 9.0	4.29 ± 0.36	DAC, XRD, P ≤ 78 GPa, B-M EOS	96
Fe ε (hcp) phase	8.30	164.8 ± 3.6	5.33 ± 0.09	DAC, XRD, P ≤ 300 GPa, B-M EOS	132

Table 1. Bulk Moduli from Static Compression Data (continued)

Chemical Formula[1]	Density (Mg/m[3])	Isothermal Bulk Modulus (GPa)[2]	dK_{0T}/dP[3]	Technique and Notes[4]	Ref.
$Fe_{0.8}Ni_{0.2}$ ε (hcp) phase	8.37	171.8 ± 2.2	4.95 ± 0.09	DAC, XRD, P ≤ 260 GPa, B-M EOS	132
Fe-5.2 wt. % Ni ε (hcp) phase	8.41	212 ± 15	4 (assumed)	DAC, XRD, P ≤ 30 GPa, B-M EOS	196
Fe-10.3 wt. % Ni ε (hcp) phase	8.42	215 ± 25	4 (assumed)	DAC, XRD, P ≤ 30 GPa, B-M EOS	196
Fe-7.2 wt. % Si ε (hcp) phase	7.76	188 ± 14	4 (assumed)	DAC, XRD, P ≤ 23 GPa, B-M EOS	188
Fe-10 wt. % Co	7.93	171 ± 6	4 (assumed)	DAC, XRD, P ≤ 30 GPa, B-M EOS	162
Fe-20 wt. % Co	7.98	169 ± 6	4 (assumed)	DAC, XRD, P ≤ 30 GPa, B-M EOS	162
Fe-40 wt. % Co	8.08	166 ± 6	4 (assumed)	DAC, XRD, P ≤ 30 GPa, B-M EOS	162
$Fe_{0.98}O$ Wüstite	5.87 (ideal)	169 ± 10	4 (assumed)	DAC, XRD, P ≤ 14 GPa, B-M EOS	221
$Fe_{0.945}O$ Wüstite	5.87 (ideal)	157 ± 10	4 (assumed)	DAC, XRD, P ≤ 12 GPa, B-M EOS	94
FeO Wüstite	5.87 (ideal)	142 ± 10	4 (assumed)	DAC, pXRD, P ≤ 26 GPa, B-M EOS	131
FeO Wüstite	5.87 (ideal)	150 ± 3	3.8	DAC, XRD, P ≤ 8.3 GPa, B-M EOS	122
$Fe_{1-x}O$ x = 0.055, 0.07, 0.10 Wüstite	5.87 (ideal)	155 ± 5	--	DAC, scXRD, P ≤ 5 GPa, linPVfit	91
FeO Wüstite	5.87 (ideal)	154.0	3.4	DP, XRD, P ≤ 30 GPa, M EOS	41
$Fe_{0.941}O$ Wüstite	5.87 (ideal)	154 ± 5	4 (assumed)	DAC, XRD, P ≤ 20 GPa, B-M EOS	213
Fe_2U	13.19	239	3	DAC, XRD, P ≤ 45 GPa, M EOS	87
FeS Troilite	4.74	82 ± 7	-5 ± 4	DAC, scXRD, P ≤ 3.4 GPa, B-M EOS	100

Table 1. Bulk Moduli from Static Compression Data (continued)

Chemical Formula[1]	Density (Mg/m^3)	Isothermal Bulk Modulus (GPa)[2]	dK$_{0T}$/dP[3]	Technique and Notes[4]	Ref.
FeS (high-pressure MnP-type structure)	4.77	35 ± 4	5 ± 2	DAC, scXRD, 3.6<P<6.4 GPa, B-M EOS	100
FeS$_2$ *Pyrite*	4.95	143	4 (assumed)	DAC, XRD, P ≤ 70 GPa, B-M EOS	95
FeS$_2$ *Pyrite*	4.95	148	5.5	DP, XRD, P ≤ 30 GPa, M EOS	41
FeS$_2$ *Pyrite*	4.95	157	--	DAC, scXRD, P ≤ 4.2 GPa, linPVfit	58
FeS$_2$ *Pyrite*	4.95	215.4 ± 0.2	5.5	DAC, XRD, P ≤ 34 GPa, B-M EOS	29
FeS$_2$ *Marcasite*	4.87	146.5 ± 0.6	4.9	DAC, XRD, P ≤ 34 GPa, B-M EOS	29
FeH	3.21	121 ± 19	5.31 ± 0.9	DAC, XRD, P ≤ 62 GPa, V EOS	18

Ices:

CO$_2$	1.40	2.93 ± 0.1	7.8	DAC, XRD, P ≤50 GPa, B-M EOS	117
NH$_3$	0.86	7.56 ± 0.06	5.29 ± 0.03	DAC, XRD, 3<P<56 GPa, B-M EOS	161
H$_2$O (cubic phase VII)	1.46	22.3 ± 1.0 (P$_0$ = 2.3 GPa)	4.9±0.7 (P$_0$ = 2.3 GPa)	DAC, XRD, 2.3<P<36 GPa, Avg. of several EOS	152
D$_2$O (cubic phase VII)	1.48	30.0 ± 1.5 (P$_0$ = 2.9 GPa)	4.1±0.5 (P$_0$ = 2.9 GPa)	DAC, XRD, 2.3<P<36 GPa, Avg. of several EOS	152
H$_2$O (cubic phase VII)	1.46	23.7 ± 0.9	4.15 ± 0.07	DAC, XRD 2.3<P<128, B-M EOS	82

Hydrides:

VH$_{0.5}$	~5.73	193.5 ± 4.0	4	DAC, XRD, P ≤ 35 GPa, B-M EOS	19
NbH$_{0.75}$	6.60	202.3 ± 3.0	4	DAC, XRD, P ≤ 35 GPa, B-M EOS	19

Table 1. Bulk Moduli from Static Compression Data (continued)

Chemical Formula[1]	Density (Mg/m³)	Isothermal Bulk Modulus (GPa)[2]	dK_{0T}/dP[3]	Technique and Notes[4]	Ref.
PdH	~10.2	130.0 ± 5.0	4.8	DAC, XRD, P ≤ 35 GPa, B-M EOS	19
PdD	~10.2	135.0 ± 5.0	4.7	DAC, XRD, P ≤ 35 GPa, B-M EOS	19
CrH	~6.23	248.0 ± 9.3	11.0	DAC, XRD, P ≤ 35 GPa, B-M EOS	19
AlH_3	~1.50	47.9 ± 1	3.3 ± 0.2	DAC, XRD, P ≤ 35 GPa, B-M EOS	19
$H(AlH_3)$	~1.28	30.9 ± 2	3.2 ± 0.4	DAC, XRD, P ≤ 35 GPa, B-M EOS	19
CuH	6.38	72.5 ± 2	2.7 ± 0.3	DAC, XRD, P ≤ 35 GPa, B-M EOS	19
H(CuH)	~5.42	22.2 ± 2	3.6 ± 0.3	DAC, XRD, P ≤ 35 GPa, B-M EOS	19

Amorphous Materials:

Chemical Formula[1]	Density (Mg/m³)	Isothermal Bulk Modulus (GPa)[2]	dK_{0T}/dP[3]	Technique and Notes[4]	Ref.
Fe_2SiO_4 liquid	3.747 (1773 K)	24.4 (1773 K)	10.1 (1773 K)	PC, sink/float measurements, 1.0<P<5.5 GPa, B-M EOS	4
SiO_2 glass	2.21	37.0 ± 5.5	-5.6 ± 6.2	DAC, length-change measurements, P ≤10 GPa, B-M EOS	143
Ca-Mg-Na glass	?	35.5 ± 3.7	-2.9 ± 4.1	DAC, length-change measurements, P ≤10 GPa, B-M EOS	143

[1] Mineral names are given in italics where applicable.
[2] All values are at ambient pressure and room temperature except where noted.
[3] A dash implies a linear fit to the pressure-volume data (implies that $K_{0T}' = 0$) or an unstated K_{0T}'.
[4] The temperature derivative of the bulk modulus (dK_{0T}/dT) is given where available.

Key: DAC = diamond-anvil cell. BA = Bridgman anvil press. DP = Drickamer press.
CAP = cubic-anvil press. TAP = tetrahedral-anvil press. MAP = mulianvil press.
TCOA = tungsten carbide opposed anvils, BCAC = boron carbide anvil cell.
XRD = x-ray diifraction. scXRD = single-crystal XRD. pXRD = powder XRD.
PD = piston displacement. DV = direct volume measurement.
ND = neutron diffraction. EXAFS = extended x-ray absorption fine structure.
P = pressure in GPa.
B-M EOS = Birch-Murngahan equation of state (includes a variety of Birch equations).
M EOS = Murngahan equation of state.
V EOS = Vinet et al. ("universal") equation of state.
linPVfit = linear fit to the pressure volume data.
lsqPVfit = least sqares fit to the pressure volume data.
 O-H method = Olinger-Halleck method of reducing P-V data (see reference).

Acknowledgements. This work was supported by NSF and the W.M. Keck Foundation. I thank T.J. Ahrens and an anonymous reviewer for helpful comments. I'm sure references were missed, for which I apologize in advance. This paper could not have been completed without the help of Quentin Williams. Institute of Tectonics (Mineral Physics Lab) contribution number 208.

REFERENCES

1. Adams, D.M, J. Haines and S. Leonard, A high-pressure x-ray powder diffraction and infrared study of copper germanate, a high-Tc model, *J. Phys.: Condens. Matter, 3,* 5183-5190, 1991.

2. Adams, D.M, A.G. Christy, J. Haines and S.M. Clark, Second-order phase transition in PbO and SnO at high pressures: Implications for the litharge-massicot phase transformation, *Phys. Rev. B, 46,* 11358-11367, 1992.

3. Adams, D.M, A.G. Christy, J. Haines and S. Leonard, The phase stability of the alkali-metal vanadates at high pressures studied by synchrotron x-ray powder diffraction and infrared spectroscopy, *J.Phys.:Condens. Matter, 3,* 6135-6144, 1991.

4. Agee, C., Isothermal compression of molten Fe_2SiO_4, *Geophys. Res. Lett., 19,* 1169-1172, 1992.

5. Alexsandrov, I.V., A.F. Goncharov, S.M. Stishov, and E.V. Yakovenko, Equation of state and Raman scattering in cubic BN and SiC at high pressures, *JETP Lett., 50,* 127-131, 1989.

6. Alexsandrov, I.V., A.F. Goncharov, A.N. Zisman and S.M. Stishov, Diamond at high pressures: Raman scattering of light, equation of state, and high-pressure diamond scale, *Sov. Phys. JETP, 66,* 384-390, 1987.

7. Akella, J., G.S. Smith and A.P. Jephcoat, High-pressure phase transformation studies in gadolinium to 106 GPa, *J. Phys. Chem. Solids, 49,* 573-576, 1988.

8. Anderson, M.S. and C.A. Swenson, Experimental compressions for normal hydrogen and normal deuterium to 25 kbar at 4.2 K, *Phys. Rev. B., 10,* 5184-5191, 1974.

9. Anderson, M.S. and C.A. Swenson, Experimental equations of state for the rare gas solids, *J. Phys. Chem. Solids, 36,* 145-161, 1975.

10. Anderson, M.S. and C.A. Swenson, Experimental compressions for sodium, potassium, and rubidium metals to 20 kbar from 4.2 K to 300 K, *Phys. Rev. B., 28,* 5395-5418, 1983.

11. Anderson, M.S. and C.A. Swenson, Experimental equations of state for cesium and lithium metals to 20 kbar and the high-pressure behavior of the alkali metals, *Phys. Rev. B., 31,* 668-680, 1985.

12. Anderson, M.S., C.A. Swenson and D.T. Peterson, Experimental equations of state for calcium, strontium and barium metals to 20 kbar from 4 to 295 K, *Phys. Rev. B., 41,* 3329-3338, 1990.

13. Angel, R., The high-pressure, high-temperature equation of state of calcium fluoride, CaF_2, *J. Phys.: Condens. Matter, 5,* L141-L144, 1993.

14. Angel, R.J., R.M. Hazen, T.C. McCormick, C.T. Prewitt and J.R. Smyth, Comparative compressibility of end-member feldspars, *Phys.Chem. Minerals., 15,* 313-318, 1988.

15. Asaumi, K., High-pressure x-ray diffraction study of solid xenon and its equation of state in relation to metallization transition, *Phys. Rev. B., 29,* 7026-7029, 1984.

16. Åsbrink, S. and M. Malinowski, A high-pressure single-crystal x-ray diffraction study of V_3O_5 including the phase transition at 6.2 GPa, *J. Appl. Cryst., 20,* 195-199, 1987.

17. Ashida, T., Y. Miyamoto and S. Kume, Heat capacity, compressibility and thermal expansion coefficient of ilmenite-type $MgGeO_3$, *Phys. Chem. Minerals, 12,* 129-131, 1985.

18. Badding, J.V., R.J. Hemley and H.K. Mao, High-pressure chemistry of hydrogen in metals: in situ study of iron hydride, *Science, 253,* 421-424, 1991.

19. Baranowski, B., M. Tkacz and S. Majchrzak, Pressure dependence of hydrogen volume in some metallic hydrides, in *Molecular Systems Under High Pressure,* edited by R. Pucci and G. Piccitto, pp. 139-156, Elsevier, 1991.

20. Barnett, J.D., S. Block and G.J. Piermarini, An optical fluorescence system for quantitative pressure measurement in the diamond-anvil cell. *Rev. Sci. Instrum., 44,* 1-9, 1973.

21. Bassett, W.A. and J.D. Barnett, Isothermal compression of stishovite and coesite up to 85 kilobars at room temperature by x-ray diffraction, *Phys. Earth Planet. Inter., 3,* 54-60, 1970.

22. Bassett, W.A. and T. Takahashi, X-ray diffraction studies up to 300 kbar, in *Advances in High Pressure Research,* edited by R.H. Wentorf, pp. 165-247, Academic Press, New York, 1974.

23. Batlogg, B., R.G. Maines, M. Greenblatt and S. DiGregorio, Novel P-V relationship in ReO_3 under pressure, *Phys. Rev. B., 29,* 3762-3766, 1984.

24. Benedict, U. and C. Dufour,

Evaluation of high pressure x-ray diffraction data from energy-dispersive conical slit equipment, *High Temp.-High Press.*, *16*, 501-505, 1984.

25. Birch, F., Finite strain isotherm and velocities for single crystal and polycrystalline NaCl at high pressures and 300 K, *J. Geophys. Res.*, *83*, 1257-1268, 1978.

26. Campbell, A.J. and D.L. Heinz, Compression of KCl in the B2 structure to 56 GPa, *J. Phys. Chem. Solids*, *52*, 495-499, 1991.

27. Campbell, A.J. and D.L. Heinz, Equation of state and hgh pressure transition of NiS in the NiAs structure, *J. Phys. Chem. Solids*, *54*, 5-7, 1993.

28. Cavaleri, M. T.G. Plymate and J.H. Stout, A pressure-volume-temperature equation of state for $Sn(\beta)$ by energy dispersive x-ray diffraction in a heated diamond anvil cell, *J. Phys. Chem. Solids*, *49*, 945-956, 1988.

29. Chattopadhyay, T. and H. G. von Schnering, High-pressure x-ray diffraction study on p-FeS_2, m-FeS_2 and MnS_2 to 340 kbar: a possible high spin-low spin transition in MnS_2, *J. Phys. Chem. Solids*, *46*, 113-116, 1985.

30. Chattopadhyay, T., H.G. von Schnering and W.A. Grosshans, High pressure x-ray diffraction study on the structural phase transition in MnS_2, *Physica 139 & 140B*, 305-307, 1986.

31. Chattopadhyay, T., A. Werner and H.G. von Schnering, Thermal expansion and compressibility of β-As_4S_3, *J. Phys. Chem. Solids*, *43*, 919-923, 1982.

32. Clendenen, R.L. and H.G. Drickamer, Lattice parameters of nine oxides and sulfides as a function of pressure, *J. Chem. Phys.*, *44*, 4223-4228, 1966.

33. Comodi, P., M. Mellini and P.F. Zanazzi, Scapolites: variation of structure with pressure and possible role in storage of fluids,

34. Comodi, P., M. Mellini and P.F. Zanazzi, Magnesiochloritoid: compressibility and high pressure structure refinement, *Phys. Chem. Minerals.*, *18*, 483-490, 1992.

35. Comodi, P., M. Mellini, L. Ungaretti and P.F. Zanazzi, Compressibility and high pressure structure refinement of tremolite, pargasite and glaucophane, *Eur. J. Mineral.*, *3*, 485-499, 1991.

36. Cullity, B.D., *Elements of X-ray Diffraction*, Addison-Wesley, Menlo Park, CA, 555 pp., 1978.

37. da Jornada, J.A., S. Block, F.A. Mauer and G.J. Piermarini, Phase transition and compression of $LiNbO_3$ under static high pressure, *J. Appl. Phys.*, *57*, 842-844, 1985.

38. Deer W.A., R.A. Howie and J. Zussman, An *Introduction to the Rock-Forming Minerals*, 2nd edition, Longman Scientific and Technical, Essex, 696 pp., 1992.

39. Desgreniers, S., Y.K. Vohra and A.L. Ruoff, Tin at high pressure: an energy dispersive x-ray diffraction study to 120 GPa, *Phys. Rev., B.*, *39*, 10359-10361, 1989.

40. d'Amour, H., D. Schiferl, W. Denner, H. Schultz and W.B. Holzapfel, High-pressure single crystal structure determinations of ruby up to 90 kbar using an automatic diffractometer, *J. Appl. Phys.*, *49*, 4411-4416, 1978.

41. Drickamer, H.G., R.W. Lynch, R.L. Clendenen and E.A. Perez-Albuerne, X-ray diffraction studies of the lattice parameters of solids under very high pressure, *Solid State Physics*, *19*, 135-229, 1966.

42. Driessen, A., E. van der Poll and I.F. Silvera, Equation of state of solid ^{4}He, *Phys. Rev. B*, *33*, 3269-3288, 1986.

43. Duclos, S.J., Y.K. Vohra, A.L.

Eur. J. Mineral., *2*, 195-202, 1990.

Ruoff, Experimental study of the crystal stability and equation of state of Si to 248 GPa, *Phys. Rev., B, 41*, 12021-12028, 1990.

44. Duclos, S.J., K. Brister, R.C. Haddon, A.R. Kortan and F.A. Thiel, Effects of pressure and stress on C_{60} fullerite to 20 GPa, *Nature, 351*, 380-381, 1991.

45. Duclos, S.J., Y.K. Vohra, A.L. Ruoff, A. Jayaraman and G.P. Espinosa, High-pressure x-ray diffraction study of CeO_2 to 70 GPa and pressure-induced phase transformation from the fluorite structure, *Phys. Rev., B, 38*, 7755-7758, 1988.

46. Dusing, E. Fr., W.A. Grosshans and W.B. Holzapfel, Equation of state of solid chlorine and bromine, *J. Physique, C8*, C8-203-206, 1984.

47. Edwards, L.R. and R.W. Lynch, The high pressure compressibility and Grüneisen parameter of strontium titanate, *J. Phys. Chem. Solids.*, *31*, 573-574, 1970.

48. Endo, S., T. Mitsui and T. Miyadai, X-ray study of metal-insulator transition in NiS_2, *Physics Letters, 46A*, 29-30, 1973.

49. Faust, J. and E. Knittle, Equation of state, amorphization and phase diagram of muscovite at high pressures, *J. Geophys. Res.*, submitted, 1993.

50. Fei. Y. and H.K. Mao, Static compression of $Mg(OH)_2$ to 78 GPa at high temperature and constraints on the equation of state of fluid-H_2O, *J. Geophys. Res., 98*, 11875-11884, 1993.

51. Fei, Y., H.K. Mao, J. Shu and J. Hu, P-V-T equation of state of magnesiowustite ($Mg_{0.6}Fe_{0.4}$)O, *Phys. Chem. Minerals.*, *18*, 416-422, 1992.

52. Fei, Y., H.K. Mao, J. Shu, G. Parthasarathy, W.A. Bassett and J. Ko, Simultaneous high-P, high T x-ray diffraction study of β-$(Mg,Fe)_2SiO_4$ to 26 GPa and 900 K, *J. Geophys. Res., 97*,

4489-4495, 1992.

53. Finger, L. and R.M. Hazen, Crystal structure and compression of ruby to 46 kbar, *J. Appl. Phys.*, *49*, 5823, 1978.

54. Finger, L. and R.M. Hazen, Crystal structure and isothermal compression of Fe_2O_3, Cr_2O_3 and V_2O_3 to 50 kbars, *J. Appl. Phys.*, *51*, 5362-5367, 1980.

55. Finger, L.W., R.M. Hazen and A. Hofmeister, High-pressure crystal chemistry of spinel ($MgAl_2O_4$) and magnetite (Fe_3O_4): comparisons with silicate spinels, *Phys. Chem. Minerals*, *13*, 215-220, 1986.

56. Finger, L.W., R.M. Hazen and T. Yagi, Crystal structures and electron densities of nickel and iron silicate spinels at elevated temperature or pressure, *Am. Mineral.*, *64*, 1002-1009, 1979.

57. Fischer, M., B. Bonello, A. Polian and J. Leger, Elasticity of $SrTiO_3$ perovskite under pressure, in *Perovskite: A Structure of Great Interest to Geophysics and Materials Science*, A. Navrotsky and D.J Weidner, Eds., AGU Press, Washington, DC, pp. 125-130, 1989.

58. Fujii, T., A. Yoshida, K. Tanaka, F. Marumo and Y. Noda, High pressure compressibilities of pyrite and cattierite, *Mineral. J.*, *13*, 202-211, 1986.

59. Graham, E.K., Recent developments in conventional high-pressure methods, *J. Geophys. Res.*, *91*, 4630-4642, 1986.

60. Grosshans, W.A., E.Fr. Dusing and W.B. Holzapfel, *High Temp.-High Press.*, *16*, 539, 1984.

61. Hanfland, M., H. Beister and K. Syassen, Graphite under pressure: equation of state and first-order Raman modes, *Phys. Rev. B.*, *39*, 12598-12603, 1989.

62. Hazen, R.M., High pressure crystal chemistry of chrysoberyl, Al_2BeO_4: insights on the origin of olivine elastic anisotropy, *Phys. Chem.*

Minerals, 14, 13-20, 1987.

63. Hazen, R.M. and A.Y. Au, High-pressure crystal chemistry of phenakite (Be_2SiO_4) and betrandite ($Be_4Si_2O_7(OH)_2$), *Phys. Chem. Minerals*, *13*, 69-78, 1986.

64. Hazen, R.M. and L.W. Finger, The crystal structures and compressibilities of layer minerals at high pressures. I. SnS_2, berndtite, *Am. Mineral.*, *63*, 289-292, 1978.

65. Hazen, R.M. and L.W. Finger, The crystal structures and compressibilities of layer minerals at high pressures. II. phlogopite and chlorite, *Am. Mineral.*, *63*, 293-296, 1978.

66. Hazen, R.M. and L.W. Finger, Crystal structures and compressibilities of pyrope and grossular to 60 kbar, *Am. Mineral.*, *63*, 297-303, 1978.

67. Hazen, R.M. and L.W. Finger, Crystal structure and compressibility of zircon at high pressure, *Am. Mineral.*, *64*, 196-201, 1979.

68. Hazen, R.M. and L.W. Finger, Linear compressibilities of $NaNO_2$ and $NaNO_3$, *J. Appl. Phys.*, *50*, 6826-6828, 1979.

69. Hazen, R.M. and L.W. Finger, Polyhedral tilting: a common type of pure dusplacive phase transition and its relationship to analcite at high pressure, *Phase Transitions*, *1*, 1-22, 1979.

70. Hazen, R.M. and L.W. Finger, Bulk moduli and high-pressure crystal structures of rutile-type compounds, *J. Phys. Chem. Solids*, *42*, 143-151, 1981.

71. Hazen, R.M. and L.W. Finger, High pressure and high temperature crystallographic study of the gillespite I-II phase transition, *Am. Mineral.*, *68*, 595-603, 1983.

72. Hazen, R.M. and L.W. Finger, Compressibilities and high-pressure phase transitions of sodium tungstate perovskites (Na_xWO_3), *J. Appl. Phys.*, *56*, 311-315, 1984.

73. Hazen, R.M. and L.W. Finger, Compressibility of zeolite 4A is dependent on the molecular size of the hydrostatic pressure medium, *J. Appl. Phys.*, *56*, 1838-1840, 1984.

74. Hazen, R.M. and L.W. Finger, High-pressure and high-temperature crystal chemistry of beryllium oxide, *J. Appl. Phys.*, *59*, 3728-3733, 1986.

75. Hazen, R.M. and L.W. Finger, High-pressue crystal chemistry of andradite and pyrope: revised procedures for high-pressure diffraction experiments, *Am. Mineral.*, *74*, 352-359, 1989.

76. Hazen, R.M. and Z.D. Sharp, Compressibility of sodalite and scapolite, *Am. Mineral.*, *73*, 1120-1122, 1988.

77. Hazen, R.M., A.Y. Au and L.W. Finger, High-pressure crystal chemistry of beryl ($Be_3Al_2Si_6O_{18}$) and euclase ($BeAlSiO_4OH$), *Am. Mineral.*, *71*, 977-984, 1986.

78. Hazen, R.M., J.M. Zhang and J. Ko, Effects of Mg/Fe on the compressibility of synthetic wadsleyite: β-$(Mg_{1-x}Fe_x)_2SiO_4$ ($x <= 0.25$), *Phys. Chem. Minerals*, *17*, 416-419, 1990.

79. Hazen. R.M., H.K. Mao, L.W. Finger and P.M. Bell, Structure and compression of crystalline methane at high pressure and room temperature, *Appl. Phys. Lett.*, *37*, 288-289, 1980.

80. Heinz, D.L. and R. Jeanloz, Compression of the B2 high-pressure phase of NaCl, *Phys. Rev. B.*, *30*, 6045-6050, 1984a.

81. Heinz, D.L. and R. Jeanloz, The equation of state of the gold calibration standard, *J. Appl. Phys.*, *55*, 885-893, 1984b.

82. Hemley, R.J., A.P. Jephcoat, H.K. Mao, C.S.Zha, L.W. Finger and D.E. Cox, Static compression of H_2O-ice to 128 GPa, *Nature*, *330*, 737-740, 1987.

83. Hemley, R.J., H.K. Mao, L.W. Finger, A.P. Jephcoat, R.M. Hazen and C.S. Zha, Equation of state of solid hydrogen and

deuterium from single-crystal x-ray diffraction to 26.5 GPa, *Phys. Rev. B, 42,* 6458-6470, 1990.

84. Hemley, R.J., C.S.Zha, A.P. Jephcoat, H.K. Mao, L.W. Finger and D.E. Cox, X-ray diffraction and equation of state of solid neon to 110 GPa, *Phys. Rev., B, 39,* 11820-11827, 1989.

85. Holloway, J.R. and B.J. Wood, *Simulating the Earth: Experimental Geochemistry,* Unwin-Hyman, Boston, 1988.

86. Houser, B., N. Alberding, R. Ingalls, E.D. Crozier, High-pressure study of α-quartz GeO_2 using extended x-ray absorption fine structure, *Phys. Rev. B, 37,* 6513-6516, 1988.

87. Itie, J.P., J.S. Olsen, L.Gerward, U. Benedict and J.C. Spirlet, High-pressure x-ray diffraction on UX_2 compounds, Physica 139 & 140B, 330-332, 1986.

88. Jayaraman, A., Diamond anvil cell and high-pressure physical investigations, *Rev. Modern Physics, 55,* 65-108, 1983.

89. Jayaraman, A., B. Batlogg, R.G. Maines and H. Bach, Effective ionic charge and bulk modulus scaling in rock-salt structured rare-earth compounds, *Phys. Rev. B, 26,* 3347-3351, 1982.

90. Jeanloz, R., Majorite: vibrational and compressional properties of a high-pressure phase, *J. Geophys. Res., 86,* 6171-6179, 1981.

91. Jeanloz, R. and R.M. Hazen, Compression, nonstoichiometry and bulk viscosity of wüstite, *Nature, 304,* 620-622, 1983.

92. Jeanloz, R. and R.M. Hazen, Finite-strain analysis of relative compressibilities: application to the high-pressure wadsleyite phase as an illustration, *Am. Mineral., 76,* 1765-1768, 1991.

93. Jeanloz, R. and A. Rudy, Static compression of MnO manganosite to 60 GPa, *J. Geophys. Res., 92,* 11433-11436, 1987.

94. Jeanloz, R. and Y. Sato-

Sorensen, Hydrostatic compression of $Fe_{1-x}O$ wustite, *J. Geophys. Res., 91,* 4665-4672, 1986.

95. Jephcoat, A. and P. Olson, Is the inner core of the Earth pure iron?, *Nature, 325,* 332-335, 1987.

96. Jephcoat, A.P., H.K. Mao and P.M. Bell, Static compression of iron to 78 GPa with rare gas solids as pressure-transmitting media, *J. Geophys. Res., 91,* 4677-4684, 1986.

97. Jephcoat, A.P., H.K. Mao and P.M. Bell, Operation of the megabar diamond-anvil cell, in *Hydrothermal Experimental Techniques,* H.E. Barnes and G.C. Ulmer, eds., pp. 469-506, Wiley Interscience, New York, 1987.

98. Jorgensen, J.D., Compression mechanisms in α-quartz structures - SiO_2 and GeO_2, *J. Appl. Phys., 49,* 5473-5478, 1978.

98a. Kikegawa, T. and H. Iwasake, An x-ray diffraction study of lattice compression and phase transition of crystalline phosphorus, *Acta. Cryst., B39,* 158-164, 1983.

99. Kim, K.Y. and A.L. Ruoff, Isothermal equation of state of potassium, *J. Appl. Phys., 52,* 245-249, 1981.

100. King, H.E. and C.T. Prewitt, High-pressure and high-temperature polymorphism of iron sulfide (FeS), *Acta Cryst., B38,* 1877-1887, 1982.

101. Knittle, E. and R. Jeanloz, High-pressure x-ray diffraction and optical absorption studies of CsI, *J. Phys. Chem. Solids, 46,* 1179-1184, 1985.

102. Knittle, E. and R. Jeanloz, Synthesis and equation of state of $(Mg,Fe)SiO_3$ perovskite to over 100 GPa, *Science, 235,* 668-670, 1987.

103. Knittle, E., A. Rudy and R. Jeanloz, High-pressure phase transition in CsBr, *Phys. Rev., B., 31,* 588-590, 1985.

104. Kudoh, Y. and H. Takeda, Single

crystal x-ray diffraction study on the bond compressibility of fayalite, Fe_2SiO_4 and rutile, TiO_2 under high pressure, *Physica B, 139-140,* 333-336, 1986.

105. Kudoh, Y. and Y. Takeuchi, The crystal structure of forsterite Mg_2SiO_4 under high pressure up to 149 kbar, *Z. Kristallogr., 171,* 291-302, 1985.

106. Kudoh, Y., E. Ito and H. Takeda, Effect of pressure on the crystal structure of perovskite-type $MgSiO_3$, *Phys. Chem. Minerals., 14,* 350-354, 1987.

107. Leger, J.M. and A. Atouf, High-pressure phase transformation in cotunnite-type $BaCl_2$, *J. Phys.: Condens. Matter, 4,* 357-365, 1992.

108. Leger, J.M., A.M. Redon and C. Chateau, Compressions of synthetic pyrope, spessartine and uvarovite garnets up to 25 GPa, *Phys. Chem. Mineral., 17,* 161-167, 1990.

109. Leger, J.M., A. Atouf, P.E Tomaszewski and A.S. Pereira, Pressure-induced phase transitions and volume changes in HfO_2 up to 50 GPa, *Phys. Rev. B., 48,* 93-98, 1993.

110. Leger, J.M., P.E Tomaszewski, A. Atouf and A.S. Pereira, Pressure-induced structural phase transitions in zirconia under high pressure, *Phys. Rev. B., 47,* 14075-14083, 1993.

111. Leinenweber, K., W. Utsumi, Y. Tsuchida, T. Yagi and K. Kurita, Unquenchable high-pressure perovskite polymorphs of $MnSnO_3$ and $FeTiO_3$, *Phys. Chem. Minerals., 18,* 244-250, 1991.

112. Levien, L. amd C.T. Prewitt, High-pressure structural study of diopside, *Am. Mineral., 66,* 315-323, 1981.

113. Levien, L. amd C.T. Prewitt, High-pressure crystal chemistry and compressibility of coesite, *Am. Mineral., 66,* 324-333, 1981.

114. Levien, L., C.T. Prewitt and D.J. Weidner, Compression of pyrope, *Am. Mineral., 64,* 805-

808, 1979.

115. Levien, L., C.T. Prewitt and D.J. Weidner, Structure and elastic properties of quartz at pressure, *Am. Mineral.*, 65, 920-930, 1980.

116. Levien, L., D.J. Weidner and C.T. Prewitt, Elasticity of diopside, *Phys. Chem. Minerals*, 4, 105-113, 1979.

117. Liu,L., Compression and phase behavior of solid CO_2 to half a megabar, *Earth Planet. Sci. Lett.*, 71, 104-110, 1984.

118. Liu, L., Compression and polymorphism of potassium to 400 kbar, *J. Phys. Chem. Solids*, 47, 1067-1072, 1986.

119. Liu, L., Polymorphism and compression of TeO_2 at high pressure, *J. Phys. Chem. Solids*, 48, 719-722, 1987.

120. Liu, L.G. and W.A. Bassett, Changes in the crystal structure and the lattice parameter of SrO at high pressure, *J. Geophys. Res.*, 78, 8470-8473, 1973.

121. Liu, M. and L. Liu, Compressions and phase transitions of tin to half a megabar, *High Temp.-High Press.*, 18, 79-85, 1986.

122. Liu, M. and L. Liu, Bulk moduli of wustite and periclase: a comparative study, *Phys. Earth Planet. Inter.*, 45, 273-279, 1987.

123. Liu, L., W.A. Bassett and T. Takahashi, Isothermal compression of a spinel phase of Co_2SiO_4 and magnesian ilmenite, *J. Geophys. Res.*, 79, 1160-1164, 1974.

124. Liu, L., M. Liu, H. Verbeek, C. Hoffner and G.Will, Comparative compressibility of Cu, Ag and Au, *J. Phys. Chem. Solids, 51*, 435-438, 1990.

125. Luo, H. and A.L. Ruoff, X-ray diffraction study of sulfur to 32 GPa: amorphization at 25 GPa, *Phys. Rev. B., 48*, 569-572, 1993.

126. Mao, H.K. and P.M. Bell, Equations of state of MgO and ε-Fe under static pressure conditions, *J. Geophys. Res.,*

84, 4533-4536, 1979.

127. Mao, H.K., T. Takahashi and W.A. Bassett, Isothermal compression of the spinel phase of Ni_2SiO_4 up to 300 kilobars at room temperature, *Phys. Earth Planet. Inter., 3*, 51-53, 1970.

128. Mao, H.K., J. Xu and P.M. Bell, Calibration of the ruby pressure gauge to 800 kbar under quasi-hydrostatic conditions, *J. Geophys. Res., 91*, 4673-4676, 1986.

129. Mao. H.K., P.M. Bell, J. Shaner and D.J. Steinberg, Specific volume measurements of Cu, Mo, Pd and Ag and calibration of the ruby R1 fluorescence pressure gauge from 0.06 to 1 Mbar, *J. Appl. Phys., 49*, 3276-3283, 1978.

130. Mao, H.K., T. Takahashi, W.A. Bassett, G.L. Kinsland and L. Merrill, Isothermal compression of magnetite up to 320 kbar and pressure induced phase transformation, *J. Geophys. Res., 79*, 1165-1170, 1974.

131. Mao, H.K., T. Takahashi, W.A. Bassett, J.S. Weaver and S. Akimoto, Effect of pressure and temperature on the molar volumes of wustite and three $(Fe,Mg)_2SiO_4$ spinel solid solutions, *J. Geophys. Res., 74*, 1061-1069, 1969.

132. Mao, H.K., Y. Wu, L.C. Chen, J. F. Shu and A.P. Jephcoat, Static compression of iron to 300 GPa and $Fe_{0.8}Ni_{0.2}$ Alloy to 260 GPa: Implication for compos-ition of the core, *J. Geophys. Res., 95*, 21737-21742, 1990.

133. Mao, H.K., L.C. Chen, R.J. Hemley, A.P. Jephcoat, Y. Wu and W.A. Bassett, Stability and equation of state of $CaSiO_3$ perovskite to 134 GPa, *J. Geophys. Res., 94*, 17889-17894, 1989.

134. Mao, H.K., Y. Wu, J.F. Shu, J.Z. Hu, R.J. Hemley and D.E. Cox, High-pressure phase transition and equation of state of lead to 238 GPa, *Solid State Commun., 74*, 1027-1029, 1990.

135. Mao, H.K., R.J. Hemley , Y. Wu, A.P. Jephcoat, L.W. Finger, C.S. Zha, and W.A. Bassett, High-pressure phase diagram and equation of state fo solid helium from single-crystal x-ray diffraction to 23.3 GPa, *Phys. Rev. Lett.*, 60, 2649-2652, 1988.

136. Mao, H.K., A.P. Jephcoat, R.J. Hemley, L.W. Finger , C.S. Zha, R.M. Hazen and D.E. Cox, Synchrotron x-ray diffraction measurements of single-crystal hydrogen to 26.5 GPa, *Science, 239*, 1131-1134, 1988.

137. Mao, H.K., Y. Wu, R.J. Hemley, L.C. Chen, J.F. Shu, L.W. Finger and D.E. Cox, High-pressure phase transition and equation of state of CsI, *Phys. Rev. Lett., 64*, 1749-1752, 1990.

138. Mao, H.K., R.J. Hemley, Y. Fei, J.F. Shu, L.C. Chen, A.P. Jephcoat, Y. Wu and W. A. Bassett, Effect of pressure, temperature and composition on lattice parameters and density of $(Fe,Mg)SiO_3$-perovskites to 30 GPa, *J. Geophys. Res., 96*, 8069-8079, 1991.

139. Martens, R., M. Rosenhauer and K.vonGehlen, Compressibilities of carbonates, *High-Pressure Researches in Geoscience*, edited by W. Schreyer, pp. 215-222, E. Schweizerbartsche Verlagsbuchhandlung, Stuttgart, 1982.

140. McCammon, C., Static compression of α-MnS at 298 K to 21 GPa, *Phys. Chem. Minerals., 17*, 636-641, 1991.

141. McCammon, C., J. Zhang, R.M. Hazen and L.W. Finger, High-pressure crystal chemistry of cubanite, $CuFe_2S_3$, *A m . Mineral., 77*, 462-473, 1992.

142. McCormick, T.C., R.M. Hazen and R.J. Angel, Compressibility of omphacite to 60 kbar: role of vacancies, *Am. Mineral., 74*, 1287-1292, 1989.

143. Meade, C. and R. Jeanloz, Frequency-dependent equation of state of fused silica to 10 GPa, *Phys. Rev. B., 35*, 236-241, 1987.

144. Meade, C. and R. Jeanloz, Static compression of Ca(OH)$_2$ at room temperature: observations of amorphization and equation of state measurements to 10.7 GPa, *Geophys. Res. Lett., 17,* 1157-1160, 1990.

145. Meng, Y., Y. Fan, R. Lu, Q. Cui and G. Zou, Structural studies of single crystals La$_6$NiSi$_2$S$_{14}$ and La$_6$CoSi$_2$S$_{14}$ under high pressure, *Physica 139 & 140B,* 337-340, 1986.

146. Merrill, L. and W.A. Bassett, Miniature diamond anvil cell for single crustal x-ray diffraction studies, *Rev. Sci. Instrum., 45,* 290-294, 1974.

147. Ming, L.C. and M. H. Manghnani, Isothermal compression of bcc transition metals to 100 kbar, *J. Appl. Phys., 49,* 208-212, 1978.

148. Ming, L.C., D. Xiong and M. H. Manghnani, Isothermal compression of Au and Al to 20 GPa *Physica 139 & 140B,* 174-176, 1986.

149. Mirwald, P.W. and G.C. Kennedy, Phase relations for SrF$_2$ to 50 kbars and 1900 C and its compression to 40 kbars at 25°C, *J. Phys. Chem. Solids, 41,* 1157-1160, 1980.

150. Mirwald, P.W., M. Malinowski and H. Schulz, Isothermal compression of low-cordierite to 30 kbar (25 °C), *Phys. Chem. Minerals, 11,* 140-148, 1984.

151. Mizukami, S., A. Ohtani, N. Kawai and E. Ito, High pressure x-ray diffraction studies of β-Mg$_2$SiO$_4$ and γ-Mg$_2$SiO$_4$, *Phys. Earth Planet. Inter., 10,* 177-182, 1975.

152. Munro, R.G., S. Block, F.A. Mauer and G. Piermarini, Isothermal equations of state for H$_2$O-VII and D$_2$O-VII, *J. Appl. Phys., 53,* 6174-6178, 1982.

153. Murnaghan, F.D., The compressibility of media under extreme pressures, *Proc. Nat. Acad. Sci., 30,* 244-247, 1944.

154. Nakagiri, N., M.H. Manghnani, L.C. Ming and S. Kimura, Crystal structure of magnetite under pressure, *Phys. Chem. Minerals, 13,* 238-244, 1986.

155. Olijnyk, H., High pressure x-ray diffraction studies on solid N$_2$ up to 43.9 GPa, *J. Chem. Phys., 93,* 8968-8972, 1990.

156. Olijnyk, H., E. Paris, C.A. Geiger and G.A. Lager, Compressional study of katoite (Ca$_3$Al$_2$(O$_4$H$_4$)$_3$) and grossular garnet, *J. Geophys. Res., 96,* 14313-14318, 1991.

157. Olinger, B. and P. Halleck, The compression of α-quartz, *J. Geophys. Res., 81,* 5711-5714, 1976.

158. Olinger, B., The compression of stishovite, *J. Geophys. Res., 81,* 5241-5248, 1976.

159. Olinger, B., Compression studies of forsterite (Mg$_2$SiO$_4$) and enstatite (MgSiO$_3$), in *High Pressure Research: Applications in Geophysics,* edited by M.H. Manghnani and S. Akimoto, pp. 325-334, Academic Press, New York, 1977.

160. Olsen, J.S., Steenstrup, S., L. Gerward, U. Benedict and J.P. Itie, High pressure structural studies of uranium and thorium compounds with the rocksalt structure, *Physica, 139-140B,* 308-310, 1986.

161. Otto, J.W., R.F. Porter and A.L. Ruoff, Equation of state of solid ammonia (NH$_3$) to 56 GPa, *J. Phys. Chem. Solids, 50,* 171-175, 1989.

162. Papantonis, D. and W.A. Bassett, Isothermal compression and bcc-hcp transition of iron-cobalt alloys up to 300 kbar at room temperature, *J. Appl. Phys., 48,* 3374-3379, 1977.

163. Paris, E., C.R.Ross and H. Olijnyk, Mn$_3$O$_4$ at high pressure: a diamond anvil cell study and a structural model, *Eur. J. Mineral., 4,* 87-93, 1992.

164. Parthasarathy, G. and W.B. Holzapfel, High-pressure structural phase transitions in tellurium, *Phys. Rev. B., 37,* 8499-8501, 1988a.

165. Parthasarathy, G. and W.B. Holzapfel, Structural phase transitions and equations of state for selenium under pressure, *Phys. Rev. B., 38,* 10105-10108, 1988b.

166. Polian, A., J.M. Besson, M. Grimsditch and W.A. Grosshans, Solid krypton: equation of state and clastic properties, *Phys. Rev. B, 39,* 1332-1336, 1989.

167. Plymate, T.G. and J.H. Stout, Pressure-volume-temperature behavior and heterogeneous equilibria of the non-quenchable body centered tetragonal poly-morph of metallic tin, *J. Phys. Chem. Solids, 49,* 1339-1348, 1988.

168. Plymate, T.G. and J.H. Stout, Pressure-volume-temperature behavior of fayalite based on static compression measurements at 400 C, *Phys. Chem. Minerals., 17,* 212-219, 1990.

169. Ralph, R.L., L.W. Finger, R.M. Hazen and S. Ghose, Compressibility and crystal structure of andalusite at high pressure, *Am. Mineral., 69,* 513-519, 1984.

170. Richet, P.. H.K. Mao and P.M. Bell, Static compression and equation of state of CaO to 1.35 Mbar, *J. Geophys. Res., 93,* 15279-15288, 1988.

171. Richet, P., J. Xu and H.K. Mao, Quasi-hydrostatic compression of ruby to 500 kbar, *Phys. Chem. Minerals., 16,* 207-211, 1988.

172. Ross, M., H.K. Mao, P.M. Bell and J.A. Xu, The equation of state of dense argon: a comparison of shock and static studies, *J. Chem. Phys., 85,* 1028-1033, 1986.

173. Ross, N.L. and R.M. Hazen, High-pressure crystal chemistry of MgSiO$_3$ perovskite, *Phys. Chem. Minerals, 17,* 228-237, 1990.

174. Ross, N.L. and R.J. Reeder, High-pressure structural study of dolomite and ankerite, *Am. Mineral., 77,* 412-421, 1992.

175. Ross, N.L., J. Ko and C.T. Prewitt, A new phase transition

in $MnTiO_3$: $LiNbO_3$-perovskite structure, *Phys. Chem. Minerals.*, 16, 621-629, 1989.

176. Ross, N.L., J.F. Shu, R.M. Hazen and T. Gasparik, High-pressure crystal chemistry of stishovite, *Am. Mineral.*, 75, 739-747, 1990.

177. Ruoff, A.L. and S. Desgreniers, Very high density solid oxygen: indications of a metallic state, in *Molecular Systems Under High Pressure*, R. Pucci and G. Piccitto, Eds., Elsevier, pp. 123-137, 1991.

178. Sato, Y., Equation of state of mantle minerals determined through high-pressure x-ray study, in *High Pressure Research: Applications in Geophysics*, edited by M.H. Manghnani and S. Akimoto, Academic Press, New York, pp. 307-324, 1977.

179. Sato, Y., Pressure-volume relationship of stishovite under hydrostatic compression, *Earth Planet. Sci. Lett.*, 34, 307-312, 1977.

180. Sato-Sorensen, Y., Phase transitions and equations of state for the sodium halides: NaF, NaCl, NaBr and NaI, *J. Geophys. Res.*, 88, 3543-3548, 1983.

181. Sato, Y. and S. Akimoto, Hydrostatic compression of four corundum-type compounds: α-Al_2O_3, V_2O_3, Cr_2O_3 and α-Fe_2O_3, *J. Appl. Phys.*, 50, 5285-5291, 1979.

182. Sato, Y. and R. Jeanloz, Phase transition in SrO, *J. Geophys. Res.*, 86, 11773-11778, 1983.

183. Sato, Y., M. Akaogi and S. Akimoto, Hydrostatic compression of the synthetic garnets pyrope and almandine, *J. Geophys. Res.*, 83, 335-338, 1978.

184. Sato, T., E. Ito and S. Akimoto, Hydrostatic compression of ilmenite phase os $ZnSiO_3$ and $MgGeO_3$, *Phys. Chem. Minerals*, 2, 171-176, 1977.

185. Schulte, O. and W.B. Holzapfel, Effect of pressure on atomic volume and crystal structure of indium to 67 GPa, *Phys. Rev. B.*,

48, 767-773, 1993.

186. Sharp, Z.D., R.M. Hazen and L.W. Finger, High-pressure crystal chemistry of monticellite, $CaMgSiO_4$, *Am. Mineral.*, 64, 748-755, 1987.

187. Shirotani, I., A. Fukizawa, H. Kawamura, T. Yagi and S. Akimoto, Pressure induced phase transitions in black phosporus, in *Solid State Physics under Pressure*, S. Minomura, Ed., Terra Scientific, Toyko, pp. 207-211, 1985.

188. Silberman, M.L., Isothermal compression of two iron-silicon alloys to 290 kbars at 23 °C, M.S. Thesis, University of Rochester, Rochester, NY, 1867.

189. Singh, A.K. and G.C. Kennedy, Compression of calcite to 40 kbar, *J. Geophys. Res.*, 79, 2615-2622, 1974.

190. Sowa, H., J. Macavei and H. Schulz, The crystal structure of berlinite $AlPO_4$ at high pressure, *Zeitschrift fur Krist.*, 192, 119-136, 1990.

191. Spain, I.L., D.R. Black, L.D. Merkle, J.Z. Hu and C.S. Menoni, X-ray diffraction in the diamond-anvil cell: techniques and application to silicon, *High Temp.-High Press.*, 16, 507-513, 1984.

192. Suzuki, T., T. Yagi and S. Akimoto, Compression behavior of CdS and BP up to 68 GPa, *J. Appl. Phys.*, 54, 748-751, 1983.

193. Syassen, K. and W.B. Holzapfel, Isothermal compression of Al and Ag to 120 kbar, *J. Appl. Phys.*, 49, 4427-4431, 1978.

194. Syassen, K., W. Winzen, H.G. Zimmer, H. Tups and J. M. Leger, Optical response of YbS and YbO at high pressures and the pressure-volume relation of YbS, *Phys. Rev., B.*, 32, 8246-8252, 1985.

195. Takahashi, T. and L.G. Liu, Compression of ferromagnesian garnets and the effect of solid solutions on the bulk modulus, *J. Geophys. Res.*, 75, 5757-5766,

1970.

196. Takahashi, T., W.A. Bassett and H.K. Mao, Isothermal compression of the alloys of iron up to 300 kbar at room temperature: iron-nickel alloys, *J. Geophys. Res.*, 73, 4717-4725, 1968.

197. Takemura, K. and K. Syassen, High pressure equation of state of rubidium, *Solid State Commun.*, 44, 1161-1164, 1982.

198. Takemura, K., O. Shimomura, K. Hase and T. Kikegawa, The high pressure equation of state of α-Mn to 42 GPa, *J. Phys.F: Met. Phys.*,18, 197-204, 1988.

199. Tamai, H. and T. Yagi, High-pressure and high-temperature phase relations in $CaSiO_3$ and $CaMgSi_2O_6$ and elasticity of perovskite-type $CaSiO_3$, *Phys. Earth Planet. Inter.*, 54, 370-377, 1989.

200. Tarrida, M. and P. Richet, Equation of state of $CaSiO_3$ perovskite to 96 GPa, *Geophys. Res. Lett.*, 16, 1351-1354, 1989.

201. Vaidya, S.N. and G.C. Kennedy, Compressibility of 18 metals to 45 kbar, *J. Phys. Chem. Solids*, 31, 2329-2345, 1970.

202. Vaidya, S.N. and G.C. Kennedy, Compressibility of 27 halides to 45 kbar, *J. Phys. Chem. Solids*, 32, 951-964, 1971.

203. Vaidya, S.N. and G.C. Kennedy, Compressibility of 22 elemental solids to 45 kbar, *J. Phys. Chem. Solids, 33*, 1377-1389, 1972.

204. Vedel, I., A.M. Redon, J.M. Leger, J. Rossat-Mignod and O. Vogt, The continuous valence transition in CeS under high pressure, *J. Phys. C: Solid State*, 19, 6297-6302, 1986.

205. Vinet, P., J. Ferrante, J.H. Rose and J.R. Smith, Compressibility of solids, *J. Geophys. Res.*, 92, 9319-9325, 1987.

206. Vohra, Y.K. and A.L. Ruoff, Static compression of metals Mo, Pb, and Pt to 272 GPa: comparison with shock data, *Phys. Rev. B*, 42, 8651-8654,

1990.

207. Wechsler, B.A. and C.T. Prewitt, Crystal structure of ilmenite (FeTiO$_3$) at high temperature and at high pressure, *Am. Mineral., 69*, 176-185, 1984.

208. Werner, A. and H.D. Hochheimer, High-pressure x-ray of Cu$_2$O and Ag$_2$O, *Phys. Rev. B., 25*, 59295935, 1982.

209. Werner, A., H.D. Hochheimer and A. Jayaraman, Pressure-induced phase transformations in the chalcopyrite-structure compounds: CuGaS$_2$ and AgGaS$_2$, *Phys. Rev. B., 23*, 3836-3838, 1981.

210. Werner, A., H.D. Hochheimer, K. Strossner and A. Jayaraman, High-pressure x-ray diffraction studies of HgTe and HgS to 20 GPa, *Phys. Rev. B., 28*, 3330-3334, 1983.

211. Wier, S.T., Y.K. Vohra and A.L. Ruoff, High-pressure phase transitions and the equations of state of BaS and BaO, *Phys. Rev. B, 33*, 4221-4226, 1986.

212. Wilburn, D.R. and W.A. Bassett, Hydrostatic compression of iron and related compounds: an overview, *Am. Mineral., 63*, 591-596, 1978.

213. Will, G. E. Hinze and W. Nuding, The compressibility of FeO measured by energy dispersive x-ray diffraction in a diamond anvil squeezer up to 200 kbar, *Phys. Chem. Minerals., 6*, 157-167, 1980.

214. Will, G., W. Hoffbauer, E. Hinze and J. Lauerjung, The compressibility of forsterite up to 300 kbar measured with synchrotron radiation, *Physica, 139 and 140 B*, 193-197, 1986.

215. Williams, Q. and R. Jeanloz, Ultra-high-pressure experimental techniques, in *Molten Salt Techniques, Vol. 4*, R.J. Gale and D.G. Lovering, Eds., Plenum Publishing, pp. 193-227, 1991.

216. Williams, Q., E. Knittle, R. Reichlin, S. Martin and R. Jeanloz, Structural and electronic properties of Fe$_2$SiO$_4$-fayalite at ultrahigh pressures: amorphization and gap closure, *J. Geophys. Res., 95*, 21549-21563, 1990.

217. Xia, H., S.J. Duclos, A.L. Ruoff and Y.K. Vohra, New high-pressure phase transition in zirconium metal, *Phys. Rev. Lett., 64*, 204-207, 1990.

218. Xiong, D.H., L.C. Ming and M.H. Manghnani, High-pressure phase transformations and isothermal compression in CaTiO$_3$ (perovskite), *Phys. Earth Planet. Inter., 43*, 244-253, 1986.

219. Yagi, T., Experimental determination of thermal expansivity of several alkali halides at high pressures, *J. Phys. Chem. Solids, 39*, 563-571, 1978.

220. Yagi, T., H.K. Mao and P.M. Bell, Hydrostatic compression of perovskite-type MgSiO$_3$ in *Advances in Physical Geochemistry*, vol. 2, S. Saxena, Ed., pp. 317-325, Springer-Verlag, New York, 1982.

221. Yagi, T., T. Suzuki and S. Akimoto, Static compression of wustite (Fe$_{0.98}$O) to 120 GPa, *J. Geophys. Res., 90*, 8784-8788, 1985.

222. Yagi, T., Y. Ida, Y. Sato and S. Akimoto, Effects of hydrostatic pressure on the lattice parameters of Fe$_2$SiO$_4$ olivine up to 70 kbar, *Phys. Earth Planet. Int., 10*, 348-354, 1975.

223. Yagi, T., Y. Uchiyama, M. Akaogi and E. Ito, Isothermal compression curve of MgSiO$_3$ tetragonal garnet, *Phys. Earth Planet. Inter., 74*, 1-7, 1992.

224. Yagi, T., M. Akaogi, O. Shimomura, H. Tamai and S. Akimoto, High pressure and high temperature equations of state of majorite, in *High-Pressure Research in Mineral Physics*, M.H. Manghnani and Y. Syono, Eds., AGU, Washington DC, pp. 141-147, 1987.

225. Yamanaka, T. and K. Ogata, Structure refinement of GeO$_2$ polymorphs at high pressures and temperatures by energy-dispersive spectra of powder diffraction, *J. Appl. Cryst., 24*, 111-118, 1991.

226. Zhao, Y.X. and I.L. Spain, X-ray diffraction data for graphite to 20 GPa, *Phys. Rev. B, 40*, 993-997, 1989.

227. Zerr, A., H. Reichmann, H. Euler and R. Boehler, Hydrostatic compression of γ-(Mg$_{0.6}$Fe$_{0.4}$)$_2$SiO$_4$ to 50.0 GPa, *Phys. Chem. Minerals, 19*, 507-509, 1993.

228. Zhang, L. and S.S. Hafner, High-pressure [57]Fe γ resonance and compressibility of Ca(Fe,Mg)Si$_2$O$_6$ clinopyroxenes, *Am. Mineral., 77*, 462-473, 1992.

229. Zhang, L., H. Ahsbahs, A. Kutoglu and S.S. Hafner, Compressiblity of grunerite, *Am. Mineral., 77*, 480-483, 1992.

230. Zhou, Y., A.J. Campbell and D.L. Heinz, Equations of state and optical properties of the high pressure phase of zinc sulfide, *J. Phys. Chem. Solids, 52*, 821-825, 1991.

231. Zimmer, H.G., K. Takemura, K. Syassen and K. Fischer, Insulator-metal transition and valence instablity in EuO near 130 kbar, *Phys. Rev., B., 29*, 2350-2352, 1984.

Shock Wave Data for Minerals

Thomas J. Ahrens and Mary L. Johnson

1. INTRODUCTION

Shock compression of the materials of planetary interiors yields data, which upon comparison with density-pressure and density-sound velocity profiles of both terrestrial planetary mantles and cores [4,5,94], as well as density profiles for the interior of the major planets [148], constrain internal composition and temperature. Other important applications of shock wave data and related properties are found in the impact mechanics of terrestrial planets and the solid satellites of the terrestrial and major planets. Significant processes which can, or have been, studied using shock wave data include: (1) the formation of planetary metallic cores during accretion [169,192], and (2) the production of a shock-melted "magma ocean" and concurrent impact volatilization versus retention of volatiles during accretion [1]. Also of interest are the shock-induced chemical reactions between meteoritic components (e.g. H_2O and Fe: [111]). The formation of primitive atmospheres, for example, containing a large fraction of H_2O and CO_2 is also addressable using shock wave and other thermodynamic data for volatile-bearing minerals (e.g. [110,112]). A related application of both shock compression and isentropic release data for minerals

T. J. Ahrens and M. L. Johnson, Seismological Laboratory, 252-21, California Institute of Technology, Pasadena, CA 91125

Present Address: M. L. Johnson, Gemological Institute of America, 1639 Stewart Street, Santa Monica, CA 90404

Mineral Physics and Crystallography
A Handbook of Physical Constants
AGU Reference Shelf 2

[13,14] is in the mechanics of both the continued bombardment and hence cratering on planetary objects through geologic time [170], as well as the effects of giant impacts on the Earth [183,185]. Finally, recovery and characterization of shock-compressed materials have provided important insights into the nature of shock deformation mechanisms and, in some cases, provided physical data on the nature of either shock-induced phase changes or phase changes which occur upon isentropic release from the high-pressure shock state (e.g., melting) [193,194].

As indicated for the data summary of Table 1, a very large data set exists describing the Hugoniot equation of state of minerals. Whereas some earlier summaries have provided raw shock data [47,121,213], the present summary provides fits to shock wave data. Earlier summaries providing fits to data are given by Al'tshuler et al. [24] and Trunin [203].

Hugoniot data specify the locus of pressure-density (or specific volume) states which can be achieved by a mineral from some initial state with a specified initial density. An analogous summary for rocks, usually described as a mixture of minerals are given in Chapter 3-4.

Three pressure units are commonly in use in the shock wave literature: kilobar (kbar), gigapascal (GPa), and megabar (Mbar). These are equal to 10^9, 10^{10}, and 10^{12} dyne/cm^2, respectively, or 10^8, 10^9, and 10^{11} pascals in SI units.

2. SHOCK WAVE EQUATION OF STATE

The propagation of a shock wave from a detonating explosive or the shock wave induced upon impact of a flyer plate accelerated, via explosives or with a gun, result

TABLE 1. Shock Wave Equation of State of Minerals and Related Materials of the Solar System

Mineral	Formula	Sample Density (Mg/m^3)	C_o (km/sec)	error ΔC_o (km/sec)	S	error ΔS	lower U_p (km/sec)	upper U_p (km/sec)	Phase*	No. of Data	References / Temp. Refs.
Gases:											
Air[c]	(mixture)	0.884	2.28 4.25	-- --	1.20 0.85	-- --	4.317 5.788	5.788 7.379	2 4	2 2	[147]
Nitrogen plus Oxygen[c]	1:1 N_2+O_2	0.945	1.83	0.11	1.26	0.03	2.235	3.785	2	6	[179]
Nitric Oxide[e]	NO	1.263	3.76	0.06	0.98	0.02	2.01	3.245	2	7	[179]
Ammonia[i]	NH_3	0.715	2.45	0.19	1.34	0.04	1.01	7.566	2	12	[83,121,140]/ [159]
Argon[k]	Ar	0.0013	0.71	0.10	1.041	0.018	1.73	7.81	2	25	[58,71,84]
Argon[f]	Ar	0.919	1.04	0.06	1.36	0.02	1.59	4.04	2	6	[214]
Argon[l]	Ar	1.026	1.1	0.2	1.45	0.07	1.42	4.10	2	7	[213]
Argon[c]	Ar	1.401	1.01 1.28 3.04	0.10 0.06 0.14	1.79 1.58 1.09	0.08 0.02 0.03	0.301 1.32 3.65	1.35 3.758 6.451	2 3 4	9 24 10	[80,121,146,180, 214] / [80,221]
Argon[c]	Ar	1.65	0.88 1.9 2.5	0.15 0.3 0.3	2.00 1.46 1.29	0.11 0.10 0.07	0.56 1.85 3.60	1.85 3.60 4.60	2 3 4	8 7 3	[64,109,121, 189]
Carbon Dioxide[i]	CO_2	1.173	1.54 3.3	0.09 0.4	1.44 1.01	0.03 0.07	1.585 4.549	3.765 6.264	2 4	16 3	[147,178]
Carbon Dioxide[h]	CO_2	1.541	1.99 3.27	0.08 --	1.56 1.21	0.03 --	1.03 3.68	3.68 4.79	2 4	3 2	[233]
Carbon Monoxide[c]	CO	0.807	1.54 2.59 1.3	-- 0.04 0.3	1.40 0.974 1.21	-- 0.010 0.04	1.692 2.471 5.608	2.471 5.608 7.92	2 3 4	2 3 4	[150]

TABLE 1. Shock Wave Equation of State of Minerals and Related Materials of the Solar System (continued)

Mineral	Formula	Sample Density (Mg/m^3)	C_o (km/sec)	error ΔC_o (km/sec)	S	error ΔS	lower U_p (km/sec)	upper U_p (km/sec)	Phase*	No. of Data	References / Temp. Refs.
Deuterium[b]	D$_2$	0.167	1.7 2.4	0.6 0.3	1.29 1.17	0.12 0.03	3.678 6.263	6.263 9.014	2 3	8 4	[63,144]
Helium[a]	He	0.123	0.674	0.011	1.366	0.002	2.47	9.39	2	3	[145]
Hydrogen[b]	H$_2$	0.071	1.128 1.49 2.38	0.006 0.08 0.19	1.829 1.51 1.23	0.013 0.03 0.03	0 1.105 3.080	1.105 3.080 9.962	1 2 4	5 3 10	[63,144,215]
Hydrogen[a]	H$_2$	0.089	1.80	0.12	1.89	0.09	0.801	1.525	2	4	[109]
Methane[d]	CH$_4$	0.423	2.19 2.87	-- 0.09	1.35 1.166	-- 0.014	2.222 3.568	3.568 8.341	2 4	2 4	[150]
Nitrogen[k]	N$_2$	0.0013	0.38	0.02	1.038	0.004	3.80	8.99	2	10	[57]
Nitrogen[c]	N$_2$	0.811	0.94 1.14 2.1 4.0	0.10 0.18 0.3 0.2	1.83 1.59 1.26 0.88	0.09 0.08 0.06 0.04	0 1.51 3.26 5.2	1.57 3.26 5.23 8.63	1 2 3 4	9 15 24 12	[60,61,62,121,146,149,179,213,233] / [149,160,221]
Oxygen[c]	O$_2$	1.202	1.60 2.35	0.16 0.10	1.45 1.22	0.06 0.02	2.06 2.91	2.98 6.766	2 4	9 16	[121,146,223]
Xenon[k]	Xe	0.012	0.2 1.5 0.04 1.8	0.4 0.5 0.3 0.5	1.15 0.74 1.14 0.93	0.13 0.13 0.05 0.05	1.58 3.13 4.11 7.69	3.33 4.11 7.69 11.1	1 3 2 4	4 6 7 3	[71,81,216]
Xenon[g]	Xe	3.006	1.33 1.7 1.49 1.94	0.16 - 0.15 0.14	1.33 1.1 1.21 1.09	0.09 - 0.05 0.03	1.185 2.51 2.7 3.82	2.51 2.7 3.82 5.502	1 3 2 4	8 2 8 3	[151,212,213] /[161,212]

TABLE 1. Shock Wave Equation of State of Minerals and Related Materials of the Solar System (continued)

Mineral	Formula	Sample Density (Mg/m³)	C_0 (km/sec)	error ΔC_0 (km/sec)	S	error ΔS	lower U_p (km/sec)	upper U_p (km/sec)	Phase*	No. of Data	References / Temp. Refs.
Elements:											
Antimony	Sb	6.695	3.2	-	-0.8	-	0	0.311	1	2	[121,126,217,227]
			2.62	0.02	0.95	0.03	0.311	0.997	2	6	
			2.03	0.07	1.61	0.04	0.989	2.699	4	13	
Bismuth	Bi	9.817	2.17	0.06	-1.0	0.5	0	0.32	1	17	[30,68,89,113, 121,131,166, 217,226]
			1.08	0.06	2.20	0.07	0.32	1.183	2	30	
			2.01	0.04	1.358	0.019	1.183	4.45	4	21	
Carbon:											
Graphite	C	0.4665	0.4	0.3	1.14	0.06	2.114	6.147	2	6	[121]
Graphite	C	1.000	0.79	0.12	1.30	0.03	0.772	5.617	2	36	[121]
Graphite	C	1.611	1.75	0.09	1.42	0.05	0.911	4.22	2	60	[82,121,133,213, 217]
Graphite	C	1.794	2.04	0.14	1.66	0.08	0	2.563	2	30	[82,121,133,213, 217]
			4.2	0.5	0.71	0.18	2.372	3.08	3	19	
			1.9	0.3	1.49	0.07	3.069	5.42	4	41	
Graphite	C	2.205	3.11	0.07	4.7	0.2	0.012	0.41	1	12	[58,65,82,121, 126,133,162, 217]
			4.19	0.05	1.83	0.04	0.404	1.9	2	77	
			7.5	0.3	0.21	0.11	1.89	3.316	3	24	
			3.92	0.06	1.331	0.008	3.119	28.38	4	22	
Diamond	C	1.90	1.2	0.2	1.73	0.05	2	6.5	2	5	[153]
Diamond	C	3.191	7.74	0.05	1.456	0.019	1.364	3.133	2	3	[121]
Diamond[q]	C	3.51	12.16	--	1	--	2	8.5	2	--	[153]
Glassy Carbon	C	1.507	2.72	0.11	1.12	0.03	0	5.8	2	45	[82,121,181]

TABLE 1. Shock Wave Equation of State of Minerals and Related Materials of the Solar System (continued)

Mineral	Formula	Sample Density (Mg/m³)	C_0 (km/sec)	error ΔC_0 (km/sec)	S	error ΔS	lower U_p (km/sec)	upper U_p (km/sec)	Phase*	No. of Data	References / Temp. Refs.
Carbon Foam	C	0.435	0.85	0.09	0.88	0.05	0.815	2.301	2	20	
			-0.35	0.12	1.34	0.02	2.119	6.734	4	83	[121]
Carbon Fibers	C	1.519	1.18	0.09	1.73	0.06	0.924	2.361	2	5	
			2.52	0.10	1.14	0.03	2.361	5.041	4	15	[121]
Cobalt	Co	2.594	-0.15	0.06	1.602	0.017	0.651	6.39	2	11	[209]
Cobalt	Co	4.15	0.05	0.03	1.863	0.014	0.615	3.63	2	8	[209]
Cobalt	Co	5.533	0.42	0.04	2.11	0.02	0.293	2.89	2	10	
			1.38	0.08	1.76	0.02	2.89	5.2	4	4	[209]
Cobalt	Co	8.82	4.53	--	1.77	--	0	0.482	1	2	
			4.77	0.02	1.285	0.014	0.482	2.297	2	17	[24,121,126,131,
			3.98	0.13	1.66	0.04	2.289	4.32	4	4	166,217,226]
Copper	Cu	1.909	0.03	0.08	1.361	0.009	0.661	26.1	2	27	[24,209]
Copper	Cu	2.887	0.37	0.08	1.406	0.015	1.15	17.25	2	28	[24,121,209]
Copper	Cu	3.57	0.03	0.02	1.675	0.008	0.63	3.96	2	6	[209]
Copper	Cu	4.475	1.35	--	-2.02	--	0	0.315	1	2	
			0.15	0.05	1.87	0.03	0.315	2.944	2	14	[56,121,134,136,
			1.9	0.3	1.33	0.05	2.944	9.56	4	8	204,209]
Copper	Cu	6.144	2.73	0.11	-1.7	0.8	0	0.534	1	14	
			0.87	0.07	1.97	0.04	0.534	3.365	2	39	[56,121,128,134, 136,172,204,
			3.2	0.4	1.27	0.07	3.327	8.77	4	5	209, 217]
Copper	Cu	7.315	3.15	0.14	-0.4	0.2	0	0.701	1	3	
			1.73	0.07	1.94	0.04	0.669	3.063	2	18	[56,121,126,128, 134,136]

TABLE 1. Shock Wave Equation of State of Minerals and Related Materials of the Solar System (continued)

Mineral	Formula	Sample Density (Mg/m³)	C_0 (km/sec)	error ΔC_0 (km/sec)	S	error ΔS	lower U_p (km/sec)	upper U_p (km/sec)	Phase*	No. of Data	References / Temp. Refs.
Copper	Cu	7.90	3.39 2.29	0.04 0.06	-0.06 1.90	0.08 0.03	0 0.627	0.646 2.969	1 2	3 18	[56,121,127,134, 136]
Copper	Cu	8.931	3.982	0.014	1.460	0.006	0	12.1	2	315	[22,24,25,28,29, 30,90,121,126, 131,134,136, 139,141,143, 166,172,184, 205,213,217, 226]
Germanium	Ge	5.328	5.93 1.98	0.19 0.10	-1.8 1.63	0.2 0.04	0.0775 1.226	1.226 3.188	1 2	17 46	[79,121,126,136, 213]
Gold	Au	19.263	2.95 3.08	0.03 0.04	1.81 1.546	0.07 0.019	0 0.71	0.71 3.52	1 2	5 11	[24,100,121,131, 166,217,226]
Indium	In	7.281	2.54 5.48	0.05 --	1.49 0.47	0.03 --	0.56 2.932	2.932 4.87	2 4	15 2	[24,121,166,217, 226]
Iodine	I₂	4.902	1.62 1.34 2.4	0.03 0.04 0.3	1.25 1.59 1.17	0.04 0.02 0.08	0.49 0.9 2.65	0.9 2.66 4.73	1 2 4	5 38 13	[125,213]
Iridium	Ir	22.54	3.81 4.37 3.36	0.05 0.07 0.09	1.76 1.15 1.76	0.10 0.05 0.04	0 0.933 1.629	0.933 1.629 3.09	1 3 4	6 4 3	[24,121,136]
Iron	Fe	2.633	-0.04 1.8	0.08 0.3	1.63 1.28	0.04 0.03	0.646 3.27	3.27 21.54	2 4	6 8	[204,209]
Iron	Fe	3.359	0.23	0.09	1.67	0.04	0.644	5.52	2	24	[121,134,136, 204]

TABLE 1. Shock Wave Equation of State of Minerals and Related Materials of the Solar System (continued)

Mineral	Formula	Sample Density (Mg/m^3)	C_0 (km/sec)	error ΔC_0 (km/sec)	S	error ΔS	lower U_p (km/sec)	upper U_p (km/sec)	Phase*	No. of Data	References / Temp. Refs.
Iron	Fe	4.547	0.57 2.4	0.09 0.3	1.88 1.38	0.04 0.05	0.591 3.58	3.59 9.1	2 4	33 6	[121,134,136, 204, 209]
Iron	Fe	5.783	3.15 1.17 0.01 2.25	0.10 0.03 0.09 0.09	-1.7 1.98 3.22 1.61	0.6 0.04 0.08 0.03	0 0.537 0.941 1.565	0.537 0.941 1.249 4.95	1 2 3 4	12 4 3 10	[31,121,134,136, 172,213]
Iron	Fe	6.972	4.1 2.4 1.2 2.77	-- 0.2 0.3 0.08	-1.5 1.3 2.8 1.71	-- 0.3 0.2 0.04	0 0.569 0.85 1.427	0.569 0.85 1.453 3.131	1 2 3 4	2 5 4 15	[121,134,136]
Iron	Fe	7.853	5.85 3.48 3.94 5.36	0.12 0.05 0.03 0.07	-1.7 1.91 1.584 1.302	0.8 0.05 0.013 0.008	0 0.763 1.413 4.50	0.573 1.433 4.55 21.73	1 2 4a 4b	16 42 97 18	[24,25,29,30,31, 33,108,121,126, 131,132,134, 136,138,162, 166,172,202, 206,213,217, 226]/[6,20,43, 44]
Iron-Nickel (see Taenite)	(Fe, Ni)										
Iron-Silicon	Fe$_{12}$Si	7.641	3.87	0.04	1.67	0.02	0.984	3.568	2	37	[42,121]
Iron-Silicon	Fe$_7$Si	7.49	4.01	0.06	1.71	0.04	0.975	2.291	2	3	[121]
Iron-Silicon (see Suessite)	Fe$_3$Si										

TABLE 1. Shock Wave Equation of State of Minerals and Related Materials of the Solar System (continued)

Mineral	Formula	Sample Density (Mg/m³)	C_0 (km/sec)	error ΔC_0 (km/sec)	S	error ΔS	lower U_p (km/sec)	upper U_p (km/sec)	Phase*	No. of Data	References / Temp. Refs.
Lead	Pb	4.71	0.31	--	1.42	--	0.607	1.016	2	2	
			0.17	0.06	1.59	0.03	1.016	2.83	3	5	
			0.78	0.12	1.37	0.03	2.83	5.77	4	3	[209]
Lead	Pb	6.79	0.555	0.016	1.726	0.015	0.55	1.44	2	4	
			0.94	0.05	1.462	0.018	1.44	3.24	3	3	
			1.36	--	1.33	--	3.24	4.85	4	2	[209]
Lead	Pb	8.40	0.71	0.07	2.10	0.14	0.26	0.73	2	3	
			1.15	0.08	1.59	0.05	0.73	2.18	3	6	
			1.84	0.05	1.306	0.016	2.18	4.91	4	5	[209]
Lead	Pb	9.51	1.20	0.07	1.87	0.08	0.46	1.16	2	3	
			1.58	0.05	1.51	0.02	1.16	2.73	3	3	[209]
Lead	Pb	11.345	1.992	0.014	1.511	0.012	0	2.36	2	93	[25,28,30,33,
			2.70	0.04	1.213	0.006	2.335	19.12	4	42	121,131,141, 163,166,206, 213,217,226]
Mercury	Hg	13.54	1.45	--	2.26	--	0	0.56	2	2	
			1.752	0.007	1.724	0.009	0.56	0.991	4	3	[121,225]
Nickel	Ni	1.644	-0.04	0.05	1.32	0.02	0.67	3.15	1	16	
			-0.4	0.3	1.47	0.06	2.86	5.73	2	19	[209]
Nickel	Ni	3.202	-0.09	0.07	1.71	0.02	1.53	4.045	2	6	
			0.87	0.17	1.46	0.02	4.045	10.31	4	3	[209]
Nickel	Ni	4.198	0.02	0.13	1.88	0.05	0.61	3.86	2	15	[209]
Nickel	Ni	5.15	0.7	--	1.9	--	1.23	1.98	2	2	
			1.40	0.09	1.606	0.019	1.98	8.91	4	7	[209]
Nickel	Ni	6.275	0.23	0.09	2.79	0.13	0.54	0.80	1	3	
			1.02	0.10	2.03	0.05	0.80	3.28	2	9	
			2.15	--	1.64	--	3.28	4.62	4	2	[209]

TABLE 1. Shock Wave Equation of State of Minerals and Related Materials of the Solar System (continued)

Mineral	Formula	Sample Density (Mg/m³)	C₀ (km/sec)	error ΔC₀ (km/sec)	S	error ΔS	lower Up (km/sec)	upper Up (km/sec)	Phase*	No. of Data	References / Temp. Refs.
Nickel	Ni	8.896	4.57	0.04	0.29	0.17	0	0.354	1	4	[24,25,90,121,131,166,209,213,217,226]
			3.83	0.17	2.5	0.3	0.349	0.635	2	28	
			4.31	0.04	1.63	0.03	0.635	2.63	4a	52	
			5.41	0.08	1.300	0.015	2.63	7.5	4b	18	
Palladium	Pd	11.996	3.83	0.02	1.83	0.04	0.00	0.825	2	7	[56,121,136,166,217,226]
			4.09	0.05	1.49	0.03	0.803	2.317	4	13	
Platinum	Pt	21.445	3.587	0.014	1.556	0.008	0.00	3.444	2	29	[86,121,136,166,217,226]
Rhenium	Re	20.53	4.12	0.05	-0.04	0.18	0.00	0.372	1	3	[121,136]
			3.56	0.08	1.63	0.08	0.372	1.441	2	7	
			4.0	0.2	1.32	0.12	1.346	2.028	4	6	
Rhenium	Re	20.984	4.16	0.04	1.40	0.06	0.00	1.127	2	7	[121,136]
Rhodium	Rh	12.422	4.28	0.12	2.7	0.4	0.00	0.426	1	4	[24,121,136,166,226]
			4.76	0.05	1.41	0.04	0.369	2.004	2	14	
			4.043	0.018	1.713	0.006	2.004	3.8	4	3	
Silver	Ag	10.49	3.23	0.04	1.59	0.03	0.00	2.149	2	16	[24,30,56,121,131,166,217,226]
			3.56	0.13	1.46	0.04	2.12	4.32	4	9	
Suessite	(Fe,Ni)3Si	6.870	5.21	--	2.25	--	0.00	0.495	1	2	[42,121]
			5.53	0.06	1.23	0.03	0.495	3.627	2	32	
Sulfur	S	2.02	3.633	0.013	0.606	0.010	0.897	1.470	2	3	[121]
			2.8	0.3	1.18	0.15	1.431	2.046	4	6	
Taenite (also Kamacite)	(Fe,Ni)	7.933	4.41	0.05	1.01	0.05	0.00	1.09	1	11	[36,121,132,136,217]
			3.79	0.051	1.65	0.02	1.019	2.777	2	41	
			4.20	0.17	1.48	0.05	2.723	4.59	4	11	

TABLE 1. Shock Wave Equation of State of Minerals and Related Materials of the Solar System (continued)

Mineral	Formula	Sample Density (Mg/m^3)	C_0 (km/sec)	error ΔC_0 (km/sec)	S	error ΔS	lower U_p (km/sec)	upper U_p (km/sec)	Phase*	No. of Data	References / Temp. Refs.
Tin	Sn	7.299	2.60	0.15	2.2	0.9	0.00	0.304	1	3	[24,25,30,121, 126,131,166, 213,217, 226]
			3.33	0.07	-0.14	0.15	0.304	0.5	3	5	
			2.48	0.03	1.57	0.03	0.5	2.15	2	66	
			3.43	0.03	1.205	0.008	2.15	8	4	33	
Wairauite	CoFe	8.091	4.64	0.04	1.63	0.08	0.00	0.647	2	12	[121]
			5.69	0.02	-0.10	0.02	0.663	1.037	3	5	
			3.78	0.08	1.62	0.04	1.038	2.801	4	10	
Zinc	Zn	6.51	3.69	0.17	0.98	0.12	0.54	2.08	2	3	[23]
			3.04	0.15	1.35	0.05	2.08	5.04	4	4	
Zinc	Zn	7.138	3.00	0.02	1.586	0.013	0.00	3.01	2	39	[24,25,30,121, 126,131,166, 217,224, 226]
			3.70	0.15	1.37	0.04	2.98	4.85	3	10	
			4.05	0.02	1.303	0.003	4.85	8	4	9	
Carbides:											
Moissanite	SiC	2.333	2.3	0.3	1.84	0.12	2.048	3.444	4	10	[121,127,136]
Moissanite	SiC	3.029	8.4	0.6	0.3	0.3	1.535	2.112	3	3	[121]
			5.6	0.3	1.62	0.14	2.112	2.842	4	4	
Moissanite	SiC	3.122	8.0	--	6.0	--	0.00	0.464	1	2	[121,127,136]
			10.29	0.13	-0.38	0.10	0.674	1.678	3	9	
			7.84	0.11	1.03	0.05	1.678	2.912	4	10	
Tantalum Carbide	TaC	12.626	3.32	0.09	1.49	0.05	0.887	2.619	2	20	[121]
Tantalum Carbide	TaC	14.110	4.34	0.05	1.36	0.03	0.435	3.76	2	21	[121,152]

TABLE 1. Shock Wave Equation of State of Minerals and Related Materials of the Solar System (continued)

Mineral	Formula	Sample Density (Mg/m³)	C_0 (km/sec)	error ΔC_0 (km/sec)	S	error ΔS	lower U_p (km/sec)	upper U_p (km/sec)	Phase*	No. of Data	References / Temp. Refs.
Tungsten Carbide[P]	WC	15.013	4.97 / 5.21	0.11 / 0.03	2.1 / 1.14	0.4 / 0.02	0.00 / 0.369	0.369 / 1.819	1 / 2	4 / 12	[121,127,136]
Tungsten Carbide	WC	15.66	4.926	0.014	1.163	0.007	0.45	3.66	2	4	[152]
Sulfides:											
Sphalerite	ZnS	3.952	5.08 / 3.10 / 0.9	0.03 / 0.10 / --	-0.09 / 1.22 / 2.0	0.03 / 0.04 / --	0.63 / 1.52 / 2.70	1.52 / 2.70 / 3.56	2 / 3 / 4	3 / 3 / 2	[186]
Pyrrhotite	$Fe_{1-x}S$	4.605	5.8 / 2.31 / 3.23	0.2 / 0.17 / 0.10	-4.7 / 2.08 / 1.49	0.5 / 0.15 / 0.03	0.235 / 0.494 / 1.496	0.547 / 1.599 / 5.361	1 / 2 / 4	3 / 10 / 14	[3,53]
Pyrite	FeS_2	4.933	8.8 / 5.3	1.0 / 0.10	-1.4 / 1.47	1.0 / 0.04	0.225 / 1.133	1.39 / 5	1 / 2	7 / 11	[10,186]
Potassium Iron Sulfide	$KFeS_2$	2.59	2.32 / 8.2 / 0.25	0.06 / -- / 0.07	1.97 / -1.0 / 1.912	0.05 / -- / 0.017	0.223 / 2.05 / 2.79	2.05 / 2.79 / 4.72	2 / 3 / 4	11 / 2 / 3	[191,230]
Halides:											
Griceite	LiF	1.27	0.74	--	1.58	--	2.4	6.59	2	2	[106]
Griceite	LiF	2.638	5.10	0.09	1.35	0.03	0.452	10.01	2	68	[38,54,121,195,213] /[177]
Villiaumite	NaF	2.792	4.08 / 6.7 / 2.83	0.14 / 0.3 / 0.04	1.42 / -0.28 / 1.635	0.13 / 0.16 / 0.013	0.5 / 1.54 / 2.027	1.54 / 2.027 / 3.982	2 / 3 / 4	8 / 4 / 8	[54,213]
Halite	NaCl	0.868	-0.15 / 7.0 / -0.4	0.19 / 1.3 / --	1.72 / -0.2 / 1.5	0.06 / 0.3 / --	1.942 / 3.62 / 4.181	3.62 / 4.181 / 5.552	2a / 3 / 2b	4 / 3 / 2	[121]

TABLE 1. Shock Wave Equation of State of Minerals and Related Materials of the Solar System (continued)

Mineral	Formula	Sample Density (Mg/m³)	C₀ (km/sec)	error ΔC₀ (km/sec)	S	error ΔS	lower Up (km/sec)	upper Up (km/sec)	Phase*	No. of Data	References / Temp. Refs.
Halite	NaCl	0.989	0.74	0.09	1.470	0.018	2.53	6.02	2	5	
			3.9	--	0.9	--	6.02	6.6	3	2	
			-17.0	--	4.1	--	6.6	6.7	4	2	[106]
Halite	NaCl	1.427	1.56	0.12	1.49	0.03	2.29	5.66	2	5	
			5.6	--	0.8	--	5.66	6	3	2	
			-5.1	--	2.5	--	6	6.11	4	2	[106]
Halite	NaCl	2.159	3.60	0.09	1.17	0.18	0.00	0.647	1	28	
			3.41	0.03	1.42	0.03	0.646	1.7	2a	113	
			4.35	0.09	0.88	0.04	1.7	2.5	3a	109	
			2.43	0.07	1.66	0.02	2.5	3.75	4a	95	
			4.4	0.3	1.11	0.08	3.73	4.356	3b	43	
			3.5	0.3	1.33	0.05	4.324	6.52	4b	8	
			19	--	-1	--	6.52	6.8	3c	2	[32,73,84,106, 121, 213] /[48, 103, 107, 176, 177]
			3.8	0.2	1.18	0.02	6.8	11.05	4c	3	
Sylvite	KCl	0.79	0.9	--	1.3	--	2.66	7.19	2	2	[106]
Sylvite	KCl	1.41	1.9	--	1.3	--	2.3	6.56	2	2	[106]
Sylvite	KCl	1.986	2.86	0.09	1.26	0.09	0.249	2.2	1	33	
			4.0	0.2	1.09	0.05	2.2	6.71	2	9	
			14	--	-0.4	--	6.71	7.1	3	2	[34,38,85,106, 213]/[48,107, 177]
			2.5	0.4	1.25	0.04	7.1	11.38	4	3	
Potassium Bromide	KBr	2.747	2.83	0.16	-0.1	0.3	0.27	0.61	1	4	
			1.88	0.05	1.50	0.03	0.57	2.9	2	13	
			2.63	0.09	1.24	0.02	2.862	5.09	4a	12	[38,106,121, 213] / [48]
			3.23	0.19	1.11	0.03	5.09	10.6	4b	6	
Cesium Iodide	CsI	2.51	0.66	0.12	1.42	0.04	1.04	4.23	2	6	
			1.37	0.12	1.22	0.02	4.23	6.75	4	4	[154]

TABLE 1. Shock Wave Equation of State of Minerals and Related Materials of the Solar System (continued)

Mineral	Formula	Sample Density (Mg/m³)	C_0 (km/sec)	error ΔC_0 (km/sec)	S	error ΔS	lower U_p (km/sec)	upper U_p (km/sec)	Phase*	No. of Data	References / Temp. Refs.
Cesium Iodide	CsI	4.51	1.57	0.17	1.66	0.17	0.56	1.32	2	7	[38,121,154, 213] /[199,209]
			3.8	--	0.1	--	1.32	1.56	3	2	
			1.95	0.05	1.302	0.019	1.56	4.3	4a	13	
			2.66	0.13	1.141	0.019	4.3	9.28	4b	3	
Fluorite	CaF$_2$	3.18	5.5	--	0.8	--	1.1	2.22	1	2	
			4.64	0.16	1.19	0.06	2.22	3.38	2	4	
			8.2	--	0.14	--	3.38	3.67	3	2	
			0.4	0.3	2.27	0.07	3.67	5.76	4	5	[35]
Cryolite	Na$_3$AlF$_6$	2.96	4.70	0.10	0.89	0.09	0.71	1.57	2	3	[186]
			3.76	0.12	1.44	0.04	1.57	3.8	4	4	
Oxides:											
Water, Ice^j	H$_2$O	0.35	0.080	0.18	1.40	0.03	2.76	6.75	2	5	[41]
Water, Ice^j	H$_2$O	0.60	0.83	0.16	1.40	0.03	2.57	6.2	2	5	[41]
Water, Ice^j	H$_2$O	0.915	4.05	0.05	-1.89	0.16	0.045	0.858	1	7	[41,75,114]
			1.43	0.11	1.48	0.03	0.858	5.67	2	9	
Water, Ice^k	H$_2$O	0.999	1.47	0.04	1.93	0.06	0.00	0.97	1	14	[19,29,116,121, 140,157,167, 213,217, 225] / [87,119]
			1.70	0.06	1.71	0.03	0.9	2.53	2	58	
			2.64	0.07	1.270	0.008	2.479	32.42	4	25	
Seawater	(mixture)	1.03	1.69	0.08	1.73	0.10	0.31	1.11	2	3	[210]
			2.07	0.09	1.38	0.03	1.11	4.76	4	4	
Bromellite	BeO	2.454	3.5	0.3	1.92	0.12	1.799	3.356	2	6	[120,121]
Bromellite	BeO	2.661	8.5	--	-1.2	--	0.577	1.25	1	2	[120,121,213]
			5.4	0.8	1.6	0.4	1.25	3.46	2	3	

156 SHOCK WAVE DATA FOR MINERALS

TABLE 1. Shock Wave Equation of State of Minerals and Related Materials of the Solar System (continued)

Mineral	Formula	Sample Density (Mg/m³)	C₀ (km/sec)	error ΔC₀ (km/sec)	S	error ΔS	lower Up (km/sec)	upper Up (km/sec)	Phase*	No. of Data	References / Temp. Refs.
Bromellite	BeO	2.797	9.10 / 6.7	0.18 / 0.5	-1.0 / 1.32	0.2 / 0.18	0.368 / 1.094	1.094 / 3.71	1 / 2	4 / 5	[120,121,213]
Bromellite	BeO	2.886	7.76 / 8.42	0.10 / 0.06	1.22 / 1.042	0.05 / 0.013	0.77 / 2.74	2.74 / 5.78	2 / 4	12 / 3	[120,121,152,213]
Bromellite	BeO	2.989	10.84 / 9.65 / 7.7	0.08 / 0.10 / 0.3	-0.78 / 0.49 / 1.53	0.13 / 0.07 / 0.11	0.317 / 0.939 / 1.825	0.939 / 1.825 / 2.822	1 / 2 / 4	4 / 4 / 4	[120,121]
Periclase	MgO	2.842	2.8	0.2	1.84	0.08	1.749	3.528	2	6	[56,121,127]
Periclase	MgO	3.00	3.10	0.14	1.88	0.06	1.259	3.362	2	13	[56,121,127]
Periclase	MgO	3.355	3.5 / 4.78 / 5.73	0.3 / 0.13 / 0.12	2.7 / 1.77 / 1.36	0.3 / 0.06 / 0.03	0.629 / 1.509 / 2.619	1.557 / 2.652 / 5.62	2a / 2b / 4	8 / 9 / 11	[37,56,121,127]
Periclase	MgO	3.583	6.09 / 6.83	0.10 / 0.12	1.75 / 1.29	0.08 / 0.04	0.626 / 1.92	1.967 / 4.44	2 / 4	25 / 16	[2,47,56,76,121,127,217,218]/[197]
Magnesio-wüstite	Mg0.6Fe0.4O	4.397	4.81 / 6.50 / 4.6	0.04 / 0.16 / 0.7	1.63 / 0.91 / 1.51	0.02 / 0.05 / 0.17	1.55 / 2.3 / 3.33	2.3 / 3.33 / 4.29	2 / 3b / 4	3 / 4 / 4	[219]
Magnesio-wüstite	Mg0.1Fe0.9O	5.191	4.3 / 6.27 / 4.96 / 5.53 / 4.76	-- / 0.11 / 0.07 / 0.12 / 0.14	5.7 / -0.3 / 1.56 / 1.14 / 1.49	-- / 0.2 / 0.07 / 0.07 / 0.06	0.00 / 0.332 / 0.705 / 1.28 / 2.169	0.332 / 0.705 / 1.28 / 2.169 / 2.721	1 / 3a / 2 / 3b / 4	2 / 3 / 3 / 4 / 3	[121]
Lime	CaO	2.980	3.56 / 6.7 / 2.61	0.11 / 0.6 / 0.16	1.79 / 0.3 / 1.67	0.07 / 0.2 / 0.05	0.882 / 2.083 / 3.01	2.083 / 3.01 / 3.558	2 / 3 / 4	6 / 3 / 5	[121]

TABLE 1. Shock Wave Equation of State of Minerals and Related Materials of the Solar System (continued)

Mineral	Formula	Sample Density (Mg/m^3)	C_0 (km/sec)	error ΔC_0 (km/sec)	S	error ΔS	lower U_p (km/sec)	upper U_p (km/sec)	Phase*	No. of Data	References / Temp. Refs.
Lime	CaO	3.324	7.4	0.8	0.5	0.3	1.801	3.285	3	5	[97] / [50]
			4.2	1.2	1.5	0.3	3.285	4.628	4	9	
Wüstite	Fe$_{1-x}$O	5.548	4.80	0.10	1.33	0.07	1.22	1.766	2	3	
			6.9	--	0.2	--	1.766	2.034	3	2	
			3.72	0.15	1.59	0.04	2.414	4.055	4	4	[97]
Wüstite	Fe$_{1-x}$O	5.662	3.4	--	2.4	--	1.56	1.73	2	2	
			6.84	0.02	0.408	0.011	1.73	2.51	3	3	
			3.3	--	1.8	--	2.51	2.62	4	2	[232]
Corundum	Al$_2$O$_3$	3.761	11.04	0.18	-4.3	0.5	0.18	0.706	1	7	
			6.61	0.16	1.35	0.07	0.706	3.282	2	15	[9,121,127]
Corundum	Al$_2$O$_3$	3.843	10.20	0.125	-1.9	0.3	0.18	0.898	1	5	
			7.08	0.11	1.36	0.05	0.898	2.979	2	10	[47,121,122,127, 217]
Corundum	Al$_2$O$_3$	3.92	8.71	0.05	0.716	0.017	0.37	5.5	2	6	[152]
Corundum	Al$_2$O$_3$	3.979	11.04	0.07	1.1	0.3	0.033	0.46	0	41	
			17.8	1.6	-14	3	0.46	0.621	1	6	
			8.83	0.06	0.93	0.03	0.555	3.064	2	52	[9,47,78,121, 122, 127,217]
Corundum	Al$_2$O$_3$	4.00	9.52	0.04	0.955	0.008	1.02	8.28	2	4	[152]
Hematite	α-Fe$_2$O$_3$	5.047	6.18	0.12	1.40	0.17	0.00	1.03	2	6	
			7.435	0.011	0.035	0.007	1.097	2.294	3	4	
			4.39	0.11	1.37	0.04	2.294	3.18	4	10	[47,121,127,200, 217]
Ilmenite	Fe^{+2}TiO$_3$	4.75	5.85	0.08	1.28	0.07	0.85	1.38	2	3	
			7.43	--	0.13	--	1.38	1.92	3	2	
			5.46	--	1.15	--	1.92	3.09	4	2	[186]
Ilmenite	Fe^{+2}TiO$_3$	4.787	6.33	0.12	1.0	0.2	0.00	0.652	2	3	
			6.86	0.06	0.17	0.05	0.626	2.009	3	5	
			4.07	0.18	1.54	0.07	2.009	3.082	4	12	[121,127]

TABLE 1. Shock Wave Equation of State of Minerals and Related Materials of the Solar System (continued)

Mineral	Formula	Sample Density (Mg/m^3)	C_0 (km/sec)	error ΔC_0 (km/sec)	S	error ΔS	lower U_p (km/sec)	upper U_p (km/sec)	Phase*	No. of Data	References / Temp. Refs.
Perovskite	$CaTiO_3$	3.86	5.25	0.11	1.48	0.07	0.59	2.35	2	5	
			6.5	0.3	0.93	0.10	2.35	3.24	3	3	
			4.0	--	1.7	--	3.24	5.42	4	2	[187]
Barium Titanate	$BaTiO_3$	5.447	6.26	0.14	-4.0	0.3	0.016	0.655	1	23	
			2.33	0.10	2.63	0.10	0.533	1.334	2	55	
			3.7	0.2	1.60	0.12	1.321	2.479	4	9	[66,121,213]
Spinel	$MgAl_2O_4$	2.991	4.05	0.19	1.43	0.07	1.727	3.63	4	9	[121,127]
Spinel	$MgAl_2O_4$	3.417	7.1	--	1.13	--	0.00	0.987	2	2	
			8.23	0.06	-0.03	0.03	0.987	2.146	3	10	[47,121,127, 217]
			5.07	0.16	1.42	0.06	2.134	3.348	4	13	
Spinel	$MgAl_2O_4$	3.514	7.26	0.07	1.48	0.17	0.00	0.688	2	3	
			8.04	0.07	0.27	0.04	0.688	2.311	3	5	
			5.7	0.5	1.35	0.16	2.311	3.507	4	8	[121,127]
Magnetite	$Fe^{+2}Fe^{+3}{}_2O_4$	5.07	7.2	--	-0.2	--	1.14	1.5	3	2	
			3.0	1.4	1.8	0.4	2.72	4.52	4	3	[186]
Magnetite	$Fe^{+2}Fe^{+3}{}_2O_4$	5.117	5.9	--	1.3	--	0.00	0.61	2	2	
			6.56	0.08	0.05	0.06	0.61	1.786	3	10	[47,121,127, 217]
			4.24	0.11	1.36	0.05	1.757	2.975	4	15	
Rutile	TiO_2	4.21	6.96	0.06	0.23	0.03	1.14	2.44	3	3	
			2.1	0.3	2.15	0.08	2.44	5.2	4	5	[35]
Rutile	TiO_2	4.245	10.3	0.5	-4.3	1.3	0.09	0.676	1	11	
			7.68	0.11	0.21	0.07	0.468	2.858	3	8	[11,47,121,123, 127,130,217]
			3.0	0.6	1.8	0.2	2.858	3.191	4	8	
Pyrolusite	$Mn^{+4}O_2$	4.318	3.77	0.14	1.46	0.07	0.769	3.263	2	16	[47,121,217]

TABLE 1. Shock Wave Equation of State of Minerals and Related Materials of the Solar System (continued)

Mineral	Formula	Sample Density (Mg/m³)	C₀ (km/sec)	error ΔC₀ (km/sec)	S	error ΔS	lower Up (km/sec)	upper Up (km/sec)	Phase*	No. of Data	References / Temp. Refs.
Cassiterite	SnO_2	6.694	6.82	0.05	-0.22	0.05	0.509	1.866	1	7	[47,121,217]
			5.15	0.06	0.68	0.03	1.866	2.501	2	4	
			2.6	0.6	1.7	0.2	2.501	2.833	4	3	
Argutite	GeO_2	6.277	9.51	0.12	0.29	0.11	0.15	2.57	2	6	[92]
Baddeleyite	ZrO_2	4.512	4.4	--	0.27	--	0.00	1.622	2	2	[121]
			1.9	0.4	1.88	0.17	1.622	2.994	4	8	
Baddeleyite	ZrO_2	5.814	5.17	0.08	1.02	0.05	0.41	2.17	2	4	[124]
			4.42	0.07	1.35	0.03	2.17	2.99	4	4	
Cerianite	$(Ce^{+4},Th)O_2$	1.133	0.2	0.3	1.20	0.07	1.925	5.437	2	7	[121]
Uraninite	UO_2	10.337	3.99	0.06	0.20	0.13	0.00	0.571	1	3	[121]
			3.59	0.04	0.91	0.03	0.568	1.983	2	12	
			1.7	0.4	1.8	0.16	1.983	2.493	4	6	
Uraninite	UO_2	6.347	0.43	0.07	1.70	0.03	1.025	3.286	4	18	[121]
Uraninite	UO_2	4.306	0.12	0.11	1.51	0.04	0.88	3.855	4	15	[121]
Uraninite	UO_2	3.111	-0.22	0.10	1.47	0.03	1.355	4.256	4	15	[121]
Hydroxides:											
Brucite	$Mg(OH)_2$	2.37	4.75	0.06	1.26	0.02	1.25	3.41	2	6	[186]
			0.9	--	2.4	--	3.41	3.96	4	2	
Brucite	$Mg(OH)_2$	2.383	5.0	0.2	1.22	0.11	0.886	3.079	2	13	[69]
Goethite	$\alpha\text{-}Fe^{+3}O(OH)$	4.0	4.4	--	1.6	--	1.02	1.34	2	2	[187]
			5.77	0.11	0.61	0.06	1.34	2.52	3	3	
			2.9	--	1.8	--	2.52	3.51	4	2	

TABLE 1. Shock Wave Equation of State of Minerals and Related Materials of the Solar System (continued)

Mineral	Formula	Sample Density (Mg/m^3)	C_o (km/sec)	error ΔC_o (km/sec)	S	error ΔS	lower U_p (km/sec)	upper U_p (km/sec)	Phase*	No. of Data	References / Temp. Refs.
Carbonates:											
Calcite	$CaCO_3$	2.701	6.9	0.2	-2.8	1.0	0.081	0.81	1	24	[11,21,102] / [103]
			3.71	0.03	1.435	0.013	0.81	3.845	2	6	
Magnesite	$MgCO_3$	2.975	6.08	0.09	1.26	0.04	0.6	3.61	2	6	[102]
Dolomite	$CaMg(CO_3)_2$	2.828	6.2	0.5	0.4	0.5	0.495	1.15	2	5	[102,184,213]
			5.30	0.10	1.16	0.03	1.12	5.32	4	19	
Aragonite	$CaCO_3$	2.928	5.82	0.11	0.78	0.12	0.11	1.83	2	12	[220]
Sulfates:											
Mascagnite	$(NH_4)_2SO_4$	1.3	0.77	0.15	2.28	0.167	0.36	1.21	2	8	[99]
			1.8	0.2	1.54	0.12	1.15	2.5	4	11	
Mascagnite	$(NH_4)_2SO_4$	1.6	1.96	0.10	2.09	0.10	0.28	1.83	2	8	[99]
Mascagnite	$(NH_4)_2SO_4$	1.73	3.71	0.14	1.34	0.12	0.2	1.87	2	5	[99]
Anhydrite	$CaSO_4$	2.97	3.60	0.06	1.75	0.05	0.73	1.55	2	3	[186]
			4.6	--	1.1	--	1.55	1.85	3	2	
			3.24	0.11	1.72	0.03	2.42	3.71	4	4	
Barite	$BaSO_4$	4.375	3.3	--	1.9	--	0.64	1.03	2	2	[186]
			4.7	0.2	0.54	0.15	1.03	1.69	3	3	
			2.3	0.4	1.86	0.16	1.69	3.29	4	5	
Gypsum	$CaSO_4$ $\cdot 2H_2O$	2.28	2.80	0.17	1.95	0.13	0.85	1.72	2	3	[186] / [103]
			5.2	--	0.5	--	1.72	2.15	3	2	
			2.49	0.12	1.79	0.04	2.15	4.06	4	5	
Borates:											
Sassolite	H_3BO_3	1.471	2.09	0.11	1.27	0.08	1.254	1.639	2	3	[121]
			1.14	--	1.85	--	1.639	2.114	4	2	

TABLE 1. Shock Wave Equation of State of Minerals and Related Materials of the Solar System (continued)

Mineral	Formula	Sample Density (Mg/m³)	C_0 (km/sec)	error ΔC_0 (km/sec)	S	error ΔS	lower U_p (km/sec)	upper U_p (km/sec)	Phase*	No. of Data	References / Temp. Refs.
Silica Polymorphs:											
Quartz	SiO_2	2.651	6.61	0.12	1.02	0.19	0.295	0.81	0a	16	[11,17,27,37,47, 72,118,121,127, 129,155,162, 208,217, 222] / [118,165]
			5.65	0.09	0.9	0.2	0.285	0.66	0b	29	
			8.14	0.11	-1.32	0.12	0.508	1.815	1a	16	
			6.44	0.07	-0.43	0.08	0.48	1.815	1b	12	
			5.29	0.08	0.20	0.04	1.803	2.48	3a	6	
			1.48	0.10	1.80	0.03	2.46	4.55	2	24	
			8.2	0.6	0.33	0.12	4.51	4.84	3b	3	
			4.0	0.2	1.283	0.018	4.84	26.76	4	10	
Porous Quartz	SiO_2	1.15	0.41	0.09	1.40	0.03	1.52	5.28	2	15	[188,207]
Porous Quartz	SiO_2	1.43	0.87	0.09	1.38	0.03	0.65	5.08	2	30	[188,207]
Porous Quartz	SiO_2	1.766	1.27	0.03	1.356	0.019	0.62	2.13	2	4	[188,207]
			3.42	0.13	0.34	0.05	2.13	2.55	3	3	
			0.25	0.05	1.589	0.011	2.55	9.0	4	9	
Porous Quartz	SiO_2	1.877	2.75	0.07	0.82	0.03	0.849	3.374	2	11	[121]
			-0.5	0.4	1.86	0.10	3.374	4.863	4	15	
Porous Quartz	SiO_2	2.151	3.02	0.16	0.84	0.08	0.799	3.199	2	11	[121,207]
			0.7	0.2	1.67	0.06	3.199	6.52	4	16	
Silicic Acid	H_4SiO_4	0.55	-0.09	0.11	1.27	0.02	3.51	6.59	4	5	[188]
Silicic Acid	H_4SiO_4	0.65	-0.01	0.03	1.275	0.006	2.94	5.93	4	6	[188]
Silicic Acid	H_4SiO_4	0.80	0.78	0.03	1.05	0.02	0.68	1.98	2	3	[188]
			0.33	0.04	1.261	0.009	1.98	5.71	4	7	

TABLE 1. Shock Wave Equation of State of Minerals and Related Materials of the Solar System (continued)

Mineral	Formula	Sample Density (Mg/m³)	C_0 (km/sec)	error ΔC_0 (km/sec)	S	error ΔS	lower U_p (km/sec)	upper U_p (km/sec)	Phase*	No. of Data	References / Temp. Refs.
Aerogel	$SiO_2{}^r$	0.172	0.21	0.07	1.01	0.02	1.494	4.01	2	6	
			-0.44	0.10	1.164	0.018	4.01	7.401	4	8	[88,188]
Aerogel	$SiO_2{}^r$	0.295	0.21	0.03	1.11	0.05	0.302	0.765	2	3	[158]
Aerogel	$SiO_2{}^r$	0.40	0.29	0.03	1.089	0.010	1.456	3.63	2	4	
			-0.214	0.013	1.224	0.003	3.63	6.37	4	4	[188]
Aerogel	$SiO_2{}^r$	0.55	0.590	0.006	0.945	0.006	0.335	1.43	2	3	
			0.15	0.04	1.219	0.008	1.43	6.1	4	7	[188]
Cristobalite	SiO_2	2.13	2.29	0.09	1.01	0.06	0.96	1.99	2	3	
			1.23	0.05	1.545	0.014	1.99	4.66	4	6	[156]
Coesite	SiO_2	1.15	0.44	0.07	1.44	0.02	1.2	4.59	2	5	[156]
Coesite	SiO_2	2.40	3.4	0.3	1.42	0.19	1.33	2.05	2	3	
			4.3	--	0.97	--	2.05	4.16	3	2	[156]
Coesite	SiO_2	2.92	5.833	0.010	0.902	0.009	0.66	1.48	2	3	
			6.92	0.02	0.168	0.011	1.48	2.86	3	3	
			2.6	0.4	1.68	0.12	2.86	4.05	4	3	[156]
Silicates:											
Forsterite	Mg_2SiO_4	3.059	5.71	0.07	0.64	0.03	0.825	2.771	2	6	[47,121,127, 217]
			3.0	0.3	1.65	0.10	2.771	3.626	4	6	
Forsterite	Mg_2SiO_4	3.212	10.5	0.2	-3.8	0.4	0.18	0.797	1a	16	[91,121,127,201, 229] / [165]
			6.8	0.3	1.4	0.6	0.00	0.795	1b	12	
			7.21	0.10	0.55	0.06	0.749	2.449	2	20	
			4.6	0.3	1.51	0.08	2.433	4.61	4	20	

TABLE 1. Shock Wave Equation of State of Minerals and Related Materials of the Solar System (continued)

Mineral	Formula	Sample Density (Mg/m^3)	C_0 (km/sec)	error ΔC_0 (km/sec)	S	error ΔS	lower U_p (km/sec)	upper U_p (km/sec)	Phase*	No. of Data	References / Temp. Refs.
Olivine	$(Mg,Fe)_2$-SiO_4; $(Mg_{0.92}$-$Fe_{0.08})_2SiO_4$	3.264	8.84	0.13	-0.99	0.18	0.272	1.43	1	4	
			5.6	--	1.3	--	1.43	2.19	2	2	
			8.1	--	0.15	--	2.19	2.8	3	2	
			6.0	0.3	0.88	0.08	2.8	4.81	4	3	[15,121]
Fayalite	Fe_2SiO_4	4.245	6.17	0.09	0.23	0.06	0.702	1.967	2	10	
			3.78	0.13	1.42	0.05	1.967	3.483	4	10	[47,121,127, 217]
Zircon	$ZrSiO_4$	4.549	8.58	0.18	-1.3	0.4	0.11	1.07	1	7	
			7.14	0.05	0.02	0.03	1.07	2.52	3	4	
			-1.6	0.9	3.5	0.3	2.52	2.84	2	4	[124]
Almandine	$(Fe_{0.79}$-$Mg_{0.14}$-$Ca_{0.04}$-$Mn_{0.03})_3$-$Al_2Si_3O_{12}$	4.181	-10	--	50	--	0.29	0.32	1	2	
			5.92	0.07	1.39	0.08	0.45	1.14	2	8	
			3.0	--	3.6	--	1.29	1.46	3	2	
			6.4	0.3	1.33	0.16	1.46	1.8	4	7	[77]
Grossular	$Ca_3Al_2Si_3O_{12}$	3.45	8.3	0.17	0.47	0.10	0.18	3.04	2	16	[121]
Mullite	$Al_6Si_2O_{13}$	2.668	2.30	0.13	1.65	0.04	1.935	4.077	2	13	[121,127]
Mullite	$Al_6Si_2O_{13}$	3.154	8.732	0.016	-0.394	0.015	0.717	1.479	1	3	
			8.29	0.04	-0.09	0.02	1.479	2.003	3	3	
			6.5	--	0.78	--	2.003	3.311	4	2	[121]
Kyanite	Al_2SiO_5	2.921	7.45	0.06	-0.58	0.07	0.608	1.157	1	3	
			7.02	0.09	-0.19	0.05	1.157	2.359	3	5	
			2.2	0.3	1.85	0.10	2.359	3.383	4	5	[121]
Kyanite	Al_2SiO_5	3.645	7.8	--	0.6	--	1.537	2.745	2	2	
			3.9	1.7	2.0	0.6	2.745	3.22	4	3	[121]
Andalusite	Al_2SiO_5	3.074	5.3	--	1.9	--	0.00	1.1	2	2	
			6.92	0.15	0.31	0.07	1.1	2.817	3	12	
			2.9	0.4	1.80	0.13	2.817	3.73	4	8	[47,121,217]

164 SHOCK WAVE DATA FOR MINERALS

TABLE 1. Shock Wave Equation of State of Minerals and Related Materials of the Solar System (continued)

Mineral	Formula	Sample Density (Mg/m^3)	C_0 (km/sec)	error ΔC_0 (km/sec)	S	error ΔS	lower U_p (km/sec)	upper U_p (km/sec)	Phase*	No. of Data	References / Temp. Refs.
Sillimanite	Al_2SiO_5	3.127	6.97	0.15	0.68	0.16	0.00	1.262	1	3	
			7.8	0.2	-0.10	0.11	1.068	2.461	3	8	
			3.8	0.2	1.57	0.08	2.461	3.611	4	10	[47,121,217]
Topaz	$Al_2SiO_4-(F,OH)_2$	3.53	8.10	0.09	0.054	0.08	0.052	1.75	1	4	
			5.3	0.4	1.722	0.15	1.75	3.59	2	6	[186]
Tourmaline	$Ca(Al,Fe,Mg)_3Al_6(BO_3)_3Si_6O_{18}(OH,F)_4$	3.179	6.2	--	1.2	--	0.824	1.555	1	2	
			8.1	0.4	0.05	0.15	1.555	2.888	3	7	
			3.6	0.5	1.62	0.16	2.888	3.695	4	8	[121]
Muscovite	$KAl_2(Si_3Al)O_{10}(OH,F)_2$	2.835	3.3	0.3	1.95	0.16	1.27	2.44	2	3	
			6.2	--	0.7	--	2.44	3.18	3	2	
			4.5	0.2	1.29	0.06	3.18	4.74	4	4	[182]
Serpentine	$Mg_3Si_2O_5(OH)_4$	2.621	5.30	0.15	0.90	0.11	0.431	2.025	2	10	
			6.5	0.4	0.20	0.18	1.719	2.561	3	10	
			3.8	0.5	1.34	0.12	2.658	5.427	4	16	[47,121,211,217]
Pyroxenes:											
Enstatite	$Mg_2Si_2O_6$	2.714	2.70	0.11	1.31	0.04	1.901	3.258	2	5	[47,121,217]
Enstatite	$Mg_2Si_2O_6$	2.814	2.74	0.14	2.04	0.09	0.746	1.956	2	5	
			6.8	--	-0.1	--	1.956	2.128	3	2	
			3.64	0.16	1.43	0.05	2.128	3.946	4	9	[121,127]
Enstatite	$Mg_2Si_2O_6$	3.067	8.11	0.18	-1.5	0.4	0.224	0.60	1	6	
			4.98	0.13	1.18	0.09	0.456	2.126	2	19	
			7.4	0.4	-0.1	0.2	1.817	2.349	3	6	
			3.7	0.3	1.54	0.07	2.349	4.54	4	22	[8,121,127,231]/[117,165]

TABLE 1. Shock Wave Equation of State of Minerals and Related Materials of the Solar System (continued)

Mineral	Formula	Sample Density (Mg/m^3)	C_0 (km/sec)	error ΔC_0 (km/sec)	S	error ΔS	lower U_p (km/sec)	upper U_p (km/sec)	Phase*	No. of Data	References / Temp. Refs.
Diopside	$CaMgSi_2O_6$	3.264	4.9	--	8.4	--	0.201	0.289	1	2	[18,196,213] / [198]
			7.14	0.03	0.626	0.018	0.289	1.85	2	4	
			6.1	0.2	1.16	0.06	1.85	4.7	4	4	
Molten Diopside[n]	$CaMgSi_2O_6$	2.61	3.30	0.12	1.44	0.08	0.73	2.24	2	5	[168]
Hedenbergite	$CaFe^{+2}\text{-}Si_2O_6$	3.42	2.8	--	2.8	--	1.1	1.41	2	2	
			6.1	--	0.47	--	1.41	2.42	3	2	
			3.30	0.05	1.620	0.018	2.42	3.62	4	3	[186]
Augite, also Salite	$(Ca_{0.80}\text{-}Na_{0.03})\text{-}(Mg_{0.76}Fe_{0.29}Ti_{0.03})\text{-}(Al_{0.20}\text{-}Si_{1.85})O_6,$ $CaMg_{0.82}\text{-}Fe_{0.18}Si_2O_6$	3.435	6.25	0.11	0.85	0.08	0.935	1.87	2	5	
			6.6	0.8	0.6	0.4	1.87	2.48	3	8	
			4.6	1.0	1.4	0.3	2.44	3.82	4	5	[18,196,213]
Jadeite	$Na(Al,Fe^{+3})\text{-}Si_2O_6$	3.335	6.41	0.06	1.30	0.08	0.00	1.005	2a	3	
			6.57	0.10	1.09	0.07	0.986	1.94	2b	8	
			7.44	0.12	0.64	0.04	1.94	3.434	4	8	[47,121,135, 217]
Spodumene	$LiAlSi_2O_6$	3.14	7.123	0.019	0.007	0.018	0.43	1.45	2	3	
			6.4	--	0.51	--	1.45	2.37	3	2	
			4.0	0.3	1.56	0.09	2.37	3.76	4	3	[187]
Wollastonite	$CaSiO_3$	2.82	2.4	0.3	1.65	0.08	1.34	4.06	4	6	[187]
Wollastonite	$CaSiO_3$	2.822	6.3	0.2	-0.15	0.15	0.94	1.799	3	5	
			4.1	0.2	1.07	0.09	1.799	2.778	4	4	[121]
Wollastonite	$CaSiO_3$	2.89	5.3	--	1.14	--	0.00	1.195	2	2	
			6.826	0.004	-0.156	0.003	1.195	1.933	3	3	
			4.52	0.15	1.03	0.06	1.933	3.282	4	5	[121]

TABLE 1. Shock Wave Equation of State of Minerals and Related Materials of the Solar System (continued)

Mineral	Formula	Sample Density (Mg/m^3)	C_o (km/sec)	error ΔC_o (km/sec)	S	error ΔS	lower U_p (km/sec)	upper U_p (km/sec)	Phase*	No. of Data	References / Temp. Refs.
Tremolite	$Ca_2(Mg,-Fe^{+2})_5-Si_8O_{22}(OH)_2$	2.901	5.20	0.14	0.91	0.07	1.15	2.82	2	9	
			3.2	0.8	1.6	0.2	2.78	3.94	4	5	[187]
Beryl	$Be_3Al_2Si_6O_{18}$	2.68	8.83	0.10	0.05	0.04	1.4	4.12	2	4	
			2.7	--	1.6	--	4.12	6.1	4	2	[187]
Feldspars:											
Orthoclase, Microcline	$KAlSi_3O_8$	2.561	7.70	0.11	-1.1	0.2	0.188	1.21	1	14	
			6.14	0.19	0.21	0.11	1.21	2.41	2	4	
			3.1	0.2	1.39	0.05	2.41	6.27	4	4	[12,16,186]
Oligoclase	$(NaAlSi_3-O_8)_{75}(CaAl_2-Si_2O_8)_{19.5} -(KAlSi_3-O_8/5.5$	2.635	7.6	0.3	-1.5	1.2	0.195	0.3	1	13	[16]
Anorthite	$CaAl_2Si_2O_8$	2.769	3.17	0.16	1.45	0.04	2.911	4.338	2	3	[96] / [173,174]
Molten Anorthite$^{\rm O}$	$CaAl_2Si_2O_8$	2.55	2.85	0.14	1.27	0.09	0.91	2.37	2	6	[168]
Nepheline	$(Na,K)AlSiO_4$	2.63	4.76	0.08	0.88	0.09	0.52	1.25	2	3	
			5.7	--	0.1	--	1.25	1.62	3	2	
			2.22	0.08	1.67	0.02	2.57	3.94	4	4	[187]
Glasses:											
Argutite Glass	GeO_2	3.655	3.6	--	-0.3	--	0.32	1.35	1	2	
			0.80	0.06	1.79	0.02	1.35	4.46	2	7	[92]
Quartz Glass (fused quartz)	SiO_2	0.145	1.8	--	0.36	--	1.789	2.309	2	2	
			-1.2	0.3	1.67	0.05	2.309	6.507	4	10	[121]
Quartz Glass	SiO_2	1.15	0.09	0.2	1.51	0.06	3.07	5.21	4	4	[156]

TABLE 1. Shock Wave Equation of State of Minerals and Related Materials of the Solar System (continued)

Mineral	Formula	Sample Density (Mg/m^3)	C_o (km/sec)	error ΔC_o (km/sec)	S	error ΔS	lower U_p (km/sec)	upper U_p (km/sec)	Phase*	No. of Data	References / Temp. Refs.
Quartz Glass	SiO$_2$	2.2	0.4	0.5	1.73	0.12	3.61	4.84	4	6	[91]
Quartz Glass	SiO$_2$	2.204	5.83	0.03	-2.33	0.17	0.00	0.306	1	7	[101,118,121,127,129,163,217,222] / [104,117,118,175,177]
			5.264	0.019	-0.114	0.014	0.306	2.106	3	32	
			3.40	0.08	0.78	0.03	2.063	2.742	2	20	
			0.80	0.08	1.70	0.02	2.703	5.175	4a	61	
			3.7	0.2	1.196	0.014	5.175	23.64	4b	3	
Anorthite Glass	CaAl$_2$Si$_2$O$_8$	2.692	6.7	0.3	-0.2	0.4	0.45	1.179	1	5	[52] / [51]
			6.43	0.14	0.01	0.07	1.179	2.731	3	3	
			1.8	0.3	1.77	0.07	2.731	4.698	4	15	
Pyrex (and soda-lime glass)	(SiO$_2$)$_{81}$-(B$_2$O$_3$)$_{12}$-(Al$_2$O$_3$)$_{2.5}$-(CaO)$_{0.4}$-(MgO)$_{0.3}$-(Na$_2$O)$_4$-(K$_2$O)$_1$-(As$_2$O$_3$)$_{0.6}^s$	2.307	5.26	0.14	-0.37	0.15	0.4	1.444	1	5	[95,121,122,136]
			3.96	0.12	0.57	0.06	1.444	2.397	2	4	
			0.6	0.3	1.92	0.09	2.397	4.43	4	11	
Soda-Copper Silicate Glass	(SiO$_2$)$_{72.2}$-(CuO)$_{12.4}$-(Na$_2$O)$_{14}$-(Al$_2$O$_3$)$_{0.5}$-(SO$_3$)$_{0.45}$-(MgO)$_{0.1}$-(Fe$_2$O$_3$)$_{0.08}^s$	2.48	3.10	0.04	1.28	0.08	0.37	0.71	2	3	[67]
			3.623	0.012	0.531	0.009	0.71	1.55	3	3	
			1.5	0.2	1.95	0.11	1.51	2.57	4	5	

TABLE 1. Shock Wave Equation of State of Minerals and Related Materials of the Solar System (continued)

Mineral	Formula	Sample Density (Mg/m^3)	C_0 (km/sec)	error ΔC_0 (km/sec)	S	error ΔS	lower U_p (km/sec)	upper U_p (km/sec)	Phase*	No. of Data	References / Temp. Refs.
Lunar Glass	$(SiO_2)_{40}$-$(TiO_2)_9$-$(Al_2O_3)_{11}$-$(FeO)_{17}$-$(MgO)_{10}$-$(CaO)_{11}$ s	1.8	0.10 / -1.01	0.19 / 0.13	1.3 / 2.39	0.2 / 0.09	0.682 / 0.985	1.02 / 1.825	2 / 4	4 / 6	[7]
Other Standards:											
Aluminum: 1100, Russian alloys	Al^r	1.345	0.00 / 2.11	0.03 / 0.19	1.988 / 1.38	0.012 / 0.04	1.323 / 3.6	3.6 / 6.76	2 / 4	10 / 4	[105,142,213]
Aluminum: 1100, Russian alloys	Al^r	1.885	0.8 / 3.07	0.4 / 0.11	2.2 / 1.39	0.2 / 0.03	1.197 / 2.70	2.77 / 6.27	2 / 4	12 / 6	[39,105,142,213]
Aluminum: 1100, Russian alloys	Al^r	2.707	6.09 / 5.44 / 6.0	0.09 / 0.04 / 0.3	-0.1 / 1.324 / 1.181	0.4 / 0.012 / 0.019	0.01 / 0.428 / 5.94	0.526 / 6.0 / 30.0	1 / 2 / 4	10 / 59 / 27	[24,26,27,28,45, 105,115,121, 139,155,162, 163,190,213, 228]
Aluminum 2024	Al^r	1.661	2.20 / 0.18 / 2.0	0.12 / 0.15 / 0.2	0.19 / 2.20 / 1.54	0.15 / 0.08 / 0.06	0.00 / 0.983 / 2.739	1.012 / 2.739 / 4.567	1 / 2 / 4	3 / 9 / 15	[56,121,134, 136]
Aluminum 2024	Al^r	1.955	2.62 / 0.71 / 2.8	0.10 / 0.11 / 0.2	0.24 / 2.30 / 1.51	0.14 / 0.06 / 0.07	0.00 / 0.921 / 2.566	0.926 / 2.566 / 4.361	1 / 2 / 4	3 / 9 / 15	[56,121,134, 136]

TABLE 1. Shock Wave Equation of State of Minerals and Related Materials of the Solar System (continued)

Mineral	Formula	Sample Density (Mg/m³)	C_0 (km/sec)	error ΔC_0 (km/sec)	S	error ΔS	lower U_p (km/sec)	upper U_p (km/sec)	Phase*	No. of Data	References / Temp. Refs.
Aluminum 2024	Al[r]	2.224	3.0	--	0.5	--	0.00	0.849	1	2	[56,121,134, 136]
			1.3	0.3	2.47	0.19	0.849	2.124	2	7	
			3.22	0.16	1.58	0.05	1.981	4.064	4	17	
Aluminum 2024	Al[r]	2.559	4.1	--	0.9	--	0.00	0.724	1	2	[56,121,134, 136]
			3.24	0.18	2.04	0.15	0.724	1.761	2	5	
			4.07	0.10	1.54	0.03	1.761	3.817	4	17	
Aluminum 2024	Al[r]	2.788	5.356	0.011	1.305	0.005	0.00	5.962	2	325	[56,90,121,134, 136,166,213, 217,224,226] / [165]
Molybdenum	Mo	1.277	-0.03	0.06	1.25	0.03	0.678	4.05	2	5	[209]
Molybdenum	Mo	1.72	-0.39	0.10	1.40	0.03	3.02	5.01	2	4	[209]
Molybdenum	Mo	2.55	-0.12	0.05	1.473	0.013	0.654	6.51	2	10	[209]
Molybdenum	Mo	2.914	-0.16	0.04	1.538	0.012	0.65	6.3	2	10	[209]
Molybdenum	Mo	4.435	0.04	0.02	1.735	0.008	0.62	5.66	2	8	[209]
Molybdenum	Mo	5.59	0.30	0.06	1.90	0.02	0.3	5.25	2	13	[209]
Molybdenum	Mo	8.146	0.94	0.04	2.79	0.04	0.47	1.13	2	4	[162,163,209]
			1.91	0.18	1.98	0.09	1.13	2.6	4a	6	
			3.8	1.6	1.18	0.10	2.6	19.99	4b	3	
Molybdenum	Mo	10.208	5.14	0.03	1.247	0.005	0.00	20.91	2	61	[24,56,74,108, 121,126,131, 136,141,162, 163,166,217, 226]
Molybdenum[m]	Mo	10.208	4.73	0.07	1.45	0.06	0.538	1.414	2	5	[137]

TABLE 1. Shock Wave Equation of State of Minerals and Related Materials of the Solar System (continued)

Mineral	Formula	Sample Density (Mg/m³)	C_0 (km/sec)	error ΔC_0 (km/sec)	S	error ΔS	lower U_p (km/sec)	upper U_p (km/sec)	Phase*	No. of Data	References / Temp. Refs.
Tantalum	Ta	2.82	0.05 / -0.33	0.04 / 0.03	1.26 / 1.451	0.03 / 0.010	0.654 / 2.06	2.06 / 4.43	2 / 4	4 / 3	[209]
Tantalum	Ta	5.41	-0.25	0.10	1.66	0.04	2.17	3.23	2	3	[209]
Tantalum	Ta	6.20	-0.01	0.02	1.649	0.012	0.601	2.37	2	5	[209]
Tantalum	Ta	8.19	0.028 / 0.54	0.018 / 0.09	1.891 / 1.61	0.012 / 0.04	1.15 / 1.8	1.8 / 3.03	2 / 4	3 / 6	[209]
Tantalum	Ta	10.92	0.37 / 1.43	0.10 / 0.08	2.23 / 1.52	0.09 / 0.04	0.7 / 1.49	1.49 / 2.64	2 / 4	3 / 5	[209]
Tantalum	Ta	16.649	3.31	0.03	1.306	0.010	0.00	5.86	2	31	[24,86,108,121, 136,139,166, 217,226]
Tungsten	W	4.6	-0.17	0.14	1.53	.04	1.45	4.095	2	8	[209]
Tungsten	W	5.5	-0.08 / 0.4 / -3.94	0.07 / -- / 0.19	1.54 / 1.4 / 2.57	0.03 / -- / 0.05	1.38 / 2.76 / 3.63	2.76 / 3.63 / 3.885	2 / 3 / 4	4 / 2 / 3	[209]
Tungsten	W	6.57	-0.011 / 1.07 / -1.05 / 1.7	0.013 / -- / 0.14 / 0.5	1.584 / 1.12 / 1.91 / 1.21	0.007 / -- / 0.04 / 0.04	1.295 / 2.35 / 2.68 / 3.685	2.35 / 2.68 / 3.685 / 16.51	2 / 3 / 4a / 4b	3 / 2 / 4 / 3	[204,209]
Tungsten	W	8.87	0.13	0.05	1.75	0.02	0.56	3.17	2	6	[209]
Tungsten	W	13.36	0.93	0.07	1.97	0.04	0.84	2.54	2	6	[209]
Tungsten	W	18.67	2.86	0.04	2.08	0.06	0.302	0.721	2	4	[121]
Tungsten	W	19.240	4.064	0.010	1.204	0.003	0.00	15.1	2	40	[100,121,131, 136, 163,217]

TABLE 1. Shock Wave Equation of State of Minerals and Related Materials of the Solar System (continued)

Mineral	Formula	Sample Density (Mg/m³)	C_0 (km/sec)	error ΔC_0 (km/sec)	S	error ΔS	lower U_p (km/sec)	upper U_p (km/sec)	Phase*	No. of Data	References / Temp. Refs.
Lexan	(polycarbonate)	1.193	1.9	--	2.4	--	0.00	0.421	1	2	[55,101,121]
			2.38	0.03	1.551	0.019	0.421	2.53	2	32	
			4.47	0.12	0.70	0.04	2.379	3.651	3	21	
			2.6	0.2	1.27	0.05	3.64	6.92	4	19	

Notes:

*Phases: 1) Elastic shock; 2) Low pressure phase; 3) Mixed region; 4) High pressure phase.

[a] Starting temperature 5K
[b] Starting temperature 20K
[c] Starting temperature 75-86K
[d] Starting temperature 111K
[e] Starting temperature 122K
[f] Starting temperature 148K; compressed gas
[g] Starting temperature 165K
[h] Starting temperature 196K
[i] Starting temperature 203-230K
[j] Starting temperature 258-263K
[k] Starting temperature 273-298K
[l] Starting temperature 300K; compressed gas
[m] Starting temperature 1673K
[n] Starting temperature 1773K
[o] Starting temperature 1923 K
[p] 5% cobalt
[q] raw data not provided
[r] with impurities
[s] composition in weight percent oxides

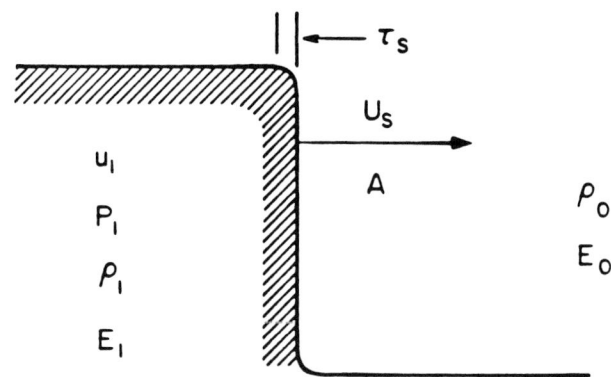

FIG. 1. Profile of a steady shock wave, rise time τ_s, imparting a particle velocity u_1 pressure P_1, and internal energy density E_1, propagating with velocity U_S into material that is at rest at density ρ_0 and internal energy density E_0.

in nearly steady waves in materials. For steady waves a shock velocity U_S with respect to the laboratory frame can be defined. Conservation of mass, momentum, and energy across a shock front can then be expressed as

$$\rho_1 = \rho_0 (U_S - u_0)/(U_S - u_1) \tag{1}$$
$$P_1 - P_0 = \rho_0 (u_1 - u_0)(U_S - u_0) \tag{2}$$
$$E_1 - E_0 = (P_1 + P_0)(1/\rho_0 - 1/\rho_1)/2 = 1/2 \ (u_1 - u_0)^2 \tag{3}$$

where ρ, u, P, and E are density, particle velocity, shock pressure, and internal energy per unit mass and, as indicated in Fig. 1, the subscripts o and 1 refer to the state in front of and behind the shock front, respectively. In Table 1, shock velocity and particle velocity are designated as U_S and u_1. Thus for a single shock $U_p = u_1$. In the case of multiple shocks, the values of U_p and u_s given in the Table are for the final (highest pressure) shock state. Equations (1)-(3) are often called the Rankine-Hugoniot equations. It should be understood that in this section pressure is used in place of stress in the indicated wave propagation direction. In actuality, stress in the wave propagation direction is specified by Eq. (2). A detailed derivation of Eqs. (1), (2), and (3) is given in Duvall and Fowles [70]. Equation (3) also indicates that the material achieves an increase in internal energy (per unit mass) which is exactly equal to the kinetic energy per unit mass.

In the simplest case, when a single shock state is achieved via a shock front, the Rankine-Hugoniot equations involve six variables (U_S, u_1, ρ_0, ρ_1, $E_1 - E_0$, and P_1); thus, measuring three, usually U_S, u_1, and ρ_0, determines the shock state variables ρ_1, $E_1 - E_0$ and P_1.

The key assumption underpinning the validity of Eqs. (1)-(3) is that the shock wave is steady, so that the rise time τ_s, is short compared to the characteristic time for which the high pressure, density, etc. are constant (see Fig. 1). Upon driving a shock of pressure P_1 into a material, a final shock state is achieved which is described by Eqs. (1)-(3). This shock state is shown in Fig. 2, in relation to other thermodynamic paths, in the pressure-volume plane. Here $V_0 = 1/\rho_0$ and $V = 1/\rho$. In the case of the isotherm and isentrope, it is possible to follow, as a thermodynamic path, the actual isothermal or isentropic curve to achieve a state on the isotherm or isentrope. A shock, or Hugoniot, state is different, however. The Hugoniot state (P_1, V_1) is achieved via a a shock front. The initial and final states are connected by a straight line called a Rayleigh line (Fig. 2). Thus successive states along the Hugoniot curve cannot be achieved, one from another, by a shock process. The Hugoniot curve itself then just represents the locus of final shock states corresponding to a given initial state.

It has long been recognized that the kinematic parameters measured in shock wave experiments U_S and U_p can empirically be described in regions where a substantial phase change in the material does not occur as:

$$U_S = C_0 + S \ U_p \tag{4}$$

As further discussed in several review articles on shock compression [22,59,136], and a recent book [40],

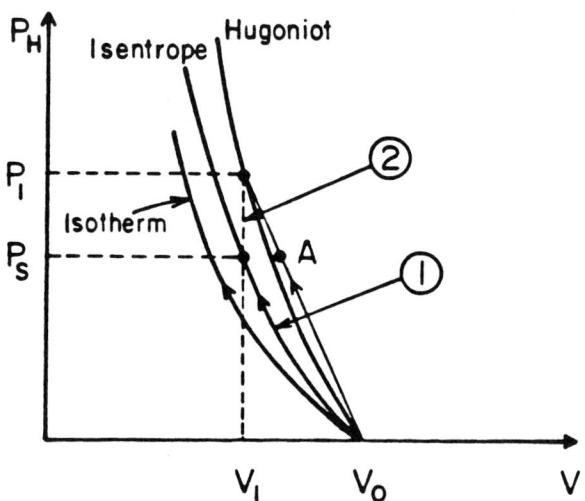

FIG. 2. Pressure-volume compression curves. For isentrope and isotherm, the thermodynamic path coincides with the locus of states, whereas for shock, the thermodynamic path is a straight line to point P_1, V_1, on the Hugoniot curve, which is the locus of shock states.

Hugoniot data for minerals and other condensed media may be described over varying ranges of pressure and density in terms of a linear relation of shock and particle velocity in Table 1. This table was assembled using the *Microsoft Excel,* version 3.0, (Redmond, WA 1993) program and the least-square fits to the shock wave data with standard errors were derived by using the LINEST function. The equations employed for line slopes and intercepts are identical to those given in Bevington [46] (Eq. 6-9, p. 104; Eq. 6-21, p. 114 for errors in slopes; and Eq. 6-22, p. 114 for errors in intercepts).

The U_S-U_p data for a wide range of minerals are given in Table 1. Here C_0 is the shock velocity at infinitesimally small particle velocity, or the ambient pressure bulk sound velocity which is given by

$$C_0 = \sqrt{K_S/\rho_0}, \tag{5}$$

where K_S is the isentropic bulk modulus, $K_S = -V\,(dP/dV)_S$ in the absence of strength effect (see Sect. 3). Upon substituting Eq. 4 into Eq. 2, and denoting the shock pressure as P_H, this is given by:

$$P_H = \rho_0 U_p\,(C_0 + S\,U_p) \tag{6}$$

Thus, from the form of Eq. 6, shock pressure is given as the sum of a linear and quadratic term in particle velocity, based on the data of Table 1. A pressure-volume relation can be obtained by combining Eq. 6 with Eq. 1 to yield:

$$P_H = \rho_0\,C_0^2\,\eta\,/(1\text{-}S\eta)^2 \tag{7}$$

where

$$\eta = 1 - V/V_0 = U_p/U_S. \tag{8}$$

Eq. 7 is often called the "shock wave equation of state" since it defines a curve in the pressure-volume plane.

The isentropic pressure can be written (e.g. [93,98,171])by an expression analogous to Eq. 7 as a series

$$P_S = \rho_0\,C_0^2\,(\eta + 2S\eta^2 + \cdots), \tag{9}$$

which upon differentiation yield the isentropic bulk modulus

$$K_S = \rho_0\,C_0^2\,(1 + (4S\text{-}1)\,\eta + \cdots) \tag{10}$$

The analogous bulk modulus along the Hugoniot is:

$$K_H = -V\,(\partial P/\partial V)_H. \tag{11}$$

The isentrope and the Hugoniot and isentropic bulk modulus are related via:

$$K_S = K_H + \left(\frac{\gamma}{2}\right)\left[P_H - K_H\ \eta\,/(1-\eta)\right] - \left[P_H - P_S\right]\left[\gamma + 1 - q_0\ (1 - q'\eta + \cdots)\right] \tag{12}$$

Here we assume a volume dependence of the Gruneisen parameter

$$\gamma = V\,(\partial P/\partial E)_V = \gamma_0\,(V/V_0)^q, \tag{13}$$

where

$$q = d\,\ell n\,\gamma/d\,\ell n\,V \text{ and } q' = d\,\ell nq/d\ell n\,V \tag{14}$$

γ_0 is the Grüneisen parameter under standard pressure and temperatures and is given by

$$\gamma_0 = \alpha\,K_T\,V_0/C_V = \alpha\,K_S V_0/C_p, \tag{15}$$

where α is the the thermal expansion coefficient, K_T is the isothermal bulk modulus and C_p and C_V are the specific heat at constant pressure and volume. We note that the P_S and P_H can be related by assuming the Mie-Grüneisen relation

$$P_H - P_S = \frac{\gamma}{V}\,(E_H\text{-}E_S), \tag{16}$$

if γ is independent of temperature, where $E_H = E_1\text{-}E_0$ is given by Eq. 3 and E_S is given by

$$E_S = -\!\!\int_{V_0}^{V} P_S dV. \tag{17}$$

Because the Grüneisen ratio relates the isentropic pressure, P_S, and bulk modulus, K_S, to the Hugoniot pressure, P_H, and Hugoniot bulk modulus, K_H, it is a key equation of state parameter.

The shock-velocity particle relation of Table 1 can be used to calculate the shock pressure when two objects impact. If (A) the flyer plate and (B) the target are known and expressed in the form of Eq. (7), the particle velocity

u_1 and pressure P_1 of the shock state produced upon impact of a flyer plate at velocity u_{fp} on a stationary target may be calculated from the solution of the equation equating the shock pressures in the flyer and driver plate:

$$\rho_{oA} (u_{fp} - u_1)(C_{oA} + S_A (u_{fp} - u_1)) = \rho_{oB} u_1 (C_{oB} + S_B u_1). \qquad (18)$$

That is;

$$u_1 = (-b - \sqrt{b^2 - 4ac})/2a, \qquad (19)$$

where

$$a = S_A \rho_{oA} - \rho_{oB} S_B, \qquad (20)$$
$$b = C_{oA} \rho_{oA} - 2S_A \rho_{oA} u_{fp} - \rho_{oB} C_{oB}, \qquad (21)$$

and

$$c = u_{fp} (C_{oA} \rho_{oA} + S_A \rho_{oA} u_{fp}). \qquad (22)$$

3. SHOCK-INDUCED DYNAMIC YIELDING AND PHASE TRANSITIONS

Both dynamic yielding and phase transitions give rise to multiple shock wave profiles when pressure or particle velocity versus time is recorded. Virtually all nonporous minerals and rocks in which dynamic compression has been studied demonstrate phenomenon related to dynamic yielding, in which materials transform from finite elastic strain states to states in which irreversible deformation has occurred. Moreover, most minerals and a large number of compounds, elements, and organic materials demonstrate shock-induced phase changes.

The dynamic yield point under shock compression, the Hugoniot elastic limit, or HEL, is defined as the maximum shock pressure a material may be subjected to without permanent, massive, microscopic rearrangement taking place at the shock front. As shown in Fig. 3a, the shock velocity of the HEL state remains nearly constant and for non-porous media is usually equal to the longitudinal elastic wave velocity. Viscoelastic polymeric media generally do not display the HEL phenomenon. We denote five regimes in Fig. 3 for the case of dynamic yielding and phase transition and the available shock wave data are separately fit to linear relations in these regimes in Table 1. For some minerals there are more than four regimes indicated, for reasons such as crystallographic control of compression at low pressures

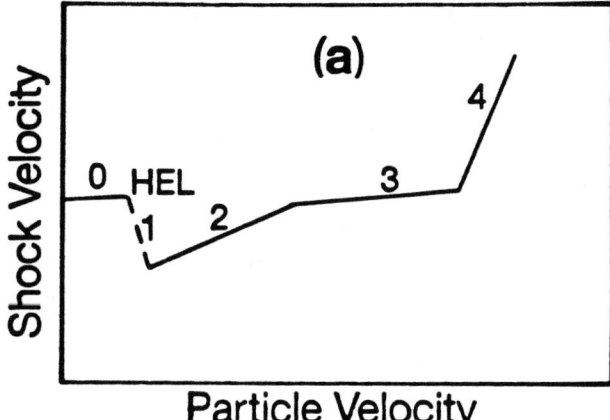

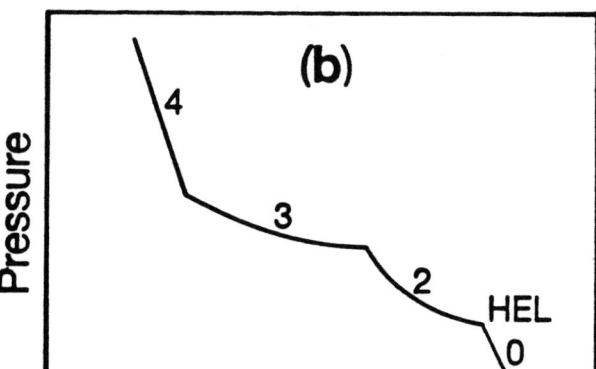

FIG. 3. Sketch of shock velocity-particle relation (a) and corresponding pressure-volume Hugoniot curves (b) for a mineral which undergoes dynamic yielding and a phase change.

    0: compression up to the Hugoniot Elastic Limit (HEL)
    1: transition via dynamic yielding to a quasi-hydrostatic state
    2: low pressure state
    3: mixed region
    4: high pressure state

(such as O_a, O_b for quartz), and for more than one high-pressure state (such as 4a, 4b, and 4c for halite).

The crystallographic or atomistic level nature of shock-induced phase changes varies from simple average coordination changes observed in various liquids, ionization and debonding in non-metallic fluids, electronic transitions in metal and non-metals, changes in crystal structure in solid materials, and transition from the solid to the fluid state.

In the case of a phase change, the pressure along the isentrope P_S at the volume V_1 corresponding to a Hugoniot state (P_1, V_1) is given by

$$\frac{P_1}{2}(V_{OO} - V_1) = -\int_{V_O}^{V_1} P\, dV + \frac{V_1}{\gamma}(P_1 - P_S) + E_{TR} \qquad (23)$$

where the left-hand side is the Rankine-Hugoniot energy, and the first and second terms on the right represent the gain in the internal energy along the paths 1 and 2 of Fig. 2 Here V_{OO} is the specific volume of the initial material and V_O the specific volume of the shock-induced high-pressure phase, or the intrinsic volume of the sample if the initial state is distended. Also E_{TR} is the energy of transition to the high-pressure phase at STP. In the case of no phase change, $E_{TR} = 0$. For zero initial porosity $V_{OO} = V_O$. The unknown parameter in Eq. 23 is P_S, which is implicit in the first integral term on the right-hand side and explicit in the second term. The second term is obtained by using the definition of the Gruneisen parameter (Eq. 13) to calculate the change in energy associated with the pressure difference $(P_1 - P_S)$ at constant volume.

4. SHOCK TEMPERATURES

For many condensed media, the Mie-Gruneisen equation of state, based on a finite-difference formulation of the Gruneisen parameter (Eq. 16), can be used to describe shock and postshock temperatures. The temperature along the isentrope [224] is given by

$$T_S = T_i \exp\left[-\int_{V_a}^{V_b}\left(\frac{\gamma}{V}\right) dV\right] \qquad (24)$$

where T_i is the initial temperature. For the principal isentrope centered at room temperature, $T_i = T_O$, $V_a = V_O$,

initial volume, and $V_b = V$, compressed volume. For the calculation of postshock temperatures $T_i = T_H$, the Hugoniot temperature, $V_a = V_H$, the volume of the shock state, and $V_b = V_{OO}{}'$, the postshock volume corresponding to the postshock temperature. For shock compression to a volume V, P_S is first obtained by using Eq. 23; then T_S, the isentropic compression temperature at volume V, may be calculated by using Eq. 24. Finally, using Eq. 16, the shock temperature T_H is given by

$$\frac{V}{\gamma}(P_H - P_S) = \int_{T_S}^{T_H} C_V\, dT \qquad (25)$$

It is useful to carry out both postshock and shock temperature measurements as they provide complementary information for the thermal equation of state, that is, γ, as well as C_V. Minerals for which shock temperatures have been measured (usually via radiative techniques) are so indicated in Table 1.

In the case of molecular fluids such as water, a formulation based on the near constancy of C_P at constant pressure is used [41,167].

Although there have been few data collected, postshock temperatures are very sensitive to the models which specify γ and its volume dependence, in the case of the Gruneisen equation of state [49,164,165]. In contrast, the absolute values of shock temperatures are sensitive to the phase transition energy E_{TR} of Eq. 23, whereas the slope of the T_H vs. pressure curve is sensitive to the specific heat.

Acknowledgments. Research supported under NSF, NASA, and DoD. We appreciate comments on this manuscript from William W. Anderson, Kathleen Gallagher, Wenbo Yang, and J. Michael Brown. Contribution number 5332, Division of Geological and Planetary Sciences.

REFERENCES

1 Abe, Y., and T. Matsui, The formation of an impact-generated H_2O atmosphere and its implications for the early thermal history of the Earth, *Proc. Lunar Planet. Sci. Conf. 15th, Part 2, J. Geophys. Res.*, 90, C545-C559, 1985.

2 Ahrens, T. J., High-pressure electrical behavior and equation

of state of magnesium oxide from shock wave measurements, *J. Appl. Phys.*, 37, 2532-2541, 1966.

3 Ahrens, T. J., Equations of state of iron sulfide and constraints on the sulfur content of the Earth, *J. Geophys. Res.*, 84, 985-998, 1979.

4 Ahrens, T. J., Dynamic

compression of earth materials, *Science*, 207, 1035-1041, 1980.

5 Ahrens, T. J., Application of shock wave data to earth and planetary science, in *Shock Waves in Condensed Matter*, edited by Y. M. Gupta, pp. 571-588, Plenum, New York, 1986.

6 Ahrens, T. J., J. D. Bass, and J. R. Abelson, Shock temperatures

in metals, in *Shock Compression of Condensed Matter - 1989*, edited by S. C. Schmidt, J. N. Johnson and L. W. Davison, pp. 851-857, Elsevier Publishers, B. V., 1990.

7   Ahrens, T. J., and D. M. Cole, Shock compression and adiabatic release of lunar fines from Apollo 17, *Proc. 5th Lunar Science Conf., Suppl. 5, Geochim. et Cosmochim. Acta, 3*, 2333-2345, 1974.

8   Ahrens, T. J., and E. S. Gaffney, Dynamic compression of enstatite, *J. Geophys. Res., 76*, 5504-5513, 1971.

9   Ahrens, T. J., W. H. Gust, and E. B. Royce, Material strength effect in the shock compression of alumina, *J. Appl. Phys., 39*, 4610-4616, 1968.

10   Ahrens, T. J., and R. Jeanloz, Pyrite: Shock compression, isentropic release, and composition of the Earth's core, *J. Geophys. Res., 92*, 10363-10375, 1987.

11   Ahrens, T. J., and R. K. Linde, Response of brittle solids to shock compression, in *Behaviour of Dense Media Under High Dynamic Pressures-Symposium on the Behavior of Dense Media under High Dynamic Pressures, Paris, September 1967*, pp. 325-336, Gordon and Breach, New York, 1968.

12   Ahrens, T. J., and H.-P. Liu, A shock-induced phase change in orthoclase, *J. Geophys. Res., 78*, 1274-1278, 1973.

13   Ahrens, T. J., and J. D. O'Keefe, Shock melting and vaporization of lunar rocks and minerals, *The Moon, 4*, 214-249, 1972.

14   Ahrens, T. J., and J. D. O'Keefe, Equation of state and impact-induced shock-wave attenuation on the Moon, in *Impact and Explosion Cratering*, edited by D. J. Roddy, R. O. Pepin and R. B. Merrill, pp. 639-656, Pergamon, New York, 1977.

15   Ahrens, T. J., and C. F. Petersen,

Shock wave data and the study of the Earth, in *The Application of Modern Physics to the Earth and Planetary Interiors*, edited by S. K. Runcorn, pp. 449-461, 1969.

16   Ahrens, T. J., C. F. Petersen, and J. T. Rosenberg, Shock compression of feldspars, *J. Geophys.Res., 74*, 2727-2746, 1969.

17   Ahrens, T. J., and J. T. Rosenberg, Shock metamorphism: Experiments on quartz and plagioclase, in *Shock Metamorphism of Natural Materials*, edited by B. M. French and N. M. Short, pp. 59-81, Mono Book Corp., Baltimore, 1968.

18   Ahrens, T. J., J. T. Rosenberg, and M. H. Ruderman, Stanford Research Institute, Menlo Park, California 94550, DASA 1868, 1966.

19   Ahrens, T. J., and M. H. Ruderman, Immersed-foil method for measuring shock wave profiles in solids, *J. Appl. Phys., 37*, 4758-4765, 1966.

20   Ahrens, T. J., H. Tan, and J. D. Bass, Analysis of shock temperature data for iron, *High Pressure Research, 2*, 145-157, 1990.

21   Ahrens, T. J., and J. V. G. Gregson, Shock compression of crustal rocks: data for quartz, calcite, and plagioclase rocks, *J. Geophys. Res., 69*, 4839-4874, 1964.

22   Al'tshuler, L. V., Use of shock waves in high-pressure physics, *Sov. Phys. Uspekhi, 8*, 52-91, 1965.

23   Al'tshuler, L. V., A. A. Bakanova, and I. P. Dudoladov, Effect of electron structure on the compressibility of metals at high pressure, *Sov. Phys. JETP, 26*, 1115-1120, 1968.

24   Al'tshuler, L. V., A. A. Bakanova, I. P. Dudoladov, E. A. Dynin, R. F. Trunin, and B. S. Chekin, Shock adiabatic curves

of metals. New data, statistical analysis, and general laws, *J. Appl. Mech. Tech. Phys., 2*, 145-169, 1981.

25   Al'tshuler, L. V., A. A. Bakanova, and R. F. Trunin, Shock adiabats and zero isotherms of seven metals at high pressures, *Sov. Phys. JETP, 15*, 65-74, 1962.

26   Al'tshuler, L. V., and S. E. Brusnikin, Equations of state of compressed and heated metals, *High Temp. USSR, 27*, 39-47, 1989.

27   Al'tshuler, L. V., N. N. Kalitkin, L. V. Kuz'mina, and B. S. Chekin, Shock adiabats for ultrahigh pressures, *Sov. Phys. JETP, 45*, 167-171, 1977.

28   Al'tshuler, L. V., S. B. Kormer, A. A. Bakanova, and R. F. Trunin, Equation of state for aluminum, copper, and lead in the high pressure region, *Sov. Phys. JETP, 11*, 573-579, 1960.

29   Al'tshuler, L. V., S. B. Kormer, M. I. Brazhnik, L. A. Vladimirov, M. P. Speranskaya, and A. I. Funtikov, The isentropic compressibility of aluminum, copper, lead, and iron at high pressures, *Sov. Phys. JETP, 11*, 766-775, 1960.

30   Al'tshuler, L. V., K. K. Krupnikov, and M. I. Brazhnik, Dynamic compressibility of metals under pressures from 400,000 to 4,000,000 atmospheres, *Sov. Phys. JETP, 34*, 614-619, 1958.

31   Al'tshuler, L. V., K. K. Krupnikov, B. N. Ledenev, V. I. Zhuchikhin, and M. I. Brazhnik, Dynamic compressibility and equation of state of iron under high pressure, *Sov. Phys. JETP, 34*, 606-614, 1958.

32   Al'tshuler, L. V., L. V. Kuleshova, and M. N. Pavlovskii, The dynamic compressibility, equation of state, and electrical conductivity of sodium chloride at high pressures, *Sov. Phys. JETP, 12*,

10-15, 1961.

33 Al'tshuler, L. V., B. N. Moiseev, L. V. Popov, G. V. Simakov, and R. F. Trunin, Relative compressibility of iron and lead at pressures of 31 to 34 Mbar, *Sov. Phys. JETP*, *27*, 420-422, 1968.

34 Al'tshuler, L. V., M. N. Pavlovskii, and V. P. Draken, Peculiarities of phase transitions in compression and rarefaction shock waves, *Sov. Phys. JETP*, *25*, 260-265, 1967.

35 Al'tshuler, L. V., M. A. Podurets, G. V. Simakov, and R. F. Trunin, High-density forms of fluorite and rutile, *Soviet Physics--Solid State*, *15*, 969-971, 1973.

36 Al'tshuler, L. V., G. V. Simakov, and R. F. Trunin, On the chemical composition of the Earth's core, *Izv., Earth Phys.*, *1*, 1-3, 1968.

37 Al'tshuler, L. V., R. F. Trunin, and G. V. Simakov, Shock-wave compression of periclase and quartz and the composition of the Earth's lower mantle, *Izv. Earth Phys.*, *10*, 657-660, 1965.

38 Al'tshuler, V. L., M. N. Pavlovskii, L. V. Kuleshova, and G. V. Simakov, Investigation of alkali-metal halides at high pressures and temperatures produced by shock compression, *Soviet Physics--Solid State*, *5*, 203-211, 1963.

39 Anderson, G. D., D. G. Doran, and A. L. Fahrenbruch, Stanford Research Institute, Menlo Park, California, AFWL-TR-65-147 and AFWL-TR-65-43, 1965.

40 Asay, J. R., and M. Shahinpoor (Ed.), *High-Pressure Shock Compression of Solids*, pp. 1-393, Springer-Verlag, New York, 1993.

41 Bakanova, A. A., V. N. Zubarev, Y. N. Sutulov, and R. F. Trunin, Thermodynamic properties of water at high pressures and temperatures, *Sov. Phys. JETP*, *41*, 544-547, 1976.

42 Balchan, A. S., and G. R. Cowan, Shock compression of two iron-silicon alloys to 2.7 Megabars, *J. Geophys. Res.*, *71*, 3577-3588, 1966.

43 Bass, J. D., T. J. Ahrens, J. R. Abelson, and H. Tan, Shock temperature measurements in metals: New results for an Fe alloy, *J. Geophys. Res.*, *95*, 21767-76, 1990.

44 Bass, J. D., B. Svendsen, and T. J. Ahrens, The temperature of shock-compressed iron, in *High Pressure Research in Mineral Physics*, edited by M. Manghnani and Y. Syono, pp. 393-402, Terra Scientific, Tokyo, 1987.

45 Belyakova, M. Y., M. V. Zhernokletov, Y. N. Sutulov, and R. F. Trunin, Shock-wave compression of metal alloys, *Isv. Earth Phys.*, *27*, 71-77, 1991.

46 Bevington, P. R., *Data reduction and Error Analysis for the Physical Sciences*, pp. 1-336, McGraw-Hill, New York, 1969.

47 Birch, F., Compressibility: Elastic Constants, in *Handbook of Physical Constants, revised edition*, edited by S. P. Clark Jr., pp. 153-159, The Geological Society of America, New York, 1966.

48 Boness, D. A., Shock-wave experiments and electronic band-structure calculations of materials at high temperature and pressure, Ph.D. thesis, 256 pp., University of Washington, Seattle, 1991.

49 Boslough, M., Postshock temperatures in silica, *J. Geophys. Res.*, *93*, 6477-6484, 1988.

50 Boslough, M. B., T. J. Ahrens, and A. C. Mitchell, Shock temperatures in CaO, *J. Geophys. Res.*, *89*, 7845-7851, 1984.

51 Boslough, M. B., T. J. Ahrens, and A. C. Mitchell, Shock temperatures in anorthite glass, *Geophys. J.R. astr. Soc.*, *84*, 475-489, 1986.

52 Boslough, M. B., S. M. Rigden, and T. J. Ahrens, Hugoniot equation of state of anorthite glass and lunar anorthosite, *Geophys. J. R. astr. Soc.*, *84*, 455-473, 1986.

53 Brown, J. M., T. J. Ahrens, and D. L. Shampine, Hugoniot data for pyrrhotite and the Earth's core, *J. Geophys. Res.*, *89*, 6041-6048, 1984.

54 Carter, W. J., Hugoniot equation of state of some alkali halides, *High Temp. - High Press.*, *5*, 313-318, 1973.

55 Carter, W. J., and S. P. Marsh, Hugoniot equations of state of polymers, Los Alamos Scientific Laboratory, unpublished report LA-UR-77-2062, 1977.

56 Carter, W. J., S. P. Marsh, J. N. Fritz, and R. G. McQueen, The equation of state of selected materials for high-pressure reference, in *Accurate Characterization of the High Pressure Environment*, edited by E. C. Boyd, pp. 147-158, NBS special pub. 326, Washington D.C., 1971.

57 Christian, R. H., R. E. Duff, and F. L. Yarger, Equation of state of gases by shock wave measurements. II. The dissociation energy of nitrogen, *J. Chem. Phys.*, *23*, 2045-2049, 1955.

58 Coleburn, N. L., Compressibility of pyrolytic graphite, *J. Chem. Phys.*, *40*, 71-77, 1964.

59 Davison, L., and R. A. Graham, Shock compression of solids, *Phys. Rep.*, *55*, 255-379, 1979.

60 Dick, R. D., Shock compression of liquid benzene, carbon disulfide, carbon tetrachloride and nitrogen, *Bull. Am. Phys. Soc.*, *13*, 579, 1968.

61 Dick, R. D., Shock wave compression of benzene, carbon disulfide, carbon tetrachloride and liquid nitrogen, *Los Alamos Scientific Laboratory report*, LA-3915, 1968.

62 Dick, R. D., Shock wave compression of benzene, carbon disulfide, carbon tetrachloride, and liquid nitrogen, *J. Chem.*

Phys., *52*, 6021-6032, 1970.

63 Dick, R. D., and G. I. Kerley, Shock compression data for liquids. II. Condensed hydrogen and deuterium, *J. Chem. Phys.*, *73*, 5264-5271, 1980.

64 Dick, R. D., R. H. Warnes, and J. J. Skalyo, Shock compression of solid argon, *J. Chem. Phys*, *53*, 1648, 1970.

65 Doran, D. G., Hugoniot equation of state of pyrolytic graphite to 300 kbars, *J. Appl. Phys.*, *34*, 844-851, 1963.

66 Doran, D. G., Shock-wave compression of barium titanate and 95/5 lead zirconate titanate, *J. Appl. Phys.*, *39*, 40-47, 1968.

67 Dremin, A. N., and G. A. Adadurov, The behavior of glass under dynamic loading, *Sov. Phys. - Solid State*, *6*, 1379-1384, 1964.

68 Duff, R. E., and F. S. Minshall, Investigation of a shock-induced transition in bismuth, *Phys. Rev.*, *108*, 1207-1212, 1957.

69 Duffy, T. S., T. J. Ahrens, and M. A. Lange, The shock wave equation of state of brucite $Mg(OH)_2$, *J. Geophys. Res.*, *96*, 14,319-14,330, 1991.

70 Duvall, G. E., and G. R. Fowles, Shock waves, in *High Pressure Physics and Chemistry*, vol. 2, edited by R. S. Bradley, pp. 209-292, Academic Press, New York, 1963.

71 Fortov, V. E., A. A. Leont'ev, A. N. Dremin, and V. K. Gryaznov, Shock-wave production of a non-ideal plasma, *Sov. Phys. JETP*, *44*, 116-122, 1976.

72 Fowles, R., Dynamic compression of quartz, *J. Geophys. Res.*, *72*, 5729-5742, 1967.

73 Fritz, J. N., S. P. Marsh, W. J. Carter, and R. G. McQueen, The Hugoniot equation of state of sodium chloride in the sodium chloride structure in, in *Accurate Characterization of the High-Pressure Environment*, *proceedings of a symposium held at the National Bureau of Standards, Gaithersburg, Maryland, Special publication*, vol. 326, NBS, Washington DC, 1971.

74 Furnish, M. D., L. C. Chhabildas, D. J. Steinberg, and G. T. G. III, Dynamic behavior of fully dense molybdenum, in *Shock Compression of Condensed Matter 1991*, edited by S. C. Schmidt, R. D. Dick, J. W. Forbes and D. G. Tasker, pp. 419-422, Elsevier Science Publishers B.V, 1992.

75 Gaffney, E. S., and T. J. Ahrens, Identification of ice VI on the Hugoniot of ice I_h, *Geophys. Res. Lett.*, *7*, 407-409, 1980.

76 Goto, T., T. J. Ahrens, G. R. Rossman, and Y. Syono, Absorption spectrum of shock-compressed Fe^{2+}-bearing MgO and the radiative conductivity of the lower mantle, *Phys. Earth Planet. Inter.*, *22*, 277-288, 1980.

77 Graham, E. K., and T. J. Ahrens, Shock wave compression of iron-silicate garnet, *J. Geophys. Res.*, *78*, 375-392, 1973.

78 Graham, R. A., and W. P. Brooks, Shock-wave compression of sapphire from 15 to 420 kbar. The effects of large anisotropic compressions, *J. Phys. Chem Solids*, *32*, 2311-2330, 1971.

79 Graham, R. A., O. E. Jones, and J. R. Holland, Physical behavior of germanium under shock wave compression, *J. Phys. Chem. Solids*, *27*, 1519-1529, 1966.

80 Grigor'ev, F. V., S. B. Kormer, O. L. Mikhailova, M. A. Mochalov, and V. D. Urlin, Shock compression and brightness temperature of a shock wave front in argon. Electron screening of radiation, *Sov. Phys. JETP*, *61*, 751-757, 1985.

81 Gryaznov, V. K., M. V. Zhernokletov, V. N. Zubarev, I. L. Ioselevskii, and V. E. Fortov, Thermodynamic properties of a nonideal argon or xenon plasma, *Sov. Phys. JETP*, *51*, 288-295, 1980.

82 Gust, W. H., Phase transition and shock-compression parameters to 120 GPa for three types of graphite and for amorphous carbon, *Phys, Rev. B.*, *22*, 4744-4756, 1980.

83 Harvey, W. B., and R. D. Dick, Shock compression of liquid ammonia, *Bull. Am. Phys. Soc.*, *20*, 48, 1975.

84 Hauver, G. E., and A. Melani, Ballistics Research Labs, Aberdeen Proving Ground, Maryland, BRL MR 2061, 1970.

85 Hayes, D. B., Polymorphic phase transformation rates in shock-loaded potassium chloride, *J. Appl. Phys.*, *45*, 1208-1217, 1974.

86 Holmes, N. C., J. A. Moriarty, G. R. Gathers, and W. J. Nellis, The equation of state of platinum to 660 GPa (6.6 Mbar), *J. Appl. Phys.*, *66*, 2962-2967, 1989.

87 Holmes, N. C., W. J. Nellis, W. B. Graham, and G. E. Walrafen, Spontaneous Raman spectroscopy of shocked H_2O, *Physica*, *139 and 140B*, 568-570, 1986.

88 Holmes, N. C., and E. F. See, Shock compression of low-density microcellular materials, in *Shock Compression of Condensed Matter 1991*, edited by S. C. Schmidt, R. D. Dick, J. W. Forbes and D. G. Tasker, pp. 91-94, Elsevier Science Publishers B.V., 1992.

89 Hughes, D. S., L. E. Gourley, and M. F. Gourley, Shock-wave compression of iron and bismuth, *J. Appl. Phys.*, *32*, 624-629, 1961.

90 Isbell, W. M., F. H. Shipman, and A. H. Jones, Hugoniot equation of state measurements for eleven materials to five Megabars, *DASA File No 15.025, MSL-68-13*, 128 pp., 1968.

91 Jackson, I., and T. J. Ahrens, Shock-wave compression of

single crystal forsterite, *J. Geophys. Res.*, *84*, 3039-3048, 1979.

92 Jackson, I., and T. J. Ahrens, Shock-wave compression of vitreous and rutile-type GeO_2: A comparative study, *Phys. Earth Planet. Int.*, *20*, 60-70, 1979.

93 Jeanloz, R., Shock wave equation of state and finite strain theory, *J. Geophys. Res.*, *94*, 5873-5886, 1989.

94 Jeanloz, R., The nature of the Earth's core, *Ann. Rev. Earth and Planet. Sci.*, *18*, 357-386, 1990.

95 Jeanloz, R., and T. J. Ahrens, Pyroxenes and olivines: Structural implications of shock-wave data for high pressure phases, in *High-Pressure Research: Applications to Geophysics*, edited by M. H. Manghnani and S. Akimoto, pp. 439-462, Academic, San Diego, Calif., 1977.

96 Jeanloz, R., and T. J. Ahrens, Anorthite, thermal equation of state to high pressures, *Geophys. J. R. astr. Soc.*, *62*, 529-549, 1980.

97 Jeanloz, R., and T. J. Ahrens, Equations of state of FeO and CaO, *Geophys. J. R. astron. Soc.*, *62*, 505-528, 1980.

98 Jeanloz, R., and R. Grover, Birch-Murnaghan and Us-Up equations of state, in *Proceedings of the American Physical Society Topical Conference on Shock Waves in Condensed Matter, Monterey, CA, 1987*, edited by S. C. Schmidt and N. C. Holmes, pp. 69-72, Plenum, New York, 1988.

99 Johnson, J. O., and J. Wackerle, Shock-wave compression of porous magnesium and ammonium sulfate, in *Behaviour of Dense Media Under High Dynamic Pressures-Symposium on the Behavior of Dense Media under High Dynamic Pressures, Paris, September 1967*, pp. 217-226, Gordon and Breach, New York, 1968.

100 Jones, A. H., W. M. Isbell, and C. J. Maiden, Measurement of the very-high-pressure properties of materials using a light-gas gun, *J. Appl. Phys.*, *37*, 3493-3499, 1966.

101 Jones, A. H., W. M. Isbell, F. H. Shipman, R. D. Perkins, S. J. Green, and C. J. Maiden, Material properties measurements for selected materials, NASA Ames, Interim Report, NAS2-3427, MSL-68-9, 55 pp., 1968.

102 Kalashnikov, N. G., M. N. Pavlovskiy, G. V. Simakov, and R. F. Trunin, Dynamic compressibilty of calcite-group minerals, *Izv. Earth Phys.*, *2*, 23-29, 1973.

103 Kondo, K.-I., and T. J. Ahrens, Heterogeneous shock-induced thermal radiation in minerals, *Phys. Chem. Minerals*, *9*, 173-181, 1983.

104 Kondo, K.-I., T. J. Ahrens, and A. Sawaoka, Shock-induced radiation spectra of fused quartz, *J. Appl. Phys.*, *54*, 4382-4385, 1983.

105 Kormer, S. B., A. I. Funtikov, V. D. Urlin, and A. N. Kolesnikova, Dynamic compression of porous metals and the equation of state with variable specific heat at high temperatures, *Sov. Phys. JETP*, *15*, 477-488, 1962.

106 Kormer, S. B., M. V. Sinitsyn, A. I. Funtikov, V. D. Urlin, and A. V. Blinov, Investigation of the compressibility of five ionic compounds at pressures up to 5 Mbar, *Sov. Phys. JETP*, *20*, 811-819, 1965.

107 Kormer, S. B., M. V. Sinitsyn, G. A. Kirilov, and V. D. Urlin, Experimental determination of temperature in shock-compressed NaCl and KCl and of their melting curves at pressures up to 700 kbar, *Soviet Phys. JETP*, *21*, 689-708, 1965.

108 Krupnikov, K. K., A. A. Bakanova, M. I. Brazhnik, and R. F. Trunin, An investigation of the shock compressibility of titanium, molybdenum, tantalum, and iron, *Sov. Phys. Doklady*, *2*, 205-208, 1963.

109 Lagus, P. L., and T. J. Ahrens, Shock wave measurements on solid hydrogen and argon, *J. Chem. Phys.*, *59*, 3517-3522, 1973.

110 Lange, M., and T. J. Ahrens, Fragmentation of ice by low velocity impact, in *12th Lunar and Planetary Science Conference*, pp. 1667-1687, Pergamon Press, Houston, TX, 1981.

111 Lange, M. A., and T. J. Ahrens, FeO and H_2O and the homogeneous accretion of the Earth, *Earth Planet. Sci. Lett.*, *71*, 111-119, 1984.

112 Lange, M. A., P. Lambert, and T. J. Ahrens, Shock effects on hydrous minerals and implications for carbonaceous meteorites, *Geochim. Cosmochim. Acta*, *49*, 1715-1726, 1985.

113 Larson, D. B., A shock-induced phase transformation in bismuth, *J. Appl. Phys.*, *38*, 1541-1546, 1967.

114 Larson, D. B., G. D. Bearson, and J. R. Taylor, Shock-wave studies of ice and two frozen soils, in *North American Contribution to the Second International Conference on Permafrost*, edited by T. L. Péwé and J. R. Mackay, pp. 318-325, 1973.

115 Lundergan, C. D., and W. Herrmann, Equation of state of 6061-T6 aluminum at low pressures, *J. Appl. Phys.*, *34*, 2046-2052, 1963.

116 Lysne, P. C., A comparison of calculated and measured low-stress Hugoniots and release adiabats of dry and water-saturated tuff, *J. Geophys. Res.*, *75*, 4375-4386, 1970.

117 Lyzenga, G. A., and T. J. Ahrens, Shock temperature measurements in Mg_2SiO_4 and SiO_2 at high pressures, *Geophys.*

Res. Lett., 7, 141-144, 1980.

118   Lyzenga, G. A., T. J. Ahrens, and A. C. Mitchell, Shock temperatures of SiO_2 and their geophysical implications, *J. Geophys. Res., 88,* 2431-2444, 1983.

119   Lyzenga, G. A., T. J. Ahrens, W. J. Nellis, and A. C. Mitchell, The temperature of shock-compressed water, *J. Chem. Phys., 76,* 6282-6286, 1982.

120   Marsh, S. P., Hugoniot equation of state of beryllium oxide, *High Temp.--High Press., 5,* 503-508, 1973.

121   Marsh, S. P. (Ed.), *LASL Shock Hugoniot Data,* pp. 1-658, University of California Press, Berkeley, 1980.

122   Mashimo, T., Y. Hanaoka, and K. Nagayama, Elastoplastic properties under shock compression of Al_2O_3 single crystal and polycrystal, *J. Appl. Phys., 63,* 327-336, 1988.

123   Mashimo, T., K. Nagayama, and A. Sawaoka, Anisotropic elastic limits and phase transitions of rutile phase TiO_2 under shock compression, *J. Appl. Phys., 54,* 5043-5048, 1983.

124   Mashimo, T., K. Nagayama, and A. Sawaoka, Shock compression of zirconia ZrO_2 and zircon $ZrSiO_4$ in the pressure range up to 150 GPa, *Phys. Chem. Minerals, 9,* 237-247, 1983.

125   McMahan, A. K., B. L. Hord, and M. Ross, Experimental and theoretical study of metallic iodine, *Phys. Rev., B 15,* 726-737, 1977.

126   McQueen, R. G., Laboratory techniques for very high pressures and the behavior of metals under dynamic loading in, in *Metallurgy at High Pressures and High Temperatures,* edited by J. K. A. Gschneider, M. T. Hepworth and N. A. D. Parlee, pp. 44-132, Gordon and Breach, New York, 1964.

127   McQueen, R. G., The equation of state of mixtures, alloys, and compounds, in *Seismic Coupling- Proceedings of a meeting sponsored by the Advanced Research Projects Agency, January 15-16, 1968,* edited by G. Simmons, Stanford Research Institute, Menlo Park, CA, 1968.

128   McQueen, R. G., W. J. Carter, J. N. Fritz, and S. P. Marsh, The solid-liquid phase line in Cu, in *Accurate Characterization of the High-Pressure Environment, proceedings of a symposium held at the National Bureau of Standards, Gaithersburg, Maryland, October 14-18, 1968, Special publication 326, NBS,* Washington DC, 1971.

129   McQueen, R. G., J. N. Fritz, and S. P. Marsh, On the equation of state of stishovite, *J. Geophys. Res., 68,* 2319-2322, 1963.

130   McQueen, R. G., J. C. Jamieson, and S. P. Marsh, Shock-wave compression and X-ray studies of titanium dioxide, *Science, 155,* 1401-1404, 1967.

131   McQueen, R. G., and S. P. Marsh, Equation of state for nineteen metallic elements from shock-wave measurements to 2 Megabars, *J. Appl. Phys., 31,* 1253-1269, 1960.

132   McQueen, R. G., and S. P. Marsh, Shock-wave compression of iron-nickel alloys and the Earth's core, *J. Geophys. Res., 71,* 1751-1756, 1966.

133   McQueen, R. G., and S. P. Marsh, Hugoniot of graphite of various initial densities and the equation of state of carbon, in *Behavior of Dense Media under High Dynamic Pressures, Symposium H. D. P., September 1967,* Gordon and Breach, New York, Paris, 1968.

134   McQueen, R. G., S. P. Marsh, and W. J. Carter, The determination of new standards for shock wave equation-of-state work, in *Behavior of Dense Media under High Dynamic Pressures, Symposium on the Behavior of Dense Media under High Dynamic Pressures, Paris, September 1967,* pp. 67-83, Gordon and Breach, New York, 1968.

135   McQueen, R. G., S. P. Marsh, and J. N. Fritz, Hugoniot equation of state of twelve rocks, *J. Geophys. Res., 72,* 4999-5036, 1967.

136   McQueen, R. G., S. P. Marsh, J. W. Taylor, J. N. Fritz, and W. J. Carter, The equation of state of solids from shock wave studies, in *High-Velocity Impact Phenomena,* edited by R. Kinslow, pp. 249-419, Academic Press, New York, 1970.

137   Miller, G. H., T. J. Ahrens, and E. M. Stolper, The equation of state of molybdenum at 1400°C, *J. Appl. Phys., 63,* 4469-4475, 1988.

138   Minshall, F. S., The dynamic response of iron and iron alloys to shock waves, *Met. Soc. Conf., 9,* 249-274, 1961.

139   Mitchell, A. C., and W. J. Nellis, Shock compression of aluminum, copper and tantalum, *J. Appl. Phys., 52,* 3363-3374, 1981.

140   Mitchell, A. C., and W. J. Nellis, Equation of state and electrical conductivity of water and ammonia shocked to the 100 GPa (1 Mbar) pressure range, *J. Chem. Phys, 76,* 6273-6281, 1982.

141   Mitchell, A. C., W. J. Nellis, J. A. Moriarty, R. A. Heinle, N. C. Holmes, R. E. Tipton, and G. W. Repp, Equation of state of Al, Cu, Mo, and Pb at shock pressures up to 2.4 TPa (24 Mbar), *J. Appl. Phys., 69,* 2981-2986, 1991.

142   Morgan, D. T., M. Rockowitz, and A. L. Atkinson, Avco Corporation, Research and Development Division, Wilmington, Massachusetts, AFWL-TR-65-117, 1965.

143   Munson, D. E., and L. M. Barker, Dynamically determined pressure-volume relationships for aluminum, copper, and lead,

J. Appl. Phys., *37*, 1652-1660, 1966.

144 Nellis, N. J., A. C. Mitchell, M. v. Thiel, G. J. Devine, and R. J. Trainor, Equation-of-state data for molecular hydrogen and deuterium at shock pressures in the range 2-76 GPa (20-760 kbar), *J. Chem. Phys.*, *79*, 1480-1486, 1983.

145 Nellis, W. J., N. C. Holmes, A. C. Mitchell, R. J. Trainor, G. K. Governo, M. Ross, and D. A. Young, Shock compression of liquid helium to 56 GPa (560 kbar), *Phys. Rev. Lett.*, *53*, 1248-1251, 1953.

146 Nellis, W. J., and A. C. Mitchell, Shock compression of liquid argon, nitrogen, and oxygen to 90 GPa (900 kbar), *J. Chem. Phys.*, *73*, 6137-6145, 1980.

147 Nellis, W. J., A. C. Mitchell, F. H. Ree, M. Ross, N. C. Holmes, R. J. Trainor, and D. J. Erskine, Equation of state of shock-compressed liquids: carbon dioxide and air, *J. Chem. Phys.*, *95*, 5268-5272, 1991.

148 Nellis, W. J., J. A. Moriarty, A. C. Mitchell, M. Ross, R. G. Dandrea, N. W. Ashcroft, N. C. Holmes, and G. R. Gathers, Metals physics at ultrahigh pressures: aluminum, copper and lead as prototypes, *Phys. Rev. Lett.*, *60*, 1414-1417, 1988.

149 Nellis, W. J., H. B. Radousky, D. C. Hamilton, A. C. Mitchell, N. C. Holmes, K. B. Christianson, and M. van Thiel, Equation of state, shock temperature, and electrical-conductivity data of dense fluid nitrogen in the region of the dissociative phase transition, *J. Chem. Phys.*, *94*, 2244-2257, 1991.

150 Nellis, W. J., F. H. Ree, M. van Thiel, and A. C. Mitchell, Shock compression of liquid carbon monoxide and methane to 90 GPa (900 kbar), *J. Chem. Phys*, *75*, 3055-3063, 1981.

151 Nellis, W. J., M. v. Thiel, and A. C. Mitchell, Shock compression of liquid xenon to 130 GPa (1.3 Mbar), *Phys. Rev. Lett.*, *48*, 816-818, 1982.

152 Pavlovskii, M. N., Shock compressibility of six very hard substances, *Soviet Physics--Solid State*, *12*, 1736-1737, 1971.

153 Pavlovskii, M. N., Shock compression of diamond, *Sov. Phys. Solid State*, *13*, 741-742, 1971.

154 Pavlovskii, M. N., V. Y. Vashchenko, and G. V. Simakov, Equation of state for cesium iodide, *Sov. Phys. - Solid State*, *7*, 972-974, 1965.

155 Podurets, M. A., L. V. Popov, A. G. Sevast'yanova, G. V. Simakov, and R. F. Trunin, On the relation between the size of studied specimens and the position of the silica shock adiabat, *Izv. Earth Phys.*, *11*, 727-728, 1976.

156 Podurets, M. A., G. V. Simakov, G. S. Telegin, and R. F. Trunin, Polymorphism of silica in shock waves and equation of state of coesite and stishovite, *Izv. Earth Phys.*, *17*, 9-15, 1981.

157 Podurets, M. A., G. V. Simakov, R. F. Trunin, L. V. Popov, and B. N. Moiseev, Compression of water by strong shock waves, *Sov. Phys. JETP*, *35*, 375-376, 1972.

158 Rabie, R., and J. J. Dick, Equation of state and crushing dynamics of low-density silica aerogels, in *Shock Compression of Condensed Matter 1991*, edited by S. C. Schmidt, R. D. Dick, J. W. Forbes and D. G. Tasker, pp. 87-90, Elsevier Science Publishers B.V., 1992.

159 Radousky, H. B., A. C. Mitchell, and W. J. Nellis, Shock temperature measurements of planetary ices: NH_3, CH_4, and "Synthetic Uranus", *Chem. Phys*, *93*, 8235-8239, 1990.

160 Radousky, H. B., W. J. Nellis, M. Ross, D. C. Hamilton, and A. C. Mitchell, Molecular dissociation and shock-induced cooling in fluid nitrogen at high densities and temperatures, *Phys.*

Rev. Let., *57*, 2419-2422, 1986.

161 Radousky, H. B., and M. Ross, Shock temperature measurements in high density fluid xenon, *Phys. Lett. A*, *129*, 43-46, 1988.

162 Ragan, C. E., III, Shock-wave experiments at threefold compression, *Phys. Rev. A*, *29*, 1391-1402, 1984.

163 Ragan, C. E. I., Shock compression measurements at 1 to 7 TPa, *Phys. Rev. A.*, *82*, 3360-3375, 1982.

164 Raikes, S. A., and T. J. Ahrens, Measurements of post-shock temperatures in aluminum and stainless steel, in *High Pressure Science and Technology*, edited by K. D. Timmerhaus and M. S. Barber, pp. 889-894, Plenum Press, New York, 1979.

165 Raikes, S. A., and T. J. Ahrens, Post-shock temperature in minerals, *Geophys. J. Roy. astr, Soc.*, *58*, 717-747, 1979.

166 Rice, M. H., R. G. McQueen, and J. M. Walsh, Compression of solids by strong shock waves, in *Solid State Phys.*, vol. 6, edited by F. Seitz and D. Turnbull, pp. 1-63, Academic Press, New York, 1958.

167 Rice, M. H., and J. M. Walsh, Equation of state of water to 250 kilobars, *J. Chem. Phys.*, *26*, 824-830, 1957.

168 Rigden, S. M., T. J. Ahrens, and E. M. Stolper, High-pressure equation of state of molten anorthite and diopside, *J. Geophys. Res.*, *94*, 9508-9522, 1989.

169 Ringwood, A. E., *Origin of the Earth and Moon*, pp. 1-295, Springer-Verlag, Berlin, New York, 1979.

170 Roddy, D. J., R. O. Pepin, and R. B. Merrill, *Impact and Explosion Cratering*, pp. 1-1301, Pergamon, New York, 1977.

171 Ruoff, A. L., Linear shock-velocity-particle-velocity relationship, *J. Appl. Phys.*, *38*, 4976-4980, 1967.

172 Schmidt, D. N., and R. K. Linde,

Response of distended copper, iron, and tungsten to shock loading, SRI, Air Force Weapons Lab Report, AF 29(601)-7236, 1968.

173 Schmitt, D. R., and T. J. Ahrens, Temperatures of shock-induced shear instabilities and their relationship to fusion curves, *Geophys. Res. Lett.*, *10*, 1077-1080, 1983.

174 Schmitt, D. R., and T. J. Ahrens, Emission spectra of shock compressed solids, in *Shock Waves in Condensed Matter-1983*, vol. Chapter VII:6, edited by J. R. Asay, R. A. Graham and G. K. Straub, pp. 513-516, Elsevier Science Publishers B.V., 1984.

175 Schmitt, D. R., and T. J. Ahrens, Shock temperatures in silica glass: Implications for modes of shock-induced deformation, phase transformation, and melting with pressure, *J. Geophys. Res.*, *94*, 5851-5871, 1989.

176 Schmitt, D. R., T. J. Ahrens, and B. Svendsen, Shock-induced melting and shear banding in single crystal NaCl, *J. Appl. Phys.*, *63*, 99-106, 1988.

177 Schmitt, D. R., B. Svendsen, and T. J. Ahrens, Shock-induced radiation from minerals, in *Shock Waves in Condensed Matter*, edited by Y. M. Gupta, pp. 261-265, Plenum Press, New York, 1986.

178 Schott, G. L., Shock-compressed carbon dioxide: liquid measurements and comparisons with selected models, *High Pressure Res.*, *6*, 187-200, 1991.

179 Schott, G. L., M. S. Shaw, and J. D. Johnson, Shocked states from initially liquid oxygen-nitrogen systems, *J. Chem. Phys.*, *82*, [4264-4275], 1982.

180 Seitz, W. L., and J. Wackerle, Reflected-shock Hugoniot for liquid argon between 0.26 and 0.74 Megabars, *Bull. Am. Phys. Soc.*, *17*, 1093, 1972.

181 Sekine, T., and T. J. Ahrens, Equation of state of heated glassy carbon, in, in *Shock Compression of Condensed Matter 1991*, edited by R. D. D. S. C. Schmidt J. W. Forbes, and D. G. Tasker, pp. 57-60, 1992.

182 Sekine, T., A. M. Rubin, and T. J. Ahrens, Shock wave equation of state of muscovite, *J. Geophys. Res.*, *96*, 19675-19680, 1991.

183 Sharpton, V. L., and P. D. Ward (Ed.), *Global Catastrophes in Earth History; An Interdisciplinary Conference on Impacts, Volcanism, and Mass Mortality*, pp. 1-631, The Geological Society of America, Inc., Special Paper 247, Boulder, Colorado, 1990.

184 Shipman, F. H., W. M. Isbell, and A. H. Jones, High pressure Hugoniot measurements for several Nevada test site rocks, DASA Report 2214, MSL-68-15, 114 pp., 1969.

185 Silver, L. T., and P. Schultz (Ed.), *Geological Implications of Impacts of Large Asteroids and Comets on the Earth*, pp. 1-528, The Geological Society of America, Special Paper 190, Boulder, Colorado, 1982.

186 Simakov, G. V., M. N. Pavlovskiy, N. G. Kalashnikov, and R. F. Trunin, Shock compressibility of twelve minerals, *Izv. Earth Phys.*, *10*, 488-492, 1974.

187 Simakov, G. V., and R. F. Trunin, Compression of minerals by shock waves, *Izv. Earth Phys.*, *16*, 134-137, 1980.

188 Simakov, G. V., and R. F. Trunin, Shockwave compression of ultraporous silica, *Izv. Earth Physics*, *26*, 952-956, 1990.

189 Skalyo, J., Jr., R. D. Dick, and R. H. Warnes, Shock compression of solid argon, *Bull. Am. Phys. Soc.*, *13*, 579, 1968.

190 Skidmore, I. C., and E. Morris, in *Thermodynamics of Nuclear Materials*, pp. 173, International Atomic Energy Agency, Vienna, Austria, 1962.

191 Somerville, M., and T. J. Ahrens, Shock compression of $KFeS_2$ and the question of potassium in the core, *J. Geophys. Res.*, *85*, 7016-7024, 1980.

192 Stevenson, D. J., Models of the earth's core, *Science*, *214*, 611-619, 1981.

193 Stöffler, D., Deformation and transformation of rock-forming minerals by natural and experimental shock processes, I., *Fortschr. Miner.*, *49*, 50-113, 1972.

194 Stöffler, D., Deformation and transformation of rock-forming minerals by natural and experimental shock processes. II. Physical properties of shocked minerals, *Fortschr. Miner.*, *51*, 256-289, 1974.

195 Stolper, E. M., D. Walker, B. H. Hager, and J. F. Hays, Melt segregation from partially molten source regions: The importance of melt density and source region size, *J. Geophys. Res.*, *86*, 6261-6271, 1981.

196 Svendsen, B., and T. J. Ahrens, Dynamic compression of diopside and salite to 200 GPa, *Geophys. Res. Lett.*, *10*, 501-504, 1983.

197 Svendsen, B., and T. J. Ahrens, Shock-induced temperatures of MgO, *Geophys. J. R. Astr. Soc.*, *91*, 667-691, 1987.

198 Svendsen, B., and T. J. Ahrens, Shock-induced temperatures of $CaMgSi_2O_6$, *J. Geophys. Res.*, *95*, 6943-6953, 1990.

199 Swenson, C. A., J. W. Shaner, and J. M. Brown, Hugoniot overtake sound-velocity measurements on CsI, *Phys. Rev. B*, *34*, 7924-7935, 1986.

200 Syono, Y., T. Goto, and Y. Nakagawa, Phase-transition pressures of Fe_3O_4 and GaAs determined from shock-compression experiments, in *High Pressure Research: Application in Geophysics*, edited by M. H. Manghnani and S. Akimoto, pp. 463-476, Academic Press, New York,

1977.

201 Syono, Y., T. Goto, J.-I. Sato, and H. Takei, Shock compression measurements of single-crystal forsterite in the pressure range 15-93 GPa, *J. Geophys. Res.*, *86*, 6181-6186, 1981.

202 Taylor, J. W., and M. H. Rice, Elastic-plastic properties of iron, *J. Appl. Phys.*, *34*, 364-371, 1963.

203 Trunin, R. F., Compressibility of various substances at high shock pressures: an overview, *Izv. Earth Phys.*, *22*, 103-114, 1986.

204 Trunin, R. F., A. B. Medvedev, A. I. Funtikov, M. A. Podurets, G. V. Simakov, and A. G. Sevast'yanov, Shock compression of porous iron, copper, and tungsten, and their equation of state in the terapascal pressure range, *Sov. Phys. JETP*, *68*, 356-361, 1989.

205 Trunin, R. F., M. A. Podurets, B. N. Moiseev, G. V. Simakov, and L. V. Popov, Relative compressibility of copper, cadmium, and lead at high pressures, *Sov. Phys. JETP*, *29*, 630-631, 1969.

206 Trunin, R. F., M. A. Podurets, G. V. Simakov, L. V. Popov, and B. N. Moiseev, An experimental verification of the Thomas-Fermi model for metals under high pressure, *Sov. Phys. JETP*, *35*, 550-552, 1972.

207 Trunin, R. F., G. V. Simakov, and M. A. Podurets, Compression of porous quartz by strong shock waves, *Izv. Earth Phys.*, *2*, 102-106, 1971.

208 Trunin, R. F., G. V. Simakov, M. A. Podurets, B. N. Moiseyev, and L. V. Popov, Dynamic compressibility of quartz and quartzite at high pressure, *Izv. Earth Phys.*, *1*, 8-12, 1971.

209 Trunin, R. F., G. V. Simakov, Y. N. Sutulov, A. B. Medvedev, B. D. Rogozkin, and Y. E. Fedorov, Compressibility of porous metals in shock waves, *Sov. Phys. JETP*, *69*, 580-588, 1989.

210 Trunin, R. F., M. V. Zhernokleto, N. F. Kuznetsov, G. V. Simakov, and V. V. Shutov, Dynamic compressibility of acqueous solutions of salts, *Izv. Earth Physics*, *23*, 992-995, 1987.

211 Tyburczy, J. A., T. S. Duffy, T. J. Ahrens, and M. A. Lange, Shock wave equation of state of serpentine to 150 GPa: Implications of the occurrence of water in the Earth's lower mantle, *J. Geophys. Res.*, *96*, 18011-18027, 1991.

212 Urlin, V. D., M. A. Machalov, and O. L. Mikhailova, Liquid xenon study under shock and quasi-isentropic compression, *High Pressure Research*, *8*, 595-605, 1992.

213 van Thiel, M., (editor), Compendium of Shock Wave Data, University of California, Lawrence Livermore Laboratory, UCRL-50801, Vol. 1, Rev. 1, 755 pp., 1977.

214 Van Thiel, M., and B. J. Alder, Shock compression of argon, *J. Chem. Phys.*, *44*, 1056-1065, 1966.

215 van Thiel, M., and B. J. Alder, Shock compression of liquid hydrogen, *Mol. Phys.*, *19*, 427-435, 1966.

216 Van Thiel, M., L. B. Hord, W. H. Gust, A. C. Mitchell, M. D'Addario, B. K, E. Wilbarger, and B. Barrett, Shock compression of deuterium to 900 kbar, *Phys. Earth Planet. Int.*, *9*, 57-77, 1974.

217 van Thiel, M., A. S. Kusubov, and A. C. Mitchell, Compendium of Shock Wave Data, Lawrence Radiation Laboratory (Livermore), UCRL-50108, 1967.

218 Vassiliou, M. S., and T. J. Ahrens, Hugoniot equation of state of periclase to 200 GPa, *Geophys. Res. Lett.*, *8*, 729-732, 1981.

219 Vassiliou, M. S., and T. J. Ahrens, The equation of state of $Mg_{0.6}Fe_{0.4}O$ to 200 GPa, *Geophys. Res. Lett.*, *9*, 127-130, 1982.

220 Vizgirda, J., and T. J. Ahrens, Shock compression of aragonite and implications for the equation of state of carbonates, *J. Geophys. Res.*, *87*, 4747-4758, 1982.

221 Voskoboinikov, I. M., M. F. Gogulya, and Y. A. Dolgoborodov, Temperatures of shock compression of liquid nitrogen and argon, *Sov. Phys. Doklady*, *24*, 375-376, 1980.

222 Wackerle, J., Shock-wave compression of quartz, *J. Appl. Phys.*, *33*, 922-937, 1962.

223 Wackerle, J., W. L. Seitz, and J. C. Jamieson, Shock-wave equation of state for high-density oxygen in, in *Behavior of Dense Media under High Dynamic Pressures Symposium, Paris, September 1967*, pp. 85, Gordon and Breach, New York, 1968.

224 Walsh, J. M., and R. H. Christian, Equation of state of metals from shock wave measurements, *Phys. Rev.*, *97*, 1544-1556, 1955.

225 Walsh, J. M., and M. H. Rice, Dynamic compression of liquids from measurements of strong shock waves, *J. Chem. Phys.*, *26*, 815-823, 1957.

226 Walsh, J. M., M. H. Rice, R. G. McQueen, and F. L. Yarger, Shock-wave compressions of twenty-seven metals. Equation of state of metals., *Phys. Rev.*, *108*, 196-216, 1957.

227 Warnes, R. H., Investigation of a shock-induced phase transition in antimony, *J. Appl. Phys.*, *38*, 4629-4631, 1967.

228 Warnes, R. H., Shock wave compression of three polynuclear aromatic compounds, *J. Chem. Phys.*, *53*, 1088, 1970.

229 Watt, J. P., and T. J. Ahrens, Shock compression of single-crystal forsterite, *J. Geophys. Res.*, *88*, 9500-9512, 1983.

230 Watt, J. P., and T. J. Ahrens, Shock wave equations of state using mixed-phase regime data,

J. Geophys. Res., *89*, 7836-7844, 1984.

231 Watt, J. P., and T. J. Ahrens, Shock-wave equation of state of enstatite, *J. Geophys. Res.*, *91*, 7495-7503, 1986.

232 Yagi, T., K. Fukuoka, and Y. Syono, Shock compression of wüstite, *J. Geophys. Res. Let.*, *15*, 816-819, 1988.

233 Zubarev, V. N., and G. S. Telegin, The impact compressibility of liquid nitrogen and solid carbon dioxide, *Sov. Phys. Doklady*, *7*, 34-36, 1962.

Electrical Properties of Minerals and Melts

James A. Tyburczy and Diana K. Fisler

Electrical properties of minerals and melts aid in the interpretation of geophysical probes of the Earth's internal electrical structure (see Hermance, Section 1-12). In addition, electrical conductivities and dielectric constants of minerals are used in studies of mineral structure, electronic and ionic transport processes, defect chemistry, and other mineral physical properties. This article contains data and references to electrical properties of minerals and related materials at elevated temperature and 1 bar pressure, and at elevated temperature and elevated pressure. Electrical properties of materials can be very sensitive to minor chemical impurities and variations. Cracks and other macroscopic defects, and microscopic defects such as voids, inclusions, twins, etc. in specimens can complicate the achievement of reliable measurements. As a result, for a given electrical property of a mineral, there are often a wide range of reported values, sometimes differing by as much as several orders of magnitude. We have striven to provide reliable data for each material listed. Because minor element content and experimental conditions are known to strongly affect electrical properties, we have attempted, where possible, to tabulate both chemical composition and experimental conditions for each measurement. Certain previous compendia of mineral electrical properties have preferred to give only ranges of values for a given mineral [17-19]. The data and tables presented here may best be used as a guide to the literature or as a source of references from which to start a search, rather than as a source of absolute values.

Recent previous compilations of data on electrical properties of rocks are given in references [17-19, 42]. Electrical properties of elements, minerals, and rocks at room temperature and 1 bar pressure, and of aqueous fluids are treated by Olhoeft in Section 3-8 of this series and in [42].

Electrical parameters. The electrical conductivity σ and the dielectric permittivity ε relate an electrical stimulus and response through Maxwell's relations. In linear systems it can be shown that

$$J_T = \sigma E + \varepsilon \partial E / \partial t \tag{1}$$

where J_T is the total current density, E is the electric field gradient, and t is time. The first term on the right hand side of equation (1) is called the conduction current and the second term is the displacement current. For a time varying electrical field of the form $E = E_0 \exp(i\omega t)$ equation (1) becomes

$$J_T = (\sigma + i\omega\varepsilon) E_0 \exp(i\omega t) \tag{2}$$

in which $i = \sqrt{-1}$ and ω = angular frequency. A total complex conductivity may be defined as

$$\sigma_T = \sigma_T{}' + i\sigma_T{}'' = \sigma + i\omega\varepsilon \tag{3}$$

and the permittivity may also be defined as a complex value

$$\varepsilon = \varepsilon' - i\varepsilon'' \tag{4}$$

ε'' is also called the loss factor, and the loss tangent δ is defined as

J. A. Tyburczy and D. K. Fisler, Department of Geology, Arizona State University, Tempe, AZ 85287-1404

Mineral Physics and Crystallography
A Handbook of Physical Constants
AGU Reference Shelf 2

185

$$\tan \delta = \varepsilon''/\varepsilon' \tag{5}$$

The complex resistivity is defined as the reciprocal of the complex total conductivity

$$\rho' - i\rho'' = (\sigma_T' + i\sigma_T'')^{-1} \tag{6}$$

The SI unit of electrical conductivity is Siemens/meter (S equivalent to $1/\Omega$). The SI unit of dielectric permittivity is Farad/meter (F/m) and is often expressed as the relative dielectric permittivity κ

$$\kappa = \varepsilon/\varepsilon_0 \tag{7}$$

where ε_0 is the dielectric permittivity of a vacuum = 8.854185×10^{-12} F/m. For general references see references [20, 22, 42, 66].

Types of conductors. Materials are classified on the basis of whether they are metallic conductors, semiconductors, or insulators.

Metallic conductors. In metallic conductors, the Fermi level (the highest filled energy level) lies within the conduction band, and the conduction electrons are not localized or bound to any particular atom. Above the Debye temperature, the conductivity decreases with increasing T, generally decreasing as $1/T$ [37, 44]. At room temperature, metallic conductors have conductivities of 10^6 S/m or more. Even at temperatures approaching 0 K metallic conductivity is very high. In addition to metals, materials such as graphite exhibit metallic conduction.

Semiconductors. Semiconductors are materials with a relatively small difference in energy E_g between the bottom of the empty conduction band and the top of the filled valence band (the band gap). Thermal promotion of electrons from the valence to the conduction band can occur, leaving positively charged holes in the valence band. Typical energy gaps for semiconductors are on the order of 1 eV. Some solid-solution impurity atoms can donate electrons to the conduction band, resulting in n-type semiconduction; others can accept electrons causing electron holes in the valence band, resulting in p-type semiconduction. Semiconductors have conductivities in the range 10^{-3} to 10^5 S/m. At elevated temperatures, many silicates and oxides fall into this category.

Insulators. In insulators, the band gap is large (often 5 eV or more) and prevents thermal elevation of electrons into the valence band. Charge transport is by ionic movement from site to site by thermally activated transport (ionic conduction) or by electronic charge transfer between ions of different valence (hopping conduction). Impurity atoms in low concentrations may have large effects on the magnitude and mechanisms of conductivity in insulators. In an impurity-containing insulator, low temperature conductivity is often dominated by the impurity-controlled mechanisms and is termed extrinsic conductivity. At high temperatures the conductivity may become dominated by the fundamental ionic conductivity of the pure crystal, which is termed intrinsic conduction. In geological materials the distinction between 'pure' materials and 'impure' or minor-element containing materials is not always clear-cut or useful. Conductivity values for insulators are generally lower than about 10^{-13} S/m at room temperature. Many earth materials, e.g., halides, are insulators.

Effects of temperature and pressure. The Nernst-Einstein relation links the electrical conductivity to the diffusion coefficient of the charge-carrying species

$$\sigma = Dz^2 e^2 n/(kT) \tag{8}$$

in which D is the diffusion coefficient, z is the valence of the conducting species, e is the charge of the electron, n is the concentration of the conducting species, k is Boltzmann's constant, and T is temperature. Diffusion is a thermally activated process so that $D = D_0 \exp(-E/kT)$, therefore the temperature dependence of the conductivity will be of the form

$$\sigma = \frac{\sigma_0}{T} \exp(-E/kT) \tag{9a}$$

or

$$\sigma T = A \exp(-E/kT) \tag{9b}$$

in which σ_0 and A are pre-exponential constants and E is the activation energy. E is often expressed in electron volts eV; Boltzmann's constant k is equal to 8.618×10^{-5} eV/deg. In practice, the $1/T$ factor in the pre exponential term on the right hand side of equation (9a) is not always used, and expressions of the form

$$\sigma = \sigma_0 \exp(-E/kT) \tag{10}$$

are employed instead. In the tables that follow, both types of expressions are used, and the reader should be careful to employ the proper form of the equation for the constants given. Plots of the logarithm of electrical conductivity versus $1/T$ that cover an extended range of temperature frequently show two or more linear regions. The variations in slope can be caused by transition from one dominant conducting species to another or by a transition from extrinsic (impurity dominated) to intrinsic conductivity. Each region is then described by an expression of the form of equation (10). However, more precise, and mechanistically more meaningful parameters are derived if the entire data set is simultaneously fit to a single expression of the form

$$\sigma = \sigma_{01} \exp(-E_1/kT) + \sigma_{02} \exp(-E_2/kT) \qquad (11)$$

In the data tables that follow, experimental data that have been fit simultaneously to multiple linear segments are indicated by braces.

The effect of pressure on conductivity can be characterized by the inclusion of an activation volume term V_σ so that

$$\sigma = \sigma_0 \exp(-\frac{U + PV_\sigma}{kT}) \qquad (12)$$

in which U is the activation internal energy. The activation volume is often interpreted as indicating the volume of the mobile species, but care should be taken when making this interpretation. When available, activation volumes are given in the tables, but more commonly measurements at elevated pressures are parameterized isobarically as functions of temperature using equation (10) or (11). In experiments at high pressure it is difficult to control all thermodynamic parameters. The values in Table 5 and 6 are the best currently available. We expect that many of these values will be revised as experimental methods improve.

Point Defects and Conductivity. The electrical conductivity of materials is determined by the point defect chemistry. Point defects are imperfections in the crystal lattice and the types of point defects possible are substitutions, vacancies, interstitial ions, electrons, and electron holes (electron deficiencies in the valence band commonly termed 'holes'). Kroger-Vink notation [23,24] was developed to describe the type of defect and its effective charge relative to a normal lattice site. In this notation, each defect is given a symbol of the form A_B^c, in which the main symbol A indicates the species of the defect (atom, electron, hole, or vacancy), the subscript B indicates the type of site (normal lattice site of a particular atom type or an interstitial site), and the superscript c indicates the net effective charge of the defect relative to the normally occupancy of that lattice (dots indicate positive relative charge, slashes indicate negative charges). Tables 1 and 2 give these definitions and show as an example the defect species in nonstoichiometric (Mg,Fe)O. Note how defects caused by aliovalent ions in different valence states are written using this notation.

The electrical conductivity of a material is the sum of the conduction of each charge carrier (or defect) type acting in parallel

$$\sigma = \Sigma \sigma_i = \Sigma c_i q_i \mu_i \qquad (13)$$

in which c_i is the concentration of the ith type of charge carrier, q_i is its effective charge, and μ_i is its mobility (in $m^2 V^{-1} sec^{-1}$). In general, all defects are present in some amount in every crystal and each contributes to the total conductivity. Usually only one or two types of defects dominate under a given set of thermodynamic conditions. The concentrations of the defect species are governed by chemical reactions that lead to the production or consumption of each defect type. For example, the removal of a positively-charged ion from its normal site resulting in the formation of a vacancy plus an interstitial ion (formation of a Frenkel defect) can be written as

$$A_A^x = V_A' + A_I^\bullet \qquad (14)$$

For such a reaction, we can write equations for chemical equilibrium constants in the usual manner

$$K = \frac{[V_A'][A_I^\bullet]}{[A_A^x]} \qquad (15)$$

in which K is the equilibrium constant and square brackets indicate site fractions. the Gibbs free energy ΔG^o for this reaction is given by

$$\Delta G^o = -RT \ln K \qquad (16)$$

in which R is the gas constant and T is temperature.

Reactions involving oxygen from the surroundings are of particular importance in oxide materials. For example, the reaction

$$O_O^x = 1/2\, O_{2(g)} + V_O^{\bullet\bullet} + 2e' \qquad (17)$$

describes the reaction of oxygen on a normal lattice site to release an oxygen molecule to the environment, leaving an oxygen vacancy and two electrons behind. Thus the mineral stoichiometry depends on the chemical potential of oxygen (or the oxygen fugacity f_{O2}) in the surrounding atmosphere. In MgO, the reaction which incorporates oxygen from the surroundings and forms a magnesium vacancy is written

$$1/2\, O_2 = V_{Mg}'' + 2h^\bullet + O_O^x \qquad (18)$$

Thus, in principle, the magnesium vacancy concentration

Table 1 . Definitions of defect species (Kröger-Vink) notation.

Symbol	Meaning	
Main symbol	Element or species of defect	Atomic symbol e (electron) h (hole) V (vacancy)
Subscript	Site	Normal Lattice Site (Atomic symbol) I (Interstitial)
Superscript	Charge relative to normal lattice site	x (zero) ' (negative) • (positive)

in even pure MgO depends on f_{O2} (at constant temperature). In the presence of even relatively small amounts of impurity ions, the concentrations of defects (and hence the conductivity) may vary widely. This extreme sensitivity of defect concentration to minor and trace element concentrations makes it very difficult to establish absolute values for the conductivities of 'pure' minerals, especially at low temperatures. The general practice has been to try to determine which defect dominates conduction under a given set of conditions by determining conductivity under varying conditions of temperature, oxygen fugacity, and trace element concentration (or some subset of these conditions). The results are then combined with defect reactions such as (17) and (18) above and defect conservation laws such as conservation of mass, charge neutrality, and conservation of lattice sites to determine the particular defect reaction that controls the concentration of the dominant defect and predicts the correct dependence on other environmental conditions (f_{O2} in particular). References such as [23-25] describe this process for many materials, and more recent references such as [9-11, 59, 60] describe defect equilibria in some important Earth materials. To properly and completely specify the thermodynamic state of a defect-containing crystal, the Gibbs phase rule must be satisfied. For a three-component system such as (Mg,Fe)O, this

constraint means that the temperature, the pressure, the oxygen fugacity, and the Mg:Fe ratio in the solid must all be specified. Thus, the conductivity will have the general form

$$\sigma_i = \sigma_0 \exp(-E/kT) (Fe^x_{Mg})^m f_{O2}^n \quad (19)$$

in which Fe^x_{Mg} is the fraction of Fe on Mg sites and n and m are constants arising from the defect reaction stoichiometries. For a four-component system such as olivine $(Mg,Fe)_2SiO_4$ an additional constraint must be specified; in olivine the most frequently specified constraint is the silica activity a_{SiO2} (or equivalently the enstatite activity a_{En}), so that the most general conductivity relation would have the form

$$\sigma_i = \sigma_0 \exp(-E/kT) (Fe^x_{Mg})^m f_{O2}^n a_{SiO2}^p \dots \quad (20)$$

in which p is also a constant. Where available, the exponents n for the dependence of conductivity on oxygen fugacity have been included in the tables that follow, but the parameters in full expressions of the form of equation (20) are still not commonly determined. See [23-25] for general discussions of defect equilibria and [8-11, 58-60] for discussions applied to Earth materials.

Table 2. Examples of defect species for (Mg,Fe)O. Note that Fe can be in +2 or +3 valence state.

Species	Symbols
Normal Lattice Sites	Mg^x_{Mg} Fe^x_{Mg} O^x_O
Substitutional Defects	Fe^x_{Mg} $Fe^•_{Mg}$
Lattice Vacancies	V''_{Mg} $V^{••}_O$
Interstitial Ions	$Mg^{••}_I$ O''_I $Fe^{••}_I$ $Fe^{•••}_I$
Electronic Defects	e' $h^•$
Defect Dimers	$(Fe^•_{Mg}\text{-}V''_{Mg})'$
Defect Trimers, etc.	$(Fe^•_{Mg}\text{-}V''_{Mg}\text{-}Fe^•_{Mg})^x$

Table 3. High temperature electrical conductivities of minerals given in the format $\sigma = A_1 \exp(-E_1/kT) f_{O_2}^{n_1}$ or $\sigma T = A_1 \exp(-E_1/kT) f_{O_2}^{n_1}$. Values surrounded by braces are fits to the 2-term expression $\sigma = A_1 \exp(-E_1/kT) f_{O_2}^{n_1} + A_2 \exp(-E_2/kT) f_{O_2}^{n_2}$ where the top line is A_1, E_1, n_1 and the bottom line is A_2, E_2 and n_2. Pressures are 1 bar total unless otherwise stated, and pressure dependences are referenced to 1 atmosphere. Oxygen activity buffer QFM refers to the quartz-fayalite-magnetite assemblage, WM refers to the wustite-magnetite assemblage.

Material	Form of equation	Temperature range K	log A1 σ in S/m, T in K	E1 eV	n1 f_{O_2} in atm	Comments	References
Calcite, CaCO3	σ	573-773	-6.15	0.1		DC, P_{CO_2} = 4 MPa	[36]
		773-998	-1.15	0.9			
		998-1073	-4.22	0.2			
		1073-1258	1.00	1.5			
		1258-1473	3.48	2.0			
‖ c	σ	120-723	$\left.\begin{matrix}-7.7\\1.04\end{matrix}\right\}$	$\left.\begin{matrix}0.15\\0.91\end{matrix}\right\}$		100 - 2.4 x 10^{10} Hz AC	[46]
⊥ c		120-723	$\left.\begin{matrix}7.9\\1.04\end{matrix}\right\}$	$\left.\begin{matrix}.011\\1.05\end{matrix}\right\}$			
Halides: Bromides, single crystal, synthetic							
AgBr	σ	248-695	5.80	0.66		DC	[28]
KBr		773-1001	8.18	1.97			
NaBr		823-1008	8.0	1.78			
	σT	600-723	4.54	0.84			[33]
		723-973	10.32	1.68			
RbBr	σ	873-954	8.25	2.03			[28]
TlBr		323-730	5.23	0.80			
Chlorides, single crystal, synthetic							
AgCl	σ	373-728	7.0	0.90		DC	[28]
KCl		823-1041	8.30	2.06			
LiCl		793-879	9.70	1.65			
NaCl		723-1073	8.0	1.90			
	σT	625-823	4.02	0.83			[33]
		823-1000	10.76	1.89			
RbCl		823-990	8.48	2.12			[28]
TlCl		293-700	5.40	0.79			

Table 3. (continued)

Material	Form of equation	Temperature range K	log A_1 σ in S/m, T in K	E_1 eV	n_1 f_{O_2} in atm	Comments	References
Halides, continued:							
Fluorides							
single crystal, synthetic							
KF	σ	973-1119	9.48	2.35		DC	[28]
LiF		993-1115	9.60	2.20			
NaF		1123-1265	8.18	2.25			
NaMgF$_3$, perovskite structure	σ	833-1173	6.26	1.51		DC	[41]
Iodides							
single crystal, synthetic							
KI	σ	773-953	7.48	1.77		DC	[28]
NaI		573-934	7.18	1.42			
Oxides:							
Aluminum oxide, Al$_2$O$_3$ single crystal (sapphire)	σ	1573-1998	5.60	2.97		10 kHz AC, DC	[43]
		1573-1873	3.96	2.62		$f_{O_2} = 10^5$ Pa	
		1898-1998	11.71	5.5		$f_{O_2} = 1$ Pa	
		1573-1873	5.01	2.84		$f_{O_2} = 1$ Pa	
		1898-1998	13.01	5.8		$f_{O_2} = 10^{-5}$ Pa	
						$f_{O_2} = 10^{-5}$ Pa	
Calcium oxide, CaO single crystal	σ	1000-1400	10.83	3.5		DC, 1.59 kHz AC	[61]
Magnesium oxide, MgO single crystal, 40 ppm Al	σ	1473-1773	{ 1.23 / 7.14	1.92 / 3.81		DC $f_{O_2} \sim 10^{-3} - 10^{-4}$ Pa	[54, 55]
		1473-1773	{ 1.23 / 6.51	1.92 / 6.51		$f_{O_2} \sim 10 - 1$ Pa	
		1473-1773	{ 1.23 / 5.87	1.92 / 3.66		$f_{O_2} \sim 1$ MPa	
320 ppm Fe	σ	1473-1773	{ 2.68 / 7.48	2.11 / 3.68		$f_{O_2} \sim 10^{-3} - 10^{-4}$ Pa	
		1473-1773	{ 2.68 / 6.57	2.11 / 3.49		$f_{O_2} \sim 10 - 1$ Pa	
		1473-1773	{ 2.68 / 5.94	2.11 / 3.08		$f_{O_2} \sim 1$ MPa	

Table 3. High temperature electrical conductivities of minerals (continued)

Material	Form of equation	Temperature range K	log A1 (σ in S/m, T in K)	E1 eV	n1 fO2 in atm	Comments	References
Oxides, continued:							
400 ppm Al	σ	1473-1773	2.97, 8.40	2.16, 4.07		$f_{O_2} \sim 10^{-3} - 10^{-4}$ Pa	[70]
		1473-1773	2.97, 5.16	2.16, 3.24		$f_{O_2} \sim 1$ MPa	
						300 - 5 x 10^4 Hz AC , DC	
Magnesiowustite, $Mg_{0.83}Fe_{0.17}O$	σ	1273-1773	2.79	0.48		f_{O_2} = QFM	
		1273-1773	3.00	0.42		f_{O_2} = 'QFM-1' to 'QFM-2'	
Magnetite, Fe_3O_4	σT	873-1673	8.30	0.15		$CO:CO_2$ = 0.003	[39]
Nickel oxide, NiO, single crystal	σ	973-1273	6.47	2.05		f_{O_2} = Ni/NiO	[63]
		973-1273	4.77	0.68		in air	
polycrystalline		973-1273	5.36	1.84		f_{O_2} = Ni/NiO	
Scheelite, $CaWO_4$, single crystal	σ	1173-1573	8.59	2.94		DC	[47]
		1173-1573	9.25	3.00		P_{H_2O}/P_{H_2} = 1.0	
						P_{H_2O}/P_{H_2} = 0.1	
Spinel, Fe-Mg-Al	σT					$CO:CO_2$ = 0.003	[39]
$Fe^{2+}Fe_{0.5}^{3+}Al_{1.5}O_4$		873-1673	6.44	0.39			
$Fe^{2+}Fe_{1.0}^{3+}Al_1O_4$		873-1673	7.74	0.25			
$Fe^{2+}Fe_{1.5}^{3+}Al_2O_4$		873-1673	7.84	0.23			
$Fe_{0.25}^{2+}Mg_{0.75}Fe_2^{3+}O_4$		873-1673	7.52	0.18			
$Fe_{0.5}^{2+}Mg_{0.5}Fe_2^{3+}O_4$		873-1673	7.74	0.18			
$Fe_{0.75}^{2+}Mg_{0.25}Fe_2^{3+}O_4$		873-1673	7.86	0.15			
$Fe_{0.25}^{2+}Mg_{0.75}Fe_{0.5}^{3+}Al_{15}O_4$		873-1673	6.52	0.39			
$Fe_{0.5}^{2+}Mg_{0.5}Fe^{3+}AlO_4$		873-1673	7.41	0.28			
$Fe_{0.75}^{2+}Mg_{0.25}Fe_{1.5}^{3+}Al_{0.5}O_4$		873-1673	7.75	0.22			
Spinel, Fe-Mg-Cr-Al	σT					$CO:CO_2$ = 0.003	[38]
$Fe^{2+}Fe_{0.5}^{3+}Cr_{1.5}O_4$		873-1673	7.97	0.47			
$Fe^{2+}Fe^{3+}CrO_4$		873-1673	7.68	0.26			
$Fe^{2+}Fe_{1.5}^{3+}Cr_{0.5}O_4$		873-1673	7.82	0.20			
$Fe_{0.25}^{2+}Fe_{0.5}^{3+}Mg_{0.25}Cr_{1.5}O_4$		873-1673	7.21	0.48			
$Fe_{0.5}^{2+}Fe_{0.5}^{3+}Mg_{0.25}CrO_4$		873-1673	7.52	0.30			
$Fe_{0.75}^{2+}Fe_{0.5}^{3+}Mg_{0.25}Cr_{0.5}O_4$		873-1673	7.78	0.19			

Table 3. (continued)

Material	Form of equation	Temperature range K	log A_1 σ in S/m, T in K	E_1 eV	n_1 f_{O_2} in atm	Comments	References
Oxides, Spinel, Fe-Mg-Cr-Al, continued:							
$Fe^{2+}Fe_{0.5}^{3+}Al_{0.75}Cr_{0.25}O_4$	σ	873-1673	7.59	0.38		$CO:CO_2 = 0.003$	[38]
$Fe^{2+}Fe^{3+}Al_{0.5}Cr_{0.5}O_4$		873-1673	7.77	0.26			
$Fe^{2+}Fe_{15}^{3+}Al_{0.25}Cr_{0.25}O_4$		873-1673	7.84	0.19			
$Fe^{2+}Fe_{0.5}^{3+}Al_{1.25}Cr_{0.25}O_4$	σ	873-1673	7.69	0.41			
$Fe^{2+}Fe^{3+}Al_{0.83}Cr_{0.17}O_4$		873-1673	7.77	0.25			
$Fe^{2+}Fe_{0.5}^{3+}Al_{0.25}Cr_{01.25}O_4$		873-1673	7.79	0.48			
$Fe^{2+}Fe^{3+}Al_{0.17}Cr_{0.83}O_4$		873-1673	7.71	0.25			
Strontium oxide, SrO, polycrystalline, synthetic	σ	860-1050	5.70	2.0		DC	[61]
Wustite, $Fe_{1-x}O$	σT					Constant thermoelectric potential	[6]
x = 0.1039 - 0.1095		1173-1573	8.062	0.161			
x = 0.0884 - 0.0967		1173-1573	8.025	0.163			
x = 0.0706 - 0.0817		1173-1573	7.967	0.167			
x = 0.0566 - 0.0706		1173-1573	7.899	0.168			
x = 0.0494 - 0.0593		1173-1573	7.869	0.182			
$Zr_{0.85}Ca_{0.15}O_{1.85}$, polycrystalline, fluorite structure	σ	1000-2025	5.18	1.26	0	$f_{O_2} = 10^{-5}$ to 10^5 Pa	[21]
Silicates: Cordierite, $Mg_{1.91}Fe_{0.08}Mn_{0.01}Al_{3.95}Si_{5.01}$ ($Na_{0.05}$, $0.56H_2O$, mCO_2)							
|| [001]	σ	473-1173	3.1	0.75		Impedance spectroscopy	[49]
$\perp$ [100]		473-1173	-0.35	0.85			
Feldspar, polycrystalline Albite, Grisons, France $NaAlSi_3O_8$	σ	673-1173	0.15	0.86		1 - 10 kHz AC in air	[34, 35]
Albite, Mewry, France, $K_{0.02}Na_{0.98}AlSi_3O_8$	σ	673-1173	0.23	0.86			
Albite, synthetic (disordered) $NaAlSi_3O_8$		673-1173	0.08	0.72			
Anorthite, synthetic, $CaAl_2Si_2O_8$,	σ	673-1173	-0.20	0.87			

Table 3. (continued)

Material	Form of equation	Temperature range K	$\log A_1$ σ in S/m, T in K	E_1 eV	n_1 f_{O_2} in atm	Comments	References
Silicates, continued:							
Anorthite, Labrador, (Ab43An57) Na$_{0.43}$Ca$_{0.57}$Al$_{1.57}$Si$_{2.43}$O$_8$		673-1173	-0.47	0.78			
Microcline, Arendal K$_{0.94}$Na$_{0.06}$AlSi$_3$O$_8$		673-1173	0.11	0.85			
Microcline, Arendal KNaAlSi$_3$O$_8$, (ion exch)		673-1173	-0.13	0.80			
Adularia, Grisons France K$_{0.84}$Na$_{0.16}$AlSi$_3$O$_8$,		673-1173	-0.62	0.86			
Microcline, St.-Gothard, Fr. K$_{0.95}$Na$_{0.05}$AlSi$_3$O$_8$		673-1173	-0.01	0.82			
Orthoclase, synthetic KAlSi$_3$O$_8$		673-1173	0.15	0.85			
Fe-Orthoclase, synthetic KFeSi$_3$O$_8$		873-1073	-0.37	0.87			
Ga-Orthoclase, synthetic, KGaSi$_3$O$_8$		823-1073	0.23	0.87			
plagioclase, synthetic Na$_{0.868}$Ca$_{0.066}$AlSi$_3$O$_8$,		673-1173	0.041	0.85			
plagioclase, synthetic Na$_{0.54}$Ca$_{0.23}$AlSi$_3$O$_8$,		673-1173	0.00	0.86			
plagioclase, synthetic Na$_{0.22}$Ca$_{0.39}$AlSi$_3$O$_8$,		673-1173	-0.20	0.87			
Kalsilite, synthetic, KAlSiO$_4$	σ	673-1173	-0.70	0.82		1 - 10 kHz AC in air	[34, 35]
Leucite, Rocca Monfina, K$_{0.96}$Na$_{0.04}$AlSi$_2$O$_6$ synthetic, KAlSi$_2$O$_6$	σ	673-898 898-1173 673-898 898-1173	3.81 0.80 3.40 1.32	0.91 0.37 0.84 0.48		1 - 10 kHz AC in air	[34, 35]
Rb-leucite, synthetic, RbAlSi$_2$O$_6$		673-1173	3.64	0.89		1 - 10 kHz AC in air	
Nepheline, synthetic NaAlSiO$_4$ Na$_3$K(SiO$_4$)$_4$	σ	673-1173 673-1173	1.58 1.11	0.93 0.98		1 - 10 kHz AC in air	[34, 35]

Table 3. (continued)

Material	Form of equation	Temperature range K	log A_1 σ in S/m, T in K	E_1 eV	n_1 fO_2 in atm	Comments	References
Silicates, continued:							
Olivine, natural, single crystal							
San Carlos, AZ (Fo90) (Mg0.9Fe0.1)2SiO4, [100] orientation	σ	1373-1473	{0.36, 2.49}	{0.55, 2.25}	{0.18, -0.18}	0.1 - 10 kHz AC $fO_2 \sim 10^{-0.5} - 10^{-6}$ Pa,	[68]
San Carlos, AZ (Fo90) (Mg0.9Fe0.1)2SiO4, [100] orientation	σ	1373-1473	{-0.75, 1.18}	{0.34, 1.30}	{0.17, 0}	0.1 - 10 kHz AC $fO_2 \sim 10^{-0.5} - 10^{-6}$ Pa, Pyroxene-buffered,	[68]
San Carlos, AZ (Fo91) (Mg0.91Fe0.09)2SiO4, [100] orientation	σ	1373-1773	{0.66, 6.30}	{1.11, 3.08}		0.1 - 10 kHz AC CO_2:CO = 10:1,	[56]
[010] orientation		1373-1773	{2.03, 22.2}	{1.52, 2.77}			
[001] orientation		1373-1773	{0.31, 6.0}	{0.97, 2.77}			
Forsterite, synthetic (Fo100) Mg2SiO4, [100] orientation	σ	1473-1523	----	1.77		0.1 - 10 kHz AC CO_2:CO = 30:1,	[51]
[010] orientation		1473-1523		2.63			
[001] orientation		1473-1523		1.63			
Co- olivine, synthetic Co2SiO4	σ	1323-1498	-0.36	0.75		10 kHz AC CO_2:CO =157.7/1	[8]
		1523-2083	5.37	1.86		CO_2:CO =157.7/1	
		1573-2083	4.18	1.41		CO_2 gas	
Olivine, polycrystalline, San Carlos olivine (Mg0.91Fe0.09)2SiO4, (Fo91)	σ	1063-1673	{1.99, 12.65}	{1.45, 4.87}		Orientationally-averaged grain interior conductivity from impedance spectroscopy on polycrystalline olivine, W-M buffer	[48]
(Mg0.91Fe0.09)2SiO4, (Fo91) Model SO2	σ	998-1773	{2.40, 9.17}	{1.60, 4.25}		Isotropic parametric model for self-buffered case, 0.1 - 10 kHz AC CO_2:CO = 10:1	[2]

Table 3. (continued)

Material	Form of equation	Temperature range K	log A1 σ in S/m, T in K	E1 eV	n1 fO2 in atm	Comments	References
Silicates, continued:							
$(Mg_{0.91}Fe_{0.09})_2SiO_4$, (Fo_{90}) Model SO1	σ	1373-1773	1.67 8.70	1.38 3.90		Isotropic parametric model for self-buffered case, 0.1 - 10 kHz AC CO_2:CO = 10:1,	[56]
Opal, amorphous SiO_2 Natural, Australia, ⊥ mineral fabric Synthetic ∥ mineral fabric ⊥ mineral fabric	σT	473-773 473-773 473-773	4.94 5.30 5.32	0.91 1.02 1.03		Air, Impedance spectroscopy Measurements made parallel and perpendicular to mineral fabric	[71]
Pyroxene, single crystal Diopside, NY $Na_3Ca_{96}Mg_{96}Fe_3Al_2Si_{200}O_{600}$ [100] orientation [010] orientation [001] orientation	σ	1273-1473 1273-1473 1273-1473	1.43 0.88 1.58	2.15 2.02 2.20	-0.18 -0.12 -0.15	$f_{O_2} = 10^{-4}$ to 10^{-9} Pa at 1273 K, and 10^{-3} to 10^{-7} Pa at 1473 K	[15]
Diopside, Brazil $Na_3Ca_{96}Mg_{85}Fe_{12}Al_2Si_{200}O_{600}$ [100] orientation [010] orientation	σ	1273-1473 1273-1473	-0.89 0.25	0.92 1.05	-0.03 -0.02	$f_{O_2} = 10^{-4}$ to 10^{-9} Pa at 1273 K, and 10^{-3} to 10^{-7} Pa at 1473 K	[15]
Orthopyroxene, North Wales, UK $Mg_{0.89}Fe_{0.11}SiO_3$ (En89) [100] orientation	σ	1190-1282 1136-1298 1163-1282 1031-1099 1190-1282 1020-1163 1111-1235 1010-1099 1235-1316	0.67 0.84 1.11 0.17 1.92 0.36 2.20 0.64 2.43	1.26 1.30 1.36 1.15 1.54 1.19 1.60 1.25 1.64		1592 Hz AC CO_2:H_2 = 99.00:1 CO_2:H_2 = 29.72:1 CO_2:H_2 = 3.643:1 CO_2:H_2 = 3.643:1 CO_2:H_2 = 0.5691 CO_2:H_2 = 0.569:1 CO_2:H_2 = 0.0733:1 CO_2:H_2 = 0.0733:1 CO_2:H_2 = 0.0420:1	[4]
[010] orientation		1190-1282 1020-1163	1.23 0.01	1.42 1.13		CO_2:H_2 = 29.72:1 CO_2:H_2 = 29.72:1	

Table 3. (continued)

Material	Form of equation	Temperature range K	log A_1 σ in S/m, T in K	E_1 eV	n_1 f_{O_2} in atm	Comments	References
Silicates, continued:							
[001] orientation		1087-1282	0.57	1.18		$CO_2:H_2$ = 29.72:1	
		1149-1282	3.20	1.76		$CO_2:H_2$ = 0.569:1	
		1042-1124	1.11	1.28		$CO_2:H_2$ = 0.569:1	
		1136-1282	3.72	1.87		$CO_2:H_2$ = 0.0733:1	
		1136-1282	3.39	1.76		Air	
		1020-1136	0.73	1.15		Air	
Orthopyroxene, Papua, New Guinea $Mg_{0.93}Fe_{0.07}SiO_3$, (En93) [010] orientation	σ					1592 Hz AC	[4]
		1176-1282	2.31	1.80		$CO_2:H_2$ = 29.72:1	
		1176-1282	2.07	1.71		$CO_2:H_2$ = 3.643:1	
		1149-1282	2.36	1.75		$CO_2:H_2$ = 0.569:1	
		1136-1282	2.13	1.66		$CO_2:H_2$ = 0.0733:1	
		1190-1282	1.21	1.50		CO_2	
		1111-1282	2.12	1.62		Air	
Orthopyroxene, $Ca_{0.8}Mg_{161.4}Fe_{31.2}Mn_{0.5}Al_{7.8}Cr_{1.4}Si_{195.6}O_{600}$	σ	973-1473	0.99	1.03		1 kHz AC $CO_2:CO$ = 1999:1 Mean of 3 orientations: [001], [010], [100]	[14]
$Ca_{0.7}Mg_{160.9}Fe_{34.2}Mn_{0.5}Al_{5.4}Cr_{0.2}Si_{197.4}O_{600}$		1098-1449	1.67	1.25			
$Ca_{0.6}Mg_{172.4}Fe_{24.8}Mn_{0.4}Al_{2.2}Cr_0Si_{199.0}O_{600}$		1111-1449	1.24	1.20			
Quartz, SiO_2, single crystal Natural, Arkansas, z-cut	σT	370-454	6.3	0.82		Air, impedance spectroscopy 200-300 x 10^{-6} Al/Si	[16]
Synthetic, Grade S4, z-cut		450-700	8.8	1.36		200 x 10^{-6} Al/Si	
Synthetic, Grade EG, z-cut		450-700	8.4	1.40		15 x 10^{-6} Al/Si	
Synthetic, Grade PQ, z-cut		450-700	8.0	1.33		1 x 10^{-6} Al/Si	

Table 4. Iron content dependence of the electrical conductivity of single crystal $(Fe,Mg)_2SiO_4$ olivine. Measurements made in the [100] direction at temperatures between 1423 - 1573 K at 1 bar total pressure. Oxygen fugacity is that of a $CO_2:CO$ 5:1 mixture. Values given for the constants in the expression $\sigma T = A_1 x^n \exp(-E/kT)$ in which x is mole fraction Fe. Valid in the composition range $0.66 \leq x \leq 0.92$ [12].

A_1 S/m	n (x in mole fraction Fe)	E eV
$10^{6.54\pm0.15}$	1.81 ± 0.02	1.35 ± 0.04

Table 5. High pressure electrical conductivity of minerals given in the form $\sigma = \sigma_o \exp(-E/kT)$.

Material	Temperature range K	Pressure GPa	σ_o S/m	E eV	Remarks	References
Magnesiowustite (wustite) (Mg,Fe)O						
$Mg_{0.78}Fe_{0.22}O$	400-700	15	2.19×10^3	0.37	Diamond cell, external heating	[32]
$Mg_{0.91}Fe_{0.09}O$	333-2000	30	4.3	0.38	Diamond cell, laser heating 1 kHz AC	[29]
$Mg_{0.725}Fe_{0.275}O$		32	5.9×10^3	0.29		
$Fe_{0.9422}O$	90-250	0	-	0.0980	Piston-cylinder	[62]
		0.26		0.102		
		0.79		0.0986		
		1.02		0.0981		
		1.36		0.0978		
		1.58		0.0960		
$Fe_{0.9216}O$	90-250	0	-	0.0993	Piston-cylinder	[62]
		0.45		0.0995		
		0.75		0.0982		
		1.00		0.0970		
		1.33		0.0960		
		1.58		0.0955		
$Fe_{0.8985}O$	90-250	0	-	0.0945	Piston-cylinder	[62]
		0.27		0.0980		
		0.45		0.0945		
		0.73		0.0925		
		1.01		0.0918		
		1.33		0.0905		
		1.55		0.0897		

Table 5. (continued)

Material	Temperature range K	Pressure GPa	σ_o S/m	E eV	Remarks	References
Olivine, $(Mg,Fe)_2SiO_4$ natural, polycrystalline						
Dreiser Weiher, Germany, $(Fo_{93.4})$	293–623	2.0	1.5×10^{-1}	0.33	DC, Ni/NiO buffer, guard ring	[50]
	623–923	2.0	8.0	0.52		
	973–1473	3.5	5.5×10^1	0.70		
	293–623	3.5	2.8×10^{-2}	0.22		
	623–923	3.5	1.0	0.42		
	973–1473	3.5	1.2×10^3	0.96		
	293–623	5.0	3.0×10^{-2}	0.21		
	623–923	5.0	4.7×10^{-1}	0.37		
	973–1473	5.0	2.6×10^2	0.83		
Dreiser Weiher, Germany, $(Fo_{88.6})$	900–1093	1.0	5.33	0.953	Piston cylinder, 1.6 kHz AC, guard ring natural paragenesis	[7]
	915–1236	1.0	6.69	0.941	QFM buffer	
	945–1189	1.0	5.41	0.976	IW buffer	
Mt. Leura, Australia, (Fo_{92})	953–1429	2.0	2.14×10^1	1.17	1592 Hz AC, 2 electrode, Al_2O_3	[51]
	953–1429	5.0	8.71	1.00	insulator	
San Carlos, AZ, $(Fo_{91.8})$	923–1223	1.0	2.04	0.85		[31]
Olivine, synthetic, polycrystalline Fo_{100} Forsterite	795–1223	1.0	2.77	0.984	Piston-cylinder, 1.6 kHz AC, 2 electrode, guard ring MgO buffer	[7]
	1223–1348	1.0	2.64×10^6	2.461	MgO buffer	
Fo_{60}	847–1331	1.0	1.92×10^1	0.479	QFM buffer	
Fo_{80}	875–1160	1.0	9.17	0.582	QFM buffer	
	800–1212	1.0	7.76	0.683	IW buffer	
Fo_{90}	823–1223	1.0	1.04	0.622	QFM buffer	
	866–1212	1.0	1.54	0.777	IW buffer	
Fo_0 (Fa_{100}) Fayalite,	636–1217	1.0	7.43×10^1	0.383	QM buffer	
	574–1091	1.0	6.75×10^1	0.523	QI buffer	
Fo_0 Fayalite	615–1273	1.0	1.21×10^2	0.38	Piston-cylinder, 1.6 kHz AC guard ring QFM buffer	[69]
			1.18×10^2	0.52	QFI buffer	

Table 5. (continued)

Material	Temperature range K	Pressure GPa	σ_0 S/m	E eV	Remarks	References
Olivine, synthetic, polycrystalline, continued:						
Fo$_0$ Fayalite	799-1500	1.0	5.82	0.843	buffered $MgSiO_3$ (SiO_2-sat)	
	915-1236	1.0	3.08	1.00	buffered $MgSiO_3$ (stoichiometric)	
	940-1283	1.0	4.35×10^0	1.05	buffered MgO	
Fo$_0$ Fayalite	423-1173	3.10	6.31×10^2	0.51	DC, 2 electrode, pyrophyllite insulator	[1]
Ni$_2$SiO$_4$	773-1100	1.0	1.38×10^{-1}	0.56	Piston-cylinder, 1.6 kHz AC guard ring, Ni/NiO buffer	[69]
	1100-1273	1.0	7.94×10^8	2.71		
α-Mg$_2$GeO$_4$	1323-1518	1.05	6.76×10^7	2.70	Piston-cylinder, Impedance spectroscopy, ERGAN buffered	[40]
	1568-1673	2.05	6.92×10^6	2.42		
Perovskite (silicate)						
Mg$_{0.89}$Fe$_{0.11}$SiO$_3$	293-673	1.2	7.0×10^0	0.48	Diamond cell, external heating	[57]
	293-673	40	7.0×10^0	0.37	$U = 0.48$ eV, $V_\sigma = -.26$ cm^3/mole	
Mg$_{0.89}$Fe$_{0.11}$SiO$_3$	295-3200	46	2.8×10^1	0.24	Diamond cell, laser heating	[30]
	295-3200	55	5.5×10^1	0.24		
Pyroxene, Orthoyroxene, synthetic, polycrystalline (Mg,Fe)SiO3 En$_{100}$ (Enstatite)	550-1000	1.0	7.73×10^1	1.13	Piston-cylinder, 1.6 kHz AC, guard ring	[53]
	550-1000	2.0	1.78×10^3	1.59	Quartz buffered	
	550-1000	1.0	1.74×10^2	1.24	Quartz buffered	
	550-1000	2.0	1.79×10^3	1.49	Forsterite buffered	
En$_{90}$Fs$_{10}$	550-1050	2.0	9.95×10^2	1.40	Forsterite buffered	
En$_{80}$Fs$_{20}$	500-1020	2.0	6.30×10^2	1.18	SiO_2/CCO-buffered	
En$_{50}$Fs$_{50}$	510-1020	2.0	1.41×10^2	0.80	SiO_2/CCO buffered	
En$_{20}$Fs$_{80}$	510-1010	2.0	1.25×10^2	0.55	SiO_2/CCO buffered	
Fs$_{100}$ (Ferrosilite)	520-950	2.0	1.12×10^2	0.42	SiO_2/CCO buffered	
Fs$_{100}$ (Ferrosilite)	500-900	2.0	1.15×10^2	0.42	SiO_2/CCO buffered	
Fs$_{100}$ (Ferrosilite)	500-970	2.0	1.17×10^2	0.37	QFsI buffered	
					QFsM buffer	
Orthopyroxene, natural polycrystalline Dreiser Weiher, Germany En$_{87.9}$Fo$_{10.1}$Wo$_{2.0}$	544-831	1.0	2.10×10^1	1.08	Piston-cylinder, 1.6 kHz AC guard-ring	[7]
	536-842	1.0	4.51	0.97	Natural paragenesis	
					QFM buffer	
Bamle, Norway En$_{86}$	1070-1323	0.5	1.08	0.98	Argon gas medium vessel, 1592 Hz AC, CO_2:CO =11:1	[5]
	1323-1673	0.5	5.13×10^2	1.78		

Table 5. (continued)

Material	Temperature range K	Pressure GPa	σ_o S/m	E eV	Remarks	References
Pyroxene, continued:						
Clinopyroxene, natural, polycrystalline Dreiser Weiher, Germany En60.7Fs7.6Wo30.7	640-847	1.0	6.24×10^{-1}	0.43	Piston-cylinder, 1.6 kHz AC, guard ring, natural paragenesis	[7]
Clinopyroxene, synthetic, polycrystalline Diopside, $CaMgSi_2O_6$	746-1064	1.0	1.24×10^2	1.28	Piston-cylinder, 1.6 kHz AC, guard ring	[7]
Spinel, γ-Mg_2GeO_4	1122-1518	2.05	7.24×10^5	1.98	Piston-cylinder Impedance spectroscopy ERGAN buffered	[40]

For oxygen buffer assemblages: Fs = ferrosilite
F = fayalite
I = iron
M = magnetite
Q = quartz
W = wüstite

Table 7. Real relative dielectric permittivity κ' (1 MHz) and temperature derivatives for forsterite Mg_2SiO_4 over the range 293 K to 1273 K [3].

Crystallographic Orientation	κ' 293 K	$1/\kappa'$ $(\partial\kappa'/\partial T)$ K^{-1}
(100)	6.97 ± 0.21	$1.22 \pm 0.14 \times 10^{-4}$
(010)	7.71 ± 0.24	$1.69 \pm 0.12 \times 10^{-4}$
(001)	7.11 ± 0.21	$1.24 \pm 0.20 \times 10^{-4}$

Table 6. Electrical conductivity of $(Fe,Mg)_2SiO_4$ olivine and spinel (γ-form) at room temperature (295 K) and pressures to 20 GPa. Values are given for the constants in the expression $\sigma_xP = \sigma_o \exp[mx + (B_o + bx)P]$, where x is mole per cent Fe, P is pressure in GPa [26, 27]. Valid in the range $50 \leq x \leq 100$.

Mineral	$\log \sigma_o$ σ_o in S/m	m (x in mole % Fe)	B_o GPa^{-1}	b, (x in mole % Fe) GPa^{-1}
olivine	-10.42	0.125	2.065×10^{-1}	4.20×10^{-4}
spinel	-5.77	0.153	1.2×10^{-1}	-

Table 8. Electrical conductivity of anhydrous naturally-occurring silicate melts at temperatures between 1200 and 1550°C. Expressed as log σ, where σ is in S/m. Data from [45, 64, 65, 67].

Melt type	Temperature, °C							
	1200	1250	1300	1350	1400	1450	1500	1550
Nephelinite (C-195)	0.59	0.91	1.15	1.27	1.35	1.43	1.48	
Basanite (C-90)	0.41	0.73	1.03	1.27	1.36	1.45	1.49	
Tholeiitic olivine basalt (C-50)	042	0.76	0.93	1.06	1.16	1.25		
Olivine tholeiite (C-214)	0.48	0.67	0.82	0.96	1.06	1.17	1.25	
Tholeiite (C-8)	0.36	0.56	0.71	0.85	0.95	1.07	1.15	
Tholeiite (70-15)	0.41	0.65	0.84	1.00	1.11	1.20	1.27	
Tholeiite (PG-16)	0.39	0.58	0.75	0.91	1.02	1.12	1.20	1.26
Quartz tholeiite (HT-1)	0.16	0.40	0.59	0.76	0.90	1.01		
Hawaiite (C-42)	0.62	0.77	0.88	1.00	1.11	1.21	1.27	
Alkali olivine basalt (C-222)	0.48	0.66	0.81	0.95	1.05	1.15	1.23	
Alkali olivine basalt (C-70)	0.53	0.69	0.82	0.95	1.04	1.16	1.22	
Alkali olivine basalt (BCR-2)	0.30	0.46	0.60	0.73	0.83	0.92	1.00	1.07
Mugearite (C-210)	0.49	0.59	0.69	0.78	0.85	0.96	1.00	
Andesite (HA)	0.39	0.48	0.56	0.64	0.71	0.79	0.86	0.92
Andesite (VC-4W)	0.34	0.42	0.50	0.58	0.65	0.71		
Latite (V-31)	0.48	0.57	0.64	0.71	0.77	0.83	0.88	0.91
Trachyte (C-128)	0.75	0.81	0.87	0.93	0.97	1.02	1.04	
Rhyodacite (HR-1)	0.46	0.52	0.60	0.66	0.72	0.76	0.81	
Rhyolite obsidian (YRO)	0.41	0.47	0.53	0.58	0.62	0.67	0.71	0.74

Table 9. Electrical conductivity of naturally occurring silicate melts at elevated pressures over the temperature range 1200-1400°C, parameters for fits to the equation $\log \sigma = \log \sigma_0 - (E_a + P\Delta V_\sigma)/2.3kT$ [64, 65].

Melt Type	P range kb	$\log \sigma_0$ σ_0 in S/m	E_a eV	ΔV_σ cm^3/mol
Yellowstone Rhyolite Obsidian, YRO	0 - 12.8	2.19	0.50	7.9
	17 - 25.5	2.77	0.76	3.2
Hawaiian Rhyodacite, HR-1	0 - 8.5	2.36	0.55	11.1
	12.8 - 25.5	2.75	0.74	5.3
Crater lake Andesite, VC-4W	0 - 8.5	3.00	0.78	17.9
	12.8 - 25.5	3.82	1.17	3.3
Hawaiin Tholeiite	0 - 4.3	5.05	1.30	4.6
	8.5 - 25.5	5.33	1.53	-0.1

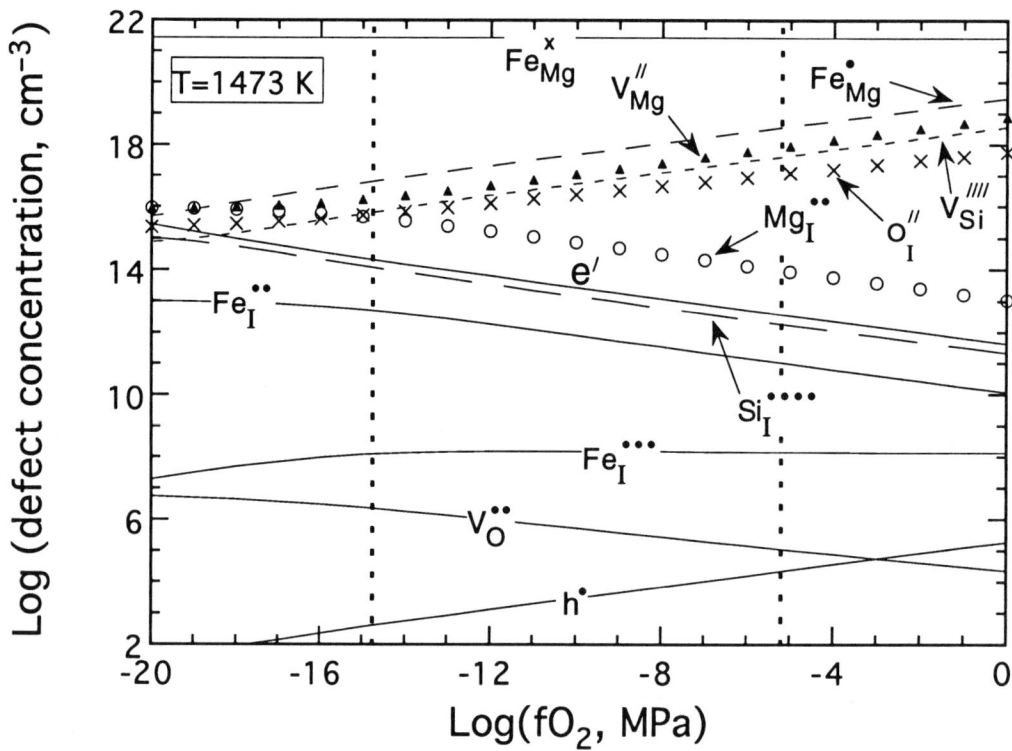

Figure 1. Calculated point defect model, plotted as log of the defect concentration versus log of oxygen fugacity, for Fo$_{90}$ olivine at 1473 K [11]. Vertical dotted lines represent stability range of olivine.

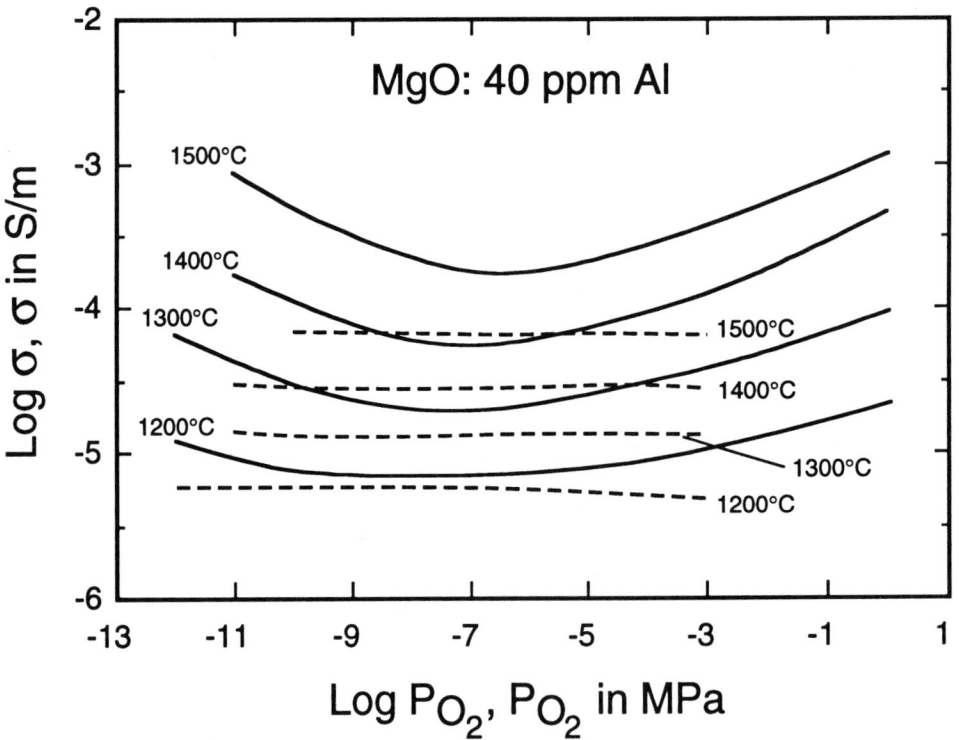

Figure 2. Log conductivity versus log f$_{O_2}$ for MgO containing 40 ppm Al. Solid lines are total conductivity, dashed lines are ionic conductivity [54, 55].

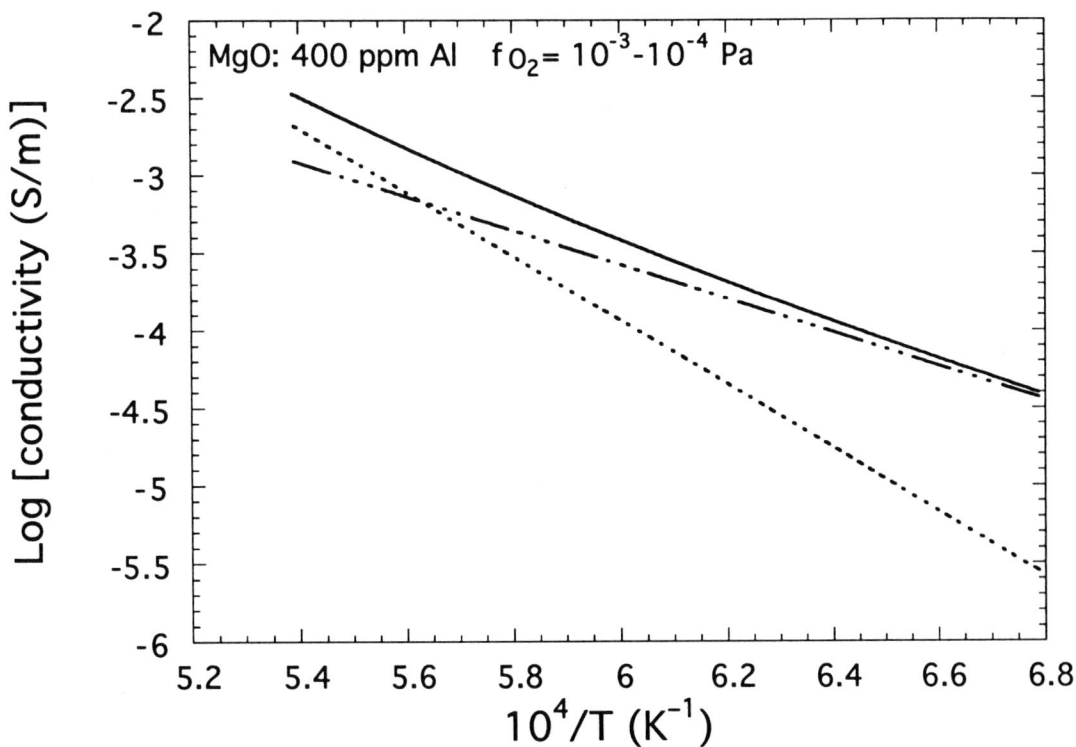

Figure 3. Log conductivity versus T^{-1} for $f_{O_2} \sim 10^{-3}$ to 10^{-4} bar for MgO containing 400 ppm Al. Solid line is total conductivity, which is the sum of ionic conductivity (dashed-dot line) and the electronic conductivity (dashed line) [54, 55].

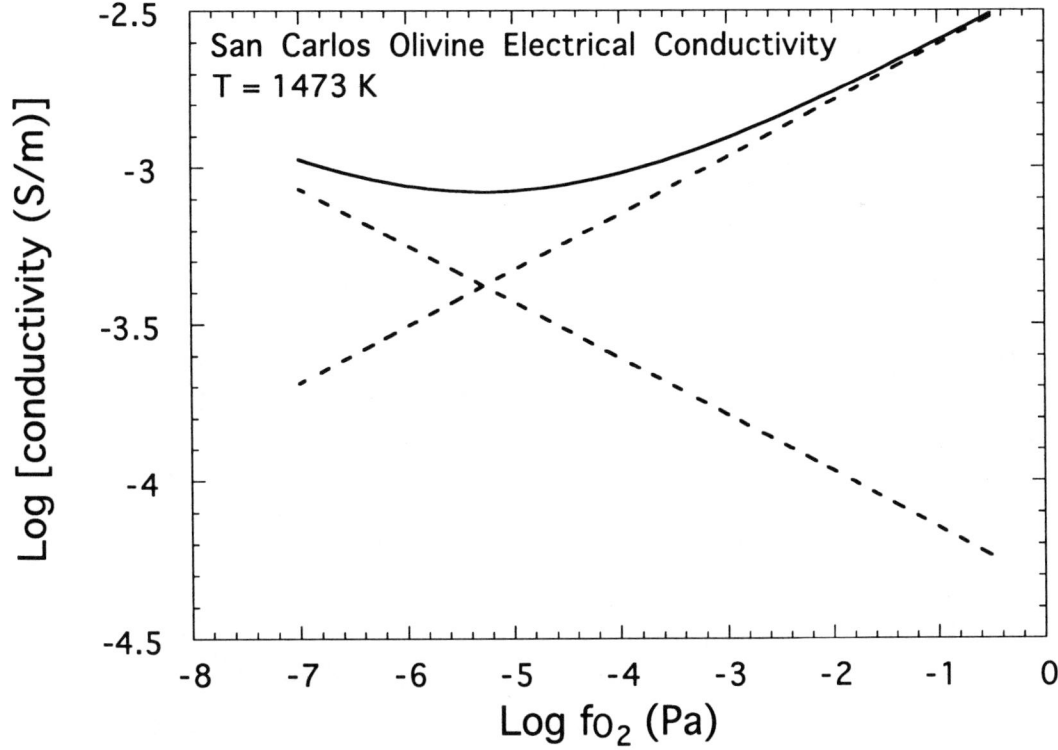

Figure 4. Log conductivity versus log f_{O_2} for San Carlos olivine (Fo90) illustrating that total conductivity is given by the sum of the two mechanisms (dashed lines) with differing oxygen fugacity dependences [68].

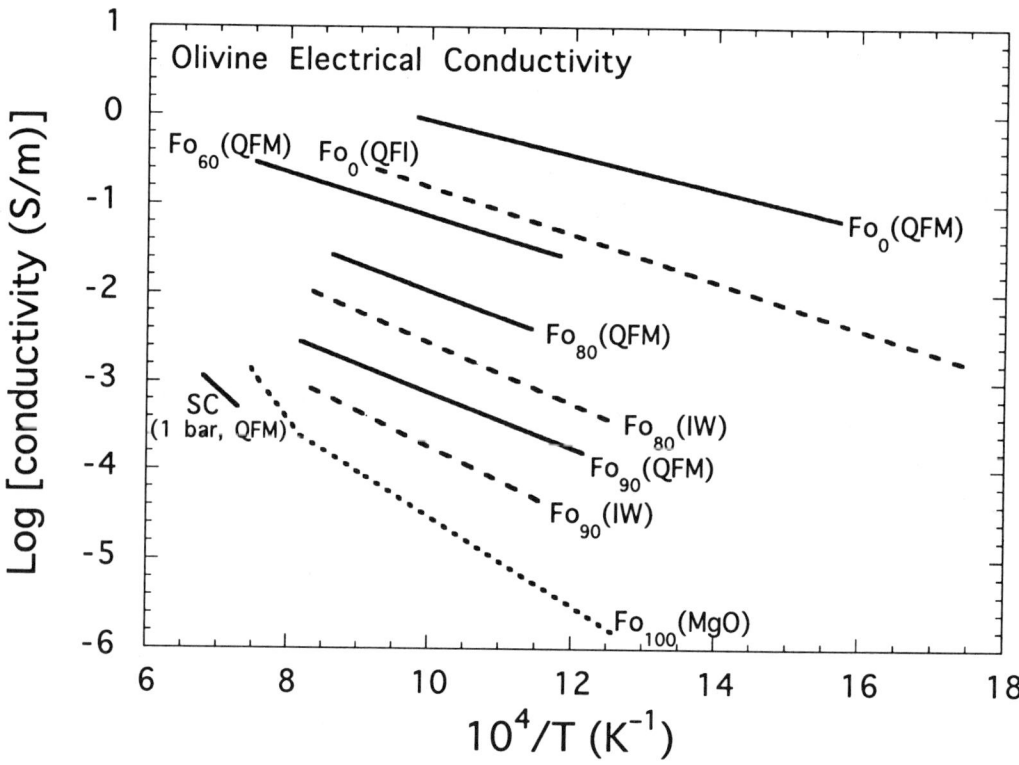

Figure 5. Log conductivity versus T^{-1} at 1 GPa for olivines of different compositions. Solid buffer assemblages controlling f_{O_2} are given in parentheses; solid lines are QFM, dashed lines at IW. Data are from [7, 69]. San Carlos olivine at 1 bar (self-buffered at QFM) is shown for comparison. [68].

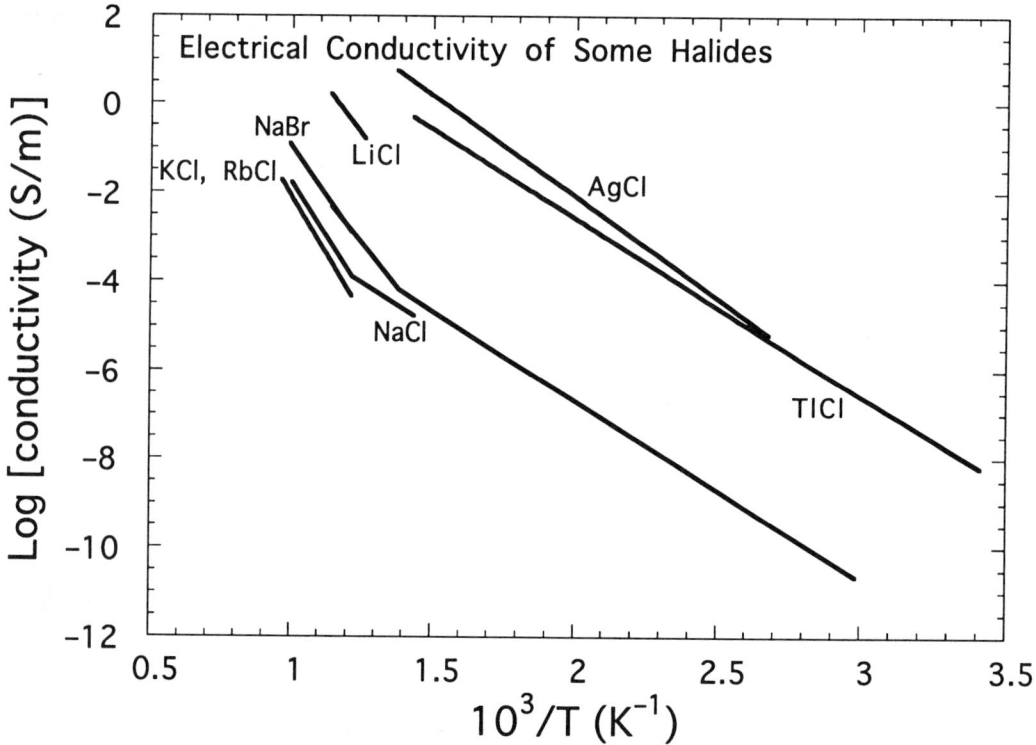

Figure 6. Log conductivity versus T^{-1} for some halides. Data from [28, 33]. The low temperature conductivity of NaCl and NaBr is impurity controlled. Conductivities for KCl and RbCl coincide.

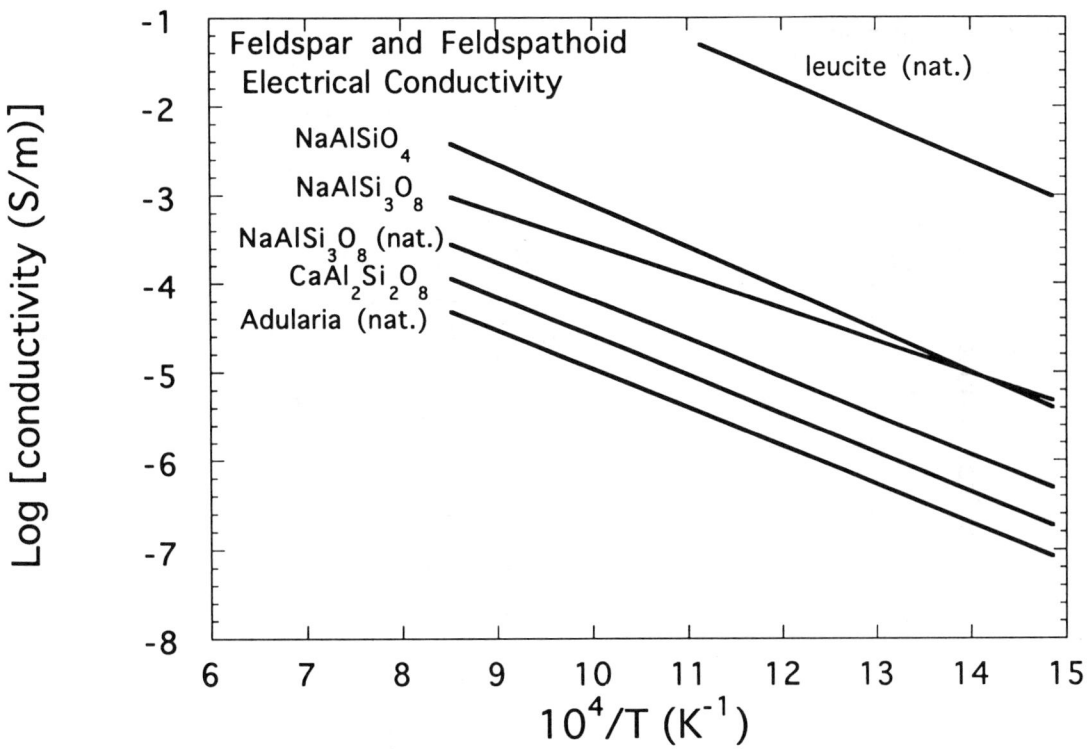

Figure 7. Log conductivity versus T^{-1} for feldspars and feldpathoids. All lines refer to intrinsic regime. Lines labeled by formula only are synthetic samples; those labeled 'nat' are natural samples. Data from [34, 35].

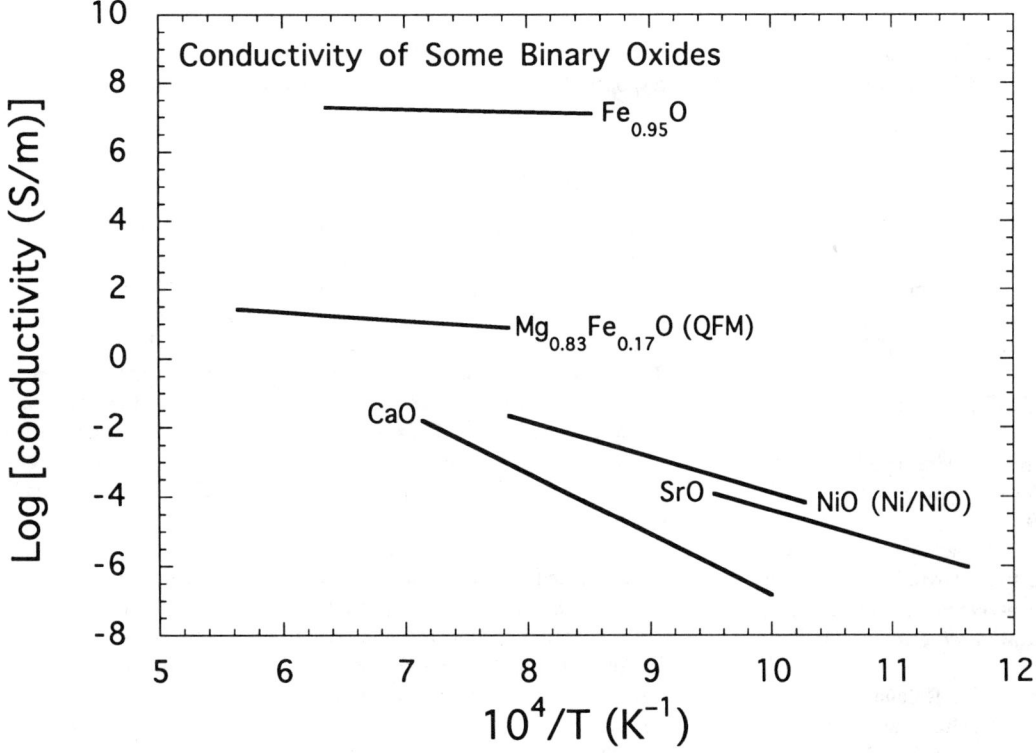

Figure 8. Log conductivity versus T^{-1} for some binary oxides. Data from [6, 61, 63, 70]. Oxygen buffers are shown in parentheses where appropriate.

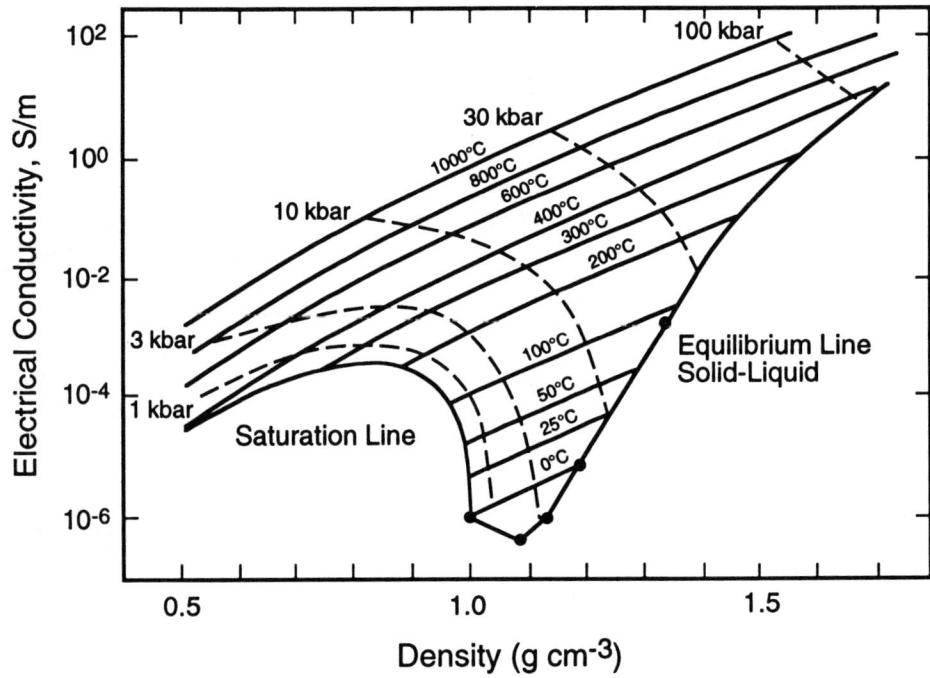

Figure 9. High pressure electrical conductivity of supercritical H_2O versus density [13].

Acknowledgements. We thank Lee Hirsch, Jeff Roberts, Tom Shankland, and Al Duba for helpful reviews. The responsibility for errors rests with the authors. This work was supported by the National Science Foundation.

REFERENCES

1. Akimoto, S. and H. Fujisawa, Demonstration of the electrical conductivity jump produced by the olivine-spinel transition, *Journal of Geophysical Research, 70*, 443-449, 1965.
2. Constable, S., T. J. Shankland and A. G. Duba, The electrical conductivity of an isotropic olivine mantle, *Journal of Geophysical Research, 97*, 3397-3404, 1992.
3. Cygan, R. T. and A. C. Lasaga, Dielectric and polarization behavior of forsterite at elevated temperatures, *American Mineralogist, 71*, 758-766, 1986.
4. Duba, A., J. N. Boland and A. E. Ringwood, The electrical conductivity of pyroxene, *Journal of Geology, 81*, 727-735, 1973.
5. Duba, A., H. C. Heard and R. N. Schock, Electrical conductivity of orthopyroxene to 1400°C and the resulting selenotherm, *Proceedings of the 7th Lunar Science Conference*, 3173-3181, 1976.
6. Gartstein, E. and T. O. Mason, Reanalysis of wustite electrical properties, *Journal of the American Ceramic Society, 65*, C-24 - C-26, 1982.
7. Hinze, E., G. Will and L. Cemic, Electrical conductivity measurements on synthetic olivines and on olivine, enstatite, and diopside from Dreiser Weiher, Eifel (Germany) under defined thermodynamic activities as a function of temperature and pressure, *Physics of the Earth and Planetary Interiors, 25*, 245-254, 1981.
8. Hirsch, L. M., Electrical conduction of Co_2SiO_4, *Physics and Chemistry of Minerals, 17*, 187-190, 1990.
9. Hirsch, L. M. and T. J. Shankland, Determination of defect equilibria in minerals, *Journal of Geophysical Research, 96*, 377-384, 1991.
10. Hirsch, L. M. and T. J. Shankland, Equilibrium point defect concentrations in MgO: Understanding the mechanisms of conduction and diffusion and the role of Fe impurities, *Journal of Geophysical Research, 96*, 385-403, 1991.
11. Hirsch, L. M. and T. J. Shankland, Quantitative olivine point defect chemical model: Insights on electricl conduction, diffusion, and the role of Fe impurities, *Geophysical Journal International, 114*, 21-35, 1993.
12. Hirsch, L. M., T. J. Shankland and A. G. Duba, Electrical conduction and polaron mobility in Fe-bearing olivine, *Geophysical Journal International, 114*, 36-44, 1993.

13. Holzapfel, W. B., Effect of pressure and temperature on the conductivity and ionic dissociation of water up to 100 kbar and 1000°C, *Journal of Chemical Physics, 50*, 4424-4428, 1969.

14. Huebner, J. S., A. Duba and L. B. Wiggins, Electrical conductivity of pyroxene which contains trivalent cations: Laboratory measurements and the lunar temperature profile, *Journal of Geophysical Research, 84*, 4652-4656, 1979.

15. Huebner, J. S. and D. E. Voigt, Electrical conductivity of diopside: Evidence for oxygen vacancies, *American Mineralogist, 73*, 1235-1254, 1988.

16. Jain, H. and S. Nowick, Electrical conductivity of synthetic and natural quartz crystals, *Journal of Applied Physics, 53*, 477-484, 1982.

17. Keller, G. V., Electrical Properties of Rocks and Minerals, in *Handbook of Physical Constants, Geological Society of America, Memoir 97*, edited by S. P. Clark Jr., pp. 553-577, Geological Society of America, New York, 1966.

18. Keller, G. V., Electrical Properties of Rocks and Minerals, in *Handbook of Physical Properties of Rocks and Minerals*, edited by R. S. Carmichael, pp. 217-293, CRC Press, Boca Raton, FL, 1981.

19. Keller, G. V., Electrical properties, in *Practical Handbook of Physical Properties of Rocks and Minerals*, edited by R. S. Carmichael, pp. 361-427, CRC Press, Inc., Boca Raton, FL, 1989.

20. Kingery, W. D., H. K. Bowen and D. R. Uhlmann, *Introduction to Ceramics*, John Wiley and Sons, Inc., New York, NY, 1976.

21. Kingery, W. D., J. Pappis, M. E. Doty and D. C. Hill, Oxygen ion mobility in cubic $Zr_{0.85}Ca_{0.15}O_{1.85}$, *Journal of the American Ceramic Society, 42*, 393-398, 1959.

22. Kittel, C., *Introduction to Solid State Physics*, John Wiley and Sons, Inc., New York, 1976.

23. Kroger, F. A. and H. H. Vink, Relations between the concentrations of imperfections in crystalline solids, in *Solid State Physics*, edited by F. Sietz and D. Turnbull, pp. 307-435, Academic Press, Inc., San Diego, CA, 1956.

24. Kroger, F. A., *Chemistry of Imperfect Crystals*, North-Holland, Amsterdam, 1974.

25. Kroger, F. A., Point defect in solids: Physics, chemistry, and thermodynamics, in *Point Defects in Minerals*, edited by R. N. Schock, pp. 1-17, American Geophysical Union, Washington, DC, 1985.

26. Lacam, A., Pressure and composition dependence of the electrical conductivity of iron-rich synthetic olivines to 200 kbar, *Physics and Chemistry of Minerals, 9*, 127-132, 1983.

27. Lacam, A., Effects of composition and high pressures on the electrical conductivity of Fe-rich $(Mg,Fe)_2SiO_4$ olivines and spinels, *Physics and Chemistry of Minerals, 12*, 23-28, 1985.

28. Lehfeldt, W., Über die elektrische Leitfähigkeit von Einkristallen, *Zeitschrift fur Physik, 85*, 717-726, 1933.

29. Li, X. and R. Jeanloz, Laboratory studies of the electrical conductivity of silicate perovskites at high temperatures and pressures, *Journal of Geophysical Research, 95*, 5067-5078, 1990.

30. Li, X. and R. Jeanloz, Effect of iron content on the electrical conductivity of perovskite and magnesiowustite assemblages at lower mantle conditions, *Journal of Geophysical Research, 96*, 6113-6120, 1991.

31. Manko, M., E. Hinze and G. Will, Die electrische leitfahigkeit und ihre frequenzabhangigkeit gemessen an synthetischen und naturlichen olivinen, in *Prot. Elektrom. Tiefenforsch.*, edited by V. Haak and J. Homilius, pp. Berlin-Lichtenrade, Berlin, 1980.

32. Mao, H. K., Observations of optical absorption and electrical conductivity in magnesiowustite at high pressure, *Annual Report of the Director, Geophysical Laboratory, Carnegie Institute of Washington, 72*, 554-557, 1973.

33. Mapother, D., H. N. Crooks and R. Maurer, Self-diffusion of sodium in sodium chloride and sodium bromide, *Journal of Chemical Physics, 18*, 1231-1236, 1950.

34. Maury, R., Conductibilite electrique des tectosilicates. I. Methode et resultats experimentaux, *Bulletin de la Societe Francaise de Mineralogie et Cristallographie, 91*, 267-278, 1968.

35. Maury, R., Conductibilite electrique des tectosilicates. II. Discussion des resultats, *Bulletin de la Societe Francaise de Mineralogie et Cristallographie, 91*, 355-366, 1968.

36. Mirwald, P. W., The electrical conductivity of calcite between 300 and 1200 C at CO_2 pressure of 40 bars, *Physics and Chemistry of Minerals, 4*, 291-297, 1979.

37. Mott, N. F. and H. Jones, *The Theory of the Properties of Metals and Alloys*, Dover Publications, Inc., New York, 1958.

38. Nell, J. and B. J. Wood, High-temperature electrical measurements and thermodynamic properties of Fe_3O_4-$FeCr_2O_4$-$MgCr_2O_4$-$FeAl_2O_4$ spinels, *American Mineralogist, 76*, 405-426, 1991.

39. Nell, J., B. J. Wood and T. O. Mason, High-temperature cation distributions in Fe_3O_4-$MgFe_2O_4$-$FeAl_2O_4$-$MgAl_2O_4$ spinels from thermopower and conductivity measurements, *American Mineralogist, 74*, 339-351, 1989.

40. Nover, G., G. Will and R. Waitz, Pressure induced phase transition in Mg_2GeO_4 as determined by frequency dependent complex electrical resistivity measurements, *Physics and Chemistry of Minerals, 19*, 133-139, 1992.

41. O'Keeffe, M. and J. O. Bovin, Solid electrolyte behavior of $NaMgF_3$: Geophysical implications, *Science, 206*, 599-600, 1979.

42. Olhoeft, G. R., Electrical properties of Rocks, in *Physical Properties of Rocks and Minerals*, edited by Y. S. Touloukian, W. R. Judd and R. F. Roy, pp. 257-329, McGraw-Hill, New York, 1981.

43. Pappis, J. and W. D. Kingery, Electrical properties of single-crystal

and polycrystalline alumina at high temperatures, *Journal of the American Ceramic Society, 44*, 459-464, 1961.

44. Pollock, D. D., *Electrical Conduction in Solids: An Introduction*, American Society for Metals, Metals Park, OH, 1985.

45. Rai, C. S. and M. H. Manghnani, Electrical conductivity of basalts to 1550°C, in *Magma Genesis: Bulletin 96, Oregon Department of Geology and Mineral Industries*, edited by H. J. B. Dick, pp. 219-232, Oregon Department of Geology and Mineral Industries, Portland, OR, 1977.

46. Rao, K. S. and K. V. Rao, Dielectric dispersion and its temperature variation in calcite single crystals, *Z. Physics, 216*, 300-306, 1968.

47. Rigdon, M. A. and R. E. Grace, Electrical charge transport in single-crystal $CaWO_4$, *Journal of the American Ceramic Society, 56*, 475-478, 1973.

48. Roberts, J. J. and J. A. Tyburczy, Frequency dependent electrical properties of polycrystalline olivine compacts, *Journal of Geophysical Research, 96*, 16205-16222, 1991.

49. Schmidbauer, E. and P. W. Mirwald, Electrical conductivity of cordierite, *Mineralogy and Petrology, 48*, 201-214, 1993.

50. Schober, M., The electrical conductivity of some samples of natural olivine at high temperatures and pressures, *Zeitschrift fur Geophysik, 37*, 283-292, 1971.

51. Schock, R. N., A. G. Duba, H. C. Heard and H. D. Stromberg, The electrical conductivity of polycrystalline olivine and pyroxene under pressure, in *High Pressure Research, Applications in Geophysics*, edited by M. H. Manghnani and S. Akimoto, pp. 39-51, Academic Press, New York, 1977.

52. Schock, R. N., A. G. Duba and T. J. Shankland, Electrical conduction in olivine, *Journal of Geophysical Research, 94*, 5829-5839, 1989.

53. Seifert, K.-F., G. Will and R. Voigt, Electrical conductivity measurements on synthetic pyroxenes $MgSiO_3$-

$FeSiO_3$ at high pressures and temperatures under defined thermodynamic conditions, in *High-Pressure researches in Geoscience*, edited by W. Schreyer, pp. 419-432, E. Schweizerbart'sche Verlagsbuchhandlung, Stuttgart, 1982.

54. Sempolinski, D. R. and W. D. Kingery, Ionic conductivity and magnesium vacancy mobility in magnesium oxide, *Journal of the American Ceramic Society, 63*, 664-669, 1980.

55. Sempolinski, D. R., W. D. Kingery and H. L. Tuller, Electronic conductivity of single crystalline magnesium oxide, *Journal of the American Ceramic Society, 63*, 669-675, 1980.

56. Shankland, T. J. and A. G. Duba, Standard electrical conductivity of isotropic, homogeneous olivine in the temperature range 1200-1500°C, *Geophysical Journal International, 103*, 25-31, 1990.

57. Shankland, T. J., J. Peyronneau and J.-P. Poirier, Electrical conductivity of the earth's lower mantle, *Nature, 344*, 453-455, 1993.

58. Smyth, D. M. and R. L. Stocker, Point defects and non-stoichiometry in forsterite, *Physics of the Earth and Planetary Interiors, 10*, 183-192, 1975.

59. Stocker, R. L. and D. M. Smyth, Effect of enstatite activity and oxygen partial pressure on the point-defect chemistry of olivine, *Physics of the Earth and Planetary Interiors, 16*, 145-156, 1978.

60. Stocker, R. L., Influence of oxygen pressure on defect concentrations in olivine with a fixed cationic ratio, *Physics of the Earth and Planetary Interiors, 17*, 118-129, 1978.

61. Surplice, N. A., The electrical conductivity of calcium and strontium oxides, *British Journal of Applied Physics, 17*, 175 - 180, 1966.

62. Tamura, S., Pressure dependence of the Neel temperature and the resistivity of $Fe_{1-y}O$ to 2 GPa, *High Temperatures-High Pressures, 22*, 399-403, 1990.

63. Tare, V. B. and J. B. Wagner, Electrical conductivity in two phase nickel-nickel oxide mixtures and at the nickel-nickel oxide phase boundary, *Journal of Applied Physics, 54*, 6459 - 6462, 1983.

64. Tyburczy, J. A. and H. S. Waff, Electrical conductivity of molten basalt and andesite to 25 kilobars pressure: Geophysical implications and significance for charge transport, *Journal of Geophysical Research, 88*, 2413-2430, 1983.

65. Tyburczy, J. A. and H. S. Waff, High pressure electrical conductivity in naturally occurring silicate liquids, in *Point Defects in Minerals, Geophysical Monograph 31*, edited by R. N. Schock, pp. 78-87, American Geophysical Union, Washington, D. C., 1985.

66. von Hippel, A. R., editor, *Dielectric Materials and Applications*, MIT Press, Cambridge, MA, 1954.

67. Waff, H. S. and D. F. Weill, Electrical conductivity of magmatic liquids: Effects of temperature, oxygen fugacity, and composition, *Earth and Planetary Science Letters, 28*, 254-260, 1975.

68. Wanamaker, B. J. and A. G. Duba, Electrical conductivity of San Carlos olivine along [100] under oxygen- and pyroxene-buffered conditions and implications for defect equilibria, *Journal of Geophysical Research, 98*, 1993.

69. Will, G., L. Cemic, E. Hinze, K.-F. Seifert and R. Voigt, Electrical conductivity measurements on olivines and pyroxenes under defined thermodynamic activities as a function of temperature and pressure, *Physics and Chemistry of Minerals, 4*, 189-197, 1979.

70. Wood, B. J. and J. Nell, High-temperature electrical conductivity of the lower-mantle phase (Mg,Fe)O, *Nature, 351*, 309-311, 1991.

71. Xu, M. Y., H. Jain and M. R. Notis, Electrical properties of opal, *American Mineralogist, 74*, 821-825, 1989.

Viscosity and Anelasticity of Melts.

Donald B. Dingwell

1. DEFINITIONS

A Newtonian fluid is one for which a linear relation exists between stress and the spatial variation of velocity. The *viscosity* of a Newtonian material is defined as the ratio of stress to strain rate

$$\eta = \frac{\sigma}{\dot{\varepsilon}} \qquad (1)$$

In contrast, the *elastic modulus* of a Hookean elastic material is defined as the ratio of stress to strain

$$M = \frac{\sigma}{\varepsilon} \qquad (2)$$

As is the case for elastic moduli, we can, for an isotropic medium like a silicate melt, speak of two components of viscosity, a volume (η_v) and a shear (η_s) component. The combination of volume and shear viscosity yields a longitudinal viscosity, (η_s)

$$\eta_l = \eta_v + \frac{4}{3}\eta_s \qquad (3)$$

A comparison of longitudinal and shear viscosities illustrates that volume and shear viscosities are subequal at very low strains [9] (see Figure 1). The ratio of Newtonian

D. B. Dingwell, Bayerisches Geoint., University Bayreuth,
Postfach 101251, D 8580 Bayreuth, Germany

Mineral Physics and Crystallography
A Handbook of Physical Constants
AGU Reference Shelf 2

viscosity to elastic modulus yields a quantity in units of time. This ratio

$$\tau = \frac{\eta}{M} \qquad (4)$$

is the Maxwell *relaxation time* [24]. This relaxation time is a convenient approximation to the timescale of deformation where the transition from purely viscous behavior to purely elastic behavior occurs. The purely viscous response of a silicate melt is termed "*liquid*" behavior whereas the purely elastic response is termed "*glassy*" behavior. The term silicate *melt* is used to describe molten silicates quite generally, regardless of their rheology. Even for the simple case of a Maxwell body, the change from liquid to glassy behavior does not occur as a sharp transition but rather describes a region of mixed liquid-like and solid-like behavior - a region of visco-elasticity. Deformation experiments performed in the viscoelastic region contain three distinct time-resolved components of deformation, instantaneous recoverable, delayed recoverable and delayed non-recoverable. Figure 2 illustrates these components for a creep experiment. Viscous flow is the delayed non-recoverable component, the term *anelasticity* refers to the instantaneous plus the delayed recoverable deformation and the entire deformation behavior falls under the heading of *viscoelasticity* [26].

Structural relaxation in silicate melts is but one example of a *relaxation mode*. It can be mechanically represented as a series combination of viscous dashpot and elastic spring (a Maxwell element, see Figure 3). For more complex materials such as crystal suspensions, foams and partial melts, several additional mechanisms of deformation are contributed at distinct timescales of deformation via additional relaxation modes within and between the added

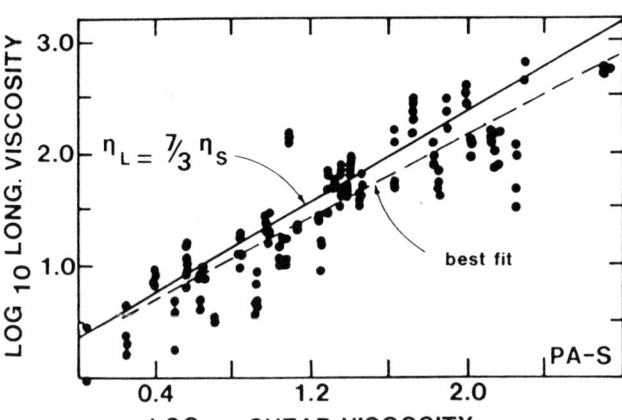

Figure 1. The relationship between longitudinal and shear viscosities in silicate melts. The solid line represents the correlation predicted from the assumption that volume and shear viscosities are equal. The dashed line is the best fit to the data and lies within 1σ of the theoretical line. Redrawn from Dingwell and Webb (1989).

phases (*e.g.*, crystal-crystal interactions, bubble deformation, matrix compaction, *etc.*). These relaxation modes involve chemical and textural equilibria and thus occur at increasingly longer timescales (*i.e.*, at lower rates of deformation). As a consequence of the multiphase nature of flowing magma the accompanying mechanical models become more complex, involving additional viscoelastic elements in the linear regime and non-linear effects as well. Inherent in the experimental study of magma rheology then, is the experimental deformation rate or its inverse, the experimental timescale. It is this experimental timescale which must most closely match the natural process in order to make secure applications of the rheological results.

2. CLASSES OF EXPERIMENTS

In principle, any experiment that records strain as a function of time and applied stress can be used to obtain viscosity. Viscosity experiments can be subdivided in several ways. We can, for example, distinguish between those in which the strain rate is controlled and the resultant stress is measured (*e.g.*, concentric cylinder) and those in which the stress is controlled and the resultant strain rate is measured (*e.g.*, fiber elongation). We can also separate those methods that involve only shear stresses (*e.g.*, concentric cylinder) from those which involve a combination of shear and volume (compressive or dilational) stress (*e.g.*, beam bending or fiber elongation). For any given geometry that employs controlled strain we

can further distinguish between the application of a step function of strain and the continuous application of strain (*e.g.*, stress relaxation versus steady-state flow). Stress relaxation involves the recording of stress versus time as the stress field decays to zero after the application of a step function of strain. Steady-state strain experiments record the equilibrium stress sustained by the liquid due to the constant strain rate being applied to the sample. We can distinguish between experiments performed at low total strains such that the perturbations from equilibrium are small and in the linear regime of stress-strain relations. Such measurements probe the response of the system to stresses without macroscopic rearrangements of the phases. In contrast nonlinear measurements can deal with the redistribution of phases during macroscopic flow. Finally, we can distinguish between time domain experiments where a stress is applied and the time variant relaxation of the system to the new equilibrium is measured (e.g., creep experiments) and frequency domain experiments where sinusoidal variations of the stress field, small in magnitude but up to very high frequencies (from mHz to MHz), are applied and the magnitude and phase shift of strain is measured (e.g., torsion, ultrasonic wave attenuation). Implicit in considerations of rheological measurements is the ratio of deformation rate to the relaxation rate of structural components in the material. The characteristic strain rate of the experiment is easily defined in most rheological experiments. For example, in Figure 4, several experimental timescales are compared with structural relaxation times of silicate melts. Ultrasonic measurements are performed on time scales of 1-50 ns, forced oscillation techniques in the range of 0.01-150 s and fiber elongation

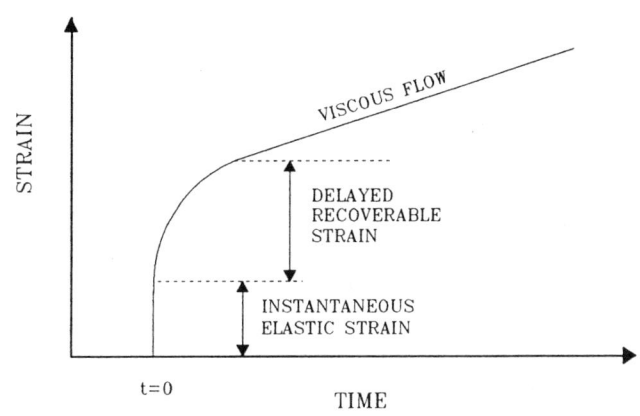

Figure 2. The instantaneous recoverable elastic strain, the delayed recoverable strain, and the delayed non-recoverable viscous flow occurring in a linear viscoelastic material upon application of a step function in stress at time, t=0.

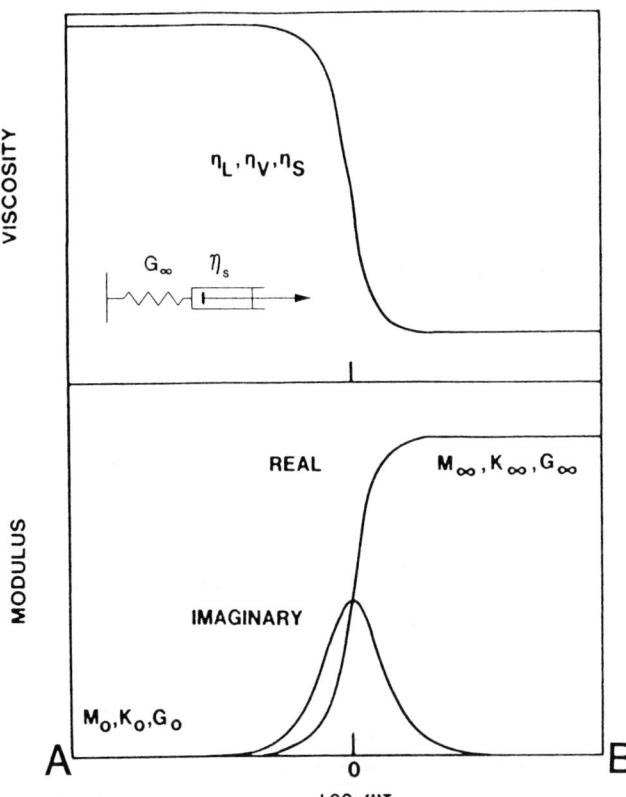

Figure 3. Calculated frequency dependent behavior of longitudinal, volume and shear viscosities (η_1, η_v, η_s) and moduli (M, K, G) of a linear viscoelastic melt with a relaxation time τ plotted as a function of $\omega\tau$ where ω is the angular frequency of the applied sinusoidal stress. The subscripts "0" and "∞" indicate zero frequency and infinite frequency values. The shear relaxation time $\tau = \eta/G$ is for the mechanical model of a spring and dash-pot in series (i.e., a Maxwell element).

from 10^2 to 10^6 s. Each structural or physical component of the system relaxes as a distinct relaxation mode at a distinct timescale of frequency which is a function of pressure and temperature. Frequency-dependent or non-Newtonian effects are seen in frequency domain rheological experiments when the measurement timescale approaches a relaxation mode timescale.

3. GEOMETRIES

The most widely used example of controlled strain rate and measured stress is the rotational (Couette) method (see Figure 5). Ryan and Blevins [34] provide a description of the physics of Couette viscometry. The strain rate is

imposed on an annulus of liquid filling the gap between an inner cylinder (or spindle) and the outer cylinder (or cup) which usually takes the form of a cylindrical crucible. The strain rate is delivered to the liquid by rotating either the inner or outer cylinder. The shear stress sustained in the liquid is usually recorded as the torque exerted by the liquid on the inner cylinder. The concentric cylinder method is commonly used for viscosity measurements in the range of 10^0-10^5 Pa s. Another form of the Couette viscometer is the cone-and-plate geometry. Viscometers based on the cone-and-plate geometry have been used in both steady state (as above) and pulse strain modes. The geometry of the cone-and-plate viscometer allows the preparation of starting materials by pre-machining to the final form. This avoids the need to pour or melt the silicate material into the cylinder and to immerse the spindle into the liquid at high temperature. The cone-and-plate method finds application at lower temperatures and higher viscosities than the concentric cylinder method. This method has also been used in the investigation of the viscoelastic behavior of silicate melts at/near the glass transition [41].

Shear viscosities in the range 10^9-10^{14} Pa s can be determined using fiber elongation techniques (see Figure 5). Glass fibers with a diameter 0.1-0.3 mm and lengths 10-18 mm are commonly used. In a vertically mounted silica glass dilatometer the silica glass holder of the dilatometer supports the beaded glass fiber in a fork. A second silica glass rod holds the lower bead of the fiber in tension. The strain-rate range is machine limited to 10^{-7}-10^{-4} s. A tensile stress (~10^7 Pa) is applied to the melt fiber and the viscosity is determined as the ratio of the applied stress to the observed strain-rate. In this geometry the observed viscosity, η_{elong}, is the elongational viscosity and is related to the shear viscosity, η_s, by

$$\eta_{elong} = \frac{\sigma}{\dot{\varepsilon}} = \frac{9\,\eta_v\,\eta_s}{3\eta_v + \eta_s} \qquad (5)$$

[11, 15, 17] Although infinite shear strains are possible in a melt, volume strain must be limited in magnitude [25]. The volume viscosity of a melt, therefore, approaches an infinite value with increasing time and Equation 5 becomes

$$\eta_{elong} = 3\,\eta_s \qquad (6)$$

[11, 25] for times greater than the relaxation time of the melt. Equation 6 is known as Trouton's rule.

Absolute shear viscosities in the range 10^9 - 10^{11} Pa s can be determined using micropenetration techniques (see Figure 5). This involves determining the rate at which an

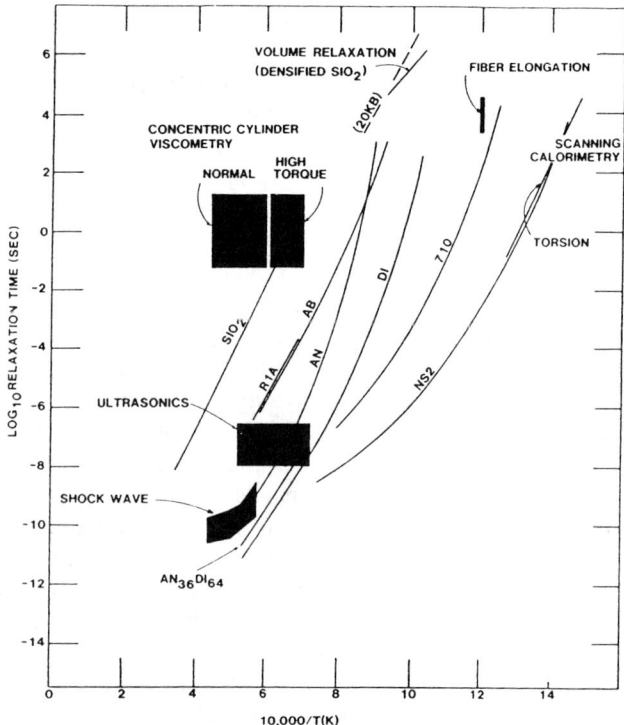

Figure 4. The liquid-glass transition as a function of temperature plotted for several silicate melts calculated from Equation 4 together with the timescales on which a range of experimental measurements are performed. The relaxation times relating to the various experimental techniques are discussed in Dingwell and Webb (1989). Redrawn from Dingwell and Webb (1989).

indenter under a fixed load moves into the melt surface. High accuracy ($\pm 0.1 \log_{10}$ Pa s) is obtained by using a silica glass sample holder and pushrod for temperatures under 1000°C. The indenter may be conical, cylindrical or spherical in shape [6]. For the case in which a spherical indenter is used [8] the absolute shear viscosity is determined from

$$\eta_s \,(\text{Pa s}) = \frac{0.1875 \, P \, t}{r^{0.5} \, l^{1.5}} \tag{7}$$

[28, 44] for the radius of the sphere, r (m), the applied force, P (N), indent distance l (m) and time, t (s), (t=0 and l =0 upon application of the force).

The shear viscosity of large cylinders of melt or partially molten material can be determined by deforming the cylinder between parallel plates moving perpendicular to their planes. Absolute viscosities in the range 10^4 - 10^8 Pa s [12, 16] and 10^7 - 10^{11} Pa s [2] can be determined.

Specimen deformation rates are measured with a linear voltage displacement transducer. For a cylindrical specimen of any thickness the shear viscosity can be determined by

$$\eta_s \,(\text{Pa s}) = \frac{2 \, \pi \, m \, g \, h^5}{3V \, \delta h/\delta t \, (2\pi \, h^3 + V)} \tag{8}$$

for the applied mass, m (kg), the acceleration due to gravity, g (m^2/s), the volume, V (m^3) of the material, the height, h (m) of the cylinder, and time, t (s) [12, 16] for the case in which the surface area of contact between the melt and the parallel plates remains constant and the cylinder bulges with increasing deformation. This is the "no-slip" condition. For the case in which the surface area between the cylinder and the plate increases with deformation and the cylinder does not bulge the viscosity is

$$\eta_s \,(\text{Pa s}) = \frac{m \, g \, h^2}{3V \, \delta h/\delta t} . \tag{9}$$

This is the "perfect slip" condition. The parallel plate method involves the uniaxial deformation of a cylinder of melt at either constant strain rate [18] or constant load [40].

The counterbalanced sphere and falling sphere viscometers (see Figure 5) are based on Stokes law. Riebling [30] has described a counterbalanced sphere viscometer which he subsequently used in viscosity determinations in the B_2O_3-SiO_2 [31] and Na_2O-Al_2O_3-SiO_2 [32] systems. Falling sphere viscometry has been employed at 1 atm and at high pressures [13, 34, 36]. The falling sphere method may be used for the simultaneous determination of density and viscosity. Alternatively the falling sphere method may be used with input values of density provided the density contrast between melt and sphere is relatively large. Maximizing the density contrast reduces errors associated with the estimation of melt density. Errors in density contrast can be reduced below the uncertainties inherent in the other variables affecting viscosity determination. Very high pressure measurements of viscosity have been made [19] by imaging the falling sphere in real time using a synchrotron radiation source and a MAX 80 superpress. This method extends the lower limit of measurable viscosity using the falling sphere method at high pressure to 10^{-3} Pa s.

The shear and longitudinal viscosities of a melt can be determined from the velocity, c(ω), and attenuation, $\alpha(\omega)$, of a shear or longitudinal wave travelling through the melt; where the amplitude of the wave is

$$A(\omega,t) = A_0 \exp^{-\alpha x} \exp^{i\omega(t-x/c)} \tag{10}$$

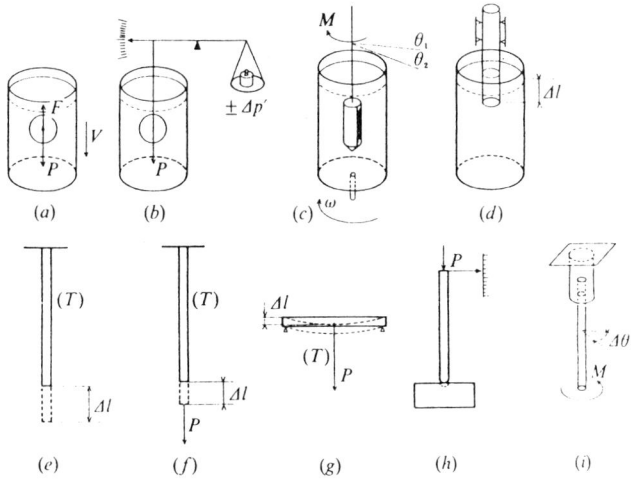

Figure 5. Methods of determining viscosity: (a) falling sphere (b) restrained sphere (c) rotational viscometer (d) sink point measurement (e) free fiber elongation (f) loaded fiber elongation (g) beam bending (h) micropenetration (i) torsion of a cylinder. Redrawn from Zarzicki (1991).

for time t and distance x [27]. Small strains ($<10^{-6}$) are used and linear stress-strain theory is applicable. The real component of the modulus, $M'(\omega) = \rho c^2$, and the imaginary component of the modulus, $M''(\omega) = M'(\omega)c\alpha/\pi f$ (where the quality factor $Q = M'(\omega)/M''(\omega) = \pi f/c\alpha$ [27]. Viscosities are calculated for time, t and distance, x [27]. The shear viscosity can be calculated from the velocity and attenuation data in that $\eta^*(\omega) = M^*(\omega)/i\omega$, or

$$M^*(\omega) = M'(\omega) + iM''(\omega) = M''(\omega) + i\eta'(\omega)\cdot\omega = \eta''(\omega)\cdot\omega + i\eta'(\omega)\cdot\omega \qquad (11)$$

and the real component of the frequency dependent viscosity is

$$\eta'(\omega) = \frac{2\rho c^3 \alpha}{\omega^2} . \qquad (12)$$

The shear viscosity can be calculated from the velocity and attenuation of the shear wave. The data from the propagating longitudinal wave results in the determination of the longitudinal viscosity from which the volume viscosity can be calculated

$$\eta^*_l(\omega) = \eta^*_k(\omega) + 4\eta^*_s(\omega)/3 . \qquad (13)$$

Most ultrasonic studies of silicate melts are conducted at

the temperature (1000-1500°C) and frequency (3-22 MHz) conditions required to observe the relaxed (frequency-independent) longitudinal viscosity of most silicate melts. In cases (basalt melts [22], basaltic andesite [35], synthetic silicate melts [33]) where the experimental conditions were approaching the relaxation frequency of the melt, frequency-dependence of the longitudinal viscosity has been observed. In studies where the frequency of the applied signal has been greater than that of the relaxation frequency of the melt shear wave propagation has been observed (B_2O_3 [43], Na_2O-B_2O_3-SiO_2 [23], $Na_2Si_2O_5$ [46]).

4. TEMPERATURE DEPENDENCE OF VISCOSITY

The temperature-dependence of silicate liquid viscosity has been described using a number of different equations. The simplest form, often valid for restricted temperature intervals, is a linear dependence of the logarithm of viscosity on reciprocal temperature, i.e. the Arrhenius equation

$$\log_{10}\eta = \log_{10}\eta_0 + 2.303\frac{E}{RT} \qquad (14)$$

where η is the viscosity at temperature, T, and η_0 and E are termed the frequency or pre-exponential factor and the activation energy, respectively. Quite often the data obtained by a single method of investigation over a restricted temperature or viscosity interval are adequately described by Equation 14. With the possible exception of SiO_2, however, sufficient data now exist to demonstrate that, in general, silicate liquids exhibit non-Ahrrenian viscosity temperature relationships. The degree of "curvature" of the viscosity temperature relationship plotted as log viscosity versus reciprocal absolute temperature varies greatly with chemical composition (see [29] for a summary). The temperature-dependence of non-Arrhenian data can be reproduced by adding a parameter to Equation 14 to yield

$$\log_{10}\eta = \log_{10}\eta_0 + 2.303\frac{E}{R(T-T_0)} \qquad (15)$$

where T_0 is a constant. This empirical description of the temperature-dependence of liquid viscosity is called the Fulcher or TVF (Tamann-Vogel-Fulcher) equation [14, 15, 42, 45].

Angell [1] describes the concept of strong and fragile liquids based on the extent of nonArrhenian temperature-dependence of viscosity. This consideration is extremely

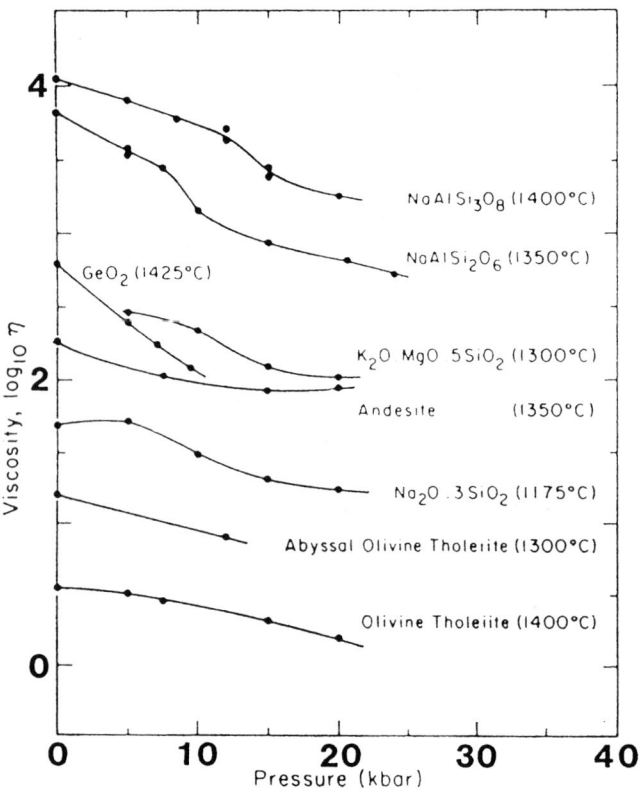

Figure 6. Viscosity as a function of pressure for various silicate melts. Redrawn from Scarfe et al. (1989).

important where extrapolations of viscosity to lower temperatures are performed. In general, silicate melts are strong liquids in comparison with nonsilicate liquids. The extent of nonArrhenian temperature dependence varies greatly with composition. Decreasing SiO_2 content generally leads to less Arrhenian, more fragile, behavior. Additionally the increase of pressure likely leads to more fragile behavior.

5. PRESSURE DEPENDENCE OF VISCOSITY

The falling sphere method has been used to study high pressure viscosities of natural and synthetic silicate liquids [10, 13, 20]. The pressure-dependence of anhydrous silicate liquids has been summarized by Scarfe et al. [36]. Figure 6 illustrates the pressure dependence of silicate liquid viscosities to 2 GPa. The viscosities of silicate and aluminosilicate melt compositions with relatively low calculated NBO/T (ratio of nonbridging oxygens to tetrahedrally coordinated cations) values decrease with increasing pressure. In contrast the viscosities of some silicate liquids increase with pressure. Two studies have

investigated the pressure-dependence of silicate melt viscosities across compositional joins exhibiting both negative and positive pressure dependence of viscosity. Kushiro [21] investigated liquids along the SiO_2-$CaAl_2O_4$ join observing a linear pressure dependence of viscosity that changed from a negative value at mole fraction $CaAl_2O_4$ = 0.15 and 0.2 through pressure invariance at $CaAl_2O_4$ = 0.33 to a positive pressure dependence at $CaAl_2O_4$. Brearley et al. [5] also observed a transition from positive to negative pressure dependence of viscosity along the join $CaMgSi_2O_6$-$NaAlSi_3O_8$. Volatile-bearing silicate liquids have also been studied at high pressure [4, 7, 38, 49]. Figure 7 summarizes the effects of H_2O and F_2O_{-1} on the viscosity of $NaAlSi_3O_8$ liquid.

6. CALCULATION SCHEMES

A number of empirical calculation schemes for the viscosity of silicate glass melts have been around for several decades [37] but calculation methods for geological melts are a more recent phenomenon. Bottinga and Weill [3] produced a method for the estimation of melt viscosities

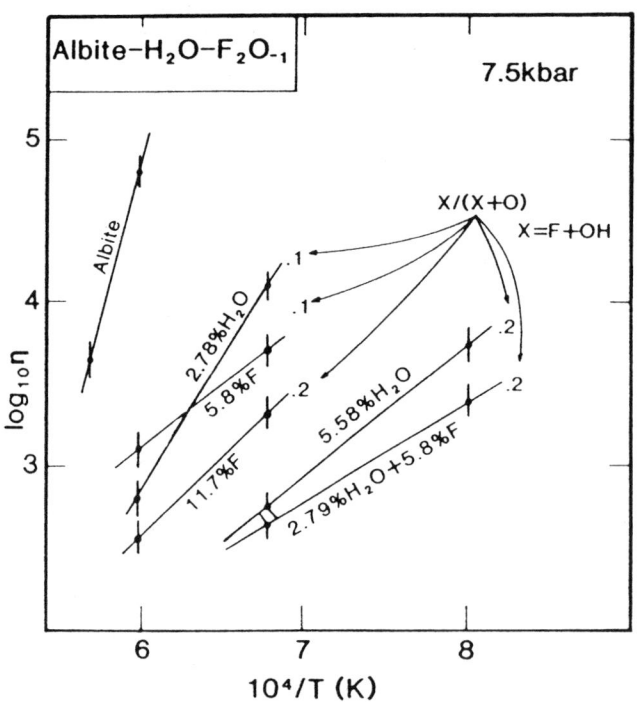

Figure 7. Viscosities of melts in the system $NaAlSi_3O_8$-H_2O-F_2O_{-1} with $(X/X+O) = 0.1$ and 0.2 plotted versus reciprocal temperature $(X = F,OH)$. Redrawn from Dingwell (1987).

at 1 atm pressure that was based on the literature data for synthetic melt compositions. Shaw [39] included water in his calculation model and with that allowed for the calculation of granitic melt viscosities at low pressures. No intrinsic pressure dependence of viscosity was included.

7. NON-NEWTONIAN MELT RHEOLOGY

The Newtonian viscosity of silicate melts is a low strain rate limiting case. With increasing strain rate the viscosity of silicate melts eventually becomes strain rate dependent as the inverse of the strain rate approaches the relaxation time of the silicate melt structure. This is the direct consequence of the relaxation mode due to structural relaxation in silicate melts. All available evidence points to a single relaxation mode being responsible for the onset of non-Newtonian

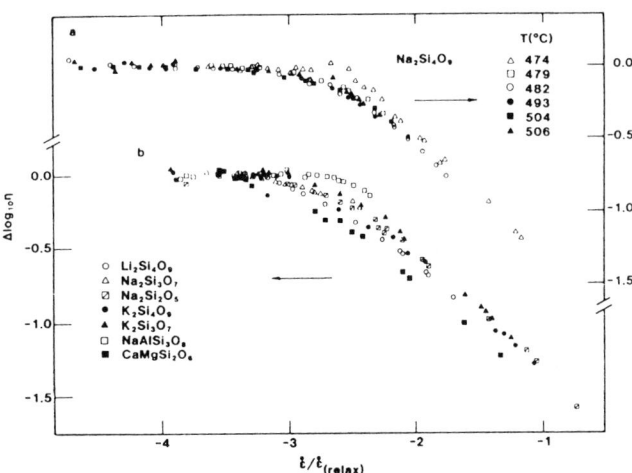

Figure 9. *(a)* A reduced plot of shear viscosity relative to Newtonian versus strain-rate normalized onto the relaxation strain-rate for $Na_2Si_4O_9$ at temperatures from 474 to 506°C. *(b)* A reduced plot of shear viscosity versus normalized strain-rate for a range of silicate melt compositions. These reduced plots based on Equation 4 remove the temperature- and composition-dependence of the onset of non-Newtonian viscosity of silicate melts. Redrawn from Webb and Dingwell (1990a).

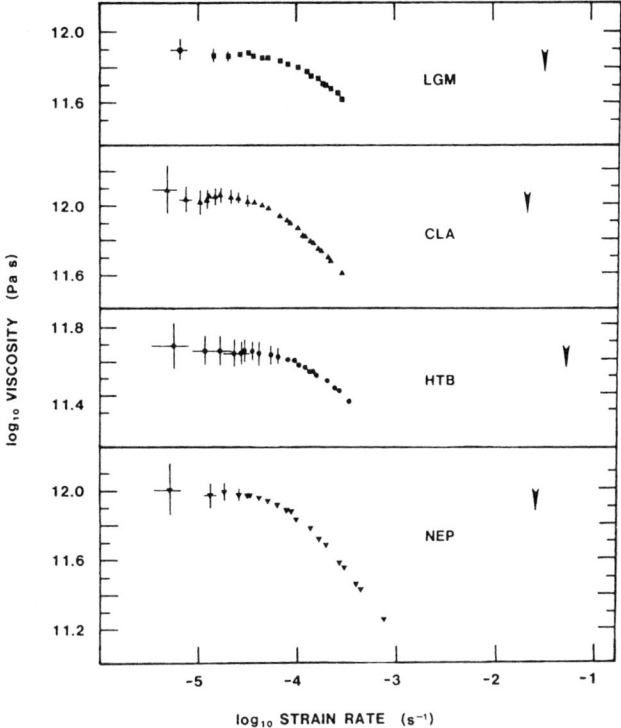

Figure 8. Shear viscosities of Little Glass Mountain rhyolite (LGM), Crater Lake andesite (CLA), Hawaiian tholeiite (HTB) and nephelinite (NEP) composition melts as a function of applied strain-rate. With increasing strain-rate the onset of non-Newtonian (strain-rate dependent) behavior is observed. The arrows indicate the calculated relaxation strain-rate for these melts. Redrawn from Webb and Dingwell (1990b).

rheology in silicate melts. The onset of non-Newtonian rheology has been extensively investigated [47, 48]. The onset of non-Newtonian rheology for four geological melts (a rhyolite, an andesite, a basalt and a nephelinite) is illustrated in Figure 8. The low-strain-rate limit of Newtonian viscosity is seen below strain rates of 10^{-4} whereas shear thinning (a decrease in viscosity with increasing shear rate) sets in at higher strain rate. The onset of non-Newtonian rheology occurs approximately 3 log_{10} units of time "above" the relaxation time of each silicate melt approximated from the Maxwell relation. A normalization of the data for several compositions and at several temperatures is illustrated in Figure 9 where the viscosities are normalized to the low-strain-rate limiting value of Newtonian viscosity and the strain rate of the experiment is normalized to the relaxation frequency (the inverse of the relaxation time), calculated from the Maxwell relation for each melt. The composition-dependence of the onset of non-Newtonian rheology can be removed in large part through this normalization. The success of this normalization indicates that the onset of non-Newtonian melt rheology in silicate melts can be easily estimated using values for the Newtonian viscosity and the shear modulus.

REFERENCES

1. Angell, C.A., Strong and fragile liquids, in *Relaxations in Complex Systems* edited by K.L. Ngai and G.B. Wright. Naval Research Lab., Arlington, VA, 3-11, 345pp., 1984.

2. Bagdassarov, N.S. and D.B. Dingwell, Rheological investigation of vesicular rhyolite, *J. Volc. Geotherm. Res.*, 50, 307-322, 1992.

3. Bottinga, Y. and D.F. Weill, The viscosity of magmatic silicate liquids: a model for calculation, *Am. J. Sci.*, 277, 438-475, 1972.

4. Brearley, M. and A. Montana, The effect of CO_2 on the viscosities of silicate liquids at high pressures, *Geochim. Cosmochim. Acta*, 53, 2609-2616, 1989.

5. Brearley, M., J.E. Dickinson Jr and C.M. Scarfe, Pressure dependence of melt viscosities on the join diopside-albite, *Geochim. Cosmochim. Acta*, 50, 2563-2570, 1986.

6. Brückner, R. and G. Demharter, Systematische Untersuchung über die Anwendbarkeit von Penetrationsviskosimetern,*Glastechn. Ber.* , 48, 12-18, 1975.

7. Dingwell, D.B., Melt viscosities in the system $NaAlSi_3O_8-H_2O-F_2O_{-1}$, *Geochemical Society Special Publication*, 1, 423-433, 1987.

8. Dingwell, D.B., R. Knoche , S. L. Webb, and M. Pichavant, The effect of B_2O_3 on the viscosity of haplogranitic liquids, *Am Mineral*, 77, 467-471, 1992.

9. Dingwell, D.B. and S.L.Webb, Structural relaxation in silicate melts and non-Newtonian melt rheology in geological processes, *Phys. Chem. Mineral.*, 16, 508-516, 1989.

10. Dunn, T. and C.M. Scarfe, Variation in the chemical diffusivity of oxygen and viscosity of an andesite melt with pressure at constant temperature, *Chem. Geol.*, 54, 203-216, 1986.

11. Ferry, J.D., *Viscoelastic properties of polymers*, 641pp., John Wiley, New York, 1980.

12. Fontana, E.H., A versatile parallel-plate viscosimeter for glass viscosity measurements to 1000°C, *Amer. Cer. Soc. Bull.*, 49, 594-597, 1970.

13. Fujii, T. and I. Kushiro, Density, viscosity and compressibility of basaltic liquid at high pressures, *Carnegie Inst. Wash. Yearb.*, 76, 419-424, 1977.

14. Fulcher, G.S., Analysis of recent measurements of the viscosity of glass, I. *J. Am. Cer. Soc.* , 12, 339-355, 1925a.

15. Fulcher, G.S., Analysis of recent measurements of the viscosity of glass, II. *J. Am. Cer. Soc.* , 12, 789-794, 1925b.

16. Gent, A.N. Theory of the parallel plate viscosimeter, *Brit. J. Appl. Phys.*, 11, 85-88, 1960.

17. Herzfeld, K.F. and T.A. Litovitz, *Absorption and Dispersion of Ultrasonic Waves*. Academic Press, 535pp., 1959.

18. Hessenkemper, H. and R. Brückner, Load-dependent relaxation behavior of various glass-melts with different structural configurations, *Glastech. Ber.*, 63, 1-6, 1989.

19. Kanzaki, M., K. Kurita, T. Fujii, T. Kato, O. Shimomura and S. Akimoto, A new technique to measure the viscosity and density of silicate melts at high pressure. in: *High-pressure research in Mineral Physics* edited by M.H. Manghnani and S. Syono , American Geophysical Union, Washington, DC, 195-200, 1987.

20. Kushiro, I., Changes in viscosity and structure of melt of $NaAlSi_2O_6$ composition at high pressures, *J.*
Geophy. Res., 81, 6347-6350, 1976.

21. Kushiro, I., Changes in viscosity with pressure of melts in the system $SiO_2-CaAl_2O_4$, Carnegie Inst. Wash. Yearb., 80, 339-341, 1981.

22. Macedo, P.B., J.H. Simmons and W. Haller, Spectrum of relaxation times and fluctuation theory: ultrasonic studies on alkali-borosilicate melt, *Phys. Chem. Glasses*, 9, 156-164, 1968.

23. Manghnani, M.H., C.S. Rai, K.W. Katahara and G.R. Olhoeft, Ultrasonic velocity and attenuation in basalt melt, *Anelasticity in the Earth* Geodynamics Series, Volume 4, American Geophysical Union, Washington, DC, 118-122, 1981.

24 Maxwell, J.C., On the dynamical theory of gases, *Philos Trans R Soc London* Ser. A.157, 49-88, 1867.

25. Mazurin, O.V., Glass relaxation, *J. Non-Cryst. Sol.*, 87, 392-407, 1986.

26. Nowick, A.S. and B.S. Berry, *Anelastic relaxation in crystalline solids*. Academic Press, New York, 1972.

27. O'Connell, R.J. and B. Budiansky, Measures of dissipation in viscoelastic media, *Geophy. Res. Lett.*, 5, 5-8, 1978.

28. Pocklington, H.C., Rough measurement of high viscosities, *Proc. Cambridge Philos. Soc.*, 36, 507-508, 1940.

29. Richet, P., Viscosity and configurational entropy of silicate melts, *Geochim. Cosmochim. Acta*, 48, 471-483, 1984.

30. Riebling, E.F., An improved counterbalanced sphere viscometer for use to 1750°C, *Rev. Sci. Instr.*, 34, 568-572, 1963.

31. Riebling, E.F., Viscosities in the $B_2O_3-SiO_2$ system, *J. Am. Cer.*

Soc., 47, 478-483, 1964.

32. Riebling, E.F., Structure of sodium aluminosilicate melts containing at least 50 mole% SiO_2 at 1500°C, *J. Chem. Phys.,* 44, 2857-2865, 1966.

33. Rivers, M.L. and I.S.E. Carmichael, Ultrasonic studies of silicate melts, *J. Geophys. Res.,* 92, 9247-9270, 1987.

34. Ryan, M.P. and J.Y.K. Blevins, *The viscosity of synthetic and natural silicate melts and glasses at high temperatures and at 1 bar (10^5 Pa) pressure and at higher pressures,* U.S. Geol. Surv. Bull., 17864, 563pp., 1987.

35. Sato, H. and M.H. Manghnani, Ultrasonic measurements of Vp and Qp: relaxation spectrum of complex modulus of basalt melt, *Phys. Earth Planet. Int.,* 41, 18-33, 1985.

36. Scarfe, C.M., B.O. Mysen and D. Virgo, Pressure dependence of the viscosity of silicate melts, *Geochemical Society Special Publication,* 1, 159-167, 1987.

37. Scholze, H., *Glas: Natur, Struktur und Eigenschaften.* 407pp. Springer, Berlin, 1988.

38. Shaw, H.R., Obsidian-H_2O viscosities at 1000 and 2000 bars in the temperature range 700 to 900°C, *J. Geophy. Res.,* 68, 6337-6343, 1963.

39. Shaw, H.R., Viscosities of magmatic silicate liquids: an empirical method of prediction, *Am. J. Sci.,* 272, 870-889, 1972.

40. Shiraishi, Y., L. Granasy, Y. Waseda and E. Matsubara, Viscosity of glassy Na_2O-B_2O_3-SiO_2 system, *J. Non-Cryst. Solid.*, 95, 1031-1038, 1987.

41. Simmons, J.H., R. Ochoa and K.D. Simmons, Non-Newtonian viscous flow in soda-lime-silica glass at forming and annealing temperatures, *J. Non-Cryst. Solid.,* 105, 313-322, 1988.

42. Tamann, G. and W. Hesse, Die Abhängigkeit der Viskosität von der Temperatur bei untergekuhlten Flüssigkeiten, *Z. anorg. allg. Chem.,* 156, 245-257, 1926.

43. Tauke, J., T.A. Litovitz and P.B. Macedo, Viscous relaxation and non-Arrhenius behavior in B_2O_3, *J. Am. Cer. Soc.,* 51, 158-163, 1968.

44. Tobolsky, A.V. and R.B. Taylor,

Viscoelastic properties of a simple organic glass, *J. Phys. Chem.,* 67, 2439-2442, 1963.

45. Vogel, H., Das Temperaturabhängigkeitsgesetz der Viskosität von Flussigkeiten, *Physik Z.,* 22, 645-646, 1921.

46. Webb, S.L., Shear and volume relaxation in $Na_2Si_2O_5$, *Am. Mineral.,* 76, 451-456, 1991.

47. Webb, S.L. and D.B. Dingwell, The onset of non-Newtonian rheology of silicate melts, *Phys. Chem. Mineral.,* 17, 125-132, 1990a.

48. Webb, S.L. and D.B. Dingwell, Non-Newtonian rheology of igneous melts at high stresses and strain-rates: experimental results for rhyolite, andesite, basalt and nephelinite, *J Geophys. Res.,* 95, 15695-15701, 1990b.

49. White, B. and A. Montana, The effect of H_2O and CO_2 on the viscosity of sanidine liquid at high pressures, *J. Geophy. Res.,* 95, 15683-15693, 1990.

50. Zarzicki, J., *Glasses and the vitreous state,* Cambridge University Press, Cambridge, 505pp., 1991.

Viscosity of the Outer Core

R. A. Secco

Estimates of outer core viscosity span 14 orders of magnitude. This wide range of values may be partially explained by the difference in type of viscosity, molecular viscosity (a rheological property of the material) vis-à-vis a modified or eddy viscosity (a property of the motion), inferred from the various observational and theoretical methods [24]. The motion associated with eddy viscosity implies the possibility of non-viscous dissipative mechanisms such as ohmic dissipation. Molecular viscosity is separated into nearly equal components of shear viscosity, η_s, and bulk or volume viscosity, η_v, depending on the type of strain involved [14,15]. η_s is a measure of resistance to isochoric flow in a shear field whereas η_v is a measure of resistance to volumetric flow in a 3-dimensional compressional field. In cases where outer core viscosity estimates are based on observations of the attenuation of longitudinal waves, both η_s and η_v play significant roles but only η_s is important for damping whole Earth torsional mode oscillation and η_v for damping radial mode oscillation [4].

The majority of estimates of outer core viscosity is based on whole Earth geodetic and seismological observations. In terms of the observation times required by a particular method, studies of p-wave attenuation benefit from their short periods (seconds) and all 7 studies cited in Table 2 give upper bound viscosity values within a very confined range of 10^8-10^9 poises. While long period (minutes to years) geodetic phenomena, such as the radial and torsional modes of free oscillation, length of day

variations, and polar motion can all be accurately measured, the outer core viscosity estimates inferred from these measurements, ranging from 10^{-1} to 10^{10} poises, suffer from long observation times which can introduce large uncertainties from additional energy sources and/or sinks affecting the observed phenomenon. It must be noted, however, that time independent factors that affect seismic wave amplitudes such as scattering, geometrical spreading, radial and lateral inhomogeneities will be embodied in derived attenuation values. Confirmation of any observation of inner core oscillation has not yet been made and therefore any so-derived viscosity estimate from geodetic observation or theory must be admitted with caution. Other methods of outer core viscosity estimation are from extrapolations launched from experimental data at low (relative to core) pressures and temperatures using various theories of liquid metals and from geomagnetic field observations at Earth's surface. Both of these suffer from large extrapolations and from the lack of any experimental viscosity data on liquid Fe at pressures of even a few kilobars. Most of the experimental data on liquid Fe and liquid Fe-Ni, Fe-S, Fe-O and Fe-Si alloys is found in the metallurgical literature and they are all restricted to measurements at a pressure of 1 atm.

The data and references are presented in 9 tables. Tables 1 and 2 are viscosity estimates based on geodetic and seismological studies, respectively. Tables 3 and 4 are viscosity estimates from geomagnetic and liquid metal theory studies, respectively. A brief description of the method used for each estimate or measurement is also given. Values of kinematic viscosity, ν, reported in some references have been converted to dynamic viscosity, η, by $\nu = \eta/\rho$, using a value of 10 g/cm^3 for ρ, the density. The data in Tables 1-4 are graphically presented in Figure 1. Table 5 contains experimental shear viscosity data for

R. A. Secco, Department of Geophysics, University of Western Ontario, London, Canada N6A 5B7

Mineral Physics and Crystallography
A Handbook of Physical Constants
AGU Reference Shelf 2

218

TABLE 1. Dynamic Viscosity Estimates of Outer Core from Geodetic Studies

Reference	Dynamic Viscosity (Poises)	Method[(o)]
Sato and Espinosa (1967)	$0.35 - 4.7 \times 10^{11}$	[o]Torsional free oscillations of whole Earth
Verhoogen (1974)	2.6×10^{-1}	[o]Chandler wobble
Yatskiv and Sasao (1975)	5×10^9	[o]Chandler wobble
Anderson (1980)	5×10^2	[o]Damping of free oscillation radial modes
Molodenskiy (1981)	$\leq 10^7$	[o]Forced nutation of the Earth (value for MCB)
Molodenskiy (1981)	$\leq 10^8$	[o]Tidal variations in the length of day
Molodenskiy (1981)	2×10^{10}	[o]Chandler wobble
Gwinn et al (1986)	$<5.4 \times 10^4$	[o]Retrograde annual Earth nutation - VLBI measurements (value for MCB)
Neuberg et al (1990)	$<3.3 \times 10^5$	[o]Viscous damping only for nearly diurnal free wobble (tidal measurements)
Smylie (1992)	7.7×10^8	[o]Damping of inner core translational modes (superconducting gravimeter data)
Bondi and Lyttleton (1948)	$<10^{12}$	[t]Theoretically required for secular deceleration of core by viscous coupling
Stewartson and Roberts (1963)	$<10^9$	[t]Theory of rotating fluids
Toomre (1966)	$>6 \times 10^5$	[t]Theoretically required for steady precession by core/mantle viscous coupling
Won and Kuo (1973)	$> 10^{-1}$	[t]Theoretical evaluation of decay time of inner core oscillation (value for ICB)
Toomre (1974)	$<10^6$	[t]18.6 year principal core nutation (value for MCB)
Aldridge and Lumb (1987)	2.9×10^7	[t]Decay of inertial waves in outer core

[t] theory
[o] observation

TABLE 2. Dynamic Viscosity Estimates of Outer Core From Seismological Studies

Reference	Dynamic Viscosity (Poises)	Method
Jeffreys (1959)	5×10^8	Attenuation of p-waves
Sato and Espinosa (1965, 1967)	8.6×10^{12}	Multiply reflected s-waves at mantle/core boundary.
Sacks (1970)	10^8	Attenuation of p-waves
Suzuki and Sato (1970)	$3\text{-}7 \times 10^{11}$	Attenuation of s-waves
Buchbinder (1971)	2×10^8	Attenuation of p-waves
Adams (1972)	4×10^8	Attenuation of p-waves
Qamar and Eisenberg (1974)	$1\text{-}2 \times 10^8$	Attenuation of p-waves
Zharkhov and Trubitsyn (1978)	$\ll 10^9$	Attenuation of p-waves
Anderson and Hart (1978)	1.4×10^9	Attenuation of body waves and checked against radial mode Q data

TABLE 3. Dynamic Viscosity Estimates of Outer Core from Geomagnetism Studies

Reference	Dynamic Viscosity (Poises)	Method
Bullard (1949)	10^{-2}	Magnetic damping of core fluid motions
Hide (1971)	10^7	Magnetohydrodynamic interactions between fluid motions and bumps on MCB
Schloessin and Jacobs (1980)	2×10^5	Decay of free dynamo action during polarity transitions
Officer (1986)	2×10^8	Value predicts correct order of magnitude of external field and westward drift

TABLE 4. Dynamic Viscosity Estimates of Outer Core from Theories of Liquid Metals

Reference	Dynamic Viscosity (Poises)	Method
Bullard (1949)	10^{-2}	From experimental values for liquid metals at STP
Miki (1952)	10^{-2} - 10^{-1}	Quantum statistical thermodynamics of liquid metals
Backus (1968)	5×10^{2}	From experimental values of liquid Hg at 10 kb and 400K
Gans (1972)	$3.7\text{-}18.5 \times 10^{-2}$	Andrade formula for melting point viscosity (value for ICB)
Leppaluoto (1972)	$1\text{-}5 \times 10^{-1}$	For a pure Fe outer core, from significant structure theory of liquids (value for MCB)
Bukowinski and Knopoff (1976)	$>10^{1}$	For a pure Fe outer core, from band structure calculations
Schloessin and Jacobs (1980)	3.4×10^{2}	From experimental values for liquid Fe and Andrade's pressure effect on viscosity
Anderson (1980)	10^{1} - 10^{4}	Two structure state theory extrapolated to core pressures
Poirier (1988)	3×10^{-2}	Thermodynamic scaling relation between melting temperature and viscosity and diffusivity of metals
Svendsen et al (1989)	2.5×10^{-2}	Liquid state model fitted to high pressure melting data on Fe

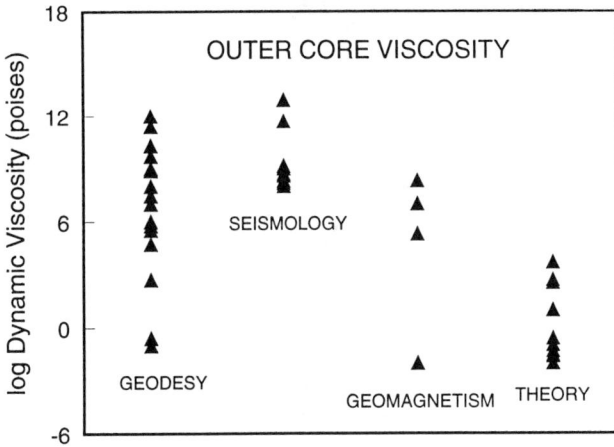

pure liquid Fe and Tables 6-9 contain experimental shear viscosity data for liquid Fe alloys, all measured by the oscillating crucible method. The viscosities in Tables 6-9 are presented as either dynamic or kinematic viscosities, as in the reference, because of insufficient density data for many of the alloy compositions.

Fig.1. The common logarithm of dynamic viscosity (poises) of the outer core plotted as a function of method used for its determination.

TABLE 5. Experimental Data of Shear Dynamic Viscosity (centipoises) of Liquid Fe at 1 atm

Reference	T(°C) 1536	1550	1600	1650	1700	1750	1800	1850
Barfield and Kitchener (1955)	7.60[e]	6.79[i]	6.41	5.89[i]	5.70	5.48[i]	5.31	5.22
Thiele (1958)	4.7	4.6	4.2	3.9	3.7	3.4		
Hoffman (1962)	5.42	5.30	4.90	4.55	3.98			
Cavalier (1963)	4.95	4.87	4.54	4.30	4.10	3.92		
Lucas (1964)	5.03	4.93	4.58	4.28	4.00	3.76		
Vostryakov et al (1964)		5.91						
Nakanishi et al (1967)	5.54[e]	5.44[i]	5.01[i]	4.69[i]	4.44[e]			
Kaplun et al (1974)		5.60	5.01					
Arkharov et al (1978)				5.96				
Steinberg et al (1981)			5.03					

[i] interpolated
[e] extrapolated

TABLE 6. Experimental Data of Shear Viscosity of Liquid Fe-Ni Alloys at 1 atm

Reference	Composition wt% Ni	Temperature (°C)	Viscosity Dynamic (centipoises)	Kinematic (millistokes)
Adachi et al (1973)	4.9	1516	6.11	
		1550	5.91	
		1604	5.69	
		1633	5.54	
		1665	5.36	
	9.7	1502	6.02	
		1513	5.95	
		1550	5.72	
		1596	5.43	
		1652	5.04	
		1693	4.81	

TABLE 6 (continued)

Reference	Composition wt% Ni	Temperature (°C)	Viscosity	
			Dynamic (centipoises)	Kinematic (millistokes)
	28.6	1471	6.15	
		1519	5.62	
		1545	5.37	
		1581	5.11	
		1615	4.91	
Arkharov et al (1978)	0.52	1600		8.48
	1.04			8.28
	1.49			8.15
	1.99			8.13
	2.46			7.98

TABLE 7. Experimental Data of Shear Viscosity of Liquid Fe-S Alloys at 1 atm

Reference	Composition wt% S	Temperature (°C)	Viscosity	
			Dynamic (centipoises)	Kinematic (millistokes)
Barfield and Kitchener (1955)	1.16 (+0.02 wt%C)	1529	7.00	
		1600	6.45	
		1700	5.84	
		1800	5.41	
Vostryakov et al (1964)	0.39	1600	6.00	
	4.70			6.61
	8.51			6.41
	18.56			3.43

TABLE 8. Experimental Data of Shear Dynamic Viscosity of Liquid Fe-O Alloys at 1 atm

Reference	Composition wt% O	Temperature (°C)	Dynamic Viscosity (centipoises)
Nakanishi et al (1967)	0.012	1600	5.02
	0.046		5.43
	0.071		5.40
	0.072		5.39

TABLE 9. Experimental Data of Shear Viscosity of Liquid Fe-Si Alloys at 1 atm

Reference	Composition wt% Si	Temperature (°C)	Viscosity Dynamic (centipoises)	Kinematic (millistokes)
Romanov and Kochegarov (1964)	0.1	1540		9.62
		1580		8.72
		1621		7.78
		1668		7.55
		1743		7.04
		1782		6.66
		1806		6.61
	0.6	1549		7.70
		1605		6.92
		1642		6.83
		1707		6.36
		1756		6.15
	2.0	1508		7.39
		1553		6.90
		1592		6.46
		1654		5.85
		1712		5.69
		1758		5.49
	5.0	1454		8.14
		1508		6.78
		1575		6.02
		1627		5.61
		1677		5.40
		1716		5.37

TABLE 9 (continued)

Reference	Composition wt% Si	Temperature (°C)	Viscosity Dynamic (centipoises)	Kinematic (millistokes)
Nakanishi et al (1967)	0.9	1615	4.13	
	2.9		3.54	
Kaplun et al (1979)	1.0	1550	5.41	
	2.0	1550	5.29	
		1600	4.89	
	3.0	1500	5.74	
		1550	5.18	
		1600	4.84	
	4.5	1500	5.51	
		1550	5.05	
		1600	4.72	
	6.0	1500	5.04	
		1550	4.65	
		1600	4.35	

Acknowledgements. I gratefully acknowledge helpful discussions with H.H. Schloessin and L. Mansinha, comments in reviews by D.L. Anderson and an anonymous reviewer, as well as financial support by the Natural Sciences and Engineering Research Council of Canada.

REFERENCES

1. Adachi, A., Morita, Z., Ogina, Y. and Ueda, M. The Viscosity of Molten Fe-Ni Alloys, in: *The Properties of Liquid Metals*, pp.561-566, ed. S. Takeuchi, Taylor and Francis, 1973.

2. Adams, R.D. Multiple Inner Core Reflections from a Novaya Zemlya Explosion, *Bull.Seis.Soc.Amer. 62*, 1063-1071, 1972.

3. Aldridge, K.D. and Lumb, L.I. Inertial Waves Identified in the Earth's Fluid Outer Core, *Nature 325,* 421-423, 1987.

4. Anderson, D.L. Bulk Attenuation in the Earth and Viscosity of the Core, *Nature 285,* 204-207, 1980.

5. Anderson, D.L. and Hart, R.S., Q of the Earth, *J.Geophys.Res. 83,* 5869-5882, 1978.

6. Arkharov, V.I., Kisun'ko, V.Z., Novokhatskii, I.A. and Khalierskii, V.P. Effect of Various Impurities on the Viscosity of Molten Iron, *High Temp. 15,* 1033-1036, 1978. Translation of: *Teplofiz.Vys.Temp.* (in Russian), *15,* 1208, 1977.

7. Backus, G.E., Kinematics of Geomagnetic Secular Variation in a Perfectly Conducting Core, *Phil.Trans.Roy.Soc.Lond., A-263,* 239-266, 1968.

8. Barfield, R.N. and Kitchener, J.A. The Viscosity of Liquid Iron and Iron-Carbon Alloys, *J.Iron Steel Inst. 180,* 324-329, 1955.

9. Bondi, H. and Lyttleton, R.A. On the Dynamical Theory of the Rotation of the Earth, *Proc.Camb.Phil.Soc. 44,* 345-359, 1948.

10. Buchbinder, G.G.R. A Velocity Structure of the Earth's Core, *Bull.Seis.Soc.Amer. 61,* 429-456, 1971.

11. Bukowinski, M.S.T. and Knopoff, L. Electronic Structure of Iron and Models of the Earth's Core, *Geophys.Res.Letts. 3,* 45-48, 1976.

12. Bullard, E.C. The Magnetic Field Within the Earth, *Proc. R. Soc. Lond., A197,* 433-453, 1949.

13. Cavalier, G. Mesure de la Viscosite du Fer, du Cobalt et du Nickel, *C.R.Acad.Sci. 256,* 1308-1311, 1963.

14. Dingwell, D.B. and Webb, S.L. Structural Relaxation in Silicate Melts and non-Newtonian Melt Rheology in Geologic Processes, *Phys.Chem.Minerals 16,* 508-516, 1989.

15. Flinn, J.M., Jarzynski, J. and Litovitz, T.A., Mechanisms of Volume Viscosity in

Molten Bismuth and Lead, *J.Chem.Phys.* *54*,(10), 4331-4340, 1971.

16. Gans, R.F. Viscosity of the Earth's Core, *J.Geophys.Res. 77*, 360-366, 1972.

17. Gwinn, C.R., Herring, T.A. and Shapiro, I.I. Geodesy by Radio Interferometry: Studies of the Forced Nutations of the Earth 2. Interpretation. *J.Geophys.Res.91*, 4755-4765, 1986.

18. Hide, R. Viscosity of the Earth's Core, *Nature Phys.Sc. 233*, 100-101, 1971.

19. Hoffmann, K. These, Saarbrucken, juillet, 1962 (cited in Lucas, 1964).

20. Jeffreys, Sir H., *The Earth, Its Origin, History and Physical Constitution*, p.255, Cambridge Univ.Press, 4th ed., 1959.

21. Kaplun, A.B. and Krut'ko, M.F. Viscosity of Fe-Si Melts with Small Silicon Contents, *Russian Metall. 3*, 71-73, 1979.

22. Leppaluoto, D. Thesis, Univ.California, Berkeley, 1972 (cited in Verhoogen, 19-74).

23. Lucas, L-D. Viscosite du Fer Pur et du Systeme Fe-C Jusqu'a 4,8% C en poids. *C.R.Acad.Sci.Paris 259*, 3760-3767, 1964.

24. Lumb, L.I. and Aldridge, K.D. On Viscosity Estimates for the Earth's Fluid Outer Core and Core-Mantle Coupling, *J.Geomag.Geoelectr. 43*, 93-110, 1991.

25. Miki, H. Physical State of the Earth's Core, *J.Phys.Earth 1*, 67, 1952.

26. Molodenskiy, S.M. Upper Viscosity Boundary of the Earth's Core, *Izv.Earth Phys. 17*, 903-909, 1981.

27. Nakanishi, K., Saito, T. and Shiraishi, Y. On the Viscosity of Molten Iron and Its Dilute Binary Alloys of Aluminum, Silicon and Oxygen, *Jap.Inst.Metals.J. 37*(7), 881-887, 1967, (in Japanese).

28. Neuberg, J., Hinderer, J. and Zurn, W. On the Complex Eigen-frequency of the "Nearly Diurnal Free Wobble" and Its Geophysical Interpretation, in: *Variations in Earth Rotation*, eds.: D.D. McCarthy and W.E. Carter, 11-16, Geophysical Monograph 59, Amer.Geophys.Union, Washington, D.C., 1990.

29. Officer, C.B. A Conceptual Model of Core Dynamics and the Earth's Magnetic Field, *J.Geophys. 59*, 89-97, 1986.

30. Poirier, J.P. Transport Properties of Liquid Metals and Viscosity of the Earth's Core, *Geophys.J. 92*, 99-105, 1988.

31. Qamar, A. and Eisenberg, A. The Damping of Core Waves, *J.Geophys Res. 79*, 758-765, 1974.

32. Romanov, A.A. and Kochegarov, V.G. Viscosity of Binary Iron-Silicon Melts in the Low Concentration Range of the Second Component, *Fiz.Metal.Metalloved. 17*(2), 300-303, 1964.

33. Sacks, I.S. Anelasticity of the Outer Core, *Carnegie Inst.Wash.Yrbk. 69*, 414-416, 1970.

34. Sato, R. and Espinosa, A.F. Dissipation in the Earth's Mantle and Rigidity and Viscosity in the Earth's Core Determined from Waves Multiply Reflected from the Mantle-Core Boundary, *Trans. Amer. Geophys. Union, 46*, 158, 1965 (also *Bull. Seis. Soc. Amer. 57*, 829-857, 1967).

35. Sato, R. and Espinosa, A.F. Dissipation Factor of the Torsional Mode $_0T_2$ for a Homogeneous-Mantle Earth with a Soft-Solid or a Viscous-Liquid Core, *J. Geophys.Res., 72*, 1761-1767, 1967.

36. Schloessin, H.H. and Jacobs, J.A. Dynamics of a Fluid Core with Inward Growing Boundaries, *Can.J.Earth Sci. 17*, 72-89, 1980.

37. Smylie, D.E. The Inner Core Transla-tional Triplet and the Density Near Earth's Center, *Science, 255*, 1678-1682, 1992.

38. Steinberg, J., Tyagi, S. and Lord, Jr., A.E. The Viscosity of Molten $Fe_{40}Ni_{40}P_{14}B_6$ and $Pd_{82}Si_{18}$, *Acta Metall. 29*, 1309-1319, 1981.

39. Stewartson, K. and Roberts, P.H. On the Motion of a Liquid in a Spheroidal Cavity of a Precessing Rigid Body, *J.Fluid Mech. 17*, 1-20, 1963.

40. Suzuki, Y. and Sato, R. Viscosity Determination in the Earth's Outer Core from ScS and SKS Phases, *J.Phys.Earth, 18*, 157-170, 1970.

41. Svendsen, B., Anderson, W.W., Ahrens, T.J. and Bass, J.D. Ideal Fe-FeS, Fe-FeO Phase Relations and Earth's Core, *Phys.Earth Planet.Int., 55*, 154-186, 1989.

42. Thiele, M. These, Berlin, juillet, 1958 (cited in Lucas, 1964).

43. Toomre, A. On the Coupling of the Earth's Core and Mantle during the 26,000 Year Precession, in: *The Earth-Moon System*, ed.: B.G. Marsden and A.G.W. Cameron, Plenum Press, New York, 33-45, 1966.

44. Toomre, A. On the "Nearly Diurnal Wobble" of the Earth, *Geophys.J.R.Astr. Soc., 38*, 335-348, 1974.

45. Verhoogen, J. Chandler Wobble and Viscosity in the Earth's Core, *Nature, 249*, 334-335, 1974.

46. Vostryakov, A.A., Vatolin, N.A. and Yesin, O.A. Viscosity and Electrical Resistivity of Molten Alloys of Iron with Phosphorus and Sulphur, *Fiz.Metal. Metalloved. 18*(3), 476-477, 1964.

47. Won, I.J. and Kuo, J.T. Oscillation of the Earth's Inner Core and Its Relation to the Generation of Geomagnetic Field, *J. Geophys.Res., 78*, 905-911, 1973.

48. Yatskiv, Ya.S. and Sasao, T. Chandler Wobble and Viscosity in the Earth's Core, *Nature, 255*, 655, 1975.

49. Zharkov, V.N. and Trubitsyn, V.P. *Physics of Planetary Interiors*, p.59, translated and edited by W.B. Hubbard, Pachart Pub.House, Tucson, 388 pp., 1978.

Models of Mantle Viscosity

Scott D. King

1. INTRODUCTION

The viscosity of the mantle is one of the most important, and least understood material properties of the Earth. Plate velocities, deep-earthquake source mechanisms, the stress distribution in subduction zones, and estimates of geochemical mixing time scales are all strongly affected by the pattern of convective flow which, in turn, is strongly influenced by the viscosity structure of the mantle. There are two approaches to understanding the viscosity structure of the Earth: using observations such as the geoid and post-glacial uplift, combined with flow models; or studying the physical deformation properties of mantle minerals in the laboratory. Both approaches have advantages and drawbacks.

Laboratory measurements of deformation indicate that the rheology of upper mantle minerals such as olivine $((Mg,Fe)_2SiO_4)$ is a strong function of temperature, grain size and stress [e.g., 3, 24, 43, 45, 46, 61, 65]. The deformation of minerals under mantle conditions generally follows a flow law of the form

$$\dot{\varepsilon} = A \left(\frac{\sigma}{\mu} \right)^n d^{-m} \exp\left(-\frac{Q}{RT} \right) \qquad (1)$$

where $\dot{\varepsilon}$ is the deformation rate, σ is the deviatoric stress, μ is the shear modulus, d is the grain size of the rock, Q is the activation energy for the deformation mechanism, T is the temperature in Kelvins, R is the gas constant and A is a

S. D. King, Purdue University, Department of Earth and Atmospheric Sciences, West Lafayette, IN 47907-1397

Mineral Physics and Crystallography
A Handbook of Physical Constants
AGU Reference Shelf 2

constant [c.f., 3, 43]. Viscosity is defined as

$$\eta = \frac{\sigma}{2\dot{\varepsilon}} \qquad (2)$$

therefore, deformation is directly related to viscosity. (Note the factor of 2 difference in equation (2) compared with the definition used by experimentalists.) For temperature changes of 100 degrees K, the viscosity changes by an order of magnitude at constant stress [c.f., 44]. Changes of deviatoric stress by a factor of 2 change the viscosity by an order of magnitude [c.f., 44]. Other factors, such as partial pressure of oxygen and water may also have important effects.

Two creep mechanisms are likely to dominate in the mantle; diffusional flow (corresponding to n=1 in equation 1) and power-law creep (corresponding to n>1 in equation 1). A rheology with a linear stress strain-rate creep mechanism, such as diffusional flow, is referred to as a Newtonian rheology. The question of which mechanism dominates in the mantle depends on the average grain size of the mantle minerals [46]. In the upper mantle, with grain sizes greater than 1mm, power-law creep should dominate at stresses greater than 1 MPa; otherwise diffusion creep dominates [46]. A deformation map (Figure 1) shows the predicted dominant deformation mechanisms for olivine with grain size 0.1mm as a function of stress and depth. It should be noted that the strain rates achieved in the lab (typically 10^{-5} - 10^{-8} s^{-1}) are much larger than those predicted in the lithosphere and mantle ($\sim 10^{-14}$ s^{-1}). While the laboratory measurements are clearly in the power-law creep field, typical mantle strain-rates lie close to the diffusional flow field at this small grain size.

The deformation of the major high pressure mantle phases perovskite $((Mg,Fe)SiO_3)$ and spinel

227

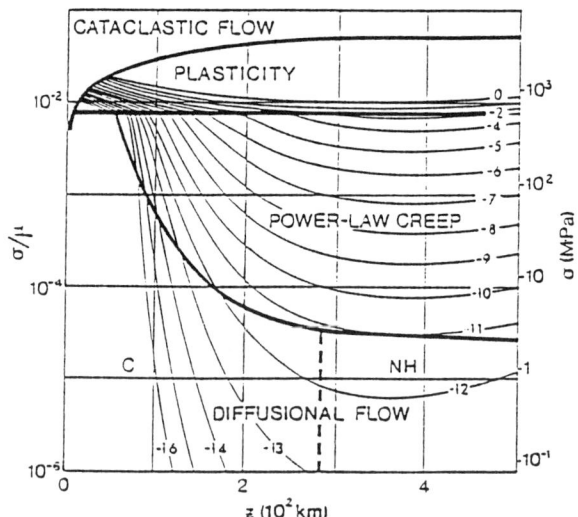

Fig. 1. (σ-z)-deformation map for polycrystalline olivine with grain size 0.1 mm. Thick lines are creep field boundaries; thin lines, constant strain rate contours (given as powers of 10). C and NH denote Coble and Nabarro-Herring creep, respectively [3].

$((Mg,Fe)_2SiO_4)$ can be studied only by analog minerals. Using high temperature creep experiments on a $CaTiO_3$ perovskite analog, Karato and Li [22] suggest the possibility of a weak zone at the top of the lower mantle (due to the grain size reduction from the spinel to perovskite phase change). Karato [21] showed that olivine, spinel and perovskite have similar normalized flow stresses, which suggests that, due to the effect of pressure, the lower mantle should have a higher viscosity than the upper mantle. Since creep parameters can differ greatly even for apparently-similar perovskites [44; Table 5], we should apply analog creep measurements to the mantle with great care.

Because of the difficulties in interpreting and applying laboratory creep measurements to mantle conditions, models of mantle viscosity based on large-scale geophysical observations continue to be important to geophysics. The post-glacial uplift problem, as described by Haskell [19], is a simple illustration. If a load is placed on the surface of a viscous fluid and allowed to deform the surface to establish hydrostatic equilibrium, the rate at which the surface deforms will depend on the viscosity of the fluid. Similarly, if the load is then removed, the rate of return is also dependent on viscosity. Viscosity models deduced from these observations are not unique, however, and require knowledge of models for the surface load (i.e., ice sheet thickness and history), which are uncertain. Also, the theoretical models are often simplified to keep

them tractable; commonly, a linear rheology (i.e. n=1 and T constant in equation (1)), which varies only with depth, is assumed.

The classic studies of post-glacial rebound illustrate the non-uniqueness of viscosity models derived from observations and flow models. Haskell [19, 20] proposed that the uplift of Fennoscandia was the result of deep flow, modeled as a half space with a uniform viscosity of 10^{21} Pa s (pascal seconds are the MKS units of dynamic viscosity; 1 Pa s = 10 poise), while van Bemmelem and Berlage [64 - see 5] proposed that the uplift of central Fennoscandia could be attributed to flow in a 100 km thick channel, with 1.3×10^{19} Pa s viscosity, overlying an effectively rigid mantle. Haskell also showed that the viscosity did not change over the interval of time of the analysis, supporting the notation of a Newtonian, rather than stress-dependent mantle.

The effects of an elastic lithosphere were first discussed by Daly [7], who appealed to the strength of the lithosphere to avoid the formation of a bulge of material squeezed out of the low viscosity channel peripheral to the ice load in his model. McConnell [28] and Cathles [4, 5] showed the strength of the elastic lithosphere as important only in considering the short wavelength harmonics. O'Connell [33] determined the viscosity of the lower mantle by looking at changes in the ice load and relating them to long wavelength rebound. He also used spherical harmonic correlation to look for the rebound signal in the geoid. The effect of phase changes in the mantle on rebound was considered by O'Connell [34]. O'Connell concluded that the effect of both the olivine-spinel and the basalt-eclogite phase change on post-glacial are rebound negligible. Peltier [37] solved for the response of a viscoelastic (Maxwell) spherical Earth using the correspondence principle, later adding the effects of varying sea level [42]. The correspondence principle asserts that one can construct the Laplace transform of the solution by solving a series of elastic problems over a range of complex frequencies and then inverting to get the time domain response for the viscoelastic problem. Cathles [4, 5] presented an alternative viscoelastic formulation which avoids the complexities of the boundary conditions for long time periods when using the correspondence principle. Cathles argues for an increase in viscosity in the lower mantle, while Peltier argues for a uniform viscosity mantle. Because no direct comparison of the two methods has been reported, it is difficult to assess whether the differences between Peltier's and Cathles' conclusions are due to their methods or ice models. An ice sheet disintegration model, including ice masses of Laurentia, Fennoscandia, and Greenland, called

ICE-1, was developed by Peltier and Andrews [40]. A summary of ice sheet models is provided by Peltier [39].

Lambeck and co-workers have used far field absolute sea level changes, rather than the rebound histories at sites near formerly-glaciated regions [26, 32]. They reason that relative sea level variations far from ice margins are less influenced by poorly-constrained ice models. The difference between Nakada and Lambeck's models based on relative sea level variations in the Pacific, and models based on Fennoscandian and Canadian Uplift could be interpreted as reflecting lateral variations in upper mantle viscosity.

Table 1 compares a number of recently published viscosity models from post-glacial rebound with geoid and plate velocity studies compiled by Hager [14]. Because of the limited power in the post-glacial rebound data set, rebound models are usually reported in very simple parameterizations. The models still range from the nearly-uniform viscosity (Haskallian) model PT to strongly layered models NL and HRPA (see Table 1).

Another constraint on mantle viscosity comes from modeling the geoid and dynamic topography of the surface and core mantle boundary, using the pattern of density anomalies inferred from seismic tomography [11, 16, 17, 18, 23, 49, 50, 51]. From early seismic tomography, it has been observed that long-wavelength (i.e., spherical harmonic degree $l = 2,3$) geoid lows are associated with long-wavelength, fast (presumably denser) regions in the lower mantle [9]. In a static Earth, this is opposite of what one predicts; geoid highs should correspond to mass excesses (Figure 2a). As shown by Pekris [36] (and others since), a mass anomaly will drive flow that deforms the surface and core mantle boundary (Figure 2b or 2c). The resulting geoid is a combination of both internal (Figure 2a) and boundary mass anomalies (Figure 2b or 2c) and can be positive or negative, depending on the viscosity structure (compare Figure 2a + 2b versus 2a + 2c).

The equations of motion of an incompressible self-gravitating spherical shell are presented in Richards and Hager [54] and Ricard et al. [50]. These equations can be solved by a separation of variables (assuming a radially-stratified viscosity). The resulting ODE's can be solved for response functions (kernels) which depend only on the viscosity structure. The geoid (δV^{lm}) can then be calculated by convolving the response functions with a distribution of density contrasts as follows

$$\delta V^{lm} = \frac{4\pi\gamma a}{2l+1} \int_c^a G^1(r,\eta(r))\delta\rho^{lm}(r)dr \qquad (3)$$

where γ is the gravitational constant, a is the radius at the

surface, c is the radius at the core, $\delta\rho^{lm}$ is the density contrast at a depth r of spherical harmonic degree l and order m, and $G^1(r,\eta(r))$ is the geoid response kernel.

The density perturbations ($\delta\rho^{lm}$) are determined by seismic velocity perturbation models from seismic tomographic inversions and/or tectonic plate and slab models from boundary layer theory and deep earthquake locations in subduction zones. To transform seismic velocity anomalies into density anomalies, we assume that changes in seismic velocity can be mapped into changes in temperature and do not represent changes in composition. Seismic velocity variations can be written in terms of elastic moduli for which the limited experimental information places reasonable bounds. Thus, the density perturbations in equation 3 can be written as

$$\delta\rho^{lm} = \frac{\delta\rho}{\delta v_s} \delta v_s^{lm} \qquad (4)$$

where $\delta\rho/\delta v_s$ is a velocity to density ratio and δv_s^{lm} is the seismic velocity perturbation model.

A simple, two-layer viscosity model, with an increase in viscosity of a factor of ten at 670 km depth or 1200 km depth, explains the longest wavelength geoid from the inferred densities in the lower mantle [11, 18]. This model is different from the two end-member, post-glacial rebound models. The differences between the models from post glacial rebound and the geoid models have led to spirited debates, but until now no consensus model.

The velocities of the Earth's plates are the surface manifestations of convective flow in the mantle. Hager and O'Connell's model [15] showed that densities from the cooling of ocean plates and subducting slabs alone provide the necessary buoyancy force to drive plates at the observed velocities. Ricard et al. [50, 51], Ricard and Vigny [47], Forte and Peltier [10], and Forte et al. [12] used observed plate velocities to deduce the radial viscosity structure of the mantle. Plate velocities constrain the absolute value of the viscosity of the mantle; the geoid does not. The plate velocity data does not have the depth resolution of the geoid, because a low viscosity zone can effectively decouple the plates from flow in the deep interior.

In all of the studies discussed, the final viscosity model is dependent upon another model. In post-glacial rebound studies, this is a model of the ice sheet, which is only crudely known. In the case of the geoid, it is the seismic velocity models (see equations 3 and 4). Most seismic tomographic inversions do not report formal uncertainties (which are difficult to perform and, because of the difficulty in defining sources of error may have little

TABLE 1. A Comparison of Recently Published Viscosity Structures
Determined by Systematic Forward Modeling (from [14])

Model	h (km)	η_{um} (Pa s)	η_{lm} (Pa s)	η_{lm}/η_{um}
PT[a]	120	10^{21}	2×10^{21}	2
MP[b]	120	10^{21}	4.5×10^{21}	4
LJN2[c]	100	3.5×10^{20}	4.7×10^{21}	15
LJN3[d]	150	3.8×10^{20}	3.4×10^{21}	8
LNA[e]	75	2×10^{20}	7.5×10^{21}	40
NLO[f]	50	10^{20}	10^{22}	100
RVGP[g]	(100)	2.6×10^{20}	1.3×10^{22}	50
HGPA[h]	(100)	2×10^{19}	6×10^{21}	300
HS[i]	(100)	2×10^{20}	6×10^{21}	30

[a]Peltier and Tushingham [41] model, based on global sea level variations, with emphasis on near-field sites.

[b]Mitrovica and Peltier [29] model, based on the assumption that the gravity anomaly over Hudson Bay is totally due to delayed rebound.

[c]Lambeck, Johnston and Nakada [26] model 2, based on European sea level variations with emphasis on relative variations in sites away from the ice margins.

[d]Lambeck, Johnston and Nakada [26] model 3, an alternative to model 2.

[e]Lambeck and Nakada [25] model for Australia, based on sea level variations with emphasis on relative variations in sites spanning the continental margin.

[f]Nakada and Lambeck [32] model for Oceanic response, based on sea level variations, with emphasis on relative variations as a function of island size.

[g]Ricard and Vigny [47] model from the Geoid and Plate velocities. Parentheses on (h) indicate the thickness of the high viscosity lid.

[h]Hager and Richards [16] model for relative mantle viscosity from the Geoid, calibrated for Plate velocities and Advected heat flux [14]. An additional layer, from 400 km to 670 km depth has a viscosity of 6×10^{20} Pa s.

[i]A modification of model HRGP that has an asthenospheric viscosity higher by a factor of 10, as might be expected for Shield regions.

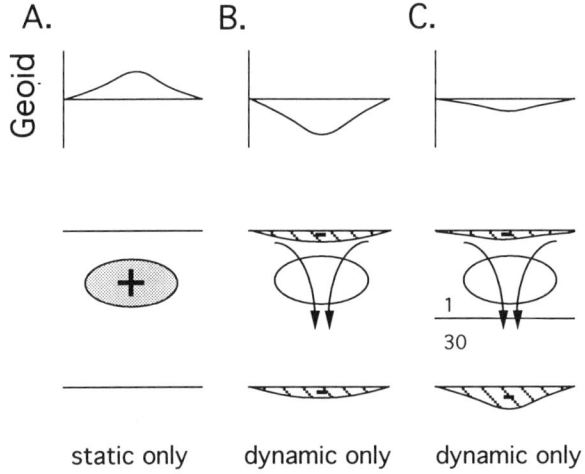

A. B. C.

Geoid

static only   dynamic only   dynamic only

Fig. 2. (a) The geoid anomaly over a positive mass anomaly (stippled) in a static earth. (b) The dynamic flow driven by the mass anomaly causes negative mass anomalies (stripped regions) at the upper and lower boundaries, hence negative geoid anomalies. The sum of (a) and (b) is a negative anomaly in a uniform viscosity medium. (c) With an increased viscosity in the lower layer, the dynamic topography of the surface is reduced and the sum of (a) plus (c) could be positive.

meaning). The use of different seismic models can produce different viscosity models [23, 51], suggesting that crude viscosity models may already be pushing the limit of the observations. Plate velocity models also have associated uncertainties; in addition, only the poloidal part of the plate velocities are driven by viscous flow without lateral variations in viscosity. Because they are ignored in the viscous flow formulations, lateral variations in viscosity themselves could also introduce a major source of error in the viscosity models.

Studies of Earth's rotation provide constraints on mantle viscosity. O'Connell [33] suggests that changing patterns of convection would change the principle moments of inertia of the Earth and that his viscosity model would permit polar wander from convection. Sabadini and Peltier [57] and Wu and Peltier [67] discuss the changes in Earth's rotation due to Pleistocene deglaciation. Ricard and Sabadini [48] discuss changes in rotation induced by density anomalies in the mantle. Polar wander also has recently been demonstrated to provide constraints on mantle viscosity [e.g., 53, 58]. A number of recent papers summarizing these results (and the post glacial uplift, geoid and plate velocity studies) can be found in *Glacial Isostasy, Sea-Level and Mantle Rheology, ed. Sabadini, Lambeck and Boschi* [59].

2. RECENT INVERSION RESULTS

Several recent studies form and solve inverse problems rather than repeatedly solving the forward problem. In theory, the inverse problem provides not only a model, but also estimates of the resolving power of the data and of the trade-offs between model parameters. The resolution and trade-off analyses are not always straight-forward. There is a surprising convergence of these results and the resulting model differs from the "traditional" models. In this paper, all of the figures will present relative viscosities. To convert to absolute viscosities (in Pa s), one should multiply the horizontal axis scale by 10^{21} Pa s. Geoid models are only sensitive to relative viscosities, so the absolute scale is chosen to be consistent with postglacial rebound and plate velocity studies.

2.1 Plate Velocity Inversion

Forte et al. [12] used the method of Bayesian inference to invert for the radial viscosity profile which best fit the observed plate velocities. Harmonic coefficients of the observed plate divergence in the degree range $l = 1-8$ were used [10]. Using a tomographic shear wave model SH425.2 - [62] as the driving force (i.e., δv_s^{lm} in equation 4) and Greens functions (kernels) for viscous flow developed in Forte and Peltier [10], they parameterized the mantle viscosity in five layers (0-100 km, 100-400 km, 400-670 km, 670-2600 km and 2600 km to the CMB). Their inversion produced a viscosity model (i.e., the best fitting uniform viscosity in each of the five layers) with a low viscosity in the transition zone and high viscosity in the 100-400 km layer with a factor of 42 jump at 670 km (see Figure 3). The plate-like divergence predicted by this model explains 48% of the variance in the observed plate divergence (in the range $l = 1-15$). Prior to the inversion, the variance reduction with an isoviscous mantle was - 770%. However, there is a significant trade-off between the top three layers (0-100 km and 400-670 km are correlated and 100-400 km is anti-correlated with the others), so a model with a low viscosity in the 100-400 km layer and higher viscosities in the 0-100 km layer and 400-670 km layers fits the data nearly as well (see also Figure 3).

Ricard and Wuming [49], using the lower mantle model of Dziewonski [8] and the upper mantle model of Woodhouse and Dziewonski [66], invert the topography, geoid, rotation poles, and angular plate velocities for mantle viscosity. Using present day plate geometries, the surface velocity boundary condition is choosen to match the stresses between a no-slip boundary condition at the surface flow driven by the internal density contrasts and

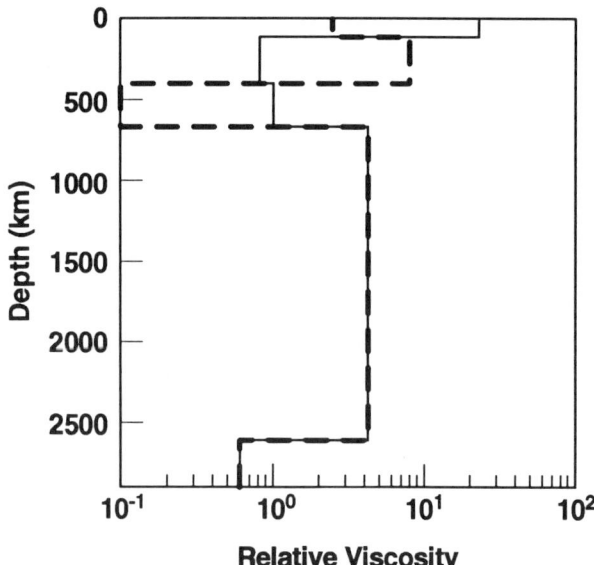

Fig. 3. 1-D viscosity models from Forte et al. [12] determined by inverting observed plate velocities for the best fitting 5 layer viscosity model. The dashed line is the preferred model, the solid line is also an acceptable model. The viscosities in this plot are scaled by a characteristic mantle viscosity ($\eta = 10^{21}$ Pa s).

flow driven by the plates. The resulting viscosity model has a continuous increase in viscosity with depth to the mid-lower mantle, then a decrease in viscosity in the lower one-third of the lower mantle, with no noticeable discontinuities or low viscosity zones. There is a peak change in viscosity of about two orders of magnitude from the surface to the maximum in the lower mantle. However, this model provides poor fits to the data; variance reductions are 44% for geoid, 58% for topography, and 19% for poliodal component of the plate velocities.

2.2 Post Glacial Uplift Inversions

There have been several attempts to form and solve an inverse problem for the viscosity of the mantle using post-glacial uplift data [35, 38]. An analysis of the relative sea level, or uplift history, over Hudson Bay was performed by Mitrovica and Peltier [30, 31]. The horizontal extent of the Laurentide ice sheet suggests that this subset of the relative sea level (RSL) data should be sensitive to the viscosity at greater depths than other data subsets [30]. They conclude that the preference of a uniform viscosity in the lower mantle of 10^{21} Pa s from other studies [e.g., 2, 4, 5, 42] is more appropriately interpreted as a constraint on the uppermost part of the lower mantle (i.e.,

670 - 1800 km), with very weak sensitivity to changes in viscosity of up to an order of magnitude below this depth or in the upper mantle. Therefore, models with large increases in viscosity with depth cannot be ruled out by the RSL data as long as the average viscosity in the 670-1800 km depth range is 10^{21} Pa s. It should be pointed out that Haskell [20] indicated that his result represented the average viscosity of the mantle.

2.3 Geoid Inversion

Ricard et al. [51] considered a three layer mantle. They used L02.56 [8] for the densities in the lower mantle, and M84C [66] and a slab model for those in the upper mantle; they also solved for the density to velocity ratio ($\delta\rho/\delta v$ - see equation 4) in the upper and lower mantle, the density coefficient for the slab model (ρ_{slab}), and viscosities in 100-300, 300-670, and 670-2900 km layers, giving them six unknowns. They used the response kernels for Newtonian viscous flow [50], and chose the viscosity value in the 0-100 km layer to be 10^{22} Pa s, because the geoid is sensitive only to relative viscosity change. They performed a Monte Carlo inversion for the viscosity model which best fit both the geoid and plate velocities. Two classes of models emerged from their study; one with an increasing viscosity with depth (Figure 4 - solid line) and another with the highest viscosity in the transition region and lower viscosity in the upper 100-300

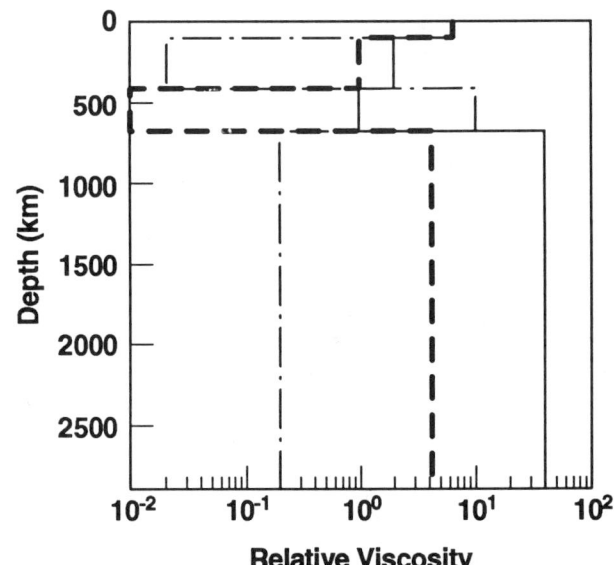

Fig. 4. Three representative 1-D viscosity models from Ricard et al. [51] from Monte Carlo inversion using geoid and plate velocities. The viscosities in this plot are scaled by a characteristic mantle viscosity ($\eta = 10^{21}$ Pa s).

km and 670-2900 km regions (Figure 4 - dash dot line). However, including the plate velocities, in addition to the geoid, a third model emerged (Figure 4 - heavy dashed line). One of their strongest conclusions is the sensitivity of their modeling to the assumed density model. As they note, Hager et al. [18] obtained better results with the same formalism using Clayton and Comers' [6] lower mantle and a boundary layer theory slab upper mantle. The more interesting result is the model that emerged when they considered both geoid and plate velocities in the inversion (Figure 4 - heavy dashed line). In addition, Ricard et al. [51] are the only investigators of the recent group to consider chemically stratified mantle models. Their results suggest that, based on the inversion study, the data are unable to discriminate between layered or whole mantle models.

A study by King and Masters [23] considered several published models for S-wave velocity for the densities providing the driving force ($\delta\rho$ in equation 3): MDLSH [63]; SH425.2 [62]; and MODSH.C [27]. Using the self-gravitating Green's functions for Newtonian viscous flow, following Richards and Hager [54] and Ricard et al. [50], they inverted for radial viscosity models that best fit the observed l = 2-8 geoid (l = 2-6 for MDLSH) using a non-negative, least-squares scheme with smoothing to inhibit

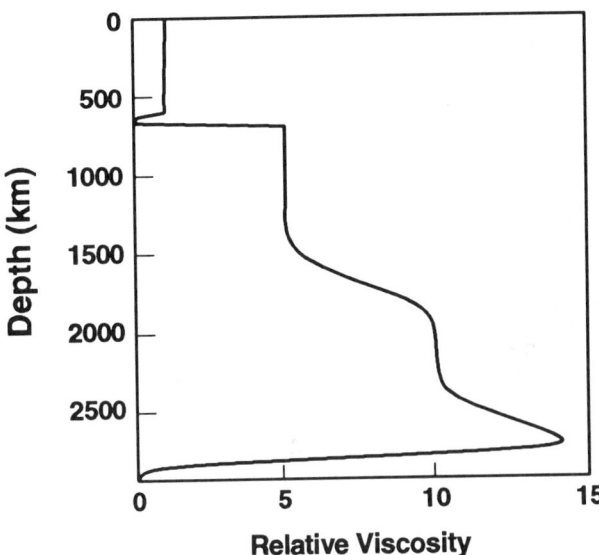

Fig. 6. 1-D viscosity model from Forte et al. [13]. This forward model provides good fits to geoid and plate velocities. It compares well with Figures 3, 4 and 5. The viscosities in this plot are scaled by a characteristic mantle viscosity ($\eta = 10^{21}$ Pa s).

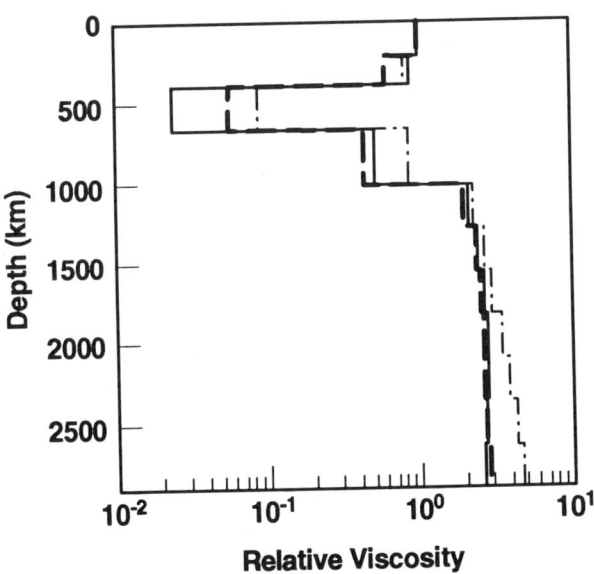

Fig. 5. 1-D viscosity models from King and Masters [23] determined by inverting the seismically determined density anomalies for the best fitting 11 layer viscosity model using the geoid. The models are normalized by the viscosity in the 1284 to 1555 region. The viscosities in this plot are scaled by a characteristic mantle viscosity ($\eta = 10^{21}$ Pa s).

wild oscillations. All three of the seismic velocity models predict a low viscosity between 400-670 km depth (Figure 5). The pattern of viscosity with depth for the three models is strikingly similar: a high viscosity from 0 to 400 km depth, a low viscosity between 400 and 670 km, and increasing viscosity below 670 km. The largest difference between the viscosity models is a factor of two difference in the viscosity of the 400-670 km layer. It is interesting to note that the viscosity in the lower mantle increases by a factor of five below 1022 km in addition to an increase at 670 km. This resembles the two-layer model of Forte and Peltier [11].

Forte et al. [13] used the recent S-wave model SH8/U4L8 which they describe. Using Frechet kernels [Forte et al., 12], they determine the viscosity profile required to fit the geoid. The model, which contains a thin, low-viscosity zone at the base of the upper mantle and an increase in viscosity in the lower mantle (Figure 6), is quite similar to those determined by King and Masters [23] (Figure 5). It may be beyond the limit of their data to constrain such a thin layer at the base of the mantle, however; because layer thickness and viscosity contrast trade-off directly, this could possibly represent a thicker, less extreme layer. Forte et al. obtain a 65% variance reduction for the observed geoid (l = 2-8), in addition to a reasonable fit to the plate velocities with these viscosity and density models. They also point out that their viscosity model is consistent with recent post-glacial uplift analyses and mineral physics.

3. COMPARISON

The viscosity model from Forte et al. [12] - determined by inversion using the plate velocity data, the viscosity models from King and Masters [23] - determined by inversion using the geoid data, and the viscosity model from Ricard et al. [51] - using a Monte Carlo inversion of geoid and plate velocity data - are remarkably similar. Furthermore, the results of the post-glacial uplift inversion by Mitrovica and Peltier [30] and the experimental data, as discussed by Ranalli [46], seem compatible with these observations. Upon considering both the uncertainty in the internal densities and surface loads and the uncertainties in the viscosity models themselves, it appears that the results from the different observations are compatible. However, if the transition zone is Ca rich, as suggested by some [1], Karato's [21] results on the strength of garnet are incompatible with this new model. Mantle models with a hard transition zone appear compatible with observations [60].

4. LATERAL VISCOSITY VARIATIONS

An area of increasing interest is the role of lateral variations in viscosity and their effect on post-glacial rebound, plate velocity and geoid predictions. Using perturbation theory on a 2-D Cartesian problem, Richards and Hager [55] suggested this effect would be small for long-wavelength structures in the lower mantle. Ritzler and Jacobi [56], however, suggest that, because radial viscosity models underpredict the geoid compared to their lateral viscosity equivalents, neglecting lateral viscosity variations leads to errors in the magnitude of radial viscosity jumps as much as a factor of two. Zhang and Christensen [68] also find that the effects of lateral viscosity variations are significant, especially when the radial viscosity is stratified. Ricard et al. [52] showed that a model with lateral viscosity variations between continental and oceanic asthenosphere is consistent with the observed global rotation of the lithosphere with respect to the hot spot reference frame (i.e., the degree 1 toroidal component of plate velocities). A number of investigations addressing lateral viscosity variations are currently underway.

Acknowledgments. The author acknowledges support from NSF grant EAR-9117406. Thanks to R. O'Connell, A. Forte, and Y. Ricard for providing theses, reprints, and preprints. Thanks also to the numerous anonymous reviewers whose comments helped improve this manuscript. Special thanks to Kathy Kincade for help in preparing this manuscript.

REFERENCES

1. Anderson, D.L., *Theory of the Earth*, pp. 57-62, Blackwell Scientific Publications, Boston, 1989.

2. Andrews, J.T., *A Geomorphological Study of Postglacial Uplift with Particular Reference to Arctic Canada*, 156 pp., Inst. British Geographers, London, 1970.

3. Ashby, M.F., and R.A. Verrall, Micromechanisms of flow and fracture, and their relevance to the rheology of the upper mantle, *Phil. Trans. R. Soc. Lond. A*, 288, 59-95, 1978.

4. Cathles, L.M., The viscosity of the Earth's mantle.   Ph.D. thesis, Princeton University, 1971.

5. Cathles, L.M., *The Viscosity of the Earth's Mantle*, 386 pp., Princeton University Press, Princeton, NJ, 1975.

6. Clayton, R.W., and R.P. Comer, A tomographic analysis of mantle heterogeneities from body wave travel time (abstract), *EOS, Trans. AGU*, 62, 776, 1983.

7. Daly, R.A., *The Changing World of the Ice Age*, 271 pp., Yale University Press, New Haven, CT, 1934.

8. Dziewonski, A.M., Mapping the lower mantle: determination of lateral heterogeneity in P velocity up to degree and order 6., *J. Geophys. Res.*, 89, 5929-5952, 1984.

9. Dziewonski, A.M., B.H. Hager, and R.J. O'Connell, Large-scale heterogeneities in the lower mantle, *J. Geophys. Res.*, 82, 239-255, 1977.

10. Forte, A.M., and W.R. Peltier, Plate tectonics and aspherical Earth structure: the importance of polidal-toroidal coupling, *J. Geophys. Res.*, 92, 3645-3679, 1987.

11. Forte, A.M., and R. Peltier, Viscous flow models of global geophysical observables 1. forward problems, *J. Geophys. Res.*, 96, 29,131-20,159, 1991.

12. Forte, A.M., W.R. Peltier, and A.M. Dziewonski, Inferences of mantle viscosity from tectonic plate velocities, *Geophys. Res. Lett.*, 18, 1747-1750, 1991.

13. Forte, A.M., A.M. Dziewonski, and R.L. Woodward, Aspherical structure of the mantle, tectonic plate motions, nonhydrostatic geoid, and topography

of the core-mantle boundary, in *Proceedings of the IUGG Symposium 6: Dynamics of the Earth's Deep Interior and Rotation,* edited by J.L. Le Mouel, in press, 1993.

14. Hager, B.H., Mantle viscosity: A comparison of models from postglacial rebound and from the geoid, plate driving forces, and advected heat flux, in *Glacial Isostasy, Sea-Level and Mantle Rheology,* edited by R. Sabadini, K. Lambeck, and E. Boschi, pp. 493-513, Kluwer Academic Publishers, London, 1991.

15. Hager, B.H., and R.J. O'Connell, A simple global model of plate dynamics and mantle convection, *J. Geophys. Res., 86,* 4843-4867, 1981.

16. Hager, B.H., and M.A. Richards, Long-wavelength variations in Earth's geoid: physical models and dynamical implications, *Phil. Trans. R. Soc. Lond. A, 328,* 309-327, 1989.

17. Hager, B.H., and R.W. Clayton, Constraints on the structure of mantle convection using seismic observations, flow models, and the geoid, in *Mantle Convection,* edited by W.R. Peltier, pp. 657-763, Gordon and Breach, New York, 1989.

18. Hager, B.H., R.W. Clayton, M.A. Richards, R.P. Comer, and A.M. Dziewonski, Lower mantle heterogeneity, dynamic topogrpahy and the geoid, *Nature, 313,* 541-545, 1985.

19. Haskell, N.A., The motion of a viscous fluid under a surface load, I, *Physics, 6,* 265-269, 1935.

20. Haskell, N.A., The motion of a viscous fluid under a surface load, II, *Physics, 7,* 56-61, 1936.

21. Karato, S-I., Plasticity-crystal structure systematics in dense oxides and its implications for the creep strength of the Earth's deep interior: A preliminary result, *Phys. Earth Planet. Int., 55,* 234-240, 1989.

22. Karato, S-I., and P. Li, Diffusive creep in perovskites: Implications for the rheology of the lower mantle, *Science, 255,* 1238-1240, 1992.

23. King, S.D., and G. Masters, An inversion for radial viscosity structure using seismic tomography, *Geophys. Res. Lett., 19,* 1551-1554, 1992.

24. Kirby, S. H., Rheology of the lithosphere, *Rev. Geophys. Space Phys., 21,* 1458-1487, 1983.

25. Lambeck, K. and M. Nakada, Late Pleistocene and Holocene sea-level change along the Australian coast, *Global Planet. Change,* 1990.

26. Lambeck, K., P. Johnston, and M. Nakada, Glacial rebound and sea-level change in northwestern Europe, *Geophys. J. Int.,* 1990.

27. Masters, G., and H. Bolton, Long-period S travel times and the three-dimensional structure of the mantle, *J. Geophys. Res.,* in press, 1993.

28. McConnell, R.K., Viscosity of the mantle from relaxation time spectra of isostatic adjustment, *J. Geophys. Res., 73,* 7089-7105, 1968.

29. Mitrovica, J.X., and W.R. Peltier, Radial resolution in the inference of mantle viscosity from observations of glacial isosatic adjustment, in *Glacial Isostasy, Sea-Level and Mantle Rheology,* edited by R. Sabadini, K. Lambeck, and E. Boschi, pp. 63-78, Kluwer Academic Publishers, London, 1991.

30. Mitrovica, J.X. and W.R. Peltier, A complete formalism for the inversion of post-glacial rebound data - Resolving power analysis, *Geophys. J. Int., 104,* 267-288, 1991.

31. Mitrovica, J.X., and W.R. Peltier, A comparison of methods for the inversion of viscoelastic relaxation spectra, *Geophys. J. Int., 108,* 410-414, 1992.

32. Nakada, M., and K. Lambeck, Late Pleistocene and Holocene sea-level change in the Australian region and mantle rheology, *J. Geophys. Res., 96,* 497-517, 1989.

33. O'Connell, R.J., Pleistocene glaciation and the viscosity of the lower mantle, *Geophys. J.R. astr. Soc., 23,* 299-327, 1971.

34. O'Connell, R.J., The effects of mantle phase changes on postglacial rebound, *J. Geophys. Res., 81,* 971-974, 1976.

35. Parsons, B.E., Changes in the Earth's shape, Ph.D. thesis, Cambridge University, Cambridge, 1972.

36. Pekris, C.L., Thermal convection in the interior of the Earth, *Mon. Not. R. Astron. Soc., Geophys. Suppl., 3,* 343-367, 1935.

37. Peltier, W.R., The impulse response of a Maxwell, *Rev. Geophys. Space Phys., 12,* 649-669, 1974.

38. Peltier, W.R., Glacial isostatic adjustment. II, The inverse problem. *Geophys. J.R. astr. Soc., 46,* 669-706, 1976.

39. Peltier, W.R., Mantle viscosity, in *Mantle Convection,* edited by W.R. Peltier, pp. 389-478, Gordon and Breach, New York, 1989.

40. Peltier, W.R., and J.T. Andrews, Glacial isostatic adjustment, I, The forward problem, *Geophys. Astr. Soc., 46,* 605-646, 1976.

41. Peltier, W.R., and A.M. Tushingham, Global sea level rise and the greenhouse effect: Might they be connected?, *Science, 244,* 806-810, 1989.

42. Peltier, W.R., W.E. Farrell, and J.A. Clark, Glacial isostasy and relative sea level: A global finite element model, *Tectonophysics, 50,* 81-110, 1978.

43. Poirier, J-P., *Creep of Crystals,* 264 pp., Cambridge University Press, Cambridge, 1985.

44. Poirier, J-P., Plastic rheology of crystals, in *AGU Handbook of Physical Constants,* this volume, 1993.

45. Ranalli, G., *Rheology of the Earth: Deformation and Flow Processes in Geophysics and Geodynamics,* 366 pp., Allen and Unwin, Boston, 1987.

46. Ranalli, G., The microphysical approach to mantle rheology, in, *Glacial Isostacy, Sea-Level, and*

Mantle Rheology, edited by R. Sabadini, K. Lambeck, and E. Boschi, pp., 343-378, Kluwer Academic Publishers, London, 1991.

47. Ricard, Y., and C. Vigny, Mantle dynamics with induced plate tectonics, *J. Geophys. Res., 94,* 17,543-17,559, 1989.

48. Ricard, Y., and R. Sabadini, Rotational instabilities of the earth induced by mantle density anomalies, *Geophys. Res. Lett., 17,* 1990.

49. Ricard, Y., and B. Wuming, Inferring the viscosity and 3-D density structure of the mantle from geoid, topography, and plate velocities, *Geophys. J. Int., 105,* 561-571, 1991.

50. Ricard, Y., L. Fleiotout, and C. Froidevaux, Geoid heights and lithospheric stresses for a dynamic Earth, *Ann. Geophys., 2,* 267-286, 1984.

51. Ricard, Y., C. Vigny, and C. Froidevaux, Mantle heterogeneities, geoid and plate motion: A Monte Carlo inversion, *J. Geophys. Res., 94,* 13,739-13.,754, 1989.

52. Ricard, Y., C. Doglioni, and R. Sabadini, Differential rotation between lithosphere and mantle: A consequence of lateral mantle viscosity variations, *J. Geophys. Res., 96,* 8407-8415, 1991.

53. Ricard, Y., G. Spada, and R. Sabadini, Polar wandering of a dynamic earth, *Geophys. J. Int.,* in press, 1993.

54. Richards, M.A., and B.H. Hager, Geoid anomalies in a dynamic earth, *J. Geophys. Res., 89,* 5987-6002, 1984.

55. Richards, M.A., and B.H. Hager, Effects of lateral viscosity variations on long-wavelength geoid anomalies and topography, *J. Geophys. Res., 94,* 10,299-10,313, 1989.

56. Ritzler, M., and W.R. Jacobi, Geoid effects in a convecting system with lateral viscosity variations, *Geophys. Res. Lett., 19,* 1547-1550, 1992.

57. Sabadini, R., and W.R. Peltier, Pleistocene deglaciation and the Earth's rotation: Implications for mantle viscosity, *Geophys. J.R. astr. Soc., 66,* 553-578, 1981.

58. Sabadini, R., and D.A. Yuen, Mantle stratification and long-term polar wander, *Nature, 339,* 373-375, 1989.

59. Sabadini, R., K. Lambeck, and E. Boschi (Eds.), *Glacial Isostacy, Sea-Level, and Mantle Rheology,* 708 pp., Kluwer Academic Publishers, London, 1991.

60. Spada, G., R. Sabadini, D.A. Yuen, and Y. Ricard, Effects on post-glacial rebound from the hard rheology in the transition zone, *Geophys. J., 109,* 1992.

61. Stocker, R.L., and M.F. Ashby, On the rheology of the upper mantle, *Rev. Geophys. Space Phys., 11,* 391-426, 1973.

62. Su, W-J., and A.M. Dziewonski, Predominance of long-wavelength heterogeneity in the mantle, *Nature, 352,* 121-126, 1991.

63. Tanimoto, T., Long-wavelength S-wave velocity structure throughout the mantle, *Geophys. J. Int., 100,* 327-336, 1990.

64. van Bemmelen, R.W., and H.P. Berlage, Versuch Einer Mathematischen Behandlung Geoteknisher Bewegungen Unter Besonderer Berucksichtegung der Undationstheorie, *Gerlands Beitrage zur Geophsysik, 43,* 19-55, 1935.

65. Weertman, J., The creep strength of the Earth's mantle, *Rev. Geophys. Space Phys., 8,* 145-168, 1970.

66. Woodhouse, J.H., and A.M. Dziewonski, Mapping the upper mantle: Three-dimensional modeling of earth structure by inversion of seismic waveforms, *J. Geophys. Res., 89,* 5953-5986, 1984.

67. Wu, P., and W.R. Peltier, Pleistocene glaciation and the Earth's rotation: A new analysis, *Geophys. J. R. astr. Soc., 76,* 753-791, 1984.

68. Zhang, S., and U. Christensen, The effect of lateral viscosity variations on geoid topography and plate motions induced by density anomalies in the mantle, *Geophys. J. Int.,* submitted, 1992.

Plastic Rheology of Crystals

1. PHENOMENOLOGY OF PLASTIC DEFORMATION

When crystals are strained above a certain limit, they cease behaving like elastic solids and become plastic. Plastic deformation is an irreversible, isovolume process, which results in a permanent shear strain after the shear stress that caused it has been removed. Hydrostatic pressure alone does not produce plastic deformation.

A *perfectly plastic* solid undergoes no plastic strain if the applied stress is lower than the *elastic limit* σ_{EL}, or *yield stress* ; if $\sigma = \sigma_{EL}$, the plastic strain can take any value. Any increase in applied stress would immediately be relaxed by strain, so that the applied stress cannot be greater than the yield stress. The yield stress usually decreases with increasing grain size and increasing temperature; in single crystals, it depends on the orientation of the stress axis with respect to the crystal lattice. In actuality, the applied stress necessary for continuing deformation usually increases with strain (*hardening*). At high temperatures, however, many crystalline materials exhibit negligible hardening and can reasonably be considered as perfectly plastic solids.

For perfectly plastic solids, the *stress-strain curve*, $\sigma = f(\varepsilon)$, is a straight line, $\sigma = \sigma_{EL}$, parallel to the strain axis (Figure 1a). Time does not explicitly appear in the constitutive equation. Standard stress-strain curves for materials are obtained by straining a sample in a testing

machine (usually in tension or compression), at constant strain-rate: $d\varepsilon/dt \equiv \dot{\varepsilon}$; when the yield stress is reached, stress remains constant and strain keeps increasing at rate $\dot{\varepsilon}$; the yield stress usually increases with $\dot{\varepsilon}$. The constitutive equation can therefore be written $\sigma = f(\dot{\varepsilon},T)$.

Plastic deformation of materials in the Earth, however, does not generally occur at constant strain-rate: at high temperatures, rocks and minerals slowly flow under constant stress. The relevant laboratory experiments are then *creep* tests [45]. In creep tests, a sample is put under constant stress (tensile or compressive) and strain is recorded as a function of time; the curve $\varepsilon(t)$ is a creep curve. In many cases, after a transient stage during which it decreases, the strain-rate becomes approximately constant, until eventual failure (Figure 1b). The creep curve is then approximately linear (*quasi steady-state creep*) and the corresponding creep-rate depends on the applied stress and on temperature: $\dot{\varepsilon} = f(\sigma,T)$. In creep tests, the crystals are free to deform, and the observed creep-rate is that for which the yield stress is equal to the applied stress. As creep tests are usually performed at temperatures high enough to obtain a measurable creep-rate, there is very little hardening, and, for most materials, a quasi steady-state can be reached (if not, the creep rate would be continuously decreasing).

It can therefore be seen that a solid deforming in steady-state creep under constant stress can be considered as a perfectly plastic solid, but, more interestingly, if the variable time is introduced, it can also be considered as a *viscous fluid* , since it flows in shear at constant rate under applied stress. The viscosity at constant stress can be defined as: $\eta = \sigma/\dot{\varepsilon}$. If creep-rate depends linearly on stress, viscosity is independent of σ and it is said to be *Newtonian* . However, creep-rate often increases faster than stress and viscosity decreases with increasing stress (*non-Newtonian viscosity*). In many cases, the creep-rate depends on stress by a power-law $\dot{\varepsilon} \propto \sigma^n$, with $1 \leq n \leq 5$.

J. P. Poirier, Institut de Physique du Globe de Paris, Depart. des Geomater., 4, Place Jussieu, 75252 Paris, Cedex 05, France

Mineral Physics and Crystallography
A Handbook of Physical Constants
AGU Reference Shelf 2

Copyright 1995 by the American Geophysical Union.

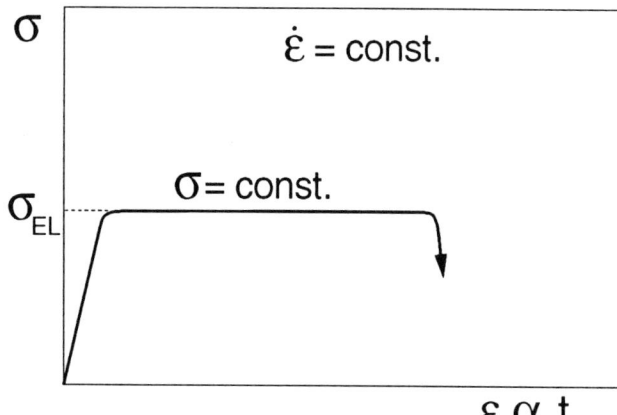

Fig. 1a   Stress-strain curve of an elastic-perfectly
            plastic solid at constant strain rate.
            Above the elastic limit σ_{EL}, stress remains
            constant.

The temperature dependence of the creep-rate usually
follows an *Arrhenius law* , i.e.: $\dot{\varepsilon} \propto \exp(-Q/RT)$, where R
is the gas constant, T is the absolute temperature, and Q is
the *apparent activation energy* , determined from the slope
of the $\ln \dot{\varepsilon}$ vs $1/T$ plot, at constant stress. The activation
energy increases with hydrostatic pressure P, through an
apparent activation volume ΔV^*. If creep is controlled by
only one physical mechanism, the activation energy can be
written: $Q = \Delta H_0(\sigma) + P\Delta V^*$, where ΔH_0 and ΔV^* are the
activation enthalpy and volume respectively of the
controlling process. In some cases, the activation enthalpy
may depend on the applied stress (decreasing as stress
increases); then, if the logarithm of the creep-rate is plotted
against the logarithm of the stress to obtain the exponent n
of a power law, the stress dependence of the activation
enthalpy may appear as a spuriously high value of n (n >
5).

The creep-rate of polycrystals (and obviously of single
crystals) does not generally depend on grain size, except for
very fine-grained polycrystals deforming by diffusion creep
(see next section).

The creep rate of pure elements (e.g., metals) is entirely
characterized by its dependence on stress, temperature,
pressure and grain-size. This, however, is not the case for
ceramics and minerals, whose thermodynamic state depends
on the activity of the components. In particular, the creep
rate of oxides (which constitute most of the minerals
relevant to deformation in the Earth) often depends on
oxygen fugacity f_{O_2}. In addition, the creep-rate of silicates
is also sensitive to the amount of water present as an
impurity, expressed as the ratio H^+/Si.

The constitutive equation (rheological equation) of pure
single crystal oxide minerals, if creep is controlled by only
one process, is usually expressed as:

$$\dot{\varepsilon} = \dot{\varepsilon}_0 \, f_{O_2}^{\ m} \, \sigma^n \, \exp\left(-\frac{\Delta H_0(\sigma) + P\Delta V^*}{RT}\right) \qquad (1)$$

where ε_0 is a constant that often depends on the orientation
of the stress axis with respect to the crystal lattice.

The values of the parameters n, m, ΔH_0, ΔV^*, are
usually obtained by fitting a curve to the experimental
values of $\dot{\varepsilon}$, determined in a range of values of the relevant
variable (σ, $1/T$, P, f_{O_2}...), all the other variables being
kept constant. More reliable results are obtained by global
inversion of the experimental results, which
simultaneously yields best values of all the parameters [51,
48].

In many cases, more than one controlling process is
active in the experimental range of σ, T, f_{O_2} etc. Several
processes, with rates $\dot{\varepsilon}_i$, may act concurrently or
sequentially. The resulting creep-rate can be expressed as:
$\dot{\varepsilon} = \Sigma_i \dot{\varepsilon}_i$, for concurrent processes, and as: $\dot{\varepsilon} = [\Sigma_i \dot{\varepsilon}_i^{-1}]^{-1}$,
for sequential processes. The Arrhenius plot $\ln \dot{\varepsilon}$ vs $1/T$ is
then usually curved and the fit of a single straight line,
when it is possible, gives only an apparent energy Q,
which does not correspond to any physical process, [e.g.,
2].

If, for some reason, it becomes easier to deform an
already deformed zone further than to initiate deformation
elsewhere, *plastic instability* occurs, and strain becomes
localized in deformation bands or shear zones. This
situation occurs when strain-softening mechanisms
predominate over strain-hardening mechanisms [27, 44].
The stress sensitivity of strain-rate n (or stress exponent in
the creep equation) is an important parameter: if n =1
(Newtonian viscous material), deformation is intrinsically
stable. If n $\approx$ 2, the solids still can be deformed to very

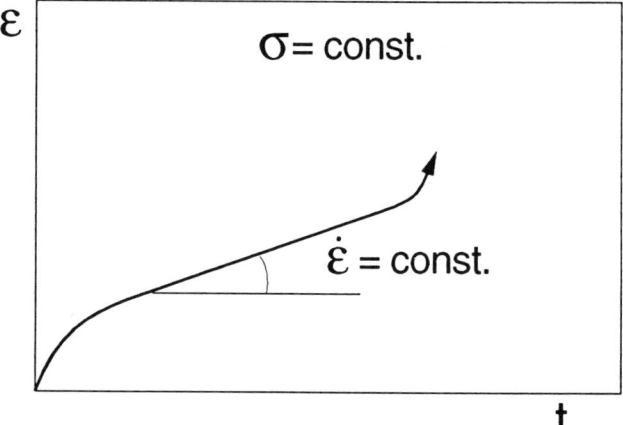

Fig. 1b   Constant stress creep curve. Strain rate is
            approximately constant during quasi steady-
            state creep.

large strains in a stable manner by the so-called *superplastic deformation* , which may occur in fine-grained materials. Non-Newtonian creep, with n ≥ 3, on the other hand, is potentially unstable and strain-induced structural changes or strain heating may cause shear localization.

Phase transformations strongly interact with plastic deformation; the resultant *transformation plasticity* [45] appears, in creep experiments, as an important transient enhancement of the creep rate during transformation (or as a stress drop in the case of deformation at constant strain-rate). Localized phase transformation under non-hydrostatic stresses may cause shear faulting instabilities [7, 30].

2. PHYSICAL MECHANISMS

Plastic deformation is the macroscopic result of transport of matter on a microscopic scale, resulting from the motion of defects: point defects, dislocations, or grain-boundaries. Applied shear stress provides the driving force for the motion of defects and the rate of motion usually increases with temperature. Deformation at low temperature $(T/T_m < 0.3)$ is due to glide of dislocations on crystallographic slip planes in the direction of their Burgers vector; plastic deformation occurs when the shear stress resolved on the slip plane in the slip direction is equal to the *critical resolved shear stress* (CRSS), characteristic of the slip system. Deformation at high temperature, in most cases, involves diffusion of point defects (usually vacancies); it may also depend on the interaction of several kinds of defects : e.g., in dislocation creep controlled by vacancy diffusion or diffusion creep accommodated by grain-boundary sliding (see [45] for a review).

In general, the shear strain rate is given by a transport equation, and is equal to the density of carriers of deformation times the strength of the carriers times their velocity. In the most frequent case of deformation by motion of dislocations, the shear strain-rate $\dot{\varepsilon}$ obeys *Orowan's equation* :

$$\dot{\varepsilon} = \rho b v \qquad (2)$$

where ρ is the density of mobile dislocations (length of dislocation line per unit volume), b is their Burgers vector (strength), and v their velocity. Orowan's equation is valid whether shear strain is caused by slip or climb of dislocations. Theoretical rheological equations such as (1) can be derived from Orowan's equation by expressing the dislocation density and the dislocation velocity in terms of the relevant physical parameters: T, σ, P, f_{O_2}, etc. [45].

Glide of dislocations is driven by the applied shear stress; to move along their slip plane dislocations have to go over the potential hills between dense rows of atoms; they also have to overcome localized obstacles, due for instance to the stress field of other dislocations. If the potential hills are high (as they are in some minerals, even at high

temperature), dislocations tend to remain straight in the deep troughs along the dense rows, and the thermally activated, stress-assisted step of passing over the hill between one row of atoms and the next (lattice friction), is the controlling process. Creep is then said to be *glide-controlled*. If the potential hills are so low, that slip between localized obstacles is easier than overcoming the obstacles, dislocations wait in front of the obstacle until they can circumvent it by climbing out of their slip planes and glide until they meet the next obstacle. As climb of edge dislocations out of their slip plane is controlled by diffusion of point defects to or from the dislocations, *climb-controlled* creep is rate-limited by thermally activated diffusion and can only occur at high temperatures; also, the activation energy does not depend on stress. In climb-controlled creep as in glide-controlled creep, the strain is still due to dislocation glide, only the controlling processes of the strain rate differ. In general, dislocation creep is grain size independent.

If one makes the reasonable assumption that, in steady-state deformation, the average distance between dislocations is determined by their long-range stress field, which balances the applied stress σ, the dislocation density ρ is then proportional to σ^2. The difference between the rheological equation for glide-controlled and climb-controlled creep essentially comes from the physics underlying the expression of the velocity.

i) In *glide-controlled creep* (Figure 2a), to move by one Burgers vector, a dislocation must locally nucleate pairs of *kinks* over the potential hill; sideways migration of the kinks then brings the dislocation to the next low-energy trough. Nucleation and migration of the kinks are thermally activated processes; the applied stress helps in overcoming the energy barrier and the activation energy decreases by an amount equal to the work done by the stress. In most cases, the dependence of the activation energy on stress can be assumed to be linear: $\Delta H (\sigma) = \Delta H_0 - B\sigma$.

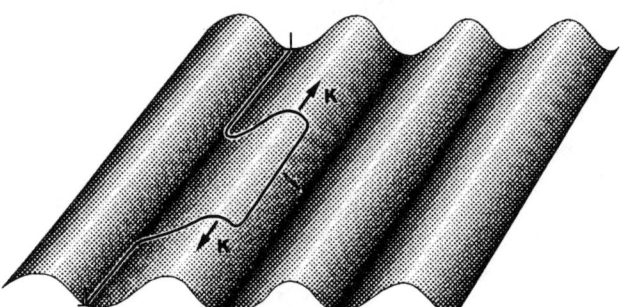

Fig. 2a   Glide-controlled creep. Dislocations tend to lie in the potential troughs, overcoming of the potential hills is achieved by nucleation and sideways spreading of kink pairs (K).

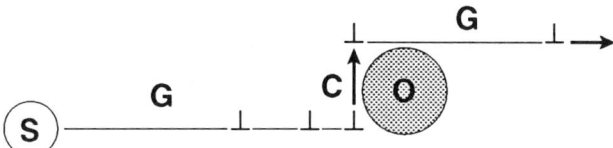

Fig. 2b   Climb-controlled creep. Edge dislocations emitted by a source (S) glide on their glide plane (G) until they are stopped in front of an obstacle (O), they have to climb (C) to clear the obstacle and resume gliding (G).

A screw dislocation can in principle glide on all crystallographic planes which contain it; if it is dissociated in one plane, it can usually easily glide in this plane. However, to *cross-slip* from on plane to another, which may be needed to avoid obstacles, the dislocation must be constricted. Constriction is thermally activated and the activation energy of cross-slip controlled creep decreases with increasing stress. Although cross-slip helps in overcoming localized obstacles, the rheological equation of cross-slip controlled creep is basically similar to that of glide-controlled creep.

ii) In *climb-controlled creep* (Figure 2b), strain results from glide of the dislocations between obstacles, but the strain-rate is controlled by the time the dislocations must wait behind the obstacles before circumventing them by climb and gliding further. Creep-rate increases with decreasing waiting time, i.e., with increasing climb velocity. Climb velocity depends on the diffusive flux of point defects to or from the dislocation, which in turn depends on the diffusion coefficient of the relevant point defects. In the case of crystals of elements (e.g., metals) the activation energy for creep is therefore independent of stress and equal to the activation energy of self-diffusion [45]; in the case of simple ionic crystals (e.g., NaCl), it is equal to the coefficient of diffusion of the slower species, usually the anion (oxygen for binary oxides). For more complicated minerals, climb involves multicomponent diffusion and the activation energy for creep is not simply related to the activation energy of diffusion of any one species (see next section and [2, 23]).

The diffusive flux of defects responsible for climb is driven by a chemical potential gradient, which is usually assumed to be proportional to the applied shear stress. The climb velocity is therefore proportional to the applied stress. Since, by Orowan's equation, the creep rate is proportional to the product of the climb velocity by the dislocation density (proportional to σ^2), and since the activation energy is stress-independent, it follows that the creep-rate at constant temperature and pressure can be theoretically expressed by a power law: $\dot{\varepsilon} \propto \sigma^3$. Although in many instances the stress exponent n of climb-controlled power-law creep is indeed $n \approx 3$, this is by no means

always the case, and for many crystals $3 \leq n \leq 5$.

iii) In *diffusion-creep* (Figure 2c), diffusion of point defects not only controls the creep rate, but also causes the creep strain: vacancies travel down the stress-induced chemical potential gradient between crystal faces in tension and in compression, thus deforming the crystal in response to the applied stress. The creep rate is proportional to the applied stress and to the self-diffusion coefficient; it is inversely proportional to the square of the grain size if diffusion occurs in the bulk (*Herring-Nabarro creep*), and inversely proportional to the cube of the grain size if diffusion occurs along the grain boundaries (*Coble creep*). Diffusion creep is a mechanism effective only at high temperatures and in polycrystals of small grain sizes. As the creep rate is linear in stress (Newtonian viscosity), diffusion creep can successfully compete with climb-controlled creep only at low stresses. Deformation of the grains of a polycrystal by diffusion creep creates incompatibilities that must be relieved by grain-boundary sliding. Diffusion creep and grain-boundary sliding are mutually accommodating processes.

Harper-Dorn creep, like diffusion creep, is characterized by a stress exponent $n \approx 1$, but exhibits no grain-size

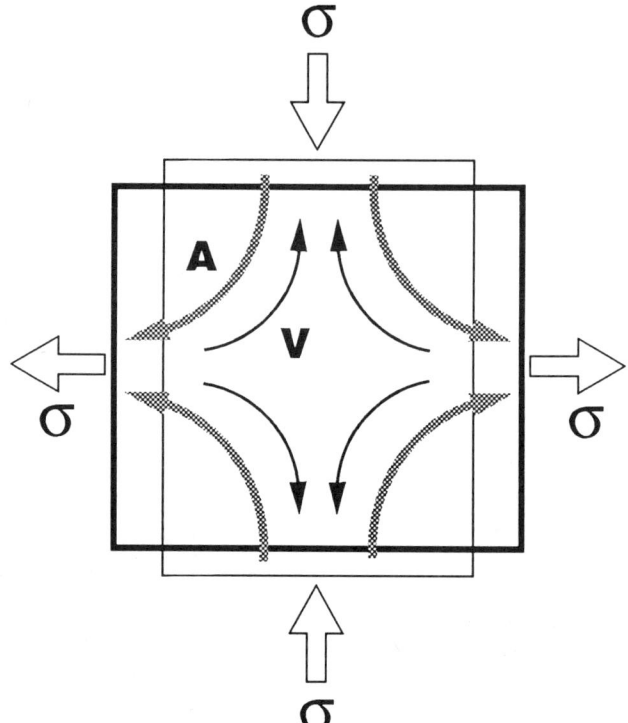

Fig. 2c   Diffusion creep. Vacancies (V) diffuse from regions of high equilibrium concentration, at surfaces normal to the extensive stress, to regions of low equilibrium concentration, at surfaces normal to the compressive stress. Atoms (A) diffuse in the opposite direction.

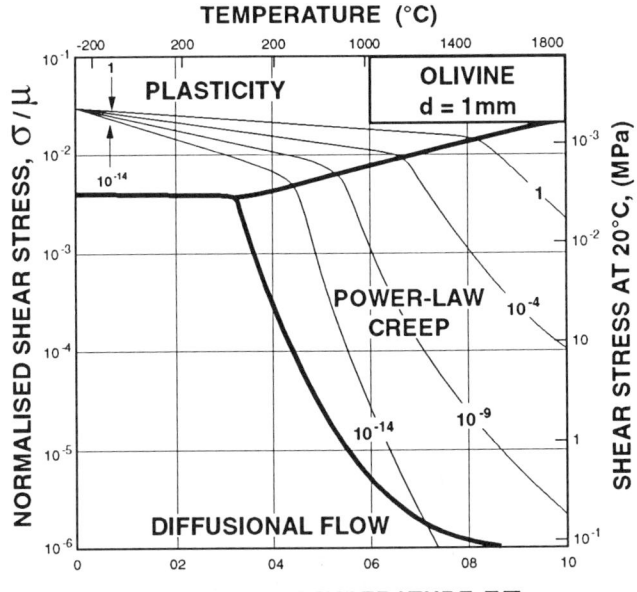

TEMPERATURE (°C)

NORMALISED SHEAR STRESS, σ/μ

SHEAR STRESS AT 20°C, (MPa)

PLASTICITY

OLIVINE
d = 1 mm

POWER-LAW
CREEP

DIFFUSIONAL FLOW

HOMOLOGOUS TEMPERATURE, T/T$_m$

Fig. 3   Deformation mechanism map for olivine,
with grain size 1 mm, showing iso-strain
rate contours (after Frost and Ashby, 1982).

dependence. It is observed only at low stresses and can probably be considered as climb-controlled creep, with a stress-independent dislocation density.

Hydrostatic pressure has little influence on plastic deformation at low temperature during glide-controlled creep. It is only in the case of diffusion-controlled creep (climb-controlled dislocation creep and diffusion creep) that pressure can cause a noticeable decrease in the creep rate through the activation volume for diffusion. The role of pressure is, of course, important in the creep of materials in the Earth's mantle and at the pressures of the lower mantle, one must consider a pressure dependence of the activation volume for creep [46].

The determination of the active mechanism of deformation cannot be done entirely on the basis of the experimental rheological equation (e.g., by comparing the stress exponent to that of the various models and the activation energy to those for diffusion of different species). It is essential, whenever possible, to examine the deformed samples by transmission electron microscopy [40] in order to characterize the dislocations and observe their configurations, looking for diagnostic features [e.g., 5, 22].

The parameters of rheological equations such as (1) can be constrained by experiments for various possible deformation mechanisms. A mechanism is dominant over a range in the variables defining the experimental conditions (T, σ, P, grain-size, f_{O_2}, etc.), if the corresponding strain-rate is greater than those of competing mechanisms. If all variables defining the experimental conditions, except two, are kept constant, *deformation-mechanism maps* can be

constructed (Figure 3), displaying in 2-dimensional space (e.g., σ/μ and T/Tm) the domains where various mechanisms are dominant (e.g., power-law climb-controlled creep, diffusion creep, etc.). The boundaries between domains are obtained by equating the rheological equations for the mechanisms dominant in each domain. Deformation-mechanism maps are useful to predict the behavior of a material only in the intervals in which the parameters have been constrained by experiments.

3. PLASTICITY OF IMPORTANT MINERALS

This section deals only with recent investigations on the minerals most widespread in the crust and mantle of the Earth, whose plastic deformation has been experimentally investigated using single crystals, thus allowing better understanding of the physical mechanisms. The plasticity of polycrystalline minerals, or monomineralic rocks (e.g., quartzite, dunite...), will be treated in the "Rock Physics" section. For a review of the plasticity of crustal rocks and minerals, see [29, 27, 17]. Data on creep of ceramics can be found in [8].

3.1. Quartz

Since the discovery of water-weakening by Griggs and Blacic [20], it has been known that the plastic behavior of quartz is very much dependent on the nature and concentration of water-related defects contained in the crystals. Reviews on the influence of water on the plasticity of quartz can be found in [40, 43]. Water-related defects might consist of 4 protons (H^+) substituted for a Si^{4+} ion or one OH_2 group substituted for an oxygen; however, recent electron microscopy work identifies water-related defects with high-pressure clusters of molecular water [41].

Natural, "dry" quartz ($H/10^6 Si < 80$) is almost impossible to deform plastically in the laboratory: it is still brittle at 1300°C, under an applied stress of 1GPa [11]. Thus, most experimental work is conducted on "wet" ($H^+/10^6 Si > 500$), often synthetic, quartz crystals.

Below a critical temperature, which decreases as water content increases, wet synthetic quartz is very strong and behaves much as dry natural quartz. Above the critical temperature, the stress-strain curves (at constant strain-rate) exhibit a sudden stress drop at the yield stress (*yield-point*) followed by a hardening stage. The creep curves correspondingly exhibit an initial incubation period during which the creep rate increases to decrease later during a hardening stage [11]. Dislocation glide usually occurs on the basal plane (0001) in the $<11\overline{2}0>$ direction (**a**), and on the prism planes $\{10\overline{1}1\}$ in the **a** or **c** [0001] , or **c+a**, directions (the indices of planes and directions are given using Miller-Bravais notation). At high enough temperatures, there may exist a quasi-steady state creep, following a power law whose parameters depend on the

TABLE 1. High-temperature compression creep of wet single-crystal quartz

OH⁻/Si (ppm)	Slip System	T (°C)	σ (MPa)	$\dot{\varepsilon}_0$ (σ in bars)	n	Q (kJ/mol)	Ref.
4300	{$2\overline{1}10$} c	400 - 570 (α)	40 - 162		3.7	109	[28]
"	"	570 - 800 (β)	"		3.7	33	"
370	{$2\overline{1}10$} c	400 - 570 (α)	80 - 200	$10^{-8.9}$	3.0	92	[35]
"	"	570 - 800 (β)	"	$10^{-8.9}$	3.0	92	"
370	{$10\overline{1}0$} a	400 - 570 (α)	80 - 200	$10^{-9.5}$	5.3	213	"
"	"	570 - 800 (β)	"	$10^{-9.7}$	3.4	117	"

orientation of the samples with respect to the stress axis (Table 1).

Observations of experimentally deformed crystals using transmission electron microscopy [11, 36, 9, 18] support the following mechanisms:

i) At low temperatures, low water fugacity, and high strain rate, deformation is controlled by lattice friction, the density of grown-in dislocations is very small, the dislocations are straight and their mobility is low; quartz is then very strong.

ii) At higher temperatures, during the incubation period, clusters of water molecules precipitate into water bubbles; the elastic strain around the bubbles is relieved by punched-out dislocation loops, which expand by climb and act as sources, thus increasing the dislocation density; plastic yield occurs suddenly with a stress drop (yield point) for an applied stress high enough to make massive multiplication of dislocations possible. FitzGerald et al. [18] also point out the role of microfractures in the nucleation of dislocations.

TABLE 2. High-temperature compression creep of single-crystal olivines

Crystal	Orientation of stress axis (pseudo-cubic)	T (°C)	σ (MPa)	Atm.	n	Q (kJ/mol)	Ref.
Fo$_{92}$	Various	1430 - 1650	5 - 150	$CO_2/H_2 = 0.1$	f(σ)	525	[31]
Fo$_{92}$	<111>	1500 - 1600	10 - 40	$CO_2/H_2 = 0.43$	3.5	523	[13]
Fo$_{92}$	<110>	1400 - 1600	10 - 60	$CO_2/H_2 = 0.43$	3.7	523	[13]
Fo$_{100}$	[101]	1550 -1650	8 - 60	$CO_2/H_2 = 0.33$	3.5	564	[14]
Fo$_{100}$	[111]	1500 - 1680	3 - 30	H_2/A	2.9	667	[16]
Fo$_{90}$	[101]	1250 - 1400	100	$CO_2/H_2 = 2.3$		536	[32]
Fo$_{90}$	[101]	1250 - 1400	100	$CO_2/H_2 = 0.25$		536	[32]
Fo$_{90}$	[101]	1250 - 1400	100	$f_{O_2} = 10^{-6}$ Pa		448	[32]
Fo$_{100}$	[110]	1400-1650	10 - 100	H_2/CO_2	2.6	460	[10]
Fo$_{100}$	[101]	1400 - 1600	20 - 100	H_2/CO_2	3.6	573	[10]
Fo$_{100}$	[011]	1500 - 1600	30 - 100	H_2/CO_2	2.7	598	[10]

TABLE 3. High-temperature creep laws for buffered single-crystal Fo91 olivine

Orientation of stress axis (pseudo-cubic)	Buffer	$\dot{\varepsilon}_0$ (σ inMPa)	n	m	Q (kJ/mol)	$\dot{\varepsilon}$
[101]	opx	0.65	3.5	0.33	250	$\dot{\varepsilon}_1$
		5.3×10^{11}	3.5	0.06	690	$\dot{\varepsilon}_2$
	mw	1.0×10^{14}	3.5	0.40	700	$\dot{\varepsilon}_1$
		0.16	3.5	0.05	300	$\dot{\varepsilon}_2$
[011]	opx	2.1×10^4	3.5	0.02	540	$\dot{\varepsilon}_1$
		5.2×10^5	3.5	0.23	540	$\dot{\varepsilon}_2$
	mw	1.0×10^{14}	3.5	0.40	750	$\dot{\varepsilon}_1$
		0.2	3.5	0.0	370	$\dot{\varepsilon}_2$
[110]	opx	0.02	3.5	0.36	230	$\dot{\varepsilon}_1$
		1.3×10^{22}	3.5	0.10	1000	$\dot{\varepsilon}_2$
		1.2	3.5	0.15	290	$\dot{\varepsilon}_3$
	mw	1.0×10^{22}	3.5	0.2	1000	$\dot{\varepsilon}_1$
		25	3.5	0.2	330	$\dot{\varepsilon}_2$

After [2]. 1200°C < T < 1525 °C ; $10^{-12} < f_{O_2} < 10^{-3}$ atm. , buffer: opx ($Mg_{0.9}Fe_{0.1}SiO_3$) or mw ($Mg_{0.7}Fe_{0.3}O$)

$\dot{\varepsilon} = \dot{\varepsilon}_0 \sigma^n f_{O_2}{}^m \exp(-Q/RT)$

Flow law: [101] opx: $\dot{\varepsilon}^{-1} = \dot{\varepsilon}_1^{-1} + \dot{\varepsilon}_2^{-1}$; [101] mw: $\dot{\varepsilon}^{-1} = \dot{\varepsilon}_1^{-1} + \dot{\varepsilon}_2^{-1}$; [011] opx: $\dot{\varepsilon} = \dot{\varepsilon}_1 + \dot{\varepsilon}_2$

[011] mw: $\dot{\varepsilon}^{-1} = \dot{\varepsilon}_1^{-1} + \dot{\varepsilon}_2^{-1}$; [110] opx: $\dot{\varepsilon} = \dot{\varepsilon}_1 + [\dot{\varepsilon}_2^{-1} + \dot{\varepsilon}_3^{-1}]^{-1}$; [110] mw: $\dot{\varepsilon}^{-1} = \dot{\varepsilon}_1^{-1} + \dot{\varepsilon}_2^{-1}$

3.2. Olivine

Olivine $(Mg, Fe)_2SiO_4$, the dominant mineral of the upper mantle, has an orthorhombic lattice and a slightly distorted hexagonal close-packed oxygen sublattice.The common slip systems are: at low temperature (100)[001], corresponding to slip on the basal plane of the hcp oxygen sublattice in the **a** direction; and at high temperature (010)[100] (prism plane of the hcp sublattice in the **c** direction), with cross-slip on other prism planes such as (001) and (011).

High temperature creep of single crystals in various orientations obeys a power-law creep equation with n ≈ 3.5 (Table 2), apparently compatible with dislocation creep controlled by climb of edge dislocations. Examination of deformed samples by transmission electron microscopy suggest that creep can be controlled by climb and/or glide, according to the orientation [15, 21]. However, when climb-control would be expected, the activation energy for creep (of the order of 500 kJ/mol) does not tally with that for diffusion of the slower species: oxygen or silicon (about 335 kJ/mol). One should however note that climb of dislocations should be controlled by *multicomponent diffusion* of all the species, rather than by diffusion of the slowest species alone [23].

The thermodynamic state of olivine with fixed Fe/Mg ratio is not entirely determined when T, P and f_{O_2} are fixed, the activity of one component (orthopyroxene or magnesiowüstite) must also be specified. Bai et al [2] performed systematic experiments on olivine buffered against orthopyroxene or magnesiowüstite for various orientations as a function of temperature, stress, and oxygen fugacity. The results (Table 3) show wide variations in activation energy and f_{O_2} dependence according to the experimental conditions: two or three power law equations are needed to describe

TABLE 4. High-temperature compression creep of single-crystal pyroxenes

Crystal	T (°C)	$\dot{\varepsilon}_0$ (σ inMPa)	n	m	Q (kJ/mol)		Ref
Enstatite (En$_{99}$)	1350 - 1450	4.6×10^{11}	3.8	0	750	Buffered by olivine	[37]
Enstatite (En$_{96}$)	1350 - 1450	2.1×10^{15}	3.9	0	880	" "	[37]
Diopside	1000 - 1050		4.3		284	P = 5-15 kbar, in talc	[1]
Diopside							
Slip on (100)	1020 - 1137	4.9×10^3	8.1		742	Ar 10% H_2-2% H_2O	[49]
	1137 - 1321	7.1×10^{-23}	8.8		85	" "	[49]
Slip on {110}	1020 - 1130	4.1×10^{-4}	6.5		442	" "	[49]
	1130 - 1321	7.0×10^{-18}	6.0		48	" "	[49]
Hedenbergite	900 - 1100	1.2×10^8	3.6		526	P = 10 kbar	[33]

phenomenologically the creep behavior over the whole range of experimental conditions, suggesting that different mechanisms have to be considered. The stress exponent, however, is equal to 3.5 in all cases (which, incidentally, is a confirmation of its limited usefulness as a criterion for the determination of the physical mechanism of creep). Examination of deformed samples by the oxidation-decoration technique [3] shows that different dislocation structures parallel the changes in power-law equations and confirms that several rate-controlling mechanisms may be operating in different experimental conditions.

As in the case of quartz, water-related defects increase olivine plasticity . For single crystals of olivine treated in a hydrous environment at 1300°C under a hydrostatic pressure of 300 MPa, and subsequently deformed in the same conditions at a constant strain rate of $10^{-5}s^{-1}$, the flow stress is reduced by a factor of 1.5 to 2.5, with respect to that of crystals treated in a dry environment [38]. Samples deformed under wet conditions and examined in transmission electron microscopy exhibited evidence of enhanced climb of dislocations (dislocation walls, etc.).

3.3. Pyroxenes

Although orthopyroxene $Mg_2Si_2O_6$ (enstatite) and clinopyroxene $MgCaSi_2O_6$ (diopside) are also important minerals of the upper mantle, their plastic properties have been much less studied than those of olivine. The easiest slip system of orthorhombic enstatite as well as of monoclinic diopside is (100)[001], with dislocation glide on the planes of the layers of chains of SiO_4 tetrahedra, and in the direction of the chains.

In orthoenstatite, the glissile dislocations with Burgers vector [001] can dissociate on (100) [52]. During high-temperature deformation, the shear stress can favor the nucleation of very fine lamellæ of clinoenstatite on (100) planes; calcium present as a minor ion in enstatite can diffuse to the clinoenstatite, thus forming thin lamellæ of calcic clinopyroxene [26], which might be confused with stacking faults. High-temperature creep of $(Mg, Fe)_2Si_2O_6$ enstatite obeys a power law (Table 4); it does not seem to be sensitive to oxygen fugacity but the creep-rate increases with increasing iron and aluminum content [37].

Mechanical twinning, primarily on the (100) plane in the [001] direction, is the dominant deformation mode for clinopyroxenes at low temperatures [33]. At high temperature, crystals oriented in such a way that twinning is impossible deform by power-law creep (Table 4). In experiments on high-temperature creep of pure diopside, Ratteron and Jaoul [49] report a drastic decrease in activation energy by a factor of almost ten above a critical temperature about 200°C below the melting point. The crystals then become much more creep-resistant than would be expected from the extrapolated lower-temperature creep law. Examination of the deformed samples by transmission electron microscopy [13] showed that above and below the critical temperature deformation took place primarily by {110}1/2<110> slip, with no detectable occurrence of climb. Deformation is therefore probably glide-controlled, which accounts for the high value of the stress exponent (n ≈ 8). Above the critical temperature, the dislocations were pinned by tiny glassy globules (<10 nm), resulting from incipient partial melting well below the currently accepted

TABLE 5. High-temperature compression creep of single-crystal oxides with perovskite structure

Crystal	Orientation of stress axis (pseudo-cubic)	T/T_m	Atm	$\ln \dot{\varepsilon}_0$ (σ in Pa)	n	Q (kJ/mol)	g	Ref.
$BaTiO_3$	<110>	0.75 - 0.92	Argon	-38	3.6	469	30	[4]
$KNbO_3$	<110>	0.84 - 0.99	Argon	-43	3.7	415	38	[5]
$KTaO_3$	<110>	0.87 - 0.99	Argon	-11	1	292	21	[5]
$CaTiO_3$	<110>	0.67 - 0.78	Air	-39	2.5	274	15	[53]
$CaTiO_3$	<100>	0.67 - 0.78	Air	-37	3.3	444	24	[53]
$NaNbO_3$	<100>	0.70 - 0.98	Air	-92	5.3	192	14	[53]

melting point of diopside. The fluid droplets may effectively act as obstacles to dislocation slip, thus causing the observed hardening.

3.4. High-Pressure Mantle Minerals

Despite their importance for the rheology of the transition zone and the lower mantle, there is no information on the plasticity of the high-pressure mantle minerals in the relevant conditions of temperature and pressure. The essential high-pressure phases are $(Mg,Fe)_2SiO_4$ β-phase and γ-spinel, and majorite garnet in the transition zone, and $(Mg,Fe)SiO_3$ perovskite and $(Mg,Fe)O$ magnesiowüstite in the lower mantle.

Magnesiowüstite has the NaCl structure and is stable at ambient pressure. There have been numerous investigations on the plastic properties of the MgO end-member (magnesia or periclase), which is easily deformable even at low temperatures, on the six {110} planes in the <110> directions. There are however no experiments at very high temperatures in controlled oxygen partial pressure conditions; experiments at about $0.5T_m$, in air or vacuum are compatible with climb-controlled dislocation creep, with n ≈ 4 and Q ≈ 400 kJ/mol [8]. Wüstite $Fe_{1-x}O$, the other end-member, is always non-stoichiometric and Fe-deficient; recent high-temperature creep experiments performed at various controlled oxygen fugacities [24] give results (n ≈ 4.8, -0.02 < m < 0.11, Q ≈ 290 kJ/mol) compatible with dislocation creep controlled by the diffusion of oxygen-related point defects whose nature varies with temperature. There are good reasons to believe that deformation of magnesiowüstite at lower mantle conditions is not very difficult: crystals formed by decomposition of olivine at high pressure in a diamond-anvil cell contain a high density of dislocations [47].

All phases other than magnesiowüstite are stable only at

high pressures and, there is as yet no high-temperature creep apparatus operating at these pressures; the only information we have so far on the plastic properties of the silicate high-pressure phases comes from experiments that were performed at room temperature, either on the metastable phases at ambient pressure or in-situ in a diamond-anvil cell. There is also qualitative information extracted from the examination of dislocation structures of the phases by transmission electron microscopy. Finally, experiments on isostructural compounds presumably belonging to isomechanical groups may provide some basis for speculation.

Karato et al. [25] performed microhardness tests on metastable $MgSiO_3$ perovskite single crystals at room temperature; they found a high value of the Vickers hardness ($H_V ≈ 18$ GPa), higher than for olivine or enstatite. Meade and Jeanloz [42] estimated the yield strength at room temperature of polycrystalline olivine, perovskite and γ-spinel, prepared in a laser-heated diamond-anvil cell, from the pressure gradient in the cell. They found that at room temperature perovskite was stronger than spinel or olivine. However, it should be emphasized that the yield strength of crystals depends on the critical resolved shear stress (CRSS) of the available slip systems; the dependence on temperature of the CRSS may differ considerably for the various slip systems of one mineral (let alone for the slip systems of different minerals), so that the dominant slip system at room temperature may not be the same at high temperature, and the ranking of minerals according to their yield strengths at room temperature cannot be assumed to remain the same at high temperature.

Although transmission electron microscopy of phases deformed at high temperature, in their stability field, does not provide quantitative information on their strength, it does at least provide reliable qualitative information on the dislocation structures, density, and dissociation from which educated guesses on the deformability can be attempted. Thus, in natural γ-spinel (ringwoodite) formed at high temperature in shocked meteorites, stacking faults and dislocations similar to those found in $MgAl_2O_4$ spinel are observed [39]. There is also evidence for the {111}<110> slip system as in $MgAl_2O_4$ [12], and dislocations are very straight, pointing to glide control of the deformation.

There is no information on the plasticity or dislocation structures of $MgSiO_3$ perovskite at high temperature; it is reasonable to assume that the slip systems are the same as in other perovskites, but crystals of perovskite structure certainly do not constitute an isomechanical group [5, 53], probably due to the fact that the distortions and the dislocation core structures are different: creep parameters can be very different even for apparently similar perovskites (Table 5). Therefore, the success of the analogue approach to try elucidating the plastic properties of $MgSiO_3$ at high temperature is contingent on the existence of a good analogue (yet to be identified) of $MgSiO_3$ perovskite.

REFERENCES

1. Avé Lallemant, H.G., Experimental deformation of diopside and websterite, *Tectonophysics, 48*, 1-27, 1978.
2. Bai, Q., S.J. Mackwell and D.L. Kohlstedt, High-temperature creep of olivine single crystals, 1. Mechanical results for buffered samples, *J. Geophys. Res., 96*, 2441-2463, 1991.
3. Bai, Q. and D.L. Kohlstedt, High-temperature creep of olivine single crystals, 3. Dislocation structures, *Tectonophysics, 206*, 1-29, 1992.
4. Beauchesne, S. and J.P. Poirier, Creep of baryum titanate perovskite: a contribution to a systematic approach to the viscosity of the lower mantle, *Phys. Earth planet. Interiors, 55*, 187-189, 1989.
5. Beauchesne, S. and J.P. Poirier, In search of a systematics for the viscosity of perovskites: creep of potassium tantalate and niobate, *Phys. Earth planet. Interiors, 61*, 182-198, 1990.
6. Blacic, J.D. and J.M. Christie, Plasticity and hydrolytic weakening of quartz single crystals, *J. Geophys. Res., 89*, 4223-4239, 1984.
7. Burnley, P.C., H.W. Green II and D.J. Prior, Faulting associated with the olivine to spinel transformation in Mg_2GeO_4 and its implications for deep-focus earthquakes, *J. Geophys. Res., 96*, 425-443, 1991.
8. Cannon, W.R. and T.G. Langdon, Creep of ceramics, *J. Materials Science, 18*, 1-50, 1983.
9. Cordier, P. and J.C. Doukhan, Water solubility in quartz and its influence on ductility, *Eur. J. Mineralogy, 1*, 221-237, 1989.
10. Darot, M. and Y. Gueguen, High-temperature creep of forsterite single crystals, *J. Geophys. Res., 86*, 6219-6234, 1981.
11. Doukhan, J.C. and L. Trépied, Plastic deformation of quartz single crystals, *Bull. Mineral., 108*, 97-123, 1985.
12. Duclos, R., N. Doukhan and B. Escaig, High-temperature creep behavior of nearly stoichiometric alumina spinel, *J. Materials Sci., 13*, 1740-1748, 1978.
13. Durham, W.B. and C. Goetze, Plastic flow of oriented single crystals of olivine, 1. Mechanical data, *J. Geophys. Res., 82*, 5737-5753, 1977a.
14. Durham, W.B. and C. Goetze, A comparison of the creep properties of pure forsterite and iron-bearing olivine, *Tectonophysics, 40*, T15-T18, 1977b.
15. Durham, W.B., C. Goetze and B. Blake, Plastic flow of oriented single crystals of olivine, 2. Observations and interpretations of the dislocation structure, *J. Geophys. Res., 82*, 5755-5770, 1977.
16. Durham, W.B., C. Froidevaux and O. Jaoul, Transient and steady-state creep of pure forsterite at low stress, *Phys. Earth planet. Interiors, 19*, 263-274. 1979.
17. Evans, B. and G. Dresen, Deformation of Earth materials: six easy pieces, *Reviews of Geophysics Supplement, 29*, 823-843, 1991.
18. FitzGerald, J.D., J.N. Boland, A.C. McLaren, A. Ord and B. Hobbs, Microstructures in water-weakened single crystals of quartz, *J. Geophys. Res., 96*, 2139-2155, 1991.
19. Frost, H.J. and M.F. Ashby, *Deformation mechanism maps*, Pergamon Press, Oxford, 1982.
20. Griggs, D.T. and J.D. Blacic, The strength of quartz in the ductile regime, *Trans. Amer. Geophys. Union, 45*, 102, 1964.
21. Guéguen, Y. and M. Darot, Les dislocations dans la forstérite déformée à haute température, *Philos. Mag., A 45*, 419-442, 1982.
22. Ingrin, J., N. Doukhan and J.C. Doukhan, High-temperature deformation of diopside single-crystal. 2.Transmission electron microscopy investigation of the defect microstructure, *J. Geophys. Res., 96*, 14,287-14,297, 1991.
23. Jaoul, O., Multicomponent diffusion and creep in olivine, *J. Geophys. Res., 95*, 17631-17642, 1990.
24. Jolles, E. and C. Monty, High temperature creep of $Fe_{1-x}O$, *Philos. Mag., 64*, 765-775, 1991.
25. Karato, S., K. Fujino and E. Ito, Plasticity of $MgSiO_3$ perovskite: the results of microhardness tests on single crystals, *Geophys. Res. Lett., 17*, 13-16, 1990.
26. Kirby, S.H. and M.A. Etheridge, Exsolution of Ca-clinopyroxene from orthopyroxene aided by deformation, *Phys. Chem. Minerals, 7*, 105-109, 1981.
27. Kirby, S.H. and A.K. Kronenberg, Rheology of the lithosphere: selected topics, *Reviews of Geophysics, 25*, 1219-1244, 1987.
28. Kirby, S.H. and J.W. McCormick, Creep of hydrolytically weakened synthetic quartz crystals oriented to promote {2$\bar{1}\bar{1}$0}<0001> slip: a brief summary of work to date, *Bull. Mineral, 102*, 124-137, 1979.

29. Kirby, S.H. and J.W. McCormick, Inelastic properties of rocks and minerals: strength and rheology, in Handbook of physical properties of rocks, edited by R.S. Carmichael, vol III, pp 139 - 280, CRC Press, Boca Raton, Florida, 1984.

30. Kirby, S.H., W.B. Durham and L.A. Stern, Mantle phase changes and deep-earthquake faulting in subducting lithosphere, *Science, 252*, 216-225, 1991.

31. Kohlstedt, D.L. and C. Goetze, Low-stress, high-temperature creep in olivine single crystals, *J. Geophys. Res., 79*, 2045-2051, 1974.

32. Kohlstedt, D.L. and P. Hornack, Effect of oxygen partial pressure on the creep of olivine, in *Anelasticity of the Earth*, edited by F.D. Stacey, M.S. Paterson and A. Nicolas, pp 101-107, American Geophysical Union, Washington D.C., 1981.

33. Kollé, J.J., and J.D. Blacic, Deformation of single-crystal clinopyroxenes: 1. Mechanical twinning in diopside and hedenbergite, *J. Geophys. Res., 87*, 4019-4034, 1982.

34. Kollé, J.J., and J.D. Blacic, Deformation of single-crystal clinopyroxenes: 2. Dislocation-controlled flow processes in hedenbergite, *J. Geophys. Res., 88*, 2381-2393, 1983.

35. Linker, M.F. and S.H. Kirby, Anisotropy in the rheology of hydrolytically weakened synthetic quartz crystals, in *Mechanical Behavior of crustal rocks (The Handin volume)*, edited by N.L. Carter, M. Friedman, J.M. Logan and D.W. Stearns, pp 29-48, American Geophysical Union, Washington, D.C., 1981.

36. Linker, M.F., S.H. Kirby, A. Ord and J.M. Christie, Effects of compression direction on the plasticity and rheology of hydrolytically weakened synthetic quartz crystals at atmospheric pressure, *J. Geophys. Res., 89*, 4241-4255, 1984.

37. Mackwell, S.J., High-temperature rheology of enstatite: implications for creep in the mantle, *Geophys. Res. Lett., 18*, 2027-2030, 1991.

38. Mackwell, S.J., D.L. Kohlstedt and M.S. Paterson, The role of water in the deformation of olivine single crystals, *J. Geophys. Res., 90*, 11319-11333, 1985.

39. Madon, M. and Poirier, J.P., Dislocations in spinel and garnet high-pressure polymorphs of olivine and pyroxene: implications for mantle rheology, *Science, 207*, 66-68, 1980.

40. McLaren, A.C., *Transmission electron microscopy of minerals and rocks*, Cambridge University Press, Cambridge, 1991.

41. McLaren, A.C., J.D. FitzGerald and J. Gerretsen, Dislocation nucleation and multiplication in synthetic quartz: relevance to water-weakening, *Phys. Chem. Minerals, 16*, 465-482, 1989.

42. Meade, C. and R. Jeanloz, The strength of metal silicates at high pressures and room temperature: implications for the viscosity of the mantle, *Nature, 348*, 533-535, 1990.

43. Paterson, M.S., The interaction of water with quartz and its influence on dislocation flow — an overview, in Rheology of solids and of the Earth, edited by S. Karato and M. Toriumi, Oxford University Press, Oxford, pp 107-142, 1989.

44. Poirier, J.P., Shear localization and shear instability in materials in the ductile field, *J.Struct. Geol., 2*, 135-142, 1980.

45. Poirier, J.P., *Creep of Crystals*, Cambridge University Press, Cambridge, 1985.

46. Poirier, J.P. and R.C.Liebermann, On the activation volume for creep and its variation with depth in the Earth's lower mantle, *Phys. Earth planet. Interiors, 35*, 283-293, 1984.

47. Poirier, J.P., J. Peyronneau, M. Madon, F. Guyot, and A. Revcolevschi, Eutectoid phase transformation of olivine and spinel into perovskite and rock salt structures, *Nature, 321*, 603-605, 1986.

48. Poirier, J.P., C.Sotin and S.Beauchesne, Experimental deformation and data processing, in *Deformation processes in minerals, ceramics and rocks*, edited by D.J. Barber and P.G. Meredith, pp 179-189, Unwin Hyman, London, 1990.

49. Ratteron, P. and O. Jaoul, High-temperature deformation of diopside single-crystal. 1.Mechanical data, *J. Geophys. Res., 96*, 14,277-14,286, 1991.

50. Ricoult, D.L. and D.L. Kohlstedt, Creep of Fe_2SiO_4 and Co_2SiO_4 single crystals in controlled thermodynamic environment, *Philos. Mag., 51*, 79-93, 1985.

51. Sotin, C. and J.P. Poirier, Analysis of high-temperature creep experiments by generalized non-linear inversion, *Mechanics of Materials, 3*, 311-317, 1984.

52. Van Duysen, J.C., N. Doukhan and J.C. Doukhan, Transmission electron microscope study of dislocations in orthopyroxene (Mg, Fe)$_2Si_2O_6$, *Phys. Chem. Minerals, 12*, 39-44, 1985.

53. Wright, K., G.D. Price and J.P. Poirier, High-temperature creep of the perovskites $CaTiO_3$ and $NaNbO_3$, *Phys. Earth planet. Interiors, 74*, 9-22, 1992.

Phase Diagrams of Earth-Forming Minerals

Dean C. Presnall

The purpose of this compilation is to present a selected and compact set of phase diagrams for the major Earth-forming minerals and to show the present state of knowledge concerning the effect of pressure on the individual mineral stabilities and their high-pressure transformation products. The phase diagrams are arranged as follows:

	Figure
Silica	1
Feldspars	2-7
Pyroxenes	8-13
Olivine	14-16
Garnet	17-20
Iron-titanium oxides	21-23
Pargasite	24
Serpentine	25
Phlogopite	25
Iron	26-27

The compilation has been compressed in three ways. (1)

For several of the mineral groups, only representative phase diagrams are shown. (2) The presentation of more complex phase diagrams that show mutual stability relationships among the various minerals and mineral groups has been minimized. (3) Many subsolidus phase diagrams important to metamorphic petrology and thermobarometry are excluded. Reviews of these subsolidus phase relationships and thermodynamic data for calculating the phase diagrams have been presented elsewhere [13, 50, 70, 122, 154]. Other useful reviews and compilations of phase diagrams are Lindsley [96] for oxides, Gilbert *et al.* [60] and Huckenholz *et al.* [74] for amphiboles, Liu and Bassett [104] for elements, oxides, and silicates at high pressures, and *Phase Diagrams for Ceramists* [129-137]. It will be noted that some diagrams are in weight percent and others are in mole percent; they have usually been left as originally published. Minor drafting errors and topological imperfections that were found on a few of the original diagrams have been corrected in the redrafted diagrams shown here.

D. C. Presnall, University of Texas, Dallas, Geoscience Program, POB 830688, Richardson, TX 75083-0688

Mineral Physics and Crystallography
A Handbook of Physical Constants
AGU Reference Shelf 2

248

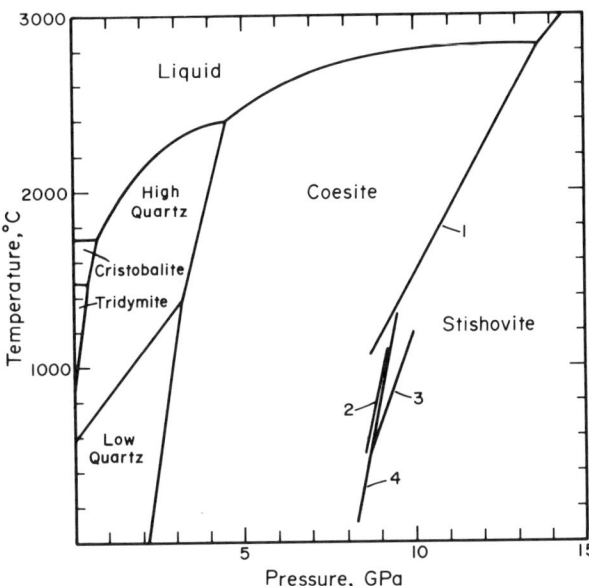

Fig. 1. Phase relationships for SiO_2. Numbers beside curves refer to the following sources: 1 - [178]; 2 - [168]; 3 - [160]; 4 - [4]. The melting curve is from Jackson [77] at pressures below 4 GPa, from Kanzaki [83] at pressures between 4 and 7 GPa, and from Zhang *et al.* [178] at pressures above 7 GPa. The temperature of the high quartz-low quartz-coesite invariant point is from Mirwald and Massonne [113]. The quartz-coesite transition is from Bohlen and Boettcher [24] but note that their curve lies toward the low-pressure side of the range of curves by others [5, 23, 31, 62, 89, 113]. The high quartz-low quartz curve is from Yoder [172]. Boundaries for the tridymite and cristobalite fields are from Tuttle and Bowen [164] except that the cristobalite-high quartz-liquid invariant point has been shifted to 0.7 GPa to accomodate the data of Jackson [77]. Silica has been synthesized in the Fe_2N structure at 35-40 GPa, T>1000°C by Liu *et al.* [105], and at 35 GPa, 500-1000°C by Togaya [162]. However, Tsuchida and Yagi [163] reported a reversible transition between stishovite and the $CaCl_2$ structure at 80-100 GPa and T>1000°C.

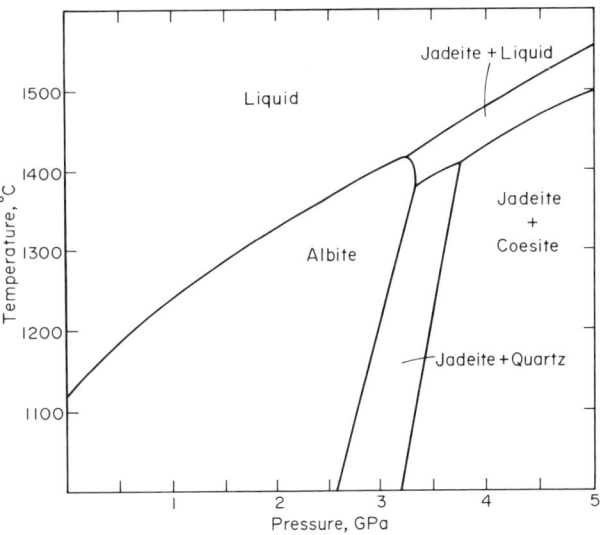

Fig. 2. Isopleth for the composition, $NaAlSi_3O_8$ [12, 16, 33]. The albite = jadeite + quartz reaction shown by Bell and Roseboom [12] and in this figure is about 0.1 GPa higher than the curve of Boettcher and Wyllie [22]. The latter passes through the "consensus" value of 1.63 GPa, 600°C for this reaction [81]. Also, the quartz-coesite curve shown by Bell and Roseboom [12] and in this figure is about 0.4 GPa higher at 1300°C than the pressure given by a linear extrapolation of the curve of Bohlen and Boettcher [24], which is shown in Figure 1. The curve of Bohlen and Boettcher would intersect the albite = jadeite + quartz curve at about 1300°C rather than the jadeite + quartz (coesite) = liquid curve. At about 1000°C, Liu [101] synthesized $NaAlSi_3O_8$ in the hollandite structure at pressures from 21 to 24 GPa, and a mixture of $NaAlSiO_4$ ($CaFe_2O_4$-type structure) + stishovite above 24 GPa. Jadeite, $NaAlSi_2O_6$; Coesite, SiO_2.

Fig. 3. Isopleth for the composition, $CaAl_2Si_2O_8$ [67, 94]. Cor, corundum; Gr, grossular; Ky, kyanite; Qz, quartz; Liq, liquid. Locations of dashed lines are inferred.

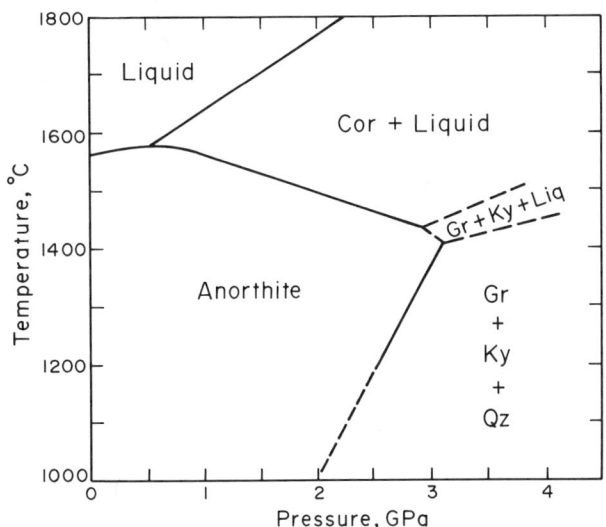

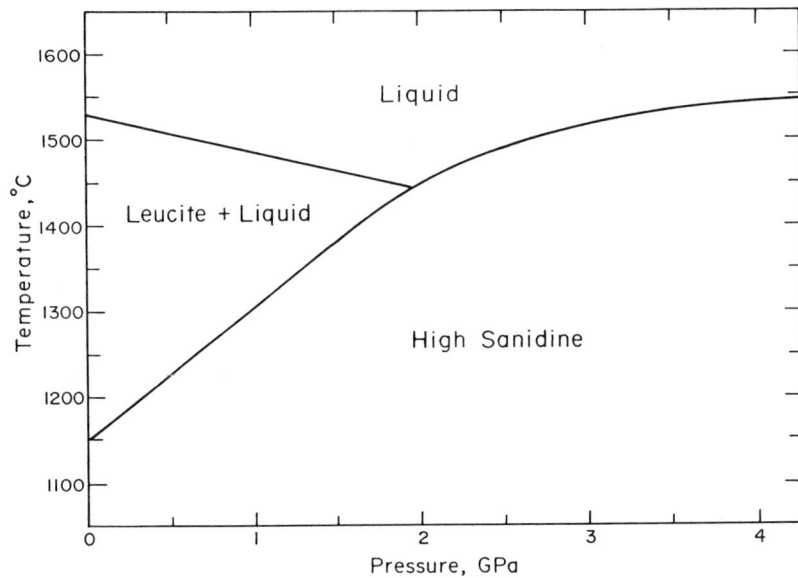

Fig. 4. Isopleth for the composition, $KAlSi_3O_8$ [93, 149]. At 12 GPa, 900°C, Ringwood *et al.* [144] synthesized $KAlSi_3O_8$ in the hollandite structure. In experiments from 8-10 GPa and 700°-1000°C, Kinomura *et al.* [88] synthesized the assemblage $K_2Si_4O_9$ (wadeite-type structure) + kyanite (Al_2SiO_5) + coesite (SiO_2) from the composition $KAlSi_3O_8$; and they synthesized the hollandite structure of $KAlSi_3O_8$ at 900°C, 12 GPa, and at 700°C, 11 and 11.5 GPa.

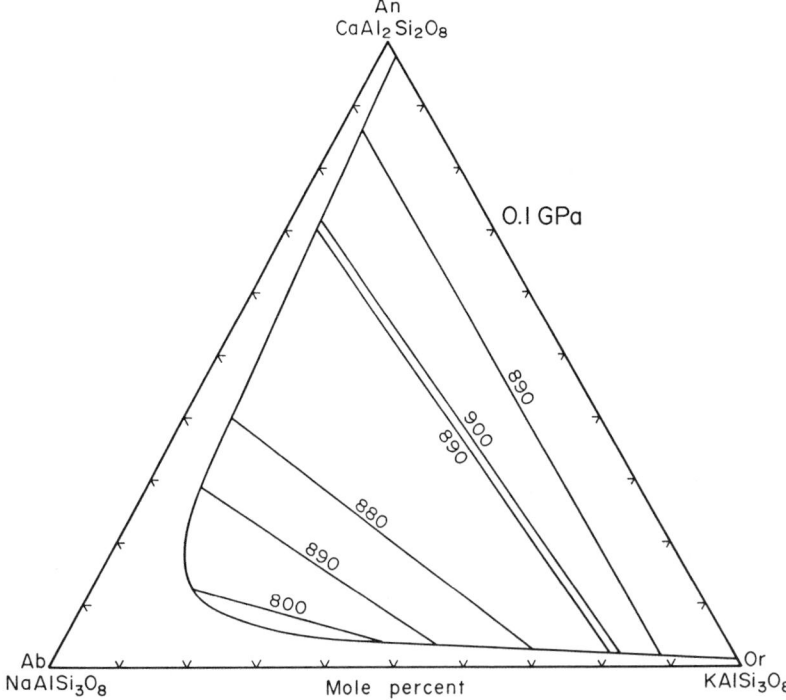

Fig. 5. Compositions of coexisting alkali feldspar and plagioclase at 0.1 GPa and temperatures from 800 to 900°C, as indicated [49]. Note that the phase boundary is essentially isothermal except in the Ab-rich portion of the diagram. Many others have discussed ternary feldspar geothermometry [10, 39, 54, 58, 63, 66, 75, 80, 139, 142, 151-153, 165] and ternary feldspar phase relationships [68, 121, 156, 164, 175]. An, anorthite; Ab, albite; Or, orthoclase.

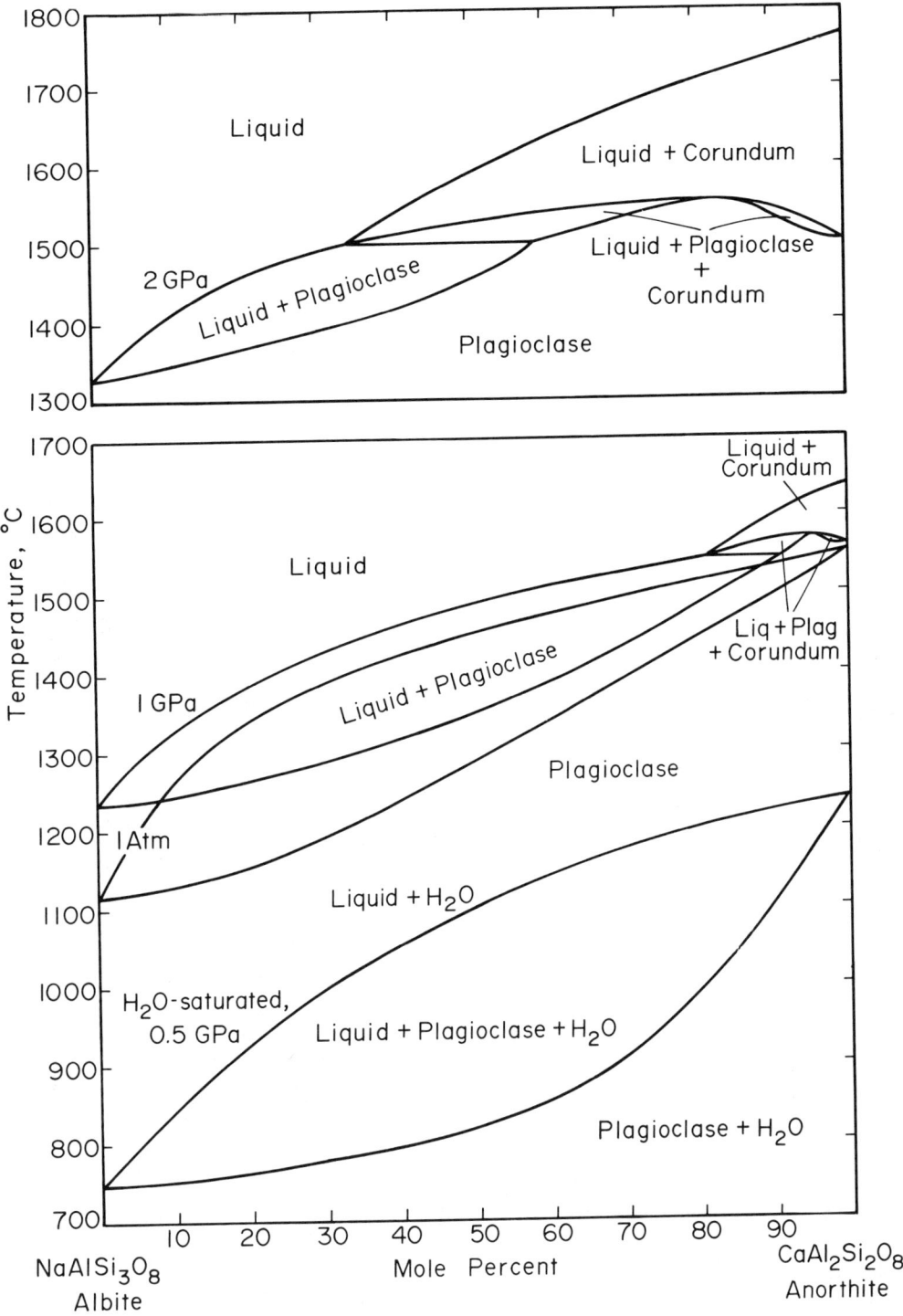

Fig. 6. Temperature-composition sections for the join NaAlSi$_3$O$_8$ (albite) - CaAl$_2$Si$_2$O$_8$ (anorthite) under anhydrous conditions at 1 atm [26, 117], 1 GPa, 2 GPa [33, 94], and under H$_2$O-saturated conditions at 0.5 GPa [79, 175].

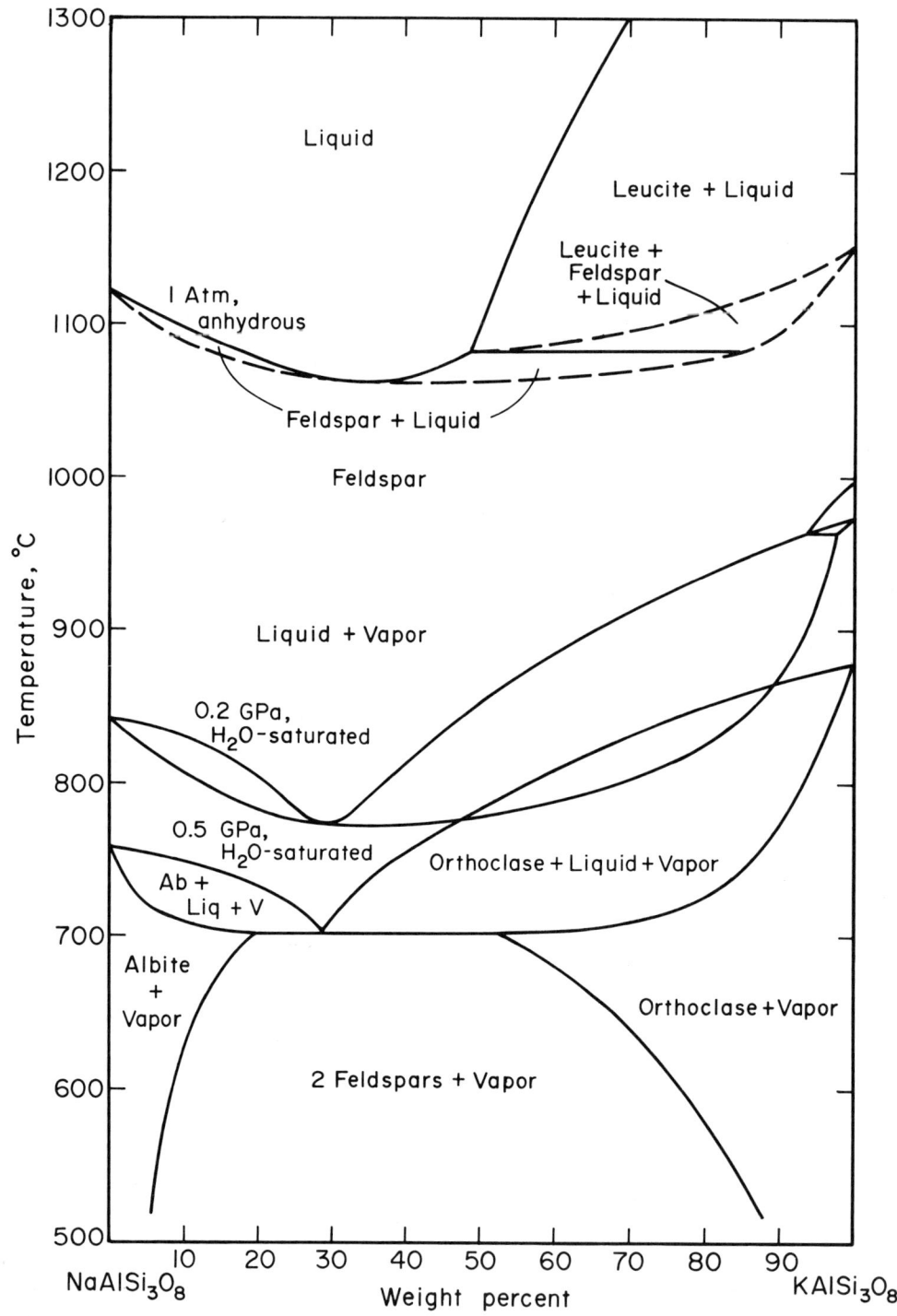

Fig. 7. Temperature-composition sections for the join NaAlSi$_3$O$_8$ (albite) - KAlSi$_3$O$_8$ (orthoclase) under anhydrous conditions at 1 atm [148], and under H$_2$O-saturated conditions at 0.2 GPa [29] and 0.5 GPa [119, 175]. Ab, albite; Liq, liquid; V, vapor. Locations of dashed lines are inferred.

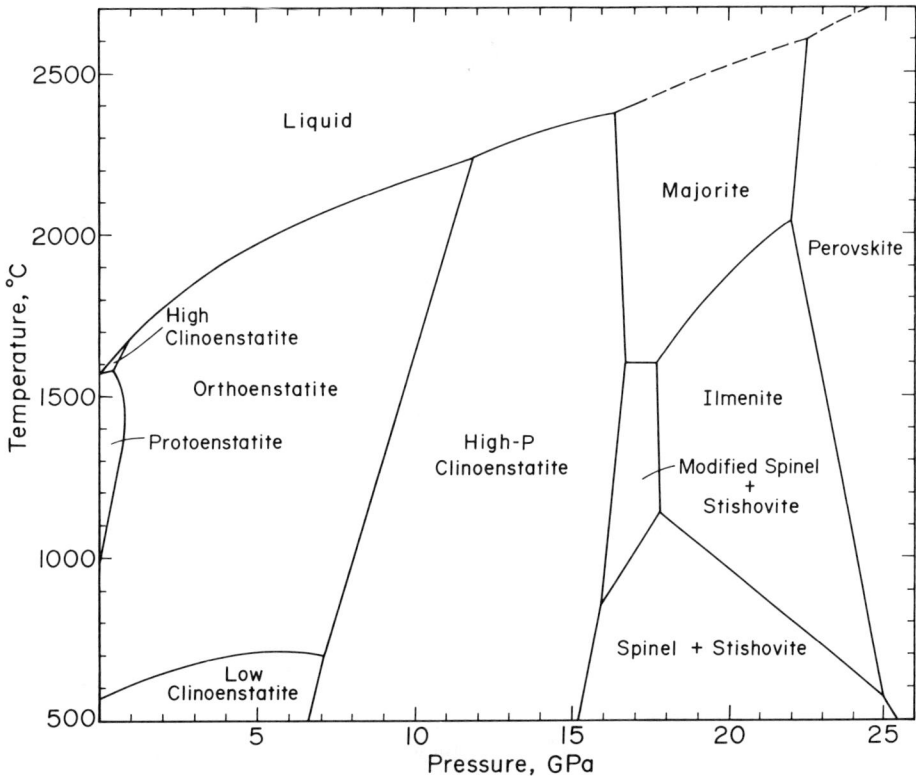

Fig. 8. Isopleth for the composition MgSiO$_3$ [7, 9, 35, 45, 56, 57, 65, 76, 92, 127, 140, 169]. For additional data at pressures above 15 GPa, see also Sawamoto [147]. Not shown is a singular point at about 0.13 GPa below which enstatite melts incongruently to forsterite + liquid [45]. Position of dashed curve is inferred. For additional data on melting temperatures up to 58 GPa, see Zerr and Boehler [177].

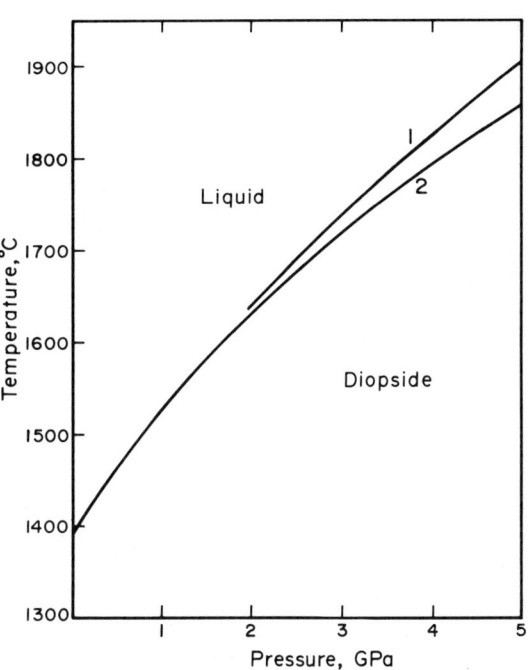

Fig. 9. Melting curve for diopside, CaMgSi$_2$O$_6$. Curve 1 is from Williams and Kennedy [166] uncorrected for the effect of pressure on thermocouple emf, and curve 2 is from Boyd and England [33]. See also Yoder [173] for data below 0.5 GPa. For CaMgSi$_2$O$_6$ composition, Mao *et al*. [110] found a mixture of perovskite (MgSiO$_3$) and glass at 21.7 and 42.1 GPa and 1000°-1200°C. They interpreted the glass to be a second perovskite phase of CaSiO$_3$ composition which inverted to glass on quenching [see also 102].

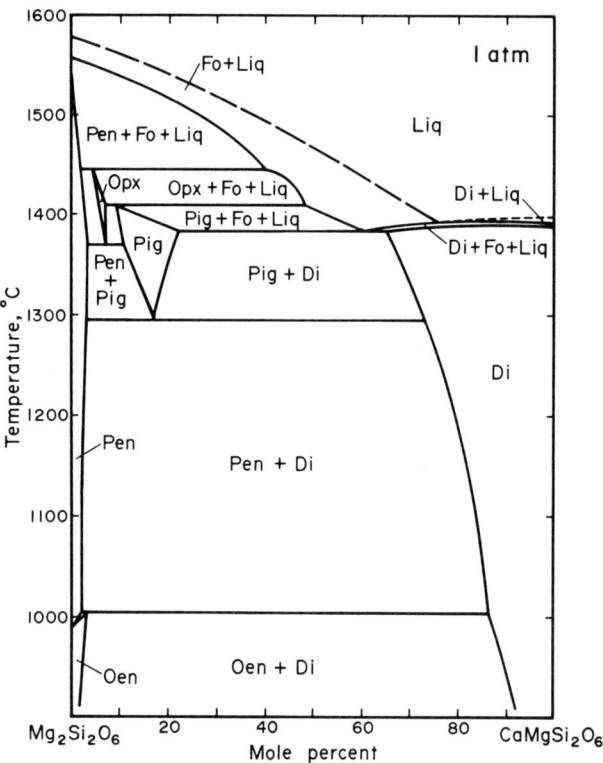

Fig. 10. The join $Mg_2Si_2O_6$ (enstatite) - $CaMgSi_2O_6$ (diopside) at 1 atm [43]. Many others have also discussed phase relationships on this join [9, 14, 36, 41, 42, 56, 78, 91, 106, 170, 171]. Fo, forsterite, Mg_2SiO_4; Liq, liquid; Pen, protoenstatite; Opx, orthopyroxene; Pig, pigeonite; Di, diopside; Oen, orthoenstatite.

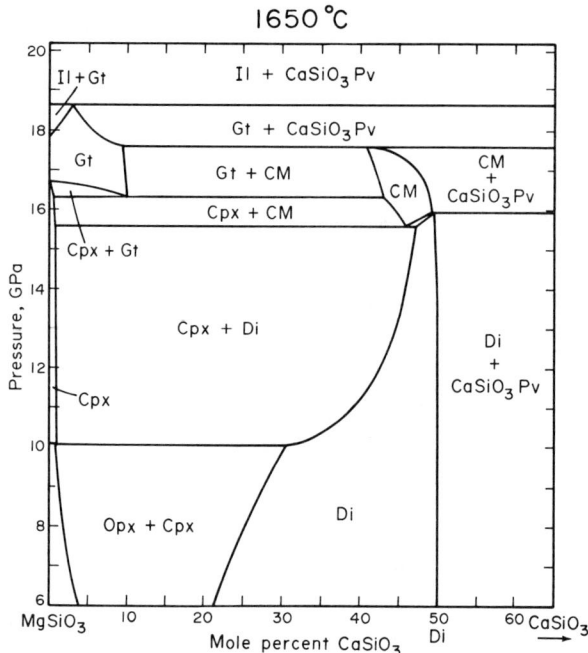

Fig. 12. Pressure-composition section for the system $MgSiO_3$-$CaSiO_3$ at 1650°C [55, 57]. Garnet and clinopyroxene, when they are free of Ca on the left-hand margin of this diagram, are the same phases, respectively, as majorite and high-P clinoenstatite on Figure 7. Il, ilmenite; Gt, garnet; Pv, perovskite; Cpx, clinopyroxene; Opx, orthopyroxene; Di, diopside; CM, a high-pressure phase of unknown structure.

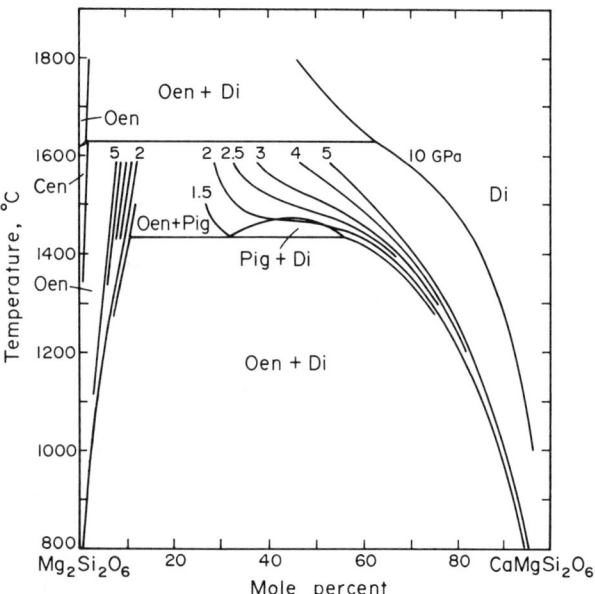

Fig. 11. Thermodynamically modeled subsolidus phase relationships for the system $Mg_2Si_2O_6$ (enstatite) - $CaMgSi_2O_6$ (diopside) from 1.5 to 10 GPa [56, 44]. The thermodynamic models are based on data from other sources [37, 98, 118, 123, 128, 150]. See also data of Biggar [15] from 1 atm to 0.95 GPa.

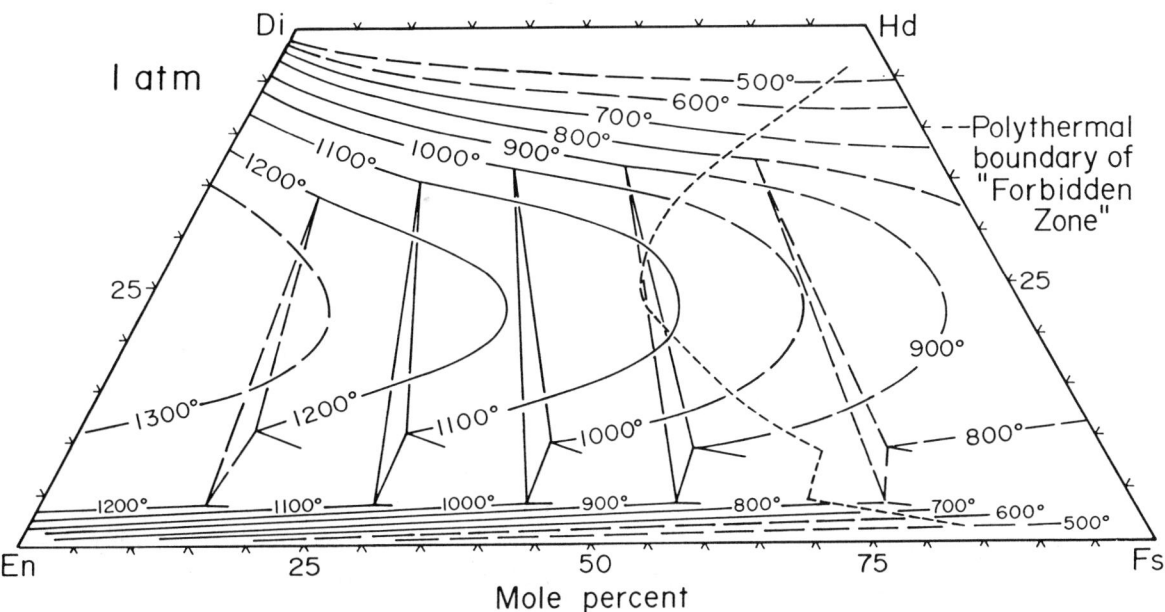

Fig. 13. Orthopyroxene + augite, orthopyroxene + augite + pigeonite, and pigeonite + augite equilibria at 1 atm and 500-1300°C [95]. Phase relationships to the right of the forbidden zone boundary are metastable relative to augite + olivine + silica. Lindsley [95] has presented three other similar diagrams at 0.5, 1, and 1.5 GPa. Lindsley and Andersen [97] should be consulted for correction procedures required before plotting pyroxenes on these diagrams for geothermometry. En, enstatite (MgSiO$_3$); Fs, ferrosilite (FeSiO$_3$); Di, diopside (CaMgSi$_2$O$_6$); Hd, hedenbergite (CaFeSi$_2$O$_6$).

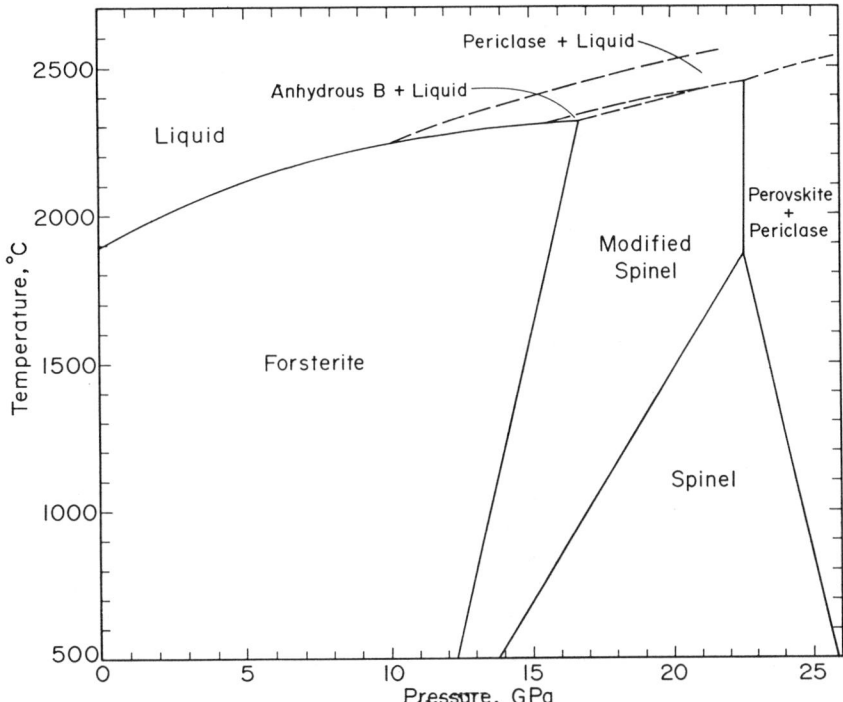

Fig. 14. Isopleth for the composition Mg$_2$SiO$_4$ [48, 57, 141]. Additional studies of the melting relationships are Ohtani and Kumazawa [126] and Kato and Kumazawa [84-86]. Locations of dashed lines are inferred.

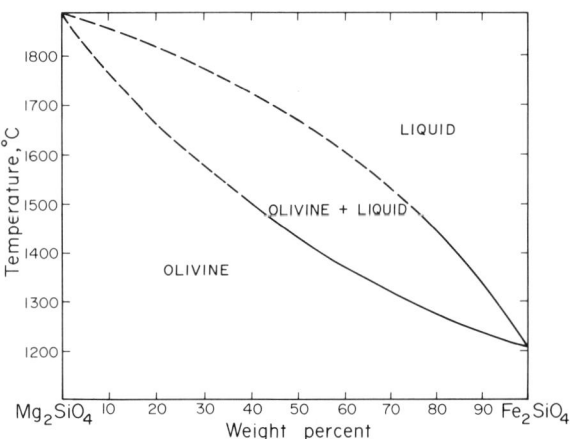

Fig. 15. Phase relationships for the system Mg_2SiO_4 (forsterite) - Fe_2SiO_4 (fayalite) in equilibrium with Fe at 1 atm [27]. Locations of dashed lines are inferred.

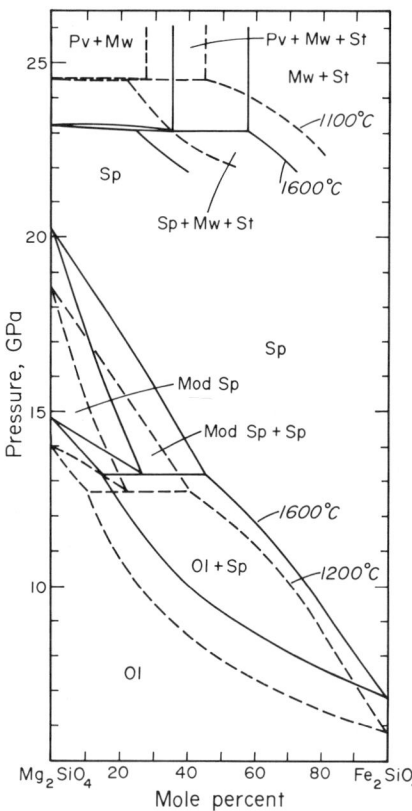

Fig. 16. Pressure-composition sections for the join Mg_2SiO_4-Fe_2SiO_4 at various temperatures. Phase relationships above 21 GPa are from Ito and Takahashi [76] and those below 21 GPa are from Akaogi et al. [3]. Other references [51, 87, 157] give additional data and discussion of these phase relationships. Pv, perovskite ($MgSiO_3$-$FeSiO_3$ solid solution); Mw, magnesiowüstite (MgO-FeO solid solution); St, stishovite (SiO_2); Sp, spinel; Mod Sp, modified spinel; Ol, olivine.

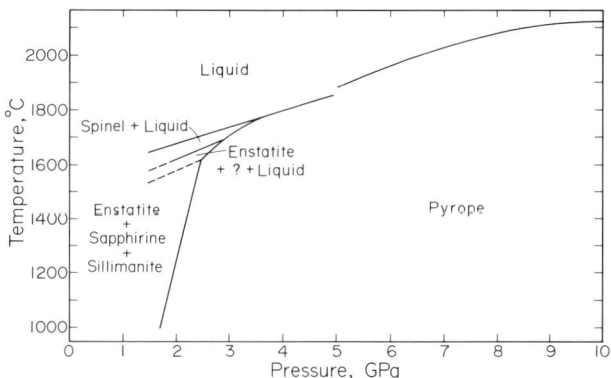

Fig. 17. Isopleth for $Mg_3Al_2Si_3O_{12}$, pyrope garnet. Phase relationships at pressures less than 5 GPa are from Boyd and England [32]. The melting curve at 5 GPa and above is from Ohtani et al. [125]. Liu [99] reported that pyrope transforms to perovskite + corundum at about 30 GPa, 200-800°C. Liu [100] subsequently revised this result and found that pyrope transforms to the ilmenite structure at about 24-25 GPa, 1000°-1400°C, and that ilmenite then transforms to perovskite at about 30 GPa. Locations of dashed lines are inferred.

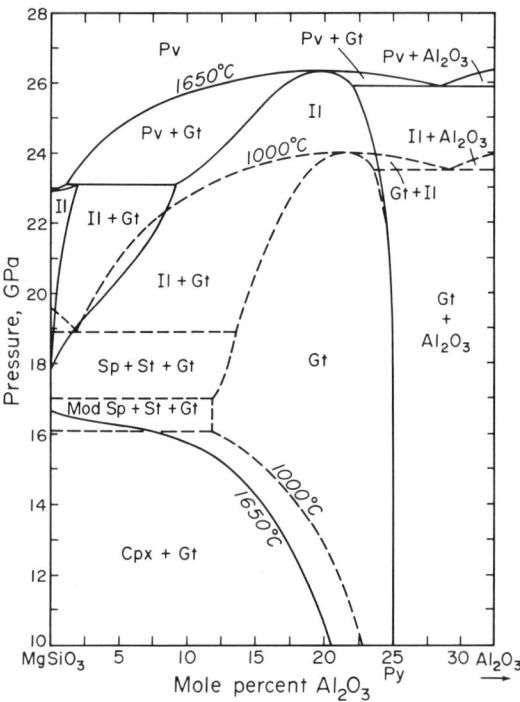

Fig. 18. Pressure-composition section for the join $MgSiO_3$-Al_2O_3 at 1000 and 1650°C [57, 82]. For additional data along the boundary between the garnet and clinopyroxene + garnet fields at 1000°C, see Akaogi and Akimoto [2]. At 1100 and 1600°C for pressures between 2 and 6.5 GPa, the Al_2O_3 content of pyroxene in equilibrium with garnet increases with decreasing pressure to at least 15 mole percent [30, 34].

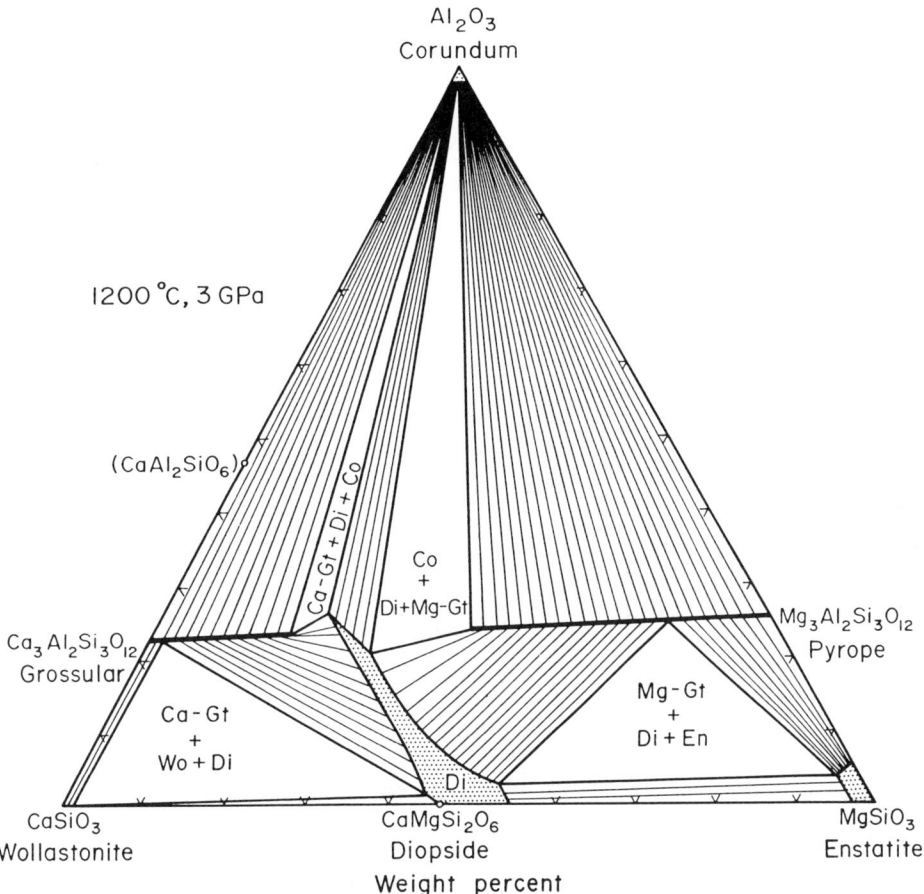

Fig. 19. The system CaSiO$_3$-MgSiO$_3$-Al$_2$O$_3$ at 1200°C, 3 GPa [30]. Co, corundum; Di, diopside; Ca-Gt, Ca-garnet; Mg-Gt, Mg-garnet; Wo, wollastonite; En, enstatite; CaAl$_2$SiO$_6$, Ca-Tschermak's molecule.

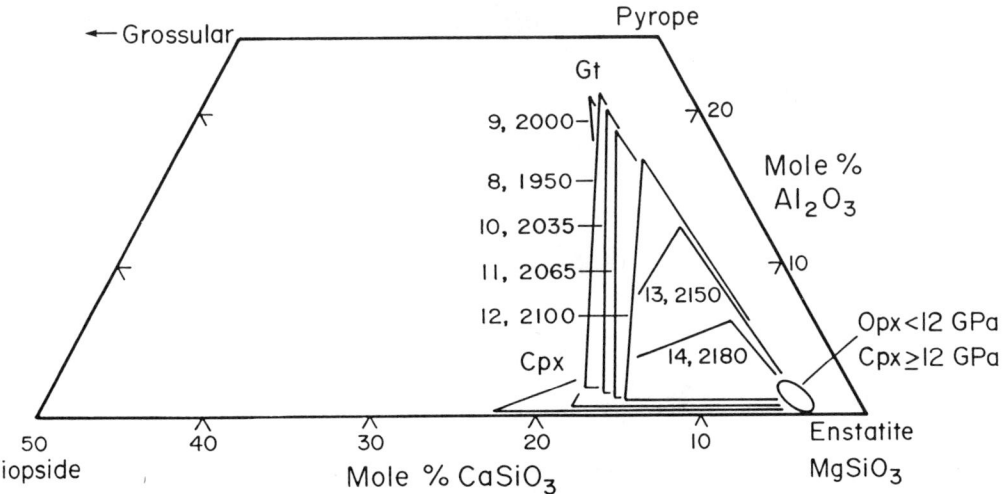

Fig. 20. Compositions (unsmoothed) of coexisting garnet (Gt), Ca-rich pyroxene (Cpx), and Ca-poor pyroxene (Opx or Cpx) at various pressures and temperatures [69]. Labels of the type, 9, 2000, indicate pressure (GPa) followed by temperature (°C). Pyrope, Mg$_3$Al$_2$Si$_3$O$_{12}$; Grossular, Ca$_3$Al$_2$Si$_3$O$_{12}$.

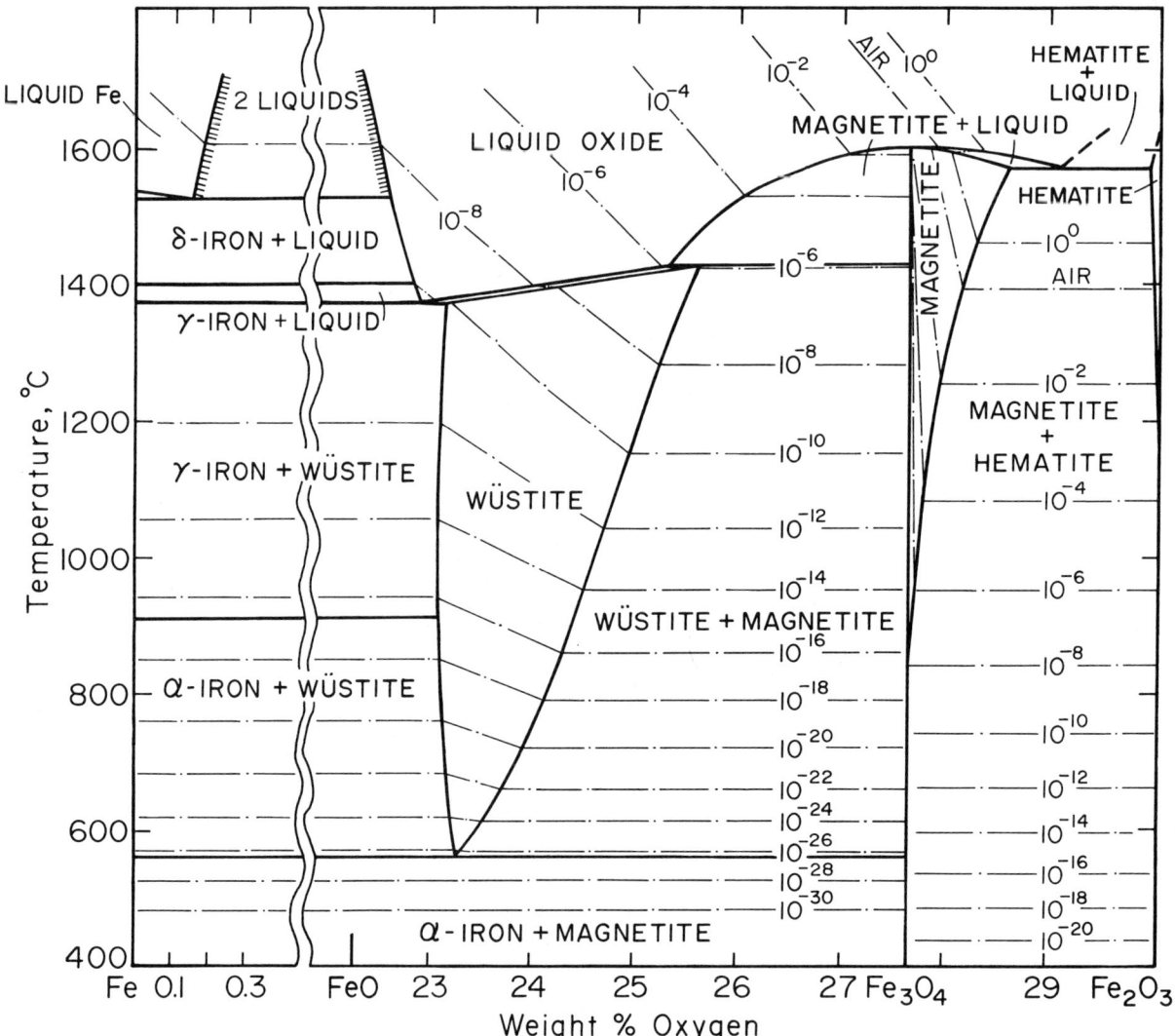

Fig. 21. Temperature-composition section for the system Fe-O at 1 atm [46, 47, 64, 120, 138]. Light dash-dot lines are oxygen isobars in atm.

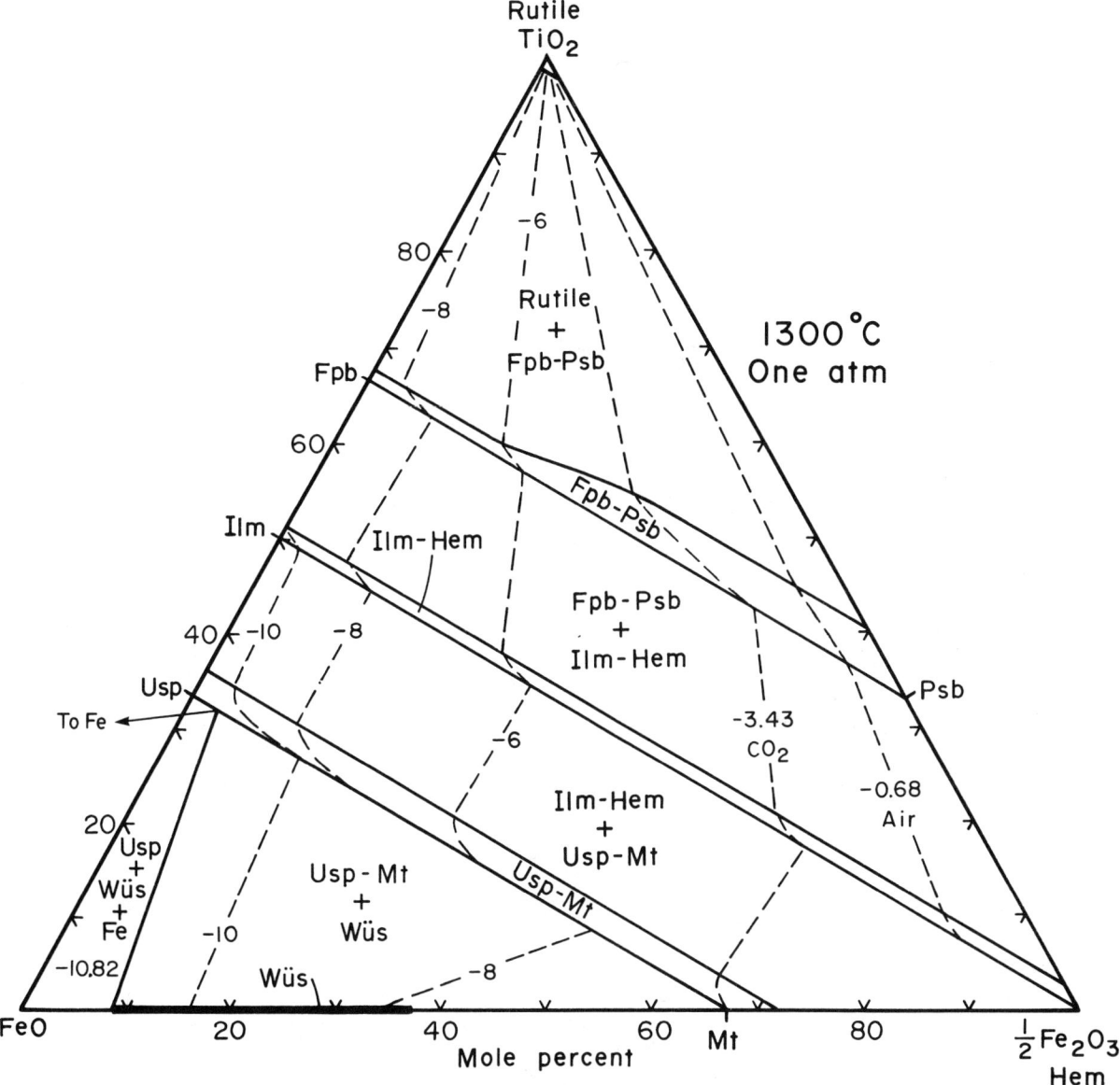

Fig. 22. The system TiO₂-FeO-Fe₂O₃ at 1300°C, 1 atm [161]. Light dashed lines are oxygen isobars labeled in log oxygen fugacity units (atm). Psb, pseudobrookite (Fe₂TiO₅); Fpb, ferropseudobrookite (FeTi₂O₅); Ilm, ilmenite (FeTiO₃); Hem, hematite (Fe₂O₃); Usp, ulvospinel (Fe₂TiO₄); Mt, magnetite (Fe₃O₄); Wüs, wüstite (Fe₁₋ₓO).

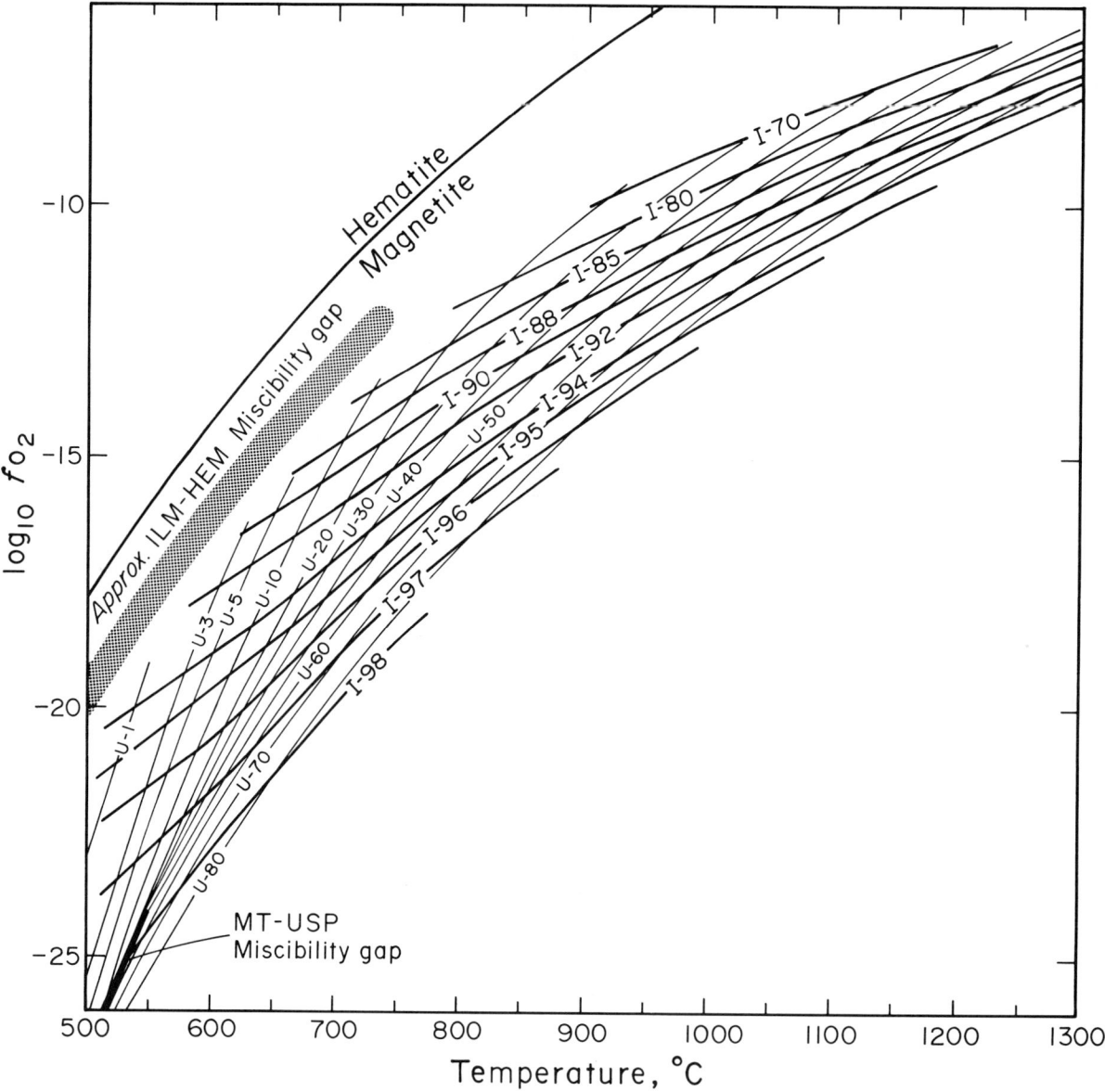

Fig. 23. Temperature-oxygen fugacity (f_{O_2}) grid for coexisting magnetite-ulvospinel solid solution and ilmenite-hematite solid solution pairs [155]. Lines with labels of the type, I-70, indicate mole % ilmenite in the ilmenite (FeTiO$_3$) - hematite (Fe$_2$O$_3$) solid solution. Lines with labels of the type, U-70, indicate mole % ulvospinel in the ulvospinel (Fe$_2$TiO$_4$) - magnetite (Fe$_3$O$_4$) solid solution. Mt, magnetite; Usp, ulvospinel; Ilm, ilmenite; Hem, hematite.

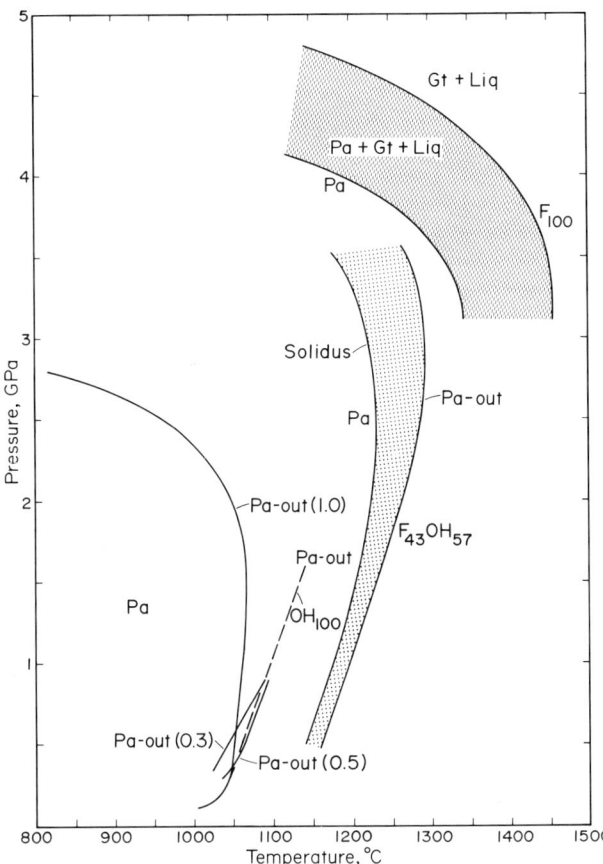

Fig. 24. Pressure-temperature projection of pargasite, (Pa) NaCa₂Mg₄Al₃Si₆O₂₂(OH,F)₂, stability limits. The univariant curves labeled Pa-out (1.0), Pa-out (0.5), and Pa-out (0.3) are from Gilbert [59], Holloway [71], and Oba [124], and give the maximum stability of pargasite during melting in the presence of a pure H_2O or H_2O-CO_2 vapor with H_2O mole fractions of 1.0, 0.5, and 0.3. Small concentrations of other constituents in the vapor are ignored. The dashed curve labeled OH_{100}, and the patterned areas labeled $F_{43}OH_{57}$ and F_{100} are from Holloway and Ford [72] and Foley [52], and show the breakdown of pargasite during vapor-absent melting for different proportions of fluorine and hydroxyl in the pargasite.

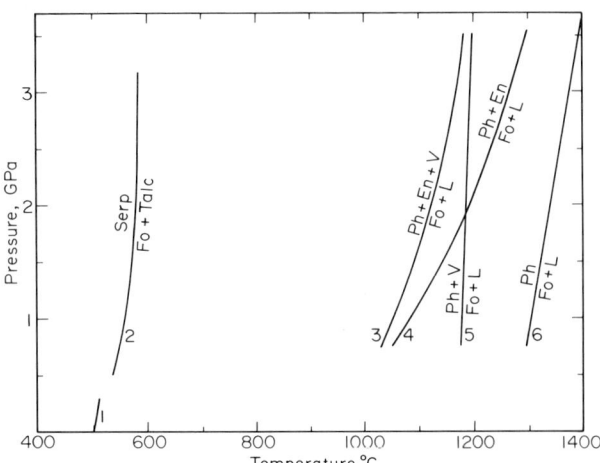

Fig. 25. Pressure-temperature projection showing the upper temperature stability limits for serpentine, $Mg_3Si_2O_5(OH)_4$ (curves 1 and 2), and phlogopite, $KMg_3AlSi_3O_{10}(OH)_2$ (curves 3-6). Numbers beside curves refer to the following sources: 1 - [28]; 2 - [90]; 3-6 - [115]. Curves 5 and 6 give the maximum stability of phlogopite in the presence (curve 5) and absence (curve 6) of vapor. Curves 3 and 4 give the corresponding maximum stability of phlogopite in the presence of forsterite and enstatite, and represent more closely the stability of phlogopite in the mantle. Curves 3-6 are not univariant [114]. See Yoder and Kushiro [174] for an earlier study of the stability of phlogopite. Montana and Brearley [116] speculated that a singular point exists at about 1.5 GPa on curve 4, so that the curve above this pressure is metastable.

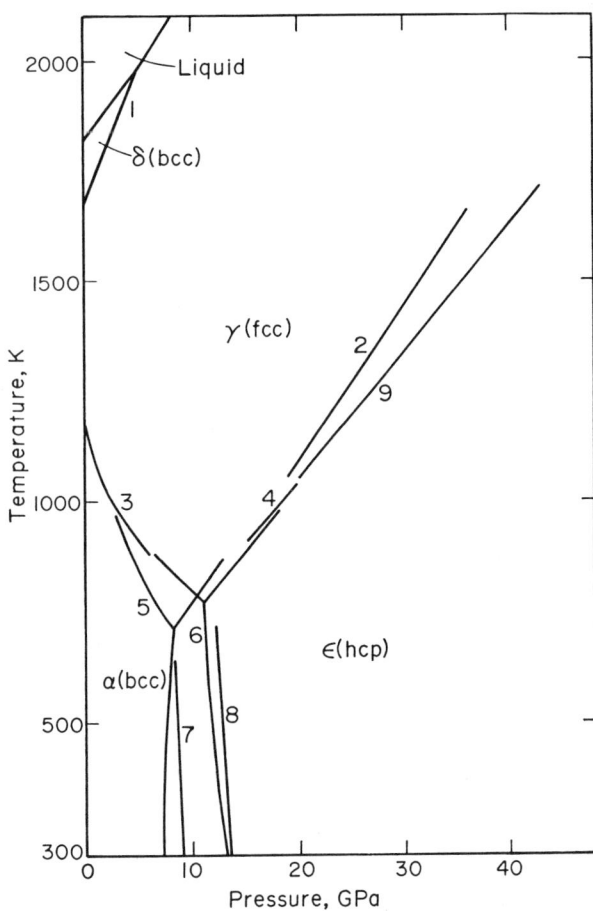

Fig. 26. Low pressure phase relationships for iron. Numbers beside curves refer to the following sources: 1 - [159]; 2 - [109]; 3 - [112]; 4 - [103]; 5 - [6]; 6 - [20, 40]; 7 - [107]; 8 - [73]; 9 - [17, 21]. Several additional references [61, 108, 179] also discuss the α-ε transition,.

Acknowledgements. Preparation of this compilation was supported by Texas Advanced Research Program Grants 009741-007 and 009741-066, and by National Science Foundation Grants EAR-8816044 and EAR-9219159. Contribution no. 738, Geosciences Program, University of Texas at Dallas.

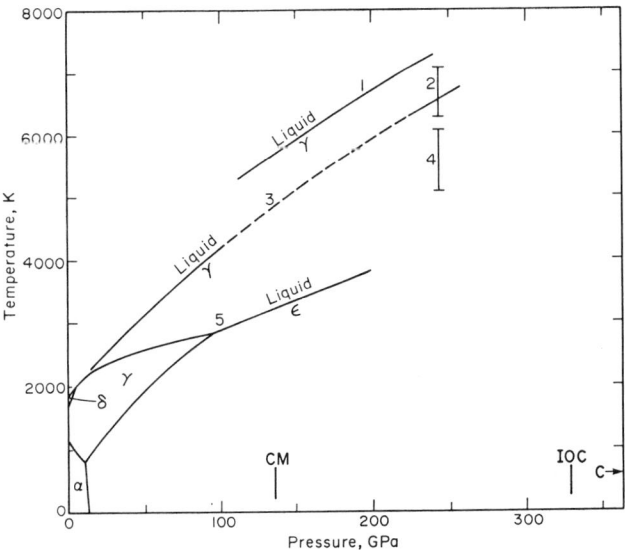

Fig. 27. High pressure melting curves for iron. For reference to Figure 26 at low pressures, the curves of Mirwald and Kennedy [112], Bundy [40], and Boehler [17] are used for the α-ε-γ phase relationships. Numbers beside melting curves and brackets refer to the following sources: 1 - [1]; 2 - [11]; 3 - [167]; 4 - [25, 38]; 5 - [17-19, 21, 103, 143, 153]. Gallagher and Ahrens [54] found that earlier shock data from their laboratory [11] are 1000°K too high, which brings the data of Bass *et al.* [11] and Brown and McQueen [38] into agreement. The shock data of Yoo *et al.* [176] (not plotted but located at 6350 K, 235 GPa and 6720 K, 300 GPa) are at slightly higher temperatures than the data of Brown and McQueen. (Ross *et al.* [145] and Anderson [8] have proposed the existence of a new phase, α'-iron, that is stable along the liquidus at pressures above about 170 GPa. On the basis of molecular dynamics calculations, Matsui [111] has also proposed the existence of a new phase at 300 GPa and temperatures above 5000 K. Saxena *et al.* [146] have suggested that α'-iron is the liquidus phase down to a pressure of 60-70 GPa. CM, core-mantle boundary (136 GPa); IOC, inner-outer core boundary (329 GPa); C, center of Earth (364 GPa).

REFERENCES

1. Ahrens, T. J., H. Tan, and J. D. Bass, Analysis of shock temperature data for iron, *High-Press. Res.*, 2, 145-157.

2. Akaogi, M., and S. Akimoto, Pyroxene-garnet solid-solution equilibria in the systems $Mg_4Si_4O_{12}$-$Mg_3Al_2Si_3O_{12}$ and $Fe_4Si_4O_{12}$-$Fe_3Al_2Si_3O_{12}$ at high pressures and temperatures, *Phys. Earth and Planet. Int., 15,* 90-106, 1977.

3. Akaogi, M., E. Ito, and A. Navrotsky, Olivine-modified spinel-spinel transitions in the system Mg_2SiO_4-Fe_2SiO_4: Calorimetric measurements, thermochemical calculation, and geo-

physical application, *J. Geophys. Res., 94*, 15,671-15,685, 1989.

4. Akaogi, M., and A. Navrotsky, The quartz - coesite - stishovite transformations: New calorimetric measurements and calculation of phase diagrams, *Phys. Earth Planet. Int., 36*, 124-134, 1984.

5. Akella, J., Quartz ↔ coesite transition and the comparative friction measurements in piston-cylinder apparatus using talc-alsimag-glass (TAG) and NaCl high-pressure cells, *Neues Jahrb. Mineral. Monatsh., 5*, 217-224, 1979.

6. Akimoto, S., T. Yagi, T. Suzuki, and O. Shimomura, Phase diagram of iron determined by high pressure X-ray diffraction using synchrotron radiation, in *High Pressure Research in Mineral Physics, Geophys. Monogr. Ser., v. 39*, edited by M. H. Manghnani and Y. Syono, pp. 149-154, Terra Scientific Pub., Tokyo, AGU, Washington, D. C., 1987.

7. Anastasiou, P., and F. Seifert, Solid solubility of Al_2O_3 in enstatite at high temperatures and 1-5 kb water pressure, *Contrib. Mineral. Petrol., 34*, 272-287, 1972.

8. Anderson, O. L., The phase diagram of iron and the temperature of the inner core, *J. Geomag. Geoelec., 45*, 1235-1248, 1993.

9. Atlas, L., The polymorphism of $MgSiO_3$ and solid-state equilibria in the system $MgSiO_3$-$CaMgSi_2O_6$, *J. Geol., 60*, 125-147, 1952.

10. Barth, T. F. W., The feldspar geologic thermometers, *Neues Jahrb. Mineral. Abhandlungen, 82*, 143-154, 1951.

11. Bass, J. D., B. Svendsen, and T. J. Ahrens, The temperature of shock compressed iron, in *High Pressure Research in Mineral Physics, Geophys. Monogr. Ser., v. 39*, edited by M. H. Manghnani and Y. Syono, pp. 393-402, Terra Scientific Pub., Tokyo, AGU, Washington, D. C., 1987.

12. Bell, P. M., and E. H. Roseboom, Jr., Melting relationships of jadeite and albite to 45 kilobars with comments on melting diagrams of binary systems at high pressures, in *Pyroxenes and Amphiboles: Crystal Chemistry and Phase Petrology, Min. Soc. Am. Spec. Paper 2*, edited by J. J. Papike, F. R. Boyd, J. R. Clark, and W. G. Ernst, pp. 151-161, Min. Soc. Am., 1969.

13. Berman, R. G., Internally-consistent thermodynamic data for minerals in the system Na_2O-K_2O-CaO-MgO-FeO-Fe_2O_3-Al_2O_3-SiO_2-TiO_2-H_2O-CO_2, *J. Petrol, 29*, 445-522, 1988.

14. Biggar, G. M., Calcium-poor pyroxenes: Phase relations in the system CaO-MgO-Al_2O_3-SiO_2, *Min. Mag., 49*, 49-58, 1985.

15. Biggar, G. M., Subsolidus equilibria between protopyroxene, orthopyroxene, pigeonite and diopside, in the $MgSiO_3$-$CaMgSi_2O_6$ at 1 bar to 9.5 kbar, and 1012 to 1450°C, *Eur. J. Mineral., 4*, 153-170, 1992.

16. Birch, F., and P. LeCompte, Temperature-pressure plane for albite composition, *Am. J. Sci., 258*, 209-217, 1960.

17. Boehler, R., The phase diagram of iron to 430 kbar, *Geophys. Res. Lett., 13*, 1153-1156, 1986.

18. Boehler, R., Melting of the Fe-FeO and the Fe-FeS systems at high pressures: Constraints on core temperatures, *Earth Planet. Sci. Lett., 111*, 217-227, 1992.

19. Boehler, R., The phase diagram of iron to 2 Mbar: New static measurements (abstract), *Eos Trans. AGU, 74, No. 16*, 305, 1993.

20. Boehler, R., M. Nicol, and M. L. Johnson, Internally-heated diamond-anvil cell: Phase diagram and P-V-T of iron, in *High Pressure Research in Mineral Physics, Geophys. Monogr. Ser., v. 39*, edited by M. H. Manghnani and Y. Syono, pp. 173-176, Terra Scientific Pub., Tokyo, AGU, Washington, D. C., 1987.

21. Boehler, R., N. von Bargen, and A. Chopelas, Melting, thermal expansion, and phase transitions of iron at high pressures, *J. Geophys. Res., 95*, 21,731-21,736, 1990.

22. Boettcher, A. L., and P. J. Wyllie, Jadeite stability measured in the presence of silicate liquids in the system $NaAlSiO_4$-SiO_2-H_2O, *Geochim. Cosmochim. Acta, 32*, 999-1012, 1968a.

23. Boettcher, A. L., and P. J. Wyllie, The quartz-coesite transition measured in the presence of a silicate liquid and calibration of piston-cylinder apparatus, *Contrib. Mineral. Petrol., 17*, 224-232, 1968b.

24. Bohlen, S. R., and A. L. Boettcher, The quartz ↔ coesite transformation: A precise determination and the effects of other components, *J. Geophys. Res., 87*, 7073-7078, 1982.

25. Boness, D. A., and J. M. Brown, The electronic band structures of iron, sulfur, and oxygen at high pressures and the Earth's core, *J. Geophys. Res., 95*, 21,721-21,730, 1990.

26. Bowen, N. L., The melting phenomena of the plagioclase feldspars, *Am. J. Sci., 35*, 577-599, 1913.

27. Bowen, N. L., and J. F. Schairer, The system MgO-FeO-SiO_2, *Am. J. Sci., 29*, 151-217, 1935.

28. Bowen, N. L., and O. F. Tuttle, The system MgO-SiO_2-H_2O, *Bull. Geol. Soc. Am., 60*, 439-460, 1949.

29. Bowen, N. L., and O. F. Tuttle, The system $NaAlSi_3O_8$-$KAlSi_3O_8$-H_2O, *J. Geol., 58*, 489-511, 1950.

30. Boyd, F. R., Garnet peridotites and the system $CaSiO_3$-$MgSiO_3$-Al_2O_3, in *Fiftieth Anniversary Symposia: Mineralogy and petrology of the upper mantle, sulfides, mineralogy and geochemistry of non-marine evaporites, Min. Soc. America Special Paper 3*, edited by B. A. Morgan, pp. 63-75, Min. Soc. Am., 1970.

31. Boyd, F. R., and J. L. England,

The quartz-coesite transition, *J. Geophys. Res., 65*, 749-756, 1960.

32. Boyd, F. R., and J. L. England, Mantle minerals, *Carnegie Inst. Washington Year Book 61*, 107-112, 1962.

33. Boyd, F. R., and J. L. England, Effect of pressure on the melting of diopside, $CaMgSi_2O_6$, and albite, $NaAlSi_3O_8$, in the range up to 50 kilobars, *J. Geophys. Res., 68*, 311-323, 1963.

34. Boyd, F. R., and J. L. England, The system enstatite-pyrope, *Carnegie Inst. Wash. Year Book 63*, 157-161, 1964.

35. Boyd, F. R., J. L. England, and B. T. C. Davis, Effects of melting and polymorphism of enstatite, $MgSiO_3$, *J. Geophys. Res., 69*, 2101-2109, 1964.

36. Boyd, F. R., and J. F. Schairer, The system $MgSiO_3$-$CaMgSi_2O_6$, *J. Petrol., 5*, 275-309, 1964.

37. Brey, G., and J. Huth, The enstatite-diopside solvus to 60 kbar, *Proc. Third Int. Kimberlite Conf., 2*, 257-264, 1984.

38. Brown, J. M., and R. G. McQueen, Phase transitions, Grüneisen parameter, and elasticity for shocked iron between 77 GPa and 400 GPa, *J. Geophys. Res., 91*, 7485-7494, 1986.

39. Brown, W. L., and I. Parsons, Towards a more practical two-feldspar geothermometer, *Contrib. Mineral Petrol., 76*, 369-377, 1981.

40. Bundy, F. P., Pressure-temperature phase diagram of iron to 200 kbar, 900°C, *J. Appl. Phys., 36*, 616-620, 1965.

41. Carlson, W. D., Evidence against the stability of orthoenstatite above 1005°C at atmospheric pressure in CaO-MgO-SiO_2, *Geophys. Res. Lett., 12*, 490-411, 1985.

42. Carlson, W. D., Reversed pyroxene phase equilibria in CaO-MgO-SiO_2 at one atmosphere pressure, *Contrib. Mineral. Petrol., 29*, 218-224, 1986.

43. Carlson, W. D., Subsolidus phase equilibria on the forsterite-saturated join $Mg_2Si_2O_6$-$CaMgSi_2O_6$ at atmospheric pressure, *Am. Mineral., 73*, 232-241, 1988.

44. Carlson, W. D., and D. H Lindsley, Thermochemistry of pyroxenes on the join $Mg_2Si_2O_6$-$CaMgSi_2O_6$, *Am. Mineral., 73*, 242-252, 1988.

45. Chen, C.-H., and D. C. Presnall, The system Mg_2SiO_4-SiO_2 at pressures up to 25 kilobars, *Am. Mineral., 60*, 398-406, 1975.

46. Darken, L. S., and R. W. Gurry, The system iron-oxygen, I. The wüstite field and related equilibria, *J. Am. Chem. Soc., 67*, 1398-1412, 1945.

47. Darken, L. S., and R. W. Gurry, The system iron-oxygen. II. Equilibrium and thermodynamics of liquid oxide and other phases, *J. Am. Chem. Soc., 68*, 798-816, 1946.

48. Davis, B. T. C., and J. L. England, The melting of forsterite up to 50 kilobars, *J. Geophys. Res., 69*, 1113-1116, 1964.

49. Elkins, L. T., and T. L. Grove, Ternary feldspar experiments and thermodynamic models, *Am. Mineral., 75*, 544-559, 1990.

50. Essene, E. J., Geologic thermometry and barometry, in *Characterization of Metamorphism Through Mineral Equilibria, Rev. Min., Min. Soc. Am., 10*, edited by J. M. Ferry, 153-206, 1982.

51. Fei, Y., H.-K. Mao, and B. O. Mysen, Experimental determination of element partitioning and calculation of phase relationships in the MgO-FeO-SiO_2 system at high pressure and high temperature, *J. Geophys. Res., 96*, 2157-2169, 1991.

52. Foley, S., High-pressure stability of the fluor- and hydroxy-endmembers of pargasite and K-richterite, *Geochim. Cosmochim. Acta, 55*, 2689-2694, 1991.

53. Fuhrman, M. L., and D. H. Lindsley, Ternary feldspar modeling and thermometry, *Am. Mineral., 73*, 201-215, 1988.

54. Gallagher, K. G., and T. J. Ahrens, First measurements of thermal conductivity in griceite and corundum at ultra high pressure and the melting point of iron (abstract), *Eos Trans. AGU, 75, No. 44*, 653, 1994.

55. Gasparik, T., Transformation of enstatite-diopside-jadeite pyroxenes to garnet, *Contrib. Mineral. Petrol., 102*, 389-405, 1989.

56. Gasparik, T., A thermodynamic model for the enstatite-diopside join, *Am. Mineral., 75*, 1080-1091, 1990a.

57. Gasparik, T., Phase relations in the transition zone, *J. Geophys. Res., 95*, 15,751-15,769, 1990b.

58. Ghiorso, M. S., Activity/composition relations in the ternary feldspars, *Contrib. Mineral. Petrol., 87*, 282-296, 1984.

59. Gilbert, M. C., Reconnaissance study of the stability of amphiboles at high pressure, *Carnegie Inst. Washington Year Book 67*, 167-170, 1969.

60. Gilbert, M. C., R. T. Helz, R. K. Popp, and F. S. Spear, Experimental studies of amphibole stability, in *Amphiboles: Petrology and Experimental Phase Relations*, edited by D. R. Veblen and P. H. Ribbe, *Rev. Mineralogy, 9B*, Chapter 2, pp. 229-353, 1982.

61. Giles, P. M., M. H. Longenbach, and A. R. Marder, High pressure martensitic transformation in iron, *J. Appl. Phys., 42*, 4290-4295, 1971.

62. Green, D. H., A. E. Ringwood, and A. Major, Friction affects and pressure calibration in a piston-cylinder apparatus at high pressure and temperature, *J. Geophys. Res., 71*, 3589-3594, 1966.

63. Green, N. L., and S. I. Udansky, Ternary-feldspar mixing relations and feldspar thermobarometry, *Am. Mineral., 71*, 1100-1108, 1986.

64. Greig, J. W., E. Posnjak, H. E. Merwin, and R. B. Sosman, Equilibrium relationships of Fe_3O_4, Fe_2O_3, and oxygen, *Am. J. Sci., 30*, 239-316, 1935.

65. Grover, J. E., The stability of low-clinoenstatite in the system Mg_2SiO_4 - $CaMgSi_2O_6$ (abstract), *Eos Trans. AGU, 53*, 539, 1972.

66. Haselton, H. T., Jr., G. L. Hovis, B. S. Hemingway, and R. A. Robie, Calorimetric investigation of the excess entropy of mixing in analbite-sanidine solid solutions: Lack of evidence for Na, K short-range order and implications for two-feldspar thermometry, *Am. Mineral., 68*, 398-413, 1983.

67. Hays, J. F., Lime-alumina-silica, *Carnegie Inst. Washington Year Book 65*, 234-239, 1966.

68. Henderson, C. M. B., Graphical analysis of phase equilibria in An-Ab-Or, in *Progress in Experimental Petrology, Nat. Environ. Res. Council Pub. 25, Series D*, edited by C. M. B. Henderson, pp. 70-78, University Press, Cambridge, 1984.

69. Herzberg, C., and T. Gasparik, Garnet and pyroxenes in the mantle: A test of the majorite fractionation hypothesis, *J. Geophys. Res., 96*, 16,263-16,274, 1991.

70. Holland, T. J. B., and R. Powell, An enlarged and updated internally consistent thermodynamic data set with uncertainties and correlations: The system K_2O - Na_2O - CaO - MgO - MnO - FeO - Fe_2O_3-Al_2O_3-TiO_2-SiO_2-C-H_2-O_2, *J. Met. Geol., 8*, 89-124, 1990.

71. Holloway, J. R., The system pargasite-H_2O-CO_2: A model for melting of a hydrous mineral with a mixed volatile fluid - I. Experimental results to 8 kbar, *Geochim. Cosmochim. Acta, 37*, 651-666, 1973.

72. Holloway, J. R., and C. E. Ford, Fluid-absent melting of the fluoro-hydroxy amphibole pargasite to 35 kilobars, *Earth Planet. Sci. Lett., 25*, 44-48, 1975.

73. Huang, E., W. A. Bassett, and P. Tao, Study of bcc-hcp iron phase transition by synchrotron radiation, in *High Pressure Research in Mineral Physics, Geophys.*

Monogr. Ser., v. 39, edited by M. H. Manghnani and Y. Syono, pp. 165-172, Terra Scientific Pub., Tokyo, AGU, Washington, D.C., 1987.

74. Huckenholz, H. G., M. C. Gilbert, and T. Kunzmann, Stability and phase relations of calcic amphiboles crystallized from magnesiohastingsite compositions in the 1 to 45 kbar pressure range, *Neues Jahrbuch Miner. Abh., 164*, 229-268, 1992.

75. Iiyama, J. T., Contribution à l'étude des équilibre sub-solidus du système ternaire orthoso-albite-anorthite à l'aide des réactions d'echange d'ions Na-K au contact d'une solution hydrothermale, *Bull. Soc Minéral. Cryst. Français, 89*, 442-454, 1966.

76. Ito, E., and E. Takahashi, Postspinel transformations in the system Mg_2SiO_4-Fe_2SiO_4 and some geophysical implications, *J. Geophys. Res., 94*, 10,637-10,646, 1989.

77. Jackson, I., Melting of the silica isotypes SiO_2, BeF_2, and GeO_2 at elevated pressures, *Phys. Earth Planet. Int., 13*, 218-231, 1976.

78. Jenner, G. A., and D. H. Green, Equilibria in the Mg-rich part of the pyroxene quadrilateral, *Min. Mag., 47*, 153-160, 1983.

79. Johannes, W., Melting of plagioclase in the system ab-an-H_2O and qz-ab-H_2O at P_{H_2O} = 5 kbars, an equilibrium problem, *Contrib. Mineral. Petrol. 66*, 295-303, 1978.

80. Johannes, W., Ternary feldspars: Kinetics and possible equilibria at 800ºC, *Contrib. Mineral. Petrol., 68*, 221-230, 1979.

81. Johannes, W., D. W. Chipman, J. F. Hays, P. M. Bell, H. K. Mao, R. C. Newton, A. L. Boettcher, and F. Seifert, An interlaboratory comparison of piston-cylinder pressure calibration using the albite-breakdown reaction, *Contrib. Mineral. Petrol., 32*, 24-38, 1971.

82. Kanzaki, M., Ultrahigh-pressure phase relations in the system

$Mg_4Si_4O_{12}$-$Mg_3Al_2Si_3O_{12}$, *Phys. Earth Planet. Inter., 49*, 168-175, 1987.

83. Kanzaki, M., Melting of silica up to 7 GPa, *J. Am. Ceram. Soc., 73*, 3706-3707, 1990.

84. Kato, T., and M. Kumazawa, Effect of high pressure on the melting relation in the system Mg_2SiO_4-$MgSiO_3$ Part I. Eutectic relation up to 7 GPa, *J. Phys. Earth, 33*, 513-524, 1985a.

85. Kato, T., and M. Kumazawa, Stability of phase B, a hydrous magnesium silicate, to 2300ºC at 20 GPa, *Geophys. Res. Lett., 12*, 534-535, 1985b.

86. Kato, T., and M. Kumazawa, Melting and phase relations in the system Mg_2SiO_4-$MgSiO_3$ at 20 GPa under hydrous conditions, *J. Geophys. Res., 91*, 9351-9355, 1986.

87. Katsura, T., and E. Ito, The system Mg_2SiO_4-Fe_2SiO_4 at high pressures and temperatures: Precise determination of stabilities of olivine, modified spinel, and spinel, *J. Geophys. Res., 94*, 15,663-15,670, 1989.

88. Kinomura, N., S. Kume, and M. Koizumi, Stability of $K_2Si_4O_9$ with wadeite type structure, in *Proc. 4th Inter. Conf. on High Pressure*, edited by J. Osugi, pp. 211-214, Phys. Chem. Soc. Japan, 1975.

89. Kitahara, S., and G. C. Kennedy, The quartz-coesite transition, *J. Geophys. Res., 69*, 5395-5400, 1964.

90. Kitahara, S., S. Takanouchi, and G. C. Kennedy, Phase relations in the system MgO-SiO_2-H_2O at high temperatures and pressures, *Am. J. Sci., 264*, 223-233, 1966.

91. Kushiro, I., Determination of liquidus relations in synthetic silicate systems with electron probe analysis: The system forsterite-diopside-silica at 1 atmosphere, *Am. Mineral., 57*, 1260-1271, 1972.

92. Kushiro, I., H. S. Yoder, Jr., and M. Nishikawa, Effect of water on the melting of enstatite, *Geol.*

Soc. Am. Bull., 79, 1685-1692, 1968.

93. Lindsley, D. H., Melting relations of KAlSi₃O₈: Effect of pressures up to 40 kilobars, *Am. Mineral, 51,* 1793-1799, 1966.

94. Lindsley, D. H., Melting relations of plagioclase at high pressures, in *New York State Museum and Science Service Memoir 18,* pp. 39-46, 1968.

95. Lindsley, D. H., Pyroxene thermometry, *Am. Mineral., 68,* 477-493, 1983.

96. Lindsley, D. H., Experimental studies of oxide minerals, in *Oxide Minerals: Petrologic and Magnetic Significance, Rev. Mineralogy, 25, Chapter 3,* edited by D. H. Lindsley, 69-106, 1991.

97. Lindsley, D. H., and D. J. Andersen, a two-pyroxene thermometer, *Proc. Thirteenth Lunar Planet. Sci. Conf., Part 2, J. Geophys. Res., 88, Supplement,* A887-A906, 1983.

98. Lindsley, D. H., and S. A. Dixon, Diopside-enstatite equilibria at 850 to 1400°C, 5 to 35 kbars, *Am. J. Sci., 276,* 1285-1301, 1976.

99. Liu, L.-G., Silicate perovskite from phase transformations of pyrope-garnet at high pressures and temperature, *Geophys. Res. Lett., 1,* 277-280, 1974.

100. Liu, L.-G., First occurrence of the garnet-ilmenite transition in silicates, *Science, 195,* 990-991, 1977.

101. Liu, L. G., High-pressure phase transformations of albite, jadeite, and nepheline, *Earth Planet. Sci. Lett., 37,* 438-444, 1978.

102. Liu, L.-G., The system enstatite-wollastonite at high pressures and temperatures with emphasis on diopside, *Phys. Earth Planet. Int., 19,* P15-P18, 1979.

103. Liu, L.-G., and W. A. Bassett, The melting of iron up to 200 kilobars, *J. Geophys. Res., 80,* 3777-3782, 1975.

104. Liu, L.-G., and W. A. Bassett, *Elements, Oxides, Silicates. High-Pressure Phases with Impli-*

cations for the Earth's Interior, 250 pp., Oxford Univ. Press, New York, 1986.

105. Liu, L.-G., W. A. Bassett, and J. Sharry, New high-pressure modifications of GeO₂ and SiO₂, *J. Geophys. Res., 83,* 2301-2305, 1978.

106. Longhi, J., and A. E. Boudreau, The orthoenstatite liquidus field in the system forsterite-diopside-silica at one atmosphere, *Am. Mineral., 65,* 563-573, 1980.

107. Manghnani, M. H., L. C. Ming, and N. Nakagiri, Investigation of the α-Fe to ε-iron phase transition by synchrotron radiation, in High Pressure Research in *Mineral Physics, Geophys. Monogr. Ser., v. 39,* edited by M. H. Manghnani and Y. Syono, pp. 155-164, Terra Scientific Pub., Tokyo, AGU, Washington, D. C., 1987.

108. Mao, H. K., W. A. Bassett, and T. Takahashi, Effects of pressure on crystal structure and lattice parameters of iron up to 300 kbar, *J. Appl. Phys., 38,* 272-276, 1967.

109. Mao, H. K., P. M. Bell, and C. Hadidiacos, Experimental phase relations of iron to 360 kbar, 1400°C, determined in an internally heated diamond-anvil apparatus, in *High Pressure Research in Mineral Physics, Geophys. Monogr. Ser., v. 39,* edited by M. H. Manghnani and Y. Syono, pp. 135-138, 1987.

110. Mao, H. K., T. Yagi, and P. M. Bell, Mineralogy of Earth's deep mantle: Quenching experiments on mineral compositions at high pressure and temperature, *Carnegie Inst. Washington Yearbook 76,* 502-504, 1977.

111. Matsui, M., Computer simulation of the structural and physical properties of iron under ultra-high pressures and high temperatures, *Central Core Earth, 2,* 79-82, 1992.

112. Mirwald, P. W., and G. C. Kennedy, The Curie point and the α-γ transition of iron to 53 kbar - a reexamination, *J. Geophys.*

Res., 84, 656-658, 1979.

113. Mirwald, P. W., and H.-J. Massonne, The low-high quartz and quartz-coesite transition to 40 kbar between 600° and 1600°C and some reconnaissance data on the effect of NaAlO₂ component on the low quartz-coesite transition, *J. Geophys. Res., 85,* 6983-7990, 1980.

114. Modreski, P. J., and A. L. Boettcher, The stability of phlogopite + enstatite at high pressures: A model for micas in the interior of the Earth, *Am. J. Sci., 272,* 852-869, 1972.

115. Modreski, P. J., and A. L. Boettcher, Phase relationships of phlogopite in the system K₂O-MgO-CaO-Al₂O₃-SiO₂-H₂O to 35 kilobars: A better model for micas in the interior of the Earth, *Am. J. Sci., 273,* 385-414, 1973.

116. Montana, A., and M. Brearley, An appraisal of the stability of phlogopite in the crust and in the mantle, *Am. Mineral., 74,* 1-4, 1989.

117. Morey, G. W., Phase-equilibrium relations of the common rock-forming oxides except water, *U. S. Geol. Survey Prof. Paper 440-L,* L1-L158, 1964.

118. Mori, T., and D. H. Green, Pyroxenes in the system Mg₂Si₂O₆-CaMgSi₂O₆ at high pressure, *Earth Planet. Sci. Lett., 26,* 277-286, 1975.

119. Morse, S. A., Alkali feldspars with water at 5 kb pressure, *J. Petrol., 11,* 221-251, 1970.

120. Muan, A., and E. F. Osborn, *Phase Equilibria Among Oxides in Steelmaking,* 236 pp., Addison-Wesley, Reading, MA, 1965.

121. Nekvasil, H., and D. H. Lindsley, Termination of the 2 feldspar + liquid curve in the system Ab-Or-An-H₂O at low H₂O contents, *Am. Mineral., 75,* 1071-1079, 1990.

122. Newton, R. C., Thermodynamic analysis of phase equilibria in simple mineral systems, in *Thermodynamic Modeling of Geological Materials: Minerals, Fluids, and Melts, Rev. Mineral.,*

Mineral Soc. Am. 17, edited by I. S. E. Carmichael and H. P. Eugster, pp. 1-33, 1987.

123. Nickel, K. G., and G. Brey, Subsolidus orthopyroxene-clinopyroxene systematics in the system CaO-MgO-SiO$_2$ to 60 kbar. A re-evaluation of the regular solution model, *Contrib. Mineral. Petrol.,* 87, 35-42, 1984.

124. Oba, T., Experimental study on the tremolite-pargasite join at variable temperatures under 10 kbar, *Proc. Indian Acad. Sci., 99,* 81-90, 1990.

125. Ohtani, E., T. Irifune, and K. Fujino, Fusion of pyrope at high-pressures and rapid crystal growth from the pyrope melt, *Nature, 294,* 62-64, 1981.

126. Ohtani, E., and M. Kumazawa, Melting of forsterite Mg$_2$SiO$_4$ up to 15 GPa, *Phys. Earth Planet. Inter., 27,* 32-38, 1981.

127. Pacalo, R. I. G., and T. Gasparik, Reversals of the orthoenstatite-clinoenstatite transition at high pressures and high temperatures, *J. Geophys. Res., 95,* 15,853-15,858, 1990.

128. Perkins, D., III, and R. C. Newton, The composition of coexisting pyroxenes and garnet in the system CaO-MgO-Al$_2$O$_3$-SiO$_2$ at 900-1100°C and high pressures, *Contrib. Mineral. Petrol., 75,* 291-300, 1980.

129. *Phase Diagrams for Ceramists, Vol. 1,* edited by E. M. Levin, C. R. Robbins, and H. F. McMurdie, 601 pp., Amer. Ceramic Soc., Columbus, Ohio, 1964.

130. *Phase Diagrams for Ceramists, Vol. 2,* edited by E. M. Levin and H. F. McMurdie, 625 pp., Amer. Ceramic Soc., Columbus, Ohio, 1969.

131. *Phase Diagrams for Ceramists, Vol. 3,* edited by E. M. Levin, C. R. Robbins, and H. F. McMurdie, 513 pp., Amer. Ceramic Soc., Columbus, Ohio, 1975.

132. *Phase Diagrams for Ceramists, Vol. 4,* edited by R. S. Roth, T. Negas, and L. P. Cook, 330 pp., Amer. Ceramic Soc., Columbus, Ohio, 1981.

133. *Phase Diagrams for Ceramists, Vol. 5,* edited by R. S. Roth, T. Negas, and L. P. Cook, 395 pp., Amer. Ceramic Soc., Columbus, Ohio, 1983.

134. *Phase Diagrams for Ceramists, Cumulative Index,* 495 pp., Amer. Ceramic Soc., Columbus, Ohio, 1984.

135. *Phase Diagrams for Ceramists, Vol. 6,* edited by R. S. Roth, J. R. Dennis, and H. F. McMurdie, 515 pp., Amer. Ceramic Soc., Columbus, Ohio, 1987.

136. *Phase Diagrams for Ceramists, Vol. 7,* 591 pp., Amer. Ceramic Soc., Columbus, Ohio, 1989.

137. *Phase Diagrams for Ceramists, Vol. 8,* edited by B. O. Mysen, 399 pp., Amer. Ceramic Soc., Columbus, Ohio, 1989.

138. Phillips, B., and A. Muan, Stability relations of iron oxides: phase equilibria in the system Fe$_3$O$_4$-Fe$_2$O$_3$ at oxygen pressures up to 45 atmospheres, *J. Phys. Chem., 64,* 1451-1453, 1960.

139. Powell, M, and R. Powell, Plagioclase-alkali feldspar geothermometry revisited, *Min. Mag., 41,* 253-256, 1977.

140. Presnall, D. C., and T. Gasparik, Melting of enstatite (MgSiO$_3$) from 10 to 16.5 GPa and the forsterite (Mg$_2$SiO$_4$) - majorite (MgSiO$_3$) eutectic at 16.5 GPa: Implications for the origin of the mantle, *J. Geophys. Res., 95,* 15,771-15,777, 1990.

141. Presnall, D. C., and M. J. Walter, Melting of forsterite, Mg$_2$SiO$_4$, from 9.7 to 16.5 GPa, *J. Geophys. Res., 98,* 19,777-19,783, 1993.

142. Price, J. G., Ideal site-mixing in solid solutions with applications to two-feldspar geothermometry, *Am. Mineral., 70,* 696-701, 1985.

143. Ringwood, A. E., and W. Hibberson, The system Fe-FeO revisited, *Phys. Chem. Miner. 17,* 313-319, 1990.

144. Ringwood, A. E., A. F. Reid, and A. D. Wadsley, High-pressure KAlSi$_3$O$_8$, an aluminosilicate with sixfold coordination, *Acta Cryst., 23,* 1093-1095, 1967.

145. Ross, M., D. A. Young, and R. Grover, Theory of the iron phase diagram at Earth core conditions, *J. Geophys. Res., 95,* 21,713-21,716, 1990.

146. Saxena, S. K., G. Shen, and P. Lazor, High pressure phase equilibrium data for iron: Discovery of a new phase with implications for Earth's core, *EOS Trans. Am. Geophys. Union, 74,* 305.

147. Sawamoto, H., Phase diagram of MgSiO$_3$ at pressures up to 24 GPa and temperatures up to 2200°C: Phase stability and properties of tetragonal garnet, in *High Pressure Research in Mineral Physics, Geophys. Monogr. Ser., v. 39,* edited by M. H. Manghnani and Y. Syono, pp. 209-219, Terra Scientific Pub., Tokyo, AGU, Washington, D. C., 1987.

148. Schairer, J. F., The alkali-feldspar join in the system NaAlSiO$_4$-KAlSiO$_4$-SiO$_2$, *J. Geol., 58,* 512-517, 1950.

149. Schairer, J. F., and N. L. Bowen, The system K$_2$O-Al$_2$O$_3$-SiO$_2$, *Am. J. Sci., 253,* 681-746, 1955.

150. Schweitzer, E., The reaction pigeonite = diopside$_{ss}$ + enstatite$_{ss}$ at 15 kbars, *Am. Mineral., 67,* 54-58, 1982.

151. Seck, H. A., Die einfluβ des drucks auf die zusammensetzung koexistierende alkalifeldspate und plagioklase im system NaAlSi$_3$O$_8$-KAlSi$_3$O$_8$-CaAl$_2$Si$_2$O$_8$ - H$_2$O, *Contrib. Mineral. Petrol., 31,* 67-86, 1971a.

152. Seck, H. A., Koexistierende alkalifeldspate und plagioklase im system NaAlSi$_3$O$_8$-KAlSi$_3$O$_8$-CaAl$_2$Si$_2$O$_8$-H$_2$O bei temperaturen von 650°C bis 900°C, *Neues Jahrb. Mineral. Abhandlungen, 115,* 315-342, 1971b.

153. Shen, G., P. Lazor, and S. K. Saxena, Laser-heated diamond-anvil cell experiments at high pressures. II. Melting of iron and

wüstite (abstract), *Eos Trans. AGU, 73, No. 14,* 368, 1992.

154. Spear, F. S. and J. T. Chaney, A petrogenetic grid for pelitic schists in the system SiO_2-Al_2O_3-FeO - MgO - K_2O - H_2O, *Contrib. Mineral. Petrol., 101,* 149-164, 1989.

155. Spencer, K. J., and D. H. Lindsley, A solution model for coexisting iron-titanium oxides, *Am. Mineral., 66,* 1189-1201, 1981.

156. Stewart, D. B., and E. H. Roseboom, Jr., Lower temperature terminations of the three-phase region plagioclase-alkali feldspar -liquid, *J. Petrol., 3,* 280-315, 1962.

157. Stixrude, L., and M. S. T. Bukowinski, Stability of $(Mg,Fe)SiO_3$ perovskite and the structure of the lowermost mantle, *Geophys. Res. Lett., 19,* 1057-1060, 1992.

158. Stormer, J. C., Jr., A practical two-feldspar geothermometer, *Am. Mineral., 60,* 667-674, 1975.

159. Strong, H. M., R. E. Tuft, and R. E. Hannemann, The iron fusion curve and the triple point, *Metall. Trans., 4,* 2657-2611, 1973.

160. Suito, K., Phase relations of pure Mg_2SiO_4 up to 200 kilobars, in *High-Pressure Research,* edited by M. H. Manghnani and S. Akimoto, pp. 255-266, Academic Press, 1977.

161. Taylor, R. W., Phase equilibria in the system FeO-Fe_2O_3-TiO_2 at 1300°C, *Am. Mineral., 49,* 1016-1030, 1964.

162. Togaya, M., High pressure generation and phase transformations of SiO_2 and GeO$_2$, in *High Pressure in Science and Technology, Mat. Res. Symp. Proc., 22,* edited by C. Homan, R. K. MacCrone, and E. Whalley, 373-376, 1984.

163. Tsuchida, Y., and T. Yagi, A new post-stishovite high-pressure polymorph of silica, *Nature, 340,* 217-220, 1989.

164. Tuttle, O. F., and N. L. Bowen, Origin of granite in the light of experimental studies in the system $NaAlSi_3O_8$-$KAlSi_3O_8$-SiO_2-H_2O, *Geol Soc. Am. Mem. 74,* 153 pp., 1958.

165. Whitney, J. A., and J. C. Stormer, Jr., The distribution of $NaAlSi_3O_8$ between coexisting microcline and plagioclase and its effect on geothermometric calculations, *Am. Mineral, 62,* 687-691, 1977.

166. Williams, D. W., and Kennedy, G. C., Melting curve of diopside to 50 kilobars, *J. Geophys. Res., 74,* 4359-5366, 1969.

167. Williams, Q., R. Jeanloz, J. Bass, B. Svendsen, and T. J. Ahrens, The melting curve of iron to 250 gigapascals: A constraint on the temperature at Earth's core, *Science, 236,* 181-183, 1987.

168. Yagi, T., and S. Akimoto, Direct determination of coesite-stishovite transition by in-situ X-ray measurements, *Tectonophys., 35,* 259-270, 1976.

169. Yamamoto, K., and S. Akimoto, The system MgO-SiO_2-H_2O at high pressures and temperatures - Stability field for hydroxyl-chondrodite, hydroxy-clinohumite and 10 Å-phase, *Am. J. Sci., 277,* 288-312, 1977.

170. Yang, H.-Y., Crystallization of iron-free pigeonite in the system anorthite-diopside-enstatite-silica at atmospheric pressure, *Am. J. Sci., 273,* 488-497, 1973.

171. Yang, H.-Y., and W. R. Foster, Stability of iron-free pigeonite at atmospheric pressure, *Am. Mineral., 57,* 1232-1241, 1972.

172. Yoder, H. S., Jr., High-low quartz inversion up to 10,000 bars, *Trans. AGU, 31,* 827-835, 1950.

173. Yoder, H. S., Jr., Change of melting point of diopside with pressure, *J. Geol., 60,* 364-374, 1952.

174. Yoder, H. S., Jr., and I. Kushiro, Melting of a hydrous phase: phlogopite, *Am. J. Sci., 267-A,* 558-582, 1969.

175. Yoder, H. S., Jr., D. B. Stewart, and J. R. Smith, Ternary feldspars, *Carnegie Inst. Washington Year Book 56,* 206-214, 1957.

176. Yoo, C. S., N. C. Holmes, M. Ross, D. J. Webb, and C. Pike, Shock temperatures and melting of iron at Earth core conditions, *Phys. Rev. Lett., 70,* 3931-3934, 1993.

177. Zerr, A., and Boehler, R., Melting of $(Mg,Fe)SiO_3$-perovskite under hydrostatic, inert conditions to 580 kbar (abstract), *Eos Trans. AGU, 74, No. 16,* 168-169, 1993.

178. Zhang, J., R. C. Liebermann, T. Gasparik, C. T. Herzberg, and Y. Fei, Melting and subsolidus relations of SiO_2 at 9-14 GPa, *J. Geophys. Res., 98,* 19,785-19,793, 1993.

179. Zou, G., P. M. Bell, and H. K. Mao, Application of the solid-helium pressure medium in a study of the α-ε Fe transition under hydrostatic pressure, *Carnegie Inst. Washington Year Book 80,* 272-274, 1981.

Diffusion Data for Silicate Minerals, Glasses, and Liquids

John B. Brady

1. INTRODUCTION

Diffusion is an integral part of many geologic processes and an increasing portion of the geologic
literature is devoted to the measurement, estimation,
and application of diffusion data. This compilation is
intended to be a guide to recent experimentally-
determined diffusion coefficients and, through the
papers cited, to important older literature. To provide
a context for the tables, a brief summary of the equations required for a phenomenological (macroscopic)
description of diffusion follows. Although the equations for well-constrained experiments are relatively
straightforward, the application of the resulting diffusion coefficients to complex geologic problems may
not be straightforward. The reader is urged to read
one or more diffusion texts [e.g., 99, 121, 32, 147,
135] before attempting to use the data presented here.

2. FORCES AND FLUXES

Diffusion is the thermally-activated, relative movement (flux) of atoms or molecules that occurs in
response to forces such as gradients in chemical
potential or temperature. Diffusion is spontaneous
and, therefore, must lead to a net decrease in free

energy. For example, the movement of Δn_i moles of
chemical component i from a region (II) of high
chemical potential ($\mu_i{}^{II}$) to a region (I) of lower
chemical potential ($\mu_i{}^{I}$) will cause the system Gibbs
energy (G) to fall since

$$\Delta G^I + \Delta G^{II} = \Delta n_i \left(\frac{\partial G^I}{\partial n_i} \right)_{P,T,n_j} - \Delta n_i \left(\frac{\partial G^{II}}{\partial n_i} \right)_{P,T,n_j} \quad (1)$$

and using the definition of μ_i [139, p.128],

$$\Delta G^{Total} = \Delta n_i \left(\mu_i^I - \mu_i^{II} \right) < 0 . \quad (2)$$

Thus, a chemical potential gradient provides a thermodynamic force for atom movement.

On a macroscopic scale, *linear* equations appear to
be adequate for relating each diffusive flux to the set
of operative forces [137, p.45]. The instantaneous,
one-dimensional, isothermal diffusive flux J_i^R (moles
of i/m^{2}s) of component i in a single-phase, n-
component system with respect to reference frame R
may be described by

$$J_i^R = \sum_{j=1}^{n} -L_{ij}^R \left(\frac{\partial \mu_j}{\partial x} \right), \quad (3)$$

where x (m) is distance and the n^2 terms L_{ij}^R (moles
of i/m·J·s) are "phenomenological diffusion coefficients" [36]. Because each component of the system
may move in response to a gradient in the chemical
potential of any other component, the complexity of
describing diffusion in multicomponent systems rises

J. B. Brady, Department of Geology, Smith College, North-
ampton, MA 01063

Mineral Physics and Crystallography
A Handbook of Physical Constants
AGU Reference Shelf 2

rapidly with the number of components. Two impor-tant results help limit this complexity. First, the iso-thermal, isobaric Gibbs-Duhem equation [139, p. 134]

$$J_i^R = \sum_{j=1}^{n} -L_{ij}^R \left(\frac{\partial \mu_j}{\partial x} \right) \qquad (4)$$

for the single phase in which the diffusion occurs reduces the number of independent gradients to (n-1) and the number of diffusion coefficients to $(n-1)^2$. Second, Onsager [132, 133] showed that if the forces, fluxes, components, and reference frame are properly chosen, the matrix of coefficients relating the forces and fluxes is symmetrical

$$L_{ij}^R = L_{ji}^R , \qquad (5)$$

reducing the number of independent diffusion coefficients to (2n-1).

Although equations (3)-(5) are theoretically satisfy-ing, they are *not* generally used to describe diffusion experiments, in part because chemical potential gradients are not directly measurable in most cases. Fisher [51], Joesten [94, 96, 97], and others [e.g., 55, 98, 16, 1] have applied these equations success-fully in modeling the diffusion evolution of coronas and other textures in some rocks. However, most workers use empirical equations related to (3) that involve measurable compositions C_j (moles of j/m^3)

$$J_i^R = \sum_{j=1}^{n-1} -D_{ij}^R \left(\frac{\partial C_j}{\partial x} \right) \qquad (6)$$

[134]. Only (n-1) compositions are independent, but unfortunately $D_{ij}^R \neq D_{ji}^R$ so that $(n-1)^2$ diffusion coefficients D_{ij}^R (m^2/s) are needed for (6).

The most obvious simplification of (6) is to limit the number of components to 2 and, therefore, the num-ber of required diffusion coefficients to 1. Most experimentalists achieve this by their experimental design. The next two sections present definitions and equations used to describe these binary (2-compo-nent) experiments. Additional ways to simplify the treatment of multicomponent systems are addressed in Section 5.

3. FICK'S LAWS

Adolf Fick's [50] empirical equations were used to describe binary diffusion experiments long before the more general equations (3) and (6) were developed. Fick's First Law

$$J_i = -D_i \left(\frac{dC_i}{dx} \right) \qquad (7)$$

relates the instantaneous flux J_i (moles of i/m^2s) of component i to the one-dimensional gradient of the concentration of i, dC_i/dx (moles of i/m^4), and defines the diffusion coefficient D_i (m^2/s). However, unless the experiment attains a steady state, time (t) is also a variable and a continuity equation (Fick's Second Law)

$$\left(\frac{\partial C_i}{\partial t} \right)_x = \left(\frac{\partial}{\partial x} \left[D_i \left(\frac{\partial C_i}{\partial x} \right)_t \right] \right)_t \qquad (8)$$

must be solved. If D_i is not a function of composition (C_i) and, therefore, not a function of position (x), then equation (4) may be simplified to

$$\left(\frac{\partial C_i}{\partial t} \right)_x = D_i \left(\frac{\partial^2 C_i}{\partial x^2} \right)_t , \qquad (9)$$

which has many analytical solutions [6, 99, 17, 32].

Some commonly used solutions to (5) for planar geometries are given in Table 1 and two of these are shown graphically in Figure 1. All of the equations (7)-(13) implicitly assume constancy of volume, for which the fixed laboratory reference frame is a mean volume reference frame, and must be modified if the sample volume does change [10]. Similar analytical solutions exist for related boundary conditions and/or other geometries, notably spherical and cylindrical cases. More complicated boundary conditions and geometries may require numerical approximation [32, Chap. 8].

It is clear from Table 1 and Figure 1 that the parameter $\sqrt{D_i t}$ may be used to characterize the extent of diffusion. In semi-infinite cases such as (10) and (11), the distance x that has attained a particular value of C_i after time t is proportional to $\sqrt{D_i t}$. In finite

Table 1. Commonly-used solutions to equation (9)

Boundary Conditions	Solution

Thin-film Solution

$$\begin{cases} C_i \to C0 \text{ for } |x| > 0 \text{ as } t \to 0 \\ C_i \to \infty \text{ for } x = 0 \text{ as } t \to 0 \end{cases}$$

$$\left(\frac{C_i - C0}{\alpha}\right) = \frac{1}{2\sqrt{\pi D_i t}} \exp\left(\frac{-x^2}{4 D_i t}\right) \quad \text{where} \quad \alpha \equiv \int_{-\infty}^{+\infty} (C_i - C0)\,dx \qquad (10)$$

Semi-infinite Pair Solution

$$\begin{cases} C_i = C0 \text{ for } x > 0 \text{ at } t = 0 \\ C_i = C1 \text{ for } x < 0 \text{ at } t = 0 \end{cases}$$

$$\left(\frac{C_i - C0}{C1 - C0}\right) = \frac{1}{2}\operatorname{erfc}\left(\frac{x}{2\sqrt{D_i t}}\right) \qquad (11)$$

Finite Pair Solution

$$\begin{cases} C_i = C1 \text{ for } 0 < x < h \text{ at } t = 0 \\ C_i = C0 \text{ for } h < x < L \text{ at } t = 0 \\ (\partial C_i / \partial x)_t = 0 \text{ for } x = 0 \,\&\, L \end{cases}$$

$$\left(\frac{C_i - C0}{C1 - C0}\right) = \frac{h}{L} + \frac{2}{\pi}\sum_{n=1}^{\infty}\frac{1}{n}\sin\left(\frac{n\pi h}{L}\right)\exp\left(\frac{-D_i n^2 \pi^2 t}{L^2}\right)\cos\left(\frac{n\pi x}{L}\right) \qquad (12)$$

Finite Sheet - Fixed Surface Composition

$$\begin{cases} C_i = C0 \text{ for } -L < x < L \text{ at } t = 0 \\ C_i = C1 \text{ for } x = 0 \,\&\, L \text{ at } t > 0 \end{cases}$$

$$\left(\frac{C_i - C0}{C1 - C0}\right) = 1 - \frac{4}{\pi}\sum_{n=0}^{\infty}\frac{-1^n}{2n+1}\exp\left(\frac{-D_i(2n+1)^2\pi^2 t}{4L^2}\right)\cos\left(\frac{(2n+1)\pi x}{2L}\right) \qquad (13)$$

cases such as (12) and (13), the fractional extent of completion of homogenization by diffusion is proportional to $\sqrt{D_i t}$. These $\sqrt{D_i t}$ relations provide important tests that experimental data must pass if diffusion is asserted as the rate-controlling process. They also provide simple approximations to the limits of diffusion when applying measured diffusion coefficients to specific problems [147].

4. DIFFUSION COEFFICIENTS

In general, one must assume that D_i *is* a function of C_i. Therefore, equations (9)-(13) may be used with confidence only if C_i does not change appreciably during the experiment. This is accomplished either (a) by using a measurement technique (typically involving radioactive tracers) that can detect very small changes in C_i, or (b) by using diffusional exchange of stable isotopes of the same element that leave the element concentration unchanged. Approach (a) yields a "tracer diffusion coefficient" for the element that is specific to the bulk composition studied. Approach (b) yields a "self-diffusion coefficient" for the isotopically doped element that is also specific to the bulk chemical composition. Both approaches generally ignore the opposite or exchange flux that must occur in dominantly ionic phases such

as silicate minerals, glasses, and liquids.

If C_i does change significantly in the experiment and D_i is a function of C_i, observed compositional profiles might not match the shape of those predicted by (9)-(13). In such cases the D_i calculated with these equations will be at best a compositional "average" and equation (8) should be considered. Experiments in which composition does change significantly are often termed "interdiffusion" or "chemical diffusion" experiments. A commonly-used analytical solution to (8), for the same boundary conditions as for equation (11), was obtained by Matano [122] using the Boltzmann [8] substitution $\left(\chi \equiv x / \sqrt{t}\right)$:

$$D_i(C2) = \frac{-1}{2t}\left(\frac{dx}{dC_i}\right)_{C_i=C2}\int_{C_i=C0}^{C_i=C2} x\,dC_i . \qquad (14)$$

Equation (14) can be evaluated numerically or graphically from a plot of $(C_i - C0)/(C1 - C0)$ versus x, where the point $x = 0$ (the Matano interface) is selected such that

$$\int_{C_i=C0}^{C_i=C1} x\,dC_i = 0 \qquad (15)$$

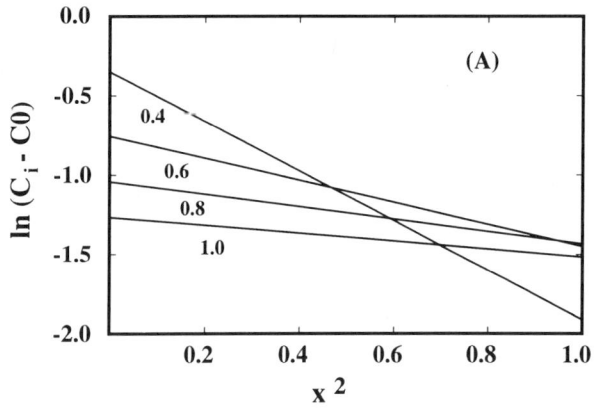 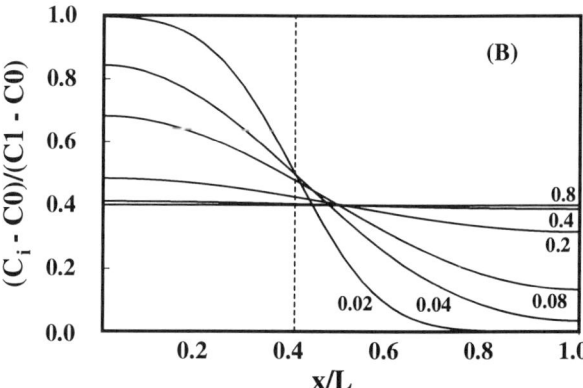

Figure 1. Graphical solutions to equation (9). (A) The "thin film" solution of equation (10) is shown with $\alpha=1$ for various values of $\sqrt{D_i t}$ (labels on lines). Plotting $\ln(C_i - C0)$ as a function of x^2 at any time yields a straight line of slope $-1/(4D_i t)$. (B) The "finite pair" solution of equation (12) is shown for $h/L=0.4$ and various values of Dt/L^2 (labels on curves). Plotting $(C_i - C0)/(C1 - C0)$ as a function of x/L normalizes all cases to a single dimensionless graph. The initial boundary between the two phases is marked by a dashed vertical line.

[32, p.230-234]. For binary, cation exchange between ionic crystals, the Matano interface is the original boundary between the two crystals. Diffusion experiments that follow the Boltzmann-Matano approach have the advantage of determining D_i as a function of C_i and the disadvantage of risking the complications of multicomponent diffusion, for which neither (8) nor (14) is correct.

Darken [35] and Hartley and Crank [83] showed that for electrically neutral species, the binary (A↔B) interdiffusion coefficients ($D_{AB_{-1}}$) obtained in a Boltzmann-Matano type experiment and the tracer diffusion coefficients (D_A^* and D_B^*) for the interdiffusing species are related by

$$D_{AB_{-1}} = \left(N_B D_A^* + N_A D_B^*\right)\left[1 + \left(\frac{\partial \ln \gamma_A}{\partial \ln N_A}\right)_{P,T}\right], \quad (16)$$

where N_A is the mole fraction and γ_A the molar activity coefficient of component A. Darken developed his analysis in response to the experiments of Smigelskas and Kirkendall [148], who studied the interdiffusion of Cu and Zn between Cu metal and $Cu_{70}Zn_{30}$ brass. Smigelskas and Kirkendall observed that Mo wires (inert markers) placed at the boundary between the Cu and brass moved in the direction of the brass during their experiments, indicating that more Zn atoms than Cu atoms crossed the boundary. Darken's analysis showed that in the presence of inert markers the

independent fluxes and, therefore, the "tracer" diffusion coefficients of Cu and Zn can be determined in addition to the interdiffusion coefficient $D_{CuZn_{-1}}$.

A similar analysis for binary cation interdiffusion (AZ ↔ BZ where AZ represents $A_z^{+a}Z_a^{-z}$ and BZ represents $B_z^{+b}Z_b^{-z}$) in appreciably ionic materials such as silicate minerals yields [5, 112, 113, 10]:

$$D_{A_b B_{-a}} = \left[\frac{\left(D_{AZ}^*\right)\left(D_{BZ}^*\right)\left(aN_{AZ} + bN_{BZ}\right)^2}{\left(a^2 N_{AZ} D_{AZ}^* + b^2 N_{BZ} D_{BZ}^*\right)}\right]\left[1 + \left(\frac{\partial \ln \gamma_{AZ}}{\partial \ln N_{AZ}}\right)_{P,T}\right], \quad (17)$$

which requires vacancy diffusion if $a \neq b$. If $a=b$, then (17) simplifies to

$$D_{AB_{-1}} = \left[\frac{\left(D_{AZ}^*\right)\left(D_{BZ}^*\right)}{\left(N_{AZ} D_{AZ}^* + N_{BZ} D_{BZ}^*\right)}\right]\left[1 + \left(\frac{\partial \ln \gamma_{AZ}}{\partial \ln N_{AZ}}\right)_{P,T}\right] \quad (18)$$

[121]. This expression permits interdiffusion coefficients to be calculated from more-easily-measured tracer diffusion coefficients. For minerals that are not

ideal solutions, the "thermodynamic factor" in brackets on the right side of (16)-(18) can significantly change the magnitude (and even the sign!) of interdiffusion coefficients from those expected for an ideal solution [12, 25].

5. MULTICOMPONENT DIFFUSION

Rarely is diffusion in geologic materials binary. Multicomponent diffusion presents the possibility that diffusive fluxes of one component may occur in response to factors not included in (7)-(18), such as gradients in the chemical potentials of other components, coupling of diffusing species, etc. In these cases, equation (6) or (3) must be used. The off-diagonal (i≠j) diffusion coefficients, D_{ij}^R or L_{ij}^R, are unknown for most materials and most workers are forced to assume (at significant peril!) that they are zero. Garnet is the one mineral for which off-diagonal diffusivities are available [19, 20] and they were found to be relatively small.

One approach used by many is to treat diffusion that is one-dimensional in real space as if it were one-dimensional in composition space. This approach was formalized by Cooper and Varshneya [28, 30] who discuss diffusion in ternary glasses and present criteria to be satisfied to obtain "effective binary diffusion coefficients" from multicomponent diffusion experiments. In general, diffusion coefficients obtained with this procedure are functions of both composition and direction in composition space. Another approach, developed by Cullinan [33] and Gupta and Cooper [75, 29], is to diagonalize the diffusion coefficient matrix for (6) through an eigenvector analysis. Although some simplification in data presentation is achieved in this way, a matrix of diffusion coefficients must be determined before the analysis can proceed.

Lasaga [105, see also 161] has generalized the relationship (17) to multicomponent minerals. The full expression is quite long, but simplifies to

$$D_{ij} = D_i^* \delta_{ij} - \left(\frac{D_i^* C_i z_i z_j}{\sum\limits_{k=1}^{n} D_k^* C_k z_k^2} \right) \left(D_j^* - D_n^* \right) \quad (19)$$

if the solid solution is thermodynamically ideal. In (19) z_i is the charge on cation i, $\delta_{ij} = 1$ if i=j, and $\delta_{ij} = 0$ if i≠j. This approach, using tracer diffusion

coefficients D_i^* to calculate multicomponent interdiffusion coefficient matrices, and its inverse offer the most hope for diffusion analysis of multicomponent problems [115, 19, 20]. It should be noted in using equations (16)-(19) that the tracer diffusivities themselves may be functions of bulk composition.

6. EXPERIMENTAL DESIGN

A few additional considerations should be mentioned regarding the collection and application of diffusion data.

• In anisotropic crystals, diffusion may be a function of crystallographic orientation and the diffusion coefficient becomes a second rank tensor. The form of this tensor is constrained by point group symmetry and the tensor can be diagonalized by a proper choice of coordinate system [131].

• Vacancies, dislocations, and other crystal defects may have profound effects on diffusion rates [e.g., 121, 164]. Therefore, it is essential that the mineral being studied is well-characterized. If the mineral, liquid, or glass contains a multivalent element like Fe, then an equilibrium oxygen fugacity should be controlled or measured because of its effect on the vacancy concentration.

• Diffusion in rocks or other polycrystalline materials may occur rapidly along grain boundaries, crystal interfaces, or surfaces and will not necessarily record intracrystalline diffusivities. Experiments involving grain-boundary diffusion [e.g., 102, 103, 11, 47, 49, 95, 154, 13, 158] are beyond the scope of this summary. If polycrystalline materials are used in experiments to measure "intrinsic" or "volume" diffusion coefficients, then the contributions of "extrinsic" grain-boundary diffusion must be shown to be negligible.

• Water can have a major effect on diffusion in many geologic materials, even in small quantities [e.g., 73, 150, 46, 167, 71]. If water is present, water fugacities are an essential part of the experimental data set.

• Because many kinetic experiments cannot be "reversed," every effort should be made to demonstrate that diffusion is the rate-controlling process. Important tests include the $\sqrt{D_i t}$ relations noted in equations (10)-(13) and the "zero time experiment." The $\sqrt{D_i t}$ test can be used if data are gathered at the same physical conditions for at least two times, preferably differing by a factor of four or more. The often-overlooked "zero time experiment," which duplicates the sample preparation, heating to the temperature of the experiment, quenching (after "zero time" at temperature), and data analysis of the other diffusion experiments, commonly reveals sources of systematic errors.

These and other important features of diffusion experiments are described in Ryerson [141].

7. DATA TABLE

Due to space limitations, the data included in this compilation have been restricted to comparatively recent experimental measurements of diffusion coefficients for silicate minerals, glasses, and liquids. Many good, older data have been left out, but can be found by following the trail of references in the recent papers that are included. Older data may also be found by consulting the compilations of Freer [56, 57], Hofmann [86], Askill [2], and Harrop [81]. Some good data also exist for important non-silicate minerals such as apatite [157, 23, 45] and magnetite-titanomagnetite [65, 136, 59], but the silicate minerals offered the clearest boundary for this paper.

Almost all of the data listed are for tracer or self-diffusion. Interdiffusion (chemical diffusion) experiments involving minerals are not numerous and interdiffusion data sets do not lend themselves to compact presentation. Interdiffusion data for silicate liquids and glasses are more abundant, but are not included [see 156, 3, 4]. Neither are non-isothermal (Soret) diffusion data [see 108]. Diffusion data listed for silicate glasses and liquids have been further restricted to bulk compositions that may be classified as either basalt or rhyolite.

Because diffusion is thermally activated, coefficients for diffusion by a single mechanism at different temperatures may be described by an Arrhenius equation

$$D = D_o \exp\left(\frac{-\Delta H}{RT}\right) \qquad (20)$$

and fit by a straight line on a graph of log D (m^2/s) as a function of $1/T$ (K^{-1}) [147, Chap. 2]. Log D_o (m^2/s) is the intercept of the line on the log D axis ($1/T = 0$). $\Delta H/(2.303 \cdot R)$ is the slope of the line where R is the gas constant (8.3143 J/mole·K) and

ΔH (J/mole) is the "activation energy." D_o and ΔH have significance in the atomic theory of diffusion [see 32, 121] and may be related for groups of similar materials [162, 82, 106].

Diffusion data are listed in terms of ΔH and log D_o and their uncertainties. In many cases data were converted to the units (kJ, m^2/s) and form (log D_o) of this compilation. Logarithms are listed to 3 decimal places for accurate conversion, even though the original data may not warrant such precision. No attempt was made to reevaluate the data, the fit, or the uncertainties given (or omitted) in the original papers. Also listed are the conditions of the experiment as appropriate including the temperature range, pressure range, oxygen fugacity, and sample geometry. Extrapolation of the data using (20) to conditions outside of the experimental range is not advisable. However, for ease of comparison log D is listed for a uniform temperature of 800°C (1200°C for the glasses and liquids), even though this temperature may be outside of the experimental range. Use these tabulated log D numbers with caution.

Finally, a sample "closure temperature" (T_c) has been calculated for the silicate mineral diffusion data. The closure temperature given is a solution of the Dodson [38] equation

$$\left(\frac{\Delta H}{RT_c}\right) = \ln\left(\frac{-55 \, RT_c^2 D_o}{a^2 \Delta H (dT/dt)}\right) \qquad (21)$$

for a sphere of radius a=0.1 mm and a cooling rate (dT/dt) of 5 K/Ma. The example closure temperatures were included because of the importance of closure temperatures in the application of diffusion data to petrologic [107] and geochronologic [123, Chap. 5] problems. Note that other closure temperatures would be calculated for different crystal sizes, cooling rates, and boundary conditions.

Acknowledgments. This paper was improved by helpful comments on earlier versions of the manuscript by D. J. Cherniak, R. A. Cooper, R. Freer, J. Ganguly, B. J. Giletti, S. J. Kozak, E. B. Watson, and an anonymous reviewer. N. Vondell provided considerable help with the library work.

Mineral, Glass, or Liquid	Orientation	Diffusing Component	Temperature Range (°C)	P (MPa)	O_2 (MPa)	ΔE_a (kJ/mole)	$\log D_0$ (or D) (m^2/s)	$\log D$ 800°C	"T_c" (°C)	Experiment/Comments	Ref.
β-quartz (SiO_2)	// c-axis	H	700-900	890-1550	UB	200 (±20)	-0.854	-10.6	188	Exchange with water-bearing fluid, "bulk" IR spectra	[104]
β-quartz (SiO_2)	// c-axis	^{18}O	600-800	100	UB	142 (±4)	-10.398(±0.272)	-17.3	282	Exchange with ^{18}O-enriched water, ion probe profiles	[68]
β-quartz (SiO_2)	⊥ c-axis	^{18}O	600-800	100	UB	234 (±8)	-8.000 (±2.239)	-19.4	498	Exchange with ^{18}O-enriched water, ion probe profiles	[68]
β-quartz (SiO_2)	// c-axis	^{18}O	700-850	100	UB/ NO	138.5(±19.1)	-10.680(±0.955)	-17.4	278	Exchange with ^{18}O-enriched water, ion probe profiles	[37]
β-quartz (SiO_2)	⊥ c-axis	^{18}O	700-850	100	UB/ NO	203.7(±2.3)	-9.413(±0.151)	-19.3	462	Exchange with ^{18}O-enriched water, ion probe profiles	[37]
β-quartz (SiO_2)	// c-axis	^{18}O	745-900	10	CO_2	159(±13)	-3.678(±0.132)	-11.4	146	Exchange with $C^{18}O_2$ gas, ion probe profiles	[146]
α-quartz (SiO_2)	// c-axis	^{18}O	450-590	100	UB	243 (±17)	-4.538	-16.4	389	Exchange with ^{18}O-enriched water, ion probe profiles	[48]
β-quartz (SiO_2)	⊥ (101)	^{30}Si	912-1028	0.1	air	230	-9.699	-20.9	570	Surface thin film of ^{30}Si; ion probe profile	[69]
β-quartz (SiO_2)	// c-axis	"^3H_2O"	720-850	0.061	UB	100 (±1.7)	-10.194(±0.099)	-15.1	115	Exchange with tritiated water vapor, serial section profiles, conc. dependent D	[145]
β-quartz (SiO_2)	// c-axis	"H_2O"	900	1500	NO	n.d.	(D=10^{-11})			Exchange with water/D_2O, bulk analysis, IR spectra	[140]
adularia ($Or_{97.6}Ab_{1.8}An_{0.5}$)	⊥ (001)	^{18}O	350-700	100	UB	107 (±5)	-11.346(±0.301)	-16.6	178	Exchange with ^{18}O-enriched water, ion probe profiles	[66]
adularia ($Or_{97.6}Ab_{1.8}An_{0.5}$)	⊥ (001)	^{18}O	650	5-1500	WM NNO	see paper - D varies w/f_{H2O}	independent of f_{O2}, f_{H2}, a_{H+}, P			Exchange with ^{18}O-enriched water, ion probe profiles	[46]
albite (low) ($Ab_{98}Or_{1.7}An_{1.2}$)	⊥ (001)	^{18}O	350-700	100	UB	89.1 (±5.0)	-12.636(±0.019)	-17.0	146	Exchange with ^{18}O-enriched water, ion probe profiles	[66]
anorthite ($An_{97}Ab_3$)	⊥ (001)	^{18}O	850-1300	0.1	O_2	236 (±8)	-9.000 (±0.349)	-20.5	553	Exchange with $^{18}O_2$ gas (+10% Ar), ion probe profile	[42]
anorthite ($An_{94}Ab_4$)	//[010]	^{18}O	1000-1300	0.1	(CO)/ (CO_2)	162 (±36)	-12.076(±1.337)	-20.0	439	Exchange with ^{18}O-enriched gas, ion probe profile	[143]
anorthite ($An_{95.3}Ab_{4.3}Or_{0.4}$)	⊥ (001)	^{18}O	350-800	100 "wet"	UB	109.6 (4.6)	-10.857(±0.021)	-16.2	172	Exchange with ^{18}O-enriched water, ion probe profiles	[66]
orthoclase ($Or_{94}Ab_6$)	powder	^{22}Na	500-800	200	UB	220.5 (±4.6)	-3.050 (±0.243)	-13.8	287	Exchange with brine, bulk analysis, cylindrical model	[53]
microcline (max) (Or_{100})	powder	^{22}Na	600-800	200	UB	80 (±8)	-9.636	-13.5	26	Exchange with $^{22}NaCl$ solution, sphere model	[109]
albite (low) ($Ab_{98}Or_{1.4}An_{0.6}$)	powder	^{22}Na	300-800	200	UB	176 (±8)	-4.903 (±0.814)	-13.5	218	Exchange with $^{22}NaCl$ solution, cylindrical model	[101] [163]

Mineral, Glass, or Liquid	Orientation	Diffusing Component	Temperature Range (°C)	P (MPa)	O_2 (MPa)	ΔE_a (kJ/mole)	log D_0 (or D) (m^2/s)	log D 800°C	"T_c" (°C)	Experiment/Comments	Ref.
orthoclase ($Or_{94}Ab_6$)	powder	^{40}Ar	500-800	200	UB	180.3 (±4.6)	-5.854 (±0.259)	-14.6	256	Ar loss into brine, bulk analysis, spherical model	[52]
microcline (max) (Or_{100})	powder	^{40}K	600-800	200	UB	293 (±8)	-1.874	-16.1	428	Exchange with ^{40}KCl solution, sphere model	[109]
albite (low) ($Ab_{98}Or_{1.4}An_{0.6}$)	powder	^{40}K	600-800	200	UB	172 (±25)	-8.125 (±0.337)	-16.5	301	Exchange with ^{40}KCl solution, cylindrical model	[101] [163]
orthoclase ($Or_{94}Ab_6$)	powder	^{41}K	600-800	200	UB	285.4 (±3.8)	-2.793 (±0.190)	-16.7	439	Exchange with brine, bulk analysis, cylindrical model	[53]
orthoclase ($Or_{94}Ab_6$)	//c-axis	^{87}Rb	625-800	100	UB	339 (±33)	-2.000 (±1.800)	-18.5	541	Exchange w/Rb-Sr-enriched water, ion probe profiles	[63]
albite (Ab_{98})	⊥(001)	^{86}Sr	570-1080	0.1	air	272	-4.509	-17.7	465	Thin film of ^{86}Sr, ion probe profile, other plag in progress	[64]
orthoclase ($Or_{94}Ab_6$)	//c-axis	^{84}Sr or ^{86}Sr	625-900	100	UB	167 (±17)	-11.000(±0.900)	-19.1	405	Exchange w/Rb-Sr-enriched water, ion probe profiles	[63]
orthoclase (Or_{93})	⊥(001)	SrAl-KSi interdiffusion	725-1075	0.1	air	284.1 (±6.7)	-6.224 (±0.302)	-20.1	568	Exchange w/Sr-Al-Si-O powder, Rutherford Back. Spec	[24]
anothoclase ($Ab_{68}Or_{27}An_5$)	⊥(001)	SrAl-KSi interdiffusion	725-1075	0.1	air	373.7(±19.2)	-1.648 (±0.913)	-19.8	609	Exchange w/Sr-Al-Si-O powder, Rutherford Back. Spec	[24]
anothoclase ($Ab_{68}Or_{27}An_5$)	⊥(010)	SrAl-KSi interdiffusion	725-1075	0.1	air	372.8(±20.1)	-2.346 (±0.951)	-20.5	634	Exchange w/Sr-Al-Si-O powder, Rutherford Back. Spec	[24]
anorthite (Ab_6An_{93})	⊥(010)	Sr-Ca interdiffusion	725-1075	0.1	air	329.7(±22.6)	-5.415 (±1.037)	-21.5	660	Exchange w/Sr-Al-Si-O powder, Rutherford Back. Spec	[24]
albite (low) ($Ab_{98}Or_{1.4}An_{0.6}$)	//c-axis	^{84}Sr	640-800	100	UB	247 (±25)	-5.600 (±1.300)	-17.6	437	Exchange w/Rb-Sr-enriched water, ion probe profiles	[63]
adularia ($Or_{89.6}$) - albite ($Ab_{98.6}$)	⊥(001) (couple)	K-Na interdiffusion	900-1000	1500	UB	n.d.	(D=10^{-17} to 10^{-15})			Microprobe profile, composition dependence, anisotropy	[25]
albite (Ab_{92}) exsolved ($An_{0/26}$)	⊥($0\bar{4}1$) (couple)	CaAl-NaSi interdiffusion	900-1050	1500 "wet"	MH	303 (±35)	-7.523 (±0.300)	-22.3	691	Average D from lamellar homogenization experiments	[114]
bytownite (An_{80}) exsolved ($An_{70/90}$)	⊥($03\bar{1}$) (couple)	CaAl-NaSi interdiffusion	1100-1400	0.1	air	516.3 (±19)	-2.959 (±0.662)	-28.1	1011	Average D from lamellar homogenization experiments	[74]
bytownite (An_{80}) exsolved ($An_{70/90}$)	⊥($03\bar{1}$) (couple)	CaAl-NaSi interdiffusion	900-975 1000-1050	1500 "wet"	MH	-317 (±35) -103	-4.959 (±0.300) -15.398	-20.4 -20.4	604 350	Average D's from lamellar homogenization experiments	[114]
K-feldspar (Or_{93})	⊥(010)	Pb-? interdiffusion	750-1050	0.1	UB	301.7(±11.3)	-6.000 (±0.519)	-20.7	609	Exchange with PbS powder, Rutherford backscattering	[21]
K-feldspar (Or_{93})	⊥(001)	Pb-? interdiffusion	750-1050	0.1	UB	306.7(±26.8)	-4.886 (±1.176)	-19.8	573	Exchange with PbS powder, Rutherford backscattering	[21]
Oligoclase (An_{23})	⊥(010)	Pb-? interdiffusion	750-1050	0.1	UB	364.5(±12.1)	-2.921 (±0.540)	-20.7	638	Exchange with PbS powder, Rutherford backscattering	[21]
Oligoclase (An_{23})	⊥(001)	Pb-? interdiffusion	750-1050	0.1	UB	226.0(±9.2)	-8.387 (±0.415)	-19.4	489	Exchange with PbS powder, Rutherford backscattering	[21]

Mineral, Glass, or Liquid	Orienta- tion	Diffusing Component	Temperature Range (°C)	P (MPa)	O_2 (MPa)	ΔE_a (kJ/mole)	log D_o (or D) (m²/s)	log D 800°C	"T_c" (°C)	Experiment/Comments	Ref.
nepheline	powder	^{18}O	1000-1300	0.1	CO_2	104.6(±10.5)	-12.229	-17.3	199	Exchange with CO_2, spherical model, bulk analysis	[27]
biotite (see paper for comp.)	powder	^{18}O	500-800	100 "wet"	UB/ NO	142 (±8)	-9.041	-16.0	233	Exchange with ^{18}O-enriched water, bulk analysis, cylindrical model, ion probe too	[54]
phlogopite (see paper for comp.)	powder	^{18}O	600-900	100 "wet"	UB/ NO	176 (±13)	-7.854	-16.4	305	Exchange with ^{18}O-enriched water, bulk analysis, cylindrical model, ion probe too	[54]
phlogopite (Ann_4) (see paper for comp.)	powder	^{40}Ar	600-900	200, 1500 "wet"	UB/ NO	242 (±11)	-4.125 (±0.514)	-15.9	373	Degassing into water, bulk analysis, cylindrical model, ($\Delta V_a \cong 0$ m³/mole)	[62] [67]
biotite (Ann_{56}) (see paper for comp. reference)	powder	^{40}Ar	600-750	100, 1400 "wet"	GM, QFM	197 (±9)	-5.114 (±0.614)	-14.7	281	Degassing into water, bulk analysis, cylindrical model ($\Delta V_a=1.4 \times 10^{-5}$ m³/mole)	[77]
biotite	powder	^{41}K	450-700	200 "wet"	UB/ NO	88				Exchange with ^{41}KCl solution, bulk analysis & ion probe, cylindrical model	[87] [88]
chlorite (sheridanite) (see paper)	powder	^2H	500-700	200 & 500	UB/ NO	171.7	-5.21	-13.6	214	Enchange with ^2H-selected water, bulk analysis, cylindrical model	[72]
mucovite	powder	^2H	450-750	200 & 400	UB	121.3	-7.98	-13.9	133	Enchange with ^2H-selected water, bulk analysis, cylindrical model	[70]
muscovite (see paper for comp.)	powder	^{18}O	512-700	100 "wet"	UB/ NO	163 (±21)	-8.114	-16.1	273	Exchange with ^{18}O-enriched water, bulk analysis, cylindrical model, ion probe too	[54]
tremolite (see paper for comp.)	//c-axis	^{18}O	650-800	100 "wet"	UB/ NO	163 (±21)	-11.699(±1.204)	-19.6	424	Exchange with ^{18}O-enriched water, ion probe profiles	[44]
hornblende (see paper for comp.)	//c-axis	^{18}O	650-800	100,20 2000 "wet"	UB/ NO	172 (±25)	-11.000(±1.322)	-19.4	421	Exchange with ^{18}O-enriched water, ion probe profiles, anisotropy, pressure depend.	[44]
richterite (see paper for comp.)	//c-axis	^{18}O	650-800	100 "wet"	UB/ NO	239 (±8)	-7.523 (±0.452)	-19.1	490	Exchange with ^{18}O-enriched water, ion probe profiles	[44]
hornblende (see paper for comp.)	powder	^{40}Ar	750-900	100 "wet"	UB or NO	268.2 (±7.1)	-5.620 (±0.506)	-18.7	497	Degassing into water, bulk analysis, spherical model	[76]
tremolite	//c-axis	Sr	800	200 "wet"	UB/ NO		$D=1 \times 10^{-21}$			Thin film solution, ion probe profile	[9]

Mineral, Glass, or Liquid	Orientation	Diffusing Component	Temperature Range (°C)	P (MPa)	O_2 (MPa)	ΔE_a (kJ/mole)	log D_0 (or D) (m^2/s)	log D 800°C	"T_c" (°C)	Experiment/Comments	Ref.
orthopyroxene	powder	Fe-Mg interdiffusion	600-800	0.1	-12.2--17.9	233	-9.432	-20.8	564	Calculated from disordering experiments [7]	[60]
diopside ($Ca_{.92}Mg_{.98}Fe_{.10}$)	powder	He "apparent diffusivity"	700-1400	0.0	UB	290 (±40)	-1.9 (±1.2)	-16.0	422	Degassing experiment, spherical model, fractures!	[151]
diopside (synthetic)	powder	^{18}O	1150-1350	0.1	CO_2	59 (±13)	-14.886 (±0.520)	-17.7	71	Exchange with CO_2, spherical model, bulk analysis	[130]
diopside ($Wo_{50.6}En_{48.3}Fs_{1.1}$)	// c-axis	^{18}O	1100-1250	0.1	NNO	-457 (±26)	-3.367 (±0.934)	-25.6	888	Exchange with ^{18}O-enriched gas, ion probe profiles	[143]
diopside ("essentially pure")	// c-axis	^{18}O	700-1250	100 "wet"	UB	226 (±21)	-9.824	-20.8	562	Exchange with ^{18}O-enriched water, ion probe profiles	[43]
diopside ("essentially pure")	⊥ c-axis	^{18}O	700-1250	100 "wet"	UB	226	-11.553	-22.6	671	Exchange with ^{18}O-enriched water, ion probe profiles	[43]
diopside ($Na_{.04}Ca_{.96}Mg_{.96}Fe_{.05}Al_{.06}Si_{1.96}O_6$)	// c-axis	Al-? interdiffusion	1180	0.1	$P(O_2)=10^{-14}$	n.d.	$(D=3.2(\pm0.7)$ $x10^{-21})$			Thin film of amorphous $CaAl_2SiO_6$, $^{27}Al(p,\gamma)^{28}Si$ nuclear reaction profile	[144]
diopside (synthetic) ($CaMgSi_2O_6$)	// c-axis	Sr	1100-1250	2000	UB	607 (+33)	2.940	-26.6	911	Thin film, sectioning, ion probe, Rutherford backscatter, anisotropy, pressure depend.	[149]
diopside ($Wo_{.99}En_{.99}Fs_{.02}$)	// c-axis	^{85}Sr	1100-1300	0.1	UB/ N_2	406	-2.268	-22.0	710	Thin film, sectioning by grinding, scintillation counts	[149]
diopside (synthetic) ($CaMgSi_2O_6$)	// c-axis	Sm	1100-1250	0.1 to 2000	UB	590 (+96)	2.146	-26.6	913	Thin film, sectioning, ion probe, Rutherford backscatter, anisotropy, pressure depend.	[149]
clinopyroxene ($Na_{.1}Ca_{.53}Mg_{1.1}Fe_{.17}Al_{.1}Si_{2.0}O_6$)	⊥ (001)	Ca-(Mg,Fe) interdiffusion	1150-1250	2500	UB/ GrPC	360.87 (±190)	-6.410	-24.0	801	Average D from lamellar homogenization experiments	[12]
wollastonite (synthetic? α-$CaSiO_3$)	sintered powder	^{45}Ca	900-1300?	0.006	UB	469	0.845	-22.0	721	Thin film, autoradiography profiles?	[110] [111]
åkermanite (syn) ($Ca_2MgSi_2O_7$)	// c-axis	^{18}O	1000-1300	0.1	CO_2	215 (±51)	-9.026	-19.5	483	Exchange with $C^{18}O_2$ atmosphere, ion probe profile	[165]
åkermanite (syn) ($Ca_2MgSi_2O_7$)	⊥ c-axis	^{18}O	800-1300	0.1	(CO)/ (CO_2)	278 (±33)	-6.328 (±1.282)	-19.9	555	Exchange with ^{18}O-enriched gass, ion probe profile	[143]
åkermanite (syn) ($Ca_2MgSi_2O_7$)	// c-axis	^{45}Ca	1100-1300	0.1	N_2	410	-0.301	-20.3	639	Thin film, sectioning by grinding, grindings counted	[127]
åkermanite (syn) ($Ca_2MgSi_2O_7$)	// c-axis	^{54}Mn	1100-1300	0.1	N_2	300	-4.569	-19.2	541	Thin film, sectioning by grinding, grindings counted	[127]
åkermanite (syn) ($Ca_2MgSi_2O_7$)	// c-axis	^{59}Fe	1100-1300	0.1	N_2	230	-7.377	-18.6	457	Thin film, sectioning by grinding, grindings counted	[127]
åkermanite (syn) ($Ca_2MgSi_2O_7$)	// c-axis	^{60}Co	1100-1300	0.1	N_2	230	-7.770	-19.0	474	Thin film, sectioning by grinding, grindings counted	[127]

Mineral, Glass, or Liquid	Orientation	Diffusing Component	Temperature Range (°C)	P (MPa)	O_2 (MPa)	ΔE_a (kJ/mole)	log D_o (or D) (m^2/s)	log D 800°C	"T_c" (°C)	Experiment/Comments	Ref.
åkermanite (syn) ($Ca_2MgSi_2O_7$)	//c-axis	^{63}Ni	1100-1300	0.1	N_2	200	-8.301	-18.0	400	Thin film, sectioning by grinding, grindings counted	[127]
åkermanite (syn) ($Ca_2MgSi_2O_7$)	//c-axis	^{85}Sr	1100-1300	0.1	N_2	380	-1.745	-20.2	627	Thin film, sectioning by grinding, grindings counted	[127]
åkermanite (syn) ($Ca_2MgSi_2O_7$)	//c-axis	^{133}Ba	1100-1300	0.1	N_2	290	-4.854	-19.0	526	Thin film, sectioning by grinding, grindings counted	[127]
gehlenite (syn) ($Ca_2Al_2SiO_7$)	//c-axis	^{18}O	1000-1300	0.1	CO_2	186 (±16)	-11.361	-20.4	498	Exchange with $C^{18}O_2$ atmosphere, ion probe profile	[165]
gehlenite (syn) ($Ca_2Al_2SiO_7$)	//a-axis	^{18}O	1000-1300	0.1	CO_2	300 (±37)	-5.157	-19.8	566	Exchange with $C^{18}O_2$ atmosphere, ion probe profile	[165]
melilite (syn) ($Ak_{50}Gh_{50}$)	powder	^{18}O	799-1300	0.1	CO_2	140.2 (±0.4)	-9.066	-15.9	227	Exchange with $C^{18}O_2$, bulk analysis, spherical model	[84]
melilite (syn) ($Ak_{75}Gh_{25}$)	powder	^{18}O	799-1300	0.1	CO_2	133.5 (±0.4)	-9.143	-15.6	206	Exchange with $C^{18}O_2$, bulk analysis, spherical model	[84]
åkermanite - gehlenite couple	//c-axis	AlAl-MgSi interdiffusion	1200 1250	0.1	N_2		D=3.9×10^{-19} D=6.9×10^{-18}			EDXA profile of cross section, D = f(composition), maximum D reported here	[127]
epidote	powder	^{2}H	450-650	200 & 400	UB	57.7	-9.48	-12.3	-56	Exchange with (^{2}H,^{1}H)$_2$O, bulk analysis, "cylinder"	[70]
zoisite	powder	^{2}H	350-650	200 & 400	UB	102.5	-8.35	-13.3	79	Exchange with (^{2}H,^{1}H)$_2$O, bulk analysis, "cylinder"	[70]
olivine ($Fo_{91}Fa_9$)	//a-axis	"H"	800-1000	300 "wet"	IW	130 (±30)	-4.222 (±0.18)	-10.5	80	Exchange with water, IR step profiles of cross section slices	[118]
olivine ($Fo_{89.2}$)	powder	He "apparent diffusivity"	700-1400	0.0	UB	420 (±20)	1.1 (±0.7)	-19.3	610	Degassing experiment, spherical model, fractures!	[151]
olivine ($Fo_{=92}Fa_{=8}$)	//c-axis	^{18}O	1200-1400	0.1	IW & NO	266 (±11)	-9.585 + 0.21x log10(f_{O_2}) (f_{O_2} in Pa)	-22.5	691	Exchange with mixed $C^{18}O_2$ gas, $^{18}O(p,\alpha)^{15}$N nuclear microanalysis profiles	[142]
olivine ($Fo_{=90}Fa_{=10}$)	//c-axis	^{18}O	1090-1500	0.1	10^{-12}-10^{-8}	318 (±17)	-5.174 + 0.34x log10(P_{O_2}) (P_{O_2} in Pa)	-20.7	616	Exchange with mixed H_2 ^{18}O gas, $^{18}O(p,\alpha)^{15}$N nuclear microanalysis profiles	[61]
olivine ($Fo_{90}Fa_{10}$)	//a-axis & //c-axis	^{30}Si	1130-1530	0.1	10^{-5}-10^{-15}	291(±15)	-12.735(±0.18)- .19xln (P_{O_2}/P_O)	-26.9	1047	Thin Fo film, Ruther-ford back-scattering profiles	[90]
olivine ($Fo_{90}Fa_{10}$)	//c-axis	Ca interdiffusion	1220-1350	0.1	10^{-9}	176	-9.155	-17.7	353	Cation exchange with basalt, microprobe profiles	[100]
olivine ($Fo_{90}Fa_{10}$)	//c-axis	Mn interdiffusion	1220-1350	0.1	10^{-9}	218	-7.167	-17.8	410	Cation exchange with basalt, microprobe profiles	[100]
olivine ($Fo_{90}Fa_{10}$)	//c-axis	Fe interdiffusion	1220-1350	0.1	10^{-9}	247	-8.000	-20.0	539	Cation exchange with basalt, microprobe profiles	[100]

Mineral, Glass, or Liquid	Orientation	Diffusing Component	Temperature Range (°C)	P (MPa)	O_2 (MPa)	ΔE_a (kJ/mole)	log D_0 (or D) (m²/s)	log D 800°C	"T_c" (°C)	Experiment/Comments	Ref.
olivine (Fo90Fa10-Fa100)	sintered powder	^{59}Fe	1130	0.1	10^{-10} - 10^{-12}	n.d.	n.d.			log D_{59Fe} = -10.143 + 0.2 a_{O2} + 2.705 [Fe/(Fe+Mg)]	[85]
olivine (Fo93.7Fa6.3)	//c-axis	Ni	1149-1234	evac. tube	UB	193 (±10)	-8.959 (±0.36)	-18.4	404	Thin Ni film, microprobe profile, anisotropy found	[26]
olivine (Fo92Fa8)-fayalite powder	//c-axis (couple)	Fe-Mg interdiffusion	1125-1200	0.1	10^{-13}	243	-5.759	-17.6	432	Microprobe profile, D varies w/direction, composition, f_{O2}	[14]
olivine (Fa97Te3)-olivine (Fo91Fa9)	//c-axis (couple)	Fe-Mg interdiffusion	900-1100	evac. tube	UB	208.5(±18.8) +9.1 x [Mg/(Mg+Fe)]	D_0= 1.5(±0.3) x10^{-4}) - 1.1x [Mg/(Mg+Fe)]			Microprobe profile, D also varies with direction and P (ΔV_a=5.5x10^{-6} m³/mole)	[124]
olivine(Mg2SiO4)-(Co2SiO4)	//c-axis (couple)	Co-Mg interdiffusion	1150-1300 / 1300-1400	0.1	air	196 / 526	-8.690 / 2.288	-18.2 / -23.3	403 / 781	Microprobe profile, D's extrapolated to pure Fo	[125]
forsterite - liebensbergite	//c-axis (couple)	Ni-Mg interdiffusion	1200-1450	0.1	air	414 to 444	-1.652	-21.8	702	Microprobe profile, other interdiffusion coefficients	[126] [128]
grossularite (Ca2.9 Fe.1Al2.0Si3.0O12)	isotropic	^{18}O	1050 / 850	800 / 200	UB	n.d. / n.d.	D=2.5x10^{-20} / D=4.8x10^{-21}			Exchange with ^{18}O-enriched water, ion probe profiles	[58]
almandine (Al67Sp28An3Py2)	isotropic	^{18}O	800-1000	100	UB	301 (±46)	-8.222 (±0.740)	-22.9	725	Exchange with ^{18}O-enriched water, ion probe profiles	[31]
pyrope (Py74Al5Gr10Ur1)	isotropic	^{25}Mg	750-900	200	UB/ MH	239 (±16)	-8.009	-19.6	514	Thin ^{25}MgO film, ion probe profiles	[34]
Alm80Pyp20-Spess94Alm6	isotropic (couple)	Fe	1300-1480	2900-4300	UB/ GrPC	275.43 (±36.49)	-7.194	-20.6	588	Calculated from interdiffusion experiments using model, (ΔV_a=5.6(±2.9)x10^{-6} m³/mole)	[20]
Alm80Pyp20-Spess94Alm6	isotropic (couple)	Mg	1300-1480	2900-4300	UB/ GrPC	284.52 (±37.55)	-6.959	-20.8	604	Calculated from interdiffusion experiments, (ΔV_a=5.6x10^{-6} m³/mole)	[20]
Alm80Pyp20-Spess94Alm6	isotropic (couple)	Mn	1300-1480	2900-4300	UB/ GrPC	253.44 (±37.19)	-7.292	-19.6	526	Calculated from interdiffusion experiments, (ΔV_a=5.3(±3.0)x10^{-6} m³/mole)	[20]
almandine (Al67Sp28An3Py2)	isotropic	^{86}Sr	800-1000	100	UB	205 (±17)	-12.000(±0.602)	-22.0	616	Exchange with ^{86}Sr water solution, ion probe profiles	[31]
almandine (Al67Sp28An3Py2)	isotropic	^{145}Nd	800-1000	100	UB	184 (±29)	-12.523(±0.602)	-21.5	562	Exchange with ^{145}Nd water solution, ion probe profiles	[31]
pyrope	powder	^{151}Sm	1300-1500	3000	UB/ GrPC	140	-11.585	-18.4	321	Exchange with silicate melt, autoradiography, "sphere"	[80]
almandine (Al67Sp28An3Py2)	isotropic	^{167}Er	800-1000	100	UB	230 (±38)	-10.301(±0.763)	-21.5	605	Exchange with ^{167}Er water solution, ion probe profiles	[31]
Alm80Pyp20-Spess94Alm6	isotropic (couple)	Fe-Mn interdiffusion	1300-1480	4000	UB/ GrPC	224.3 (±20.5)	-10.086	-21.0	571	Microprobe profiles, model fits of alm-rich composition	[41]

Mineral, Glass, or Liquid	Orientation	Diffusing Component	Temperature Range (°C)	P (MPa)	O₂ (MPa)	ΔE_a (kJ/mole)	log D_0 (or D) (m²/s)	log D 1200°C	"T_c" (°C)	Experiment/Comments	Ref.
titanite	//c-axis	^{18}O	700-900	100	UB/NO	301	-5.638	-20.3	591	Exchange with ^{18}O-enriched water, ion probe profiles	[129]
titanite	//c-axis	^{86}Sr	700-900	100	UB/NO	234	-9.420	-20.8	570	Exchange with ^{18}O-enriched water, ion probe profiles	[129]
titanite	//(100)	Pb	650-1027	0.1	air	328.5(±11.3)	-3.955(±0.315)	-19.9	591	Exchange with PbS powder, Rutherford backscattering	[22]
zircon	n.d.	Pb	550-800	0.1	air	142(±8)	-11.699	-18.6	337	Ion implantation of Pb, Rutherford backscattering	[23]
rhyolite	glass	Li	297-909	0.1	air	92.1(±1.3)	-5.599(±0.079)	-8.9		Thin film of LiNO$_3$, ion probe profile on cross section	[92]
rhyolite (obsidian) Iceland	glass	^{24}Na	140-850	0.1	air	84.5(±1.3)	-5.91(±0.18)	-8.9		Thin film, serial sectioning by etching, counting surface	[91]
"haplogranite"	melt	"B-Si" interdiffusion	1200-1600	0.1	n.d.	288.5(±20.4)	-4.864(±0.640)	-15.1		Ion probe profile of cross section	[18]
rhyolite(obsidian) Lake County, OR	melt	P	1200-1500	800	UB/GrPC	600.9(±11.7)	-12.652(±0.334)	-34.0		Apatite dissolution, microprobe profile, effect of water measured	[79]
rhyolite (obsidian) Lake County, OR	melt with 8% water	^{36}Cl	1100	1000	UB/GrPC		D=1.29x10^{-11}			Thin film of Na^{36}Cl, β-track profiles of cross section	[155]
rhyolite (obsidian) Iceland	glass	^{42}K	350-850	0.1	air	106.3(±3.8)	-6.46(±0.24)	-10.2		Thin film, serial sectioning by etching, counting surface	[91]
rhyolite (obsidian) Iceland	glass	Ca	630-930	0.1	air	283.7(±4.6)	-0.69(±0.22)	-10.7		Thin film, serial sectioning by etching, counting surface	[91]
rhyolite (obsidian) Iceland	glass	^{86}Rb	400-950	0.1	air	127.2(±0.8)	-6.86(±0.05)	-11.4		Thin film, serial sectioning by etching, counting surface	[91]
rhyolite (dehydrated) NM	glass	^{85}Sr	650-950	0.1	air	178.7(±3.3)	-5.260(±0.175)	-11.6		Thin film, serial sectioning by grinding, counting surface	[119]
rhyolite (obsidian) Lake County, OR	melt(dry)	Zr	1020-1500	800	UB/GrPC	408.8(±11.7)	-1.009(±0.386)	-15.5		Zircon dissolution, microprobe profiles	[78]
rhyolite (obsidian) Lake County, OR	melt(wet)	Zr	1020-1385	800	UB/NO	197.9(±8.0)	-5.523(±0.301)	-12.5		Zircon dissolution, microprobe profiles	[78]
rhyolite (obsidian) Iceland	glass	^{134}Cs	600-920	0.1	air	208.4(±8.4)	-5.04(±0.44)	-12.4		Thin film, serial sectioning by etching, counting surface	[91]
rhyolite (dehydrated) NM	glass and melt	^{134}Cs	790-1300	0.1	air	201.3(±12.1)	-6.01(±0.45)	-13.1		Thin film, serial sectioning by etching, counting surface	[91]
rhyolite (obsidian) Lake County, OR	melt	^{134}Cs	700-800	210	UB/NO	81.68	-8.143	-11.0		Thin film, β-track profiles of cross section	[153]
rhyolite (dehydrated) NM	glass	^{133}Ba	650-950	0.1	air	188.7(±6.3)	-5.42(±0.30)	-12.1		Thin film, serial sectioning by grinding, counting surface	[119]

Mineral, Glass, or Liquid	Orientation	Diffusing Component	Temperature Range (°C)	P (MPa)	O_2 (MPa)	ΔE_a (kJ/mole)	log D_o (or D) (m²/s)	log D 1200°C	"T_c" (°C)	Experiment/Comments	Ref.
rhyolite (dehydrated) NM	glass and melt	Ce	875-1100	0.1	air	490.4 (±23.9)	2.72 (±0.99)	-14.7		Thin film, serial sectioning by etching, counting surface	[91]
rhyolite (dehydrated) NM	glass and melt	Eu	700-1050	0.1	air	288.7 (±5.0)	-3.11 (0.22)	-13.3		Thin film, serial sectioning by etching, counting surface	[91]
rhyolite (obsidian) Lake County, OR	melt 6% water	LREE	1000-1400	800	UB/GrPC	251.5(±42.3)	-4.638 (±1.436)	-13.6		Monazite dissolution, microprobe profiles	[138]
rhyolite (obsidian) Lake County, OR	melt 1% water	LREE	1000-1400	800	UB/GrPC	510.9(±59.0)	3.362 (±0.629)	-14.8		Monazite dissolution, microprobe profiles	[138]
rhyolite (obsidian) Mono Craters	glass	H_2O	400-850	0.1	N_2	103 (±5)	-14.59 (±1.59)	-18.2		Dehydration in N_2, FTIR profile, equilibrium model	[166]
rhyolite (obsidian) New Mexico	glass	H_2O	510-980	0.1	air	46.48 (±11.40)	-10.90 (±0.56)	-12.5		Dehydration in air, bulk weight loss, low water	[93]
rhyolite (obsidian) Lake County, OR	melt with 8% water	$^{14}CO_2$	800-1100	1000	UB/GrPC	75 (±21)	-7.187	-9.9		Thin film of Na_2 $^{14}CO_3$, β-track profiles of cross section	[155]
basalt (alkali)	melt	^6Li	1300-1400	0.1	air	115.5	-5.125	-9.2		Thin film of ^6LiCl, ion probe profile of cross section	[116]
basalt Goose Island	melt	O	1160-1360	0.1	IW to CO_2	215.9 (±13.4)	-2.439	-10.1		Oxidation/reduction of bead, thermo-gravimetric balance	[160]
basalt (alkali olivine) BC	melt	O	1280-1400	400	UB/GrPC	293 (±29)	-0.790 (±2.51)	-11.2		Reduction by graphite, bulk FeO analysis by titration	[40]
basalt (alkali olivine) BC	melt	O	1280-1450	1200	UB/GrPC	360 (±25)	1.450 (±0.081)	-11.3		Reduction by graphite, bulk FeO analysis by titration	[40]
basalt (alkali olivine) BC	melt	O	1350-1450	2000	UB/GrPC	297 (±59)	-0.770 (±1.87)	-11.3		Reduction by graphite, bulk FeO analysis by titration	[40]
basalt (tholeiite) 1921 Kilauea	melt	O	1300-1450	1200	UB/GrPC	213 (±17)	-3.010 (±0.59)	-10.6		Reduction by graphite, bulk FeO analysis by titration	[40]
basalt (FeTi) Galapagos	melt	^{18}O	1320-1500	0.1	CO_2 & O_2	251 (±29)	-2.854	-11.8		Exchange with ^{18}O-selected gas, bulk analysis of sphere	[15]
"basalt" (Fe-free) (synthetic)	melt	Ar	1300-1450	1000-3000		113.2 (±7.5)	-6.140 (±0.068)	-10.2		Method not described	[39]
"basalt" ("alkali") Tenerife	melt	^{24}Na	1300-1400	0.1	air	163 (±13)	-4.02 (±0.46)	-9.8		Thin film, serial sectioning by grinding, counting surface	[116]
basalt (tholeiite) 1921 Kilauea	melt	^{45}Ca	1260-1440	0.1	air	184.1	-4.272	-10.8		Thin film, β-track profiles of cross-section on film	[89]
"basalt" (Fe-free) (synthetic)	melt	^{45}Ca	1100-1400	0.1	UB	106.3	-6.301	-10.1		Thin film of $^{45}CaCl$, β-track profiles of cross section	[152]
"basalt" (Fe-free) (synthetic)	melt	^{45}Ca	1100-1400	1000	UB	141.0	-5.284	-10.3		Thin film of $^{45}CaCl$, β-track profiles of cross section	[152]
"basalt" (Fe-free) (synthetic)	melt	^{45}Ca	1100-1400	2000	UB	208.4	-3.211	-10.6		Thin film of $^{45}CaCl$, β-track profiles of cross section	[152]

Mineral, Glass, or Liquid	Orientation	Diffusing Component	Temperature Range (°C)	P (MPa)	O_2 (MPa)	ΔE_a (kJ/mole)	log D_o (or D) (m^2/s)	log D 1200°C	"T_c" (°C)	Experiment/Comments	Ref.
basalt ("alkali") Tenerife	melt	^{46}Sc	1300-1400	0.1	air	197 ($\pm$8)	-4.55 ($\pm$0.31)	-11.5		Thin film, serial sectioning by grinding, counting surface	[116]
basalt ("alkali") Tenerife	melt	^{54}Mn	1300-1400	0.1	air	201 ($\pm$25)	-3.80 ($\pm$0.81)	-10.9		Thin film, serial sectioning by grinding, counting surface	[116]
basalt ("alkali") Tenerife	melt	^{59}Fe	1300-1400	0.1	air	264 ($\pm$17)	-2.20 ($\pm$0.59)	-11.5		Thin film, serial sectioning by grinding, counting surface	[116]
basalt (tholeiite) 1921 Kilauea	melt	^{60}Co	1260-1440	0.1	air	151.9	-5.276	-10.7		Thin film, serial sectioning by grinding, counting surface	[89]
basalt ("alkali") Tenerife	melt	^{60}Co	1300-1400	0.1	air	201 ($\pm$21)	-3.83 ($\pm$0.61)	-11.0		Thin film, serial sectioning by grinding, counting surface	[116]
basalt (tholeiite) 1921 Kilauea	melt	^{85}Sr	1260-1440	0.1	air	182.0	-4.556	-11.0		Thin film, serial sectioning by grinding, counting surface	[89]
basalt ("alkali") Tenerife	melt	^{85}Sr	1300-1400	0.1	air	213 ($\pm$25)	-3.46 ($\pm$0.83)	-11.0		Thin film, serial sectioning by grinding, counting surface	[116]
basalt (tholeiite) 1921 Kilauea	melt	^{133}Ba	1260-1440	0.1	air	164.9	-5.229	-11.1		Thin film, serial sectioning by grinding, counting surface	[89]
basalt ("alkali") Tenerife	melt	^{133}Ba	1300-1400	0.1	air	172 ($\pm$17)	-5.00 ($\pm$0.54)	-11.1		Thin film, serial sectioning by grinding, counting surface	[116]
basalt ("alkali") Tenerife	melt	^{134}Cs	1300-1400	0.1	air	272 ($\pm$17)	-2.00 ($\pm$0.60)	-11.6		Thin film, serial sectioning by grinding, counting surface	[116]
basalt (tholeiite) 1921 Kilauea	melt	^{152}Eu, ^{153}Gd	1320-1440 1320-1210	0.1	air	169.9	-5.237 D=1.4x10^{-11}			Thin film, serial sectioning by grinding, counting surface	[120]
"basalt" (Fe-free) (synthetic)	melt	$^{14}CO_2$	1350-1500	1500	UB/ GrPC	195.0	-3.449	-10.4		Thin film of Na$^{14}CO_3$, β-track profiles of cross section, pressure dependence	[159]

Key to oxygen fugacity or atmosphere abbreviations in data table

CO$_2$ -   Pure carbon dioxide atmosphere
GM -   Graphite-methane buffer
IW -   Iron-wustite buffer
MH -   Magnetite-hematite buffer
N$_2$ -   Pure nitrogen atmosphere
NO -   Nickel-nickel oxide
O$_2$ -   Pure oxygen atmosphere
QFM -   Quartz-fayalite-magnetite buffer
UB -   Unbuffered oxygen fugacity
UB/GrPC -   Unbuffered, but f$_{O2}$ limited by the graphite-bearing piston-cylinder assembly
UB/NO -   Unbuffered, but near nickel-nickel oxide due to the cold seal pressure vessel

REFERENCES

1. Ashworth, J. R., Birdi, J. J., and Emmett, T. F., Diffusion in coronas around clinopyroxene: modelling with local equilibrium and steady state, and a non-steady-state modification to account for zoned actinolite-hornblende, *Contrib. Mineral. Petrol., 109*, 307-325, 1992.

2. Askill, J., *Tracer Diffusion Data for Metals, Alloys, and Simple Oxides,* 97 pp., Plenum Press, New York, 1970.

3. Baker, D. R., Interdiffusion of hydrous dacitic and rhyolitic melts and the efficacy of rhyolite contamination of dacitic enclaves, *Contrib. Mineral. Petrol., 106*, 462-473, 1991.

4. Baker, D. R., The effect of F and Cl on the interdiffusion of peralkaline intermediate and silicic melts, *Am. Mineral., 78*, 316-324, 1993.

5. Barrer, R. M., Bartholomew, R. F., and Rees, L. V. C., Ion exchange in porous crystals. Part II. The relationship between self- and exchange-diffusion coefficients, *J. Phys. Chem. Solids, 24*, 309-317, 1963.

6. Barrer, Richard M., *Diffusion in and through Solids,* 464 pp., Cambridge University Press, Cambridge, England, 1951.

7. Besancon, J. R., Rate of disordering in orthopyroxenes, *Am. Mineral., 66*, 965-973, 1981.

8. Boltzmann, L., Zur Integration der Diffusionsgleichung bei variabeln Diffusion-coefficienten *Ann. Physik Chem., 53*, 959-964, 1894.

9. Brabander, D. J., and Giletti, B. J., Strontium diffusion kinetics in amphiboles (abstract), *AGU 1992 Fall Meeting,* supplement to *Eos Trans. AGU, 73,* 641, 1992.

10. Brady, J. B., Reference frames and diffusion coefficients, *Am. J. Sci., 275,* 954-983, 1975.

11. Brady, J. B., Intergranular diffusion in metamorphic rocks, *Am. J. Sci., 283-A,* 181-200, 1983.

12. Brady, J. B., and McCallister, R. H., Diffusion data for clinopyroxenes from homogenization and self-diffusion experiments, *Am. Mineral., 68,* 95-105, 1983.

13. Brenan, J. M., Diffusion of chlorine in fluid-bearing quartzite: Effects of fluid composition and total porosity, *Contrib. Mineral. Petrol., 115,* 215-224, 1993.

14. Buening, D. K., and Buseck, P. R., Fe-Mg lattice diffusion in olivine, *J. Geophys. Res., 78,* 6852-6862, 1973.

15. Canil, D., and Muehlenbachs, K., Oxygen diffusion in an Fe-rich basalt melt, *Geochim. Cosmochim. Acta, 54,* 2947-2951, 1990.

16. Carlson, W. D., and Johnson, C. D., Coronal reaction textures in garnet amphibolites of the Llano Uplift, *Am. Mineral., 76,* 756-772, 1991.

17. Carslaw, H. S., Jaeger, J. C., *Conduction of Heat in Solids,* 510 pp., Oxford, Clarendon Press, 1959.

18. Chakraborty, S., Dingwell, D., and Chaussidon, M., Chemical diffusivity of boron in melts of haplogranitic composition, *Geochim. Cosmochim. Acta, 57,* 1741-1751, 1993.

19. Chakraborty, S., and Ganguly, J., Compositional zoning and cation diffusion in garnets, in *Diffusion, Atomic Ordering, and Mass Transport,* edited by Ganguly, J., pp. 120-175, Springer-Verlag, New York,

1991.

20. Chakraborty, S., and Ganguly, J., Cation diffusion in aluminosilicate garnets: experimental determination in spessartine-almandine diffusion couples, evaluation of effective binary diffusion coefficients, and applications, *Contrib. Mineral. Petrol., 111,* 74-86, 1992.

21. Cherniak, D. J., Diffusion of Pb in feldspar measured by Rutherford backscatteing spectroscopy, *AGU 1992 Fall Meeting,* supplement to *Eos Trans. AGU, 73,* 641, 1992.

22. Cherniak, D., Lead diffusion in titanite and preliminary results on the effects of radiation damage on Pb transport, *Chem. Geol., 110,* 177-194, 1993.

23. Cherniak, D. J., Lanford, W. A., and Ryerson, R. J., Lead diffusion in apatite and zircon using ion implantation and Rutherford backscattering techniques, *Geochim. Cosmochim. Acta, 55,* 1663-1673, 1991.

24. Cherniak, D. J., and Watson, E. B., A study of strontium diffusion in K-feldspar, Na-K feldspar and anorthite using Rutherford backscattering spectroscopy, *Earth Planet. Sci. Lett., 113,* 411-425, 1992.

25. Christoffersen, R., Yund, R. A., and Tullis, J., Interdiffusion of K and Na in alkali feldspars, *Am. Mineral., 68,* 1126-1133, 1983.

26. Clark, A. M., and Long, J. V. P., The anisotropic diffusion of nickel in olivine, in *Thomas Graham Memorial Symposium on Diffusion Processes,* edited by Sherwood, J. N., Chadwick, A. V., Muir, W. M., and Swinton, F. L., pp. 511-521, Gordon and Breach, New York, 1971.

27. Connolly, C., and Muehlenbachs, K., Contrasting oxygen diffusion in nepheline, diopside and other silicates and their relevance to isotopic systematics in meteorites, *Geochim. Cosmochim. Acta, 52,* 1585-1591, 1988.

28. Cooper, A. R., The use and limitations of the concept of an effective binary diffusion coefficient for multi-component diffusion, in *Mass Transport in Oxides,* edited by Wachtman, J. B., Jr., and Franklin, A. D., pp. 79-84, National Bureau of Standards, Washington, 1968.

29. Cooper, A. R., Jr. Vector space treatment of multicomponent diffusion, in *Geochemical Transport and Kinetics,* edited by Hofmann, A. W., Giletti, B. J., Yoder, H. S., Jr., and Yund, R. A., pp. 15-30, Carnegie Institution of Washington, Washington, 1974.

30. Cooper, A. R., and Varshneya, A. K., Diffusion in the system K_2O-SrO-SiO$_2$, effective binary diffusion coefficients, *Am. Ceram. Soc. J., 51,* 103-106, 1968.

31. Coughlan, R. A. N., Studies in diffusional transport: grain boundary transport of oxygen in feldspars, strontium and the REE's in garnet, and thermal histories of granitic intrusions in south-central Maine using oxygen isotopes, Ph.D. thesis, Brown University, Providence, Rhode Island, 1990.

32. Crank, J., *The Mathematics of Diffusion,* 414 pp., Clarendon Press, Oxford, 1975.

33. Cullinan, H. T. , Jr., Analysis of the flux equations of multi-component diffusion, *Ind. Engr. Chem. Fund., 4,* 133-139, 1965.

34. Cygan, R. T., and Lasaga, A. C., Self-diffusion of magnesium in garnet at 750-900°C, *Am. J. Sci., 285,* 328-350, 1985.

35. Darken, L. S., Diffusion, mobility and their interrelation through free energy in binary metallic systems, *Am. Inst. Min. Metal. Engrs. Trans., 175,* 184-201, 1948.

36. de Groot, S. R., Mazur, P., *Non-equilibrium Thermodynamics,* 510 pp., North Holland, Amsterdam, 1962.

37. Dennis, P. F., Oxygen self-diffusion in quartz under hydrothermal conditions, *J. Geophys. Res., 89,* 4047-4057, 1984.

38. Dodson, M. H., Closure temperature in cooling geo-chronological and petrological problems, *Contrib. Mineral. Petrol., 40,* 259-274, 1973.

39. Draper, D. S., and Carroll, M. R., Diffusivity of Ar in haplobasaltic liquid at 10 to 30 kbar (abstract), *AGU 1992 Fall Meeting,* supplement to *Eos Trans. AGU, 73,* 642, 1992.

40. Dunn, T., Oxygen chemical diffusion in three basaltic liquids at elevated temperatures and pressures, *Geochim. Cosmochim. Acta, 47,* 1923-1930, 1983.

41. Elphick, S. C., Ganguly, J., and Loomis, T. P., Experimental determination of cation diffusivities in aluminosilicate garnets, I. Experimental methods and interdiffusion data, *Contrib. Mineral. Petrol., 90,* 36-44, 1985.

42. Elphick, S. C., Graham, C. M., and Dennis, P. F., An ion microprobe study of anhydrous oxygen diffusion in anorthite: a comparison with hydrothermal data and some geological implications, *Contrib. Mineral. Petrol., 100,* 490-495, 1988.

43. Farver, J. R., Oxygen self-diffusion in diopside with application to cooling rate determinations, *Earth Planet. Sci. Lett., 92,* 386-396, 1989.

44. Farver, J. R., and Giletti, B. J., Oxygen diffusion in amphiboles, *Geochim. Cosmochim. Acta, 49,* 1403-1411, 1985.

45. Farver, J. R., and Giletti, B. J., Oxygen and strontium diffusion kinetics in apatite and potential applications to thermal history determinations, *Geochim. Cosmochim. Acta, 53,* 1621-1631, 1989.

46. Farver, J. R., and Yund, R. A., The effect of hydrogen, oxygen, and water fugacity on oxygen diffusion in alkali feldspar, *Geochim. Cosmochim. Acta, 54,* 2953-2964, 1990.

47. Farver, J. R., and Yund, R. A., Measurement of oxygen grain boundary diffusion in natural, fine-grained quartz aggregates, *Geochim. Cosmochim. Acta, 55,* 1597-1607, 1991a.

48. Farver, J. R., and Yund, R. A., Oxygen diffusion in quartz: Dependence on temperature and water fugacity, *Chem. Geol., 90,* 55-70, 1991b.

49. Farver, J. R., and Yund, R. A., Oxygen diffusion in a fine-grained quartz aggregate with wetted and nonwetted micro-structures, *J. Geophys. Res., 97,* 14017-14029, 1992.

50. Fick, A., Über diffusion, *Ann. Physik Chem., 94,* 59, 1855.

51. Fisher, G. W., Nonequilibrium thermodynamics as a model for diffusion-controlled metamor−phic processes, *Am. J. Sci., 273,* 897-924, 1973.

52. Foland, K. A., [40]Ar diffusion in homogeneous orthoclase and an interpretation of Ar diffusion in K-feldspars, *Geochim. Cosmochim. Acta, 38,* 151-166, 1974a.

53. Foland, K. A., Alkali diffusion in orthoclase, in *Geochemical Transport and Kinetics,* edited by Hofmann, A. W., Giletti, B. J., Yoder, H. S., Jr., and Yund, R. A., pp. 77-98, Carnegie Institution of Washington, Washington, 1974b.

54. Fortier, S. M., and Giletti, B. J., Volume self-diffusion of oxygen in biotite, muscovite, and phlogopite micas, *Geochim. Cosmochim. Acta, 55,* 1319-1330, 1991.

55. Foster, C. T., A thermodynamic model of mineral segregations in the lower sillimanite zone near Rangeley, Maine, *Am. Mineral., 66,* 260-277, 1981.

56. Freer, R., Self-diffusion and impurity diffusion in oxides, *J. Materials Sci., 15,* 803-824, 1980.

57. Freer, R., Diffusion in silicate minerals and glasses: a data digest and guide to the literature, *Contrib. Mineral. Petrol., 76,* 440-454, 1981.

58. Freer, R., and Dennis, P. F., Oxygen diffusion studies. I. A preliminary ion microprobe investigation of oxygen diffusion in some rock-forming minerals, *Mineral. Mag., 45,* 179-192, 1982.

59. Freer, R., and Hauptman, Z., An experimental study of magnetite-titanomagnetite interdiffusion, *Phys. Earth Planet. Inter., 16,* 223-231, 1978.

60. Ganguly, J., and Tazzoli, V., Fe+2-Mg interdiffusion in orthopyroxene: constraints from cation ordering and structural data and implications for cooling rates of meteorites (abstract), in *Lunar and Planet. Science XXIV,* pp. 517-518, Lunar and Planet. Institute, Houston, 1992.

61. Gérard, O., and Jaoul, O.,

Oxygen diffusion in San Carlos olivine, *J. Geophys. Res., 94,* 4119-4128, 1989.

62. Giletti, B. J., Studies in diffusion I: argon in phlogopite mica, in *Geochemical Transport and Kinetics,* edited by Hofmann, A. W., Giletti, B. J., Yoder, H. S., Jr., and Yund, R. A., pp. 107-115, Carnegie Institution of Washington, Washington, 1974.

63. Giletti, B. J., Rb and Sr diffusion in alkali feldspars, with implications for cooling histories of rocks, *Geochim. Cosmochim. Acta, 55,* 1331-1343, 1991.

64. Giletti, B. J., Diffusion kinetics of Sr in plagioclase (abstract), *AGU 1992 Spring Meeting,* supplement to *Eos Trans. AGU, 73,* 373, 1992.

65. Giletti, B. J., and Hess, K. C., Oxygen diffusion in magnetite, *Earth Planet. Sci. Lett., 89,* 115-122, 1988.

66. Giletti, B. J., Semet, M. P., and Yund, R. A., Studies in diffusion -- III. Oxygen in feldspars: an ion microprobe determination, *Geochim. Cosmochim. Acta, 42,* 45-57, 1978.

67. Giletti, B. J., and Tullis, J., Studies in diffusion, IV. Pressure dependence of Ar Diffusion in Phlogopite mica, *Earth Planet. Sci. Lett., 35,* 180-183, 1977.

68. Giletti, B. J., and Yund, R. A., Oxygen diffusion in quartz, *J. Geophys. Res., 89,* 4039-4046, 1984.

69. Giletti, B. J., Yund, R. A., and Semet, M., Silicon diffusion in quartz, *Geol. Soc. Am. Abstr. Prog., 8,* 883-884, 1976.

70. Graham, C. M., Experimental hydrogen isotope studies III: diffusion of hydrogen in hydrous

minerals, and stable isotope exchange in metamorphic rocks, *Contrib. Mineral. Petrol., 76,* 216-228, 1981.

71. Graham, C. M., and Elphick, S. C., Some experimental constraints on the role of hydrogen in oxygen and hydrogen diffusion and Al-Si interdiffusion in silicates, in *Diffusion, Atomic Ordering, and Mass Transport: Selected Problems in Geochemistry,* edited by Ganguly, J., pp. 248-285, Springer-Verlag, New York, 1991.

72. Graham, C. M., Viglino, J. A., and Harmon, R. S., Experimental study of hydrogen-isotope exchange between aluminous chlorite and water and of hydrogen diffusion in chlorite, *Am. Mineral., 72,* 566-579, 1987.

73. Griggs, D. T., Hydrolytic weakening of quartz and other silicates, *Geophys. J., 14,* 19-31, 1967.

74. Grove, T. L., Baker, M. B., and Kinzler, R. J., Coupled CaAl-NaSi diffusion in plagioclase feldspar: experiments and applications to cooling rate speedometry, *Geochim. Cosmochim. Acta, 48,* 2113-2121, 1984.

75. Gupta, P. K., and Cooper, A. R., Jr., The [D] matrix for multicomponent diffusion, *Physica, 54,* 39-59, 1971.

76. Harrison, T. M., Diffusion of 40Ar in hornblende, *Contrib. Mineral. Petrol., 78,* 324-331, 1981.

77. Harrison, T. M., Duncan, I., and McDougall, I., Diffusion of 40Ar in biotite: temperature, pressure and compositional effects, *Geochim. Cosmochim. Acta, 49,* 2461-2468, 1985.

78. Harrison, T. M., and Watson,

E. B., Kinetics of zircon dissolution and zirconium diffusion in granitic melts of variable water content, *Contrib. Mineral. Petrol., 84*, 66-72, 1983.

79. Harrison, T. M., and Watson, E. B., The behavior of apatite during crustal anatexis: equilibrium and kinetic considerations, *Geochim. Cosmochim. Acta, 48*, 1467-1477, 1984.

80. Harrison, W. J., and Wood, B. J., An experimental investigation of the partitioning of REE between garnet and liquid with reference to the role of defect equilibria, *Contrib. Mineral. Petrol., 72*, 145-155, 1980.

81. Harrop, P. J., Self-diffusion in simple oxides (a bibliography), *J. Materials Sci., 3*, 206-222, 1968.

82. Hart, S. R., Diffusion compensation in natural silicates, *Geochim. Cosmochim. Acta, 45*, 279-291, 1981.

83. Hartley, G. S., and Crank, J., Some fundamental definitions and concepts in diffusion processes, *Trans. Faraday Soc., 45*, 801-818, 1949.

84. Hayashi, T., and Muehlenbachs, K., Rapid oxygen diffusion in melilite and its relevance to meteorites, *Geochim. Cosmochim. Acta, 50*, 585-591, 1986.

85. Hermeling, J., and Schmalzried, H., Tracerdiffusion of the Fe-cations in Olivine $(Fe_x Mg_{1-x})_2 SiO_4$ (III), *Phys. Chem. Miner., 11*, 161-166, 1984.

86. Hofmann, A. W., Diffusion in natural silicate melts: a critical review, in *Physics of Magmatic Processes,* edited by Hargraves, R. B., pp. 385-417, Princeton Univ. Press, 1980.

87. Hofmann, A. W., and Giletti, B. J., Diffusion of geochronologically important nuclides in minerals under hydrothermal conditions, *Eclogae Geol. Helv., 63*, 141-150., 1970.

88. Hofmann, A. W., Giletti, B. J., Hinthorne, J. R., Andersen, C. A., and Comaford, D., Ion microprobe analysis of a potassium self-diffusion experiment in biotite, *Earth Planet. Sci. Lett., 24*, 48-52, 1974.

89. Hofmann, A. W., and Magaritz, M., Diffusion of Ca, Sr, Ba, and Co in a basalt melt, *J. Geophys. Res., 82*, 5432-5440, 1977.

90. Houlier, B., Cheraghmakani, M., and Jaoul, O., Silicon diffusion in San Carlos olivine, *Phys. Earth Planet. Inter., 62*, 329-340, 1990.

91. Jambon, A., Tracer diffusion in granitic melts, *J. Geophys. Res., 87*, 10797-10810, 1982.

92. Jambon, A., and Semet, M. P., Lithium diffusion in silicate glasses of albite, orthoclase, and obsidian composition: an ion-microprobe determination, *Earth Planet. Sci. Lett., 37*, 445-450, 1978.

93. Jambon, A., Zhang, Y., and Stolper, E. M., Experimental dehydration of natural obsidian and estimation of D_{H2O} at low water contents, *Geochim. Cosmochim. Acta, 56*, 2931-2935, 1992.

94. Joesten, R., Evolution of mineral assemblage zoning in diffusion metasomatism, *Geochim. Cosmochim. Acta, 41*, 649-670, 1977.

95. Joesten, R., Grain-boundary diffusion kinetics in silicates and oxide minerals, in *Diffusion, Atomic Ordering, and Mass Transport: Selected Problems in Geochemistry,* edited by Ganguly, J., pp. 345-395, Springer-Verlag, New York, 1991a.

96. Joesten, R., Local equilibrium in metasomatic processes revisited. Diffusion-controlled growth of chert nodule reaction rims in dolomite, *Am. Mineral., 76*, 743-755, 1991b.

97. Joesten, R., and Fisher, G., Kinetics of diffusion-controlled mineral growth in the Christmas Mountains (Texas) contact aureole, *Geol. Soc. Am. Bull., 100*, 714-732, 1988.

98. Johnson, C. D., and Carlson, W. D., The origin of olivine-plagioclase coronas in meta-gabbros from the Adirondack Mountains, New York, *J. Metamorph. Geol., 8*, 697-717, 1990.

99. Jost, W., *Diffusion in Solids, Liquids, and Gases,* 558 pp., Academic Press, New York, 1960.

100. Jurewicz, A. J. G., and Watson, E. B., Cations in olivine, part 2: diffusion in olivine xenocrysts, with applications to petrology and mineral physics, *Contrib. Mineral. Petrol., 99*, 186-201, 1988.

101. Kasper, R. B., Cation and oxygen diffusion in albite, Ph.D. thesis, Brown University, Providence, Rhode Island, 1975.

102. Kingery, W. D., Plausible concepts necessary and sufficient for interpretation of grain boundary phenomena: I, grain boundary characteristics, structure, and electrostatic potential, *Am. Ceram. Soc. J., 57*, 1-8, 1974a.

103. Kingery, W. D., Plausible concepts necessary and sufficient for interpretation of grain boundary phenomena: II, solute segregation, grain boundary diffusion, and general discussion, *Am. Ceram. Soc. J., 57*, 74-83, 1974b.

104. Kronenberg, A. K., Kirby, S. H., Aines, R. D., and Rossman, G. R., Solubility and diffusional uptake of hydrogen in quartz at high water pressures: implications for hydrolytic weakening, *J. Geophys. Res., 91*, 12723-12744, 1986.

105. Lasaga, A. C., Multicomponent exchange and diffusion in silicates, *Geochim. Cosmochim. Acta, 43*, 455-469, 1979.

106. Lasaga, A. C., The atomistic basis of diffusion: defects in minerals, in *Kinetics of Geochemical Processes*, edited by Lasaga, A. C., and Kirkpatrick, R. J., pp. 261-320, Mineralogical Society of America, Washington, 1981.

107. Lasaga, A. C., Geospeedometry: an extension of geothermometry, in *Kinetics and Equilibrium in Mineral Reactions*, edited by Saxena, S. K., pp. 81-114, Springer-Verlag, New York, 1983.

108. Lesher, C. E., and Walker, D., Thermal diffusion in petrology, in *Diffusion, Atomic Ordering, and Mass Transport: Selected Problems in Geochemistry*, edited by Ganguly, J., pp. 396-451, Springer-Verlag, New York, 1991.

109. Lin, T. H., and Yund, R. A., Potassium and sodium self-diffusion in alkali feldspar, *Contrib. Mineral. Petrol., 34*, 177-184, 1972.

110. Lindner, R., Studies on solid state reactions with radiotracers, *J. Chem. Phys., 23*, 410-411, 1955.

111. Lindner, R., Silikatbildung durch Reaktion im festen Zustand, *Z. Physik. Chem., 6*, 129-142, 1956.

112. Lindström, R., Chemical diffusion in alkali halides, *J. Phys. Chem. Solids, 30*, 401-405, 1969.

113. Lindström, R., Chemical interdiffusion in binary ionic solid solutions and metal alloys with changes in volume, *J. Phys. C: Solid State, 7*, 3909-3929, 1974.

114. Liu, M., and Yund, R. A., NaSi-CaAl interdiffusion in plagioclase, *Am. Mineral., 77*, 275-283, 1992.

115. Loomis, T. P., Ganguly, J., and Elphick, S. C., Experimental determination of cation diffusivities in aluminosilicate garnets II. Multicomponent simulation and tracer diffusion coefficients, *Contrib. Mineral. Petrol., 90*, 45-51, 1985.

116. Lowry, R. K., Henderson, P., and Nolan, J., Tracer diffusion of some alkali, alkaline-earth and transition element ions in a basaltic and an andesitic melt, and the implications concerning melt structure, *Contrib. Mineral. Petrol., 80*, 254-261, 1982.

117. Lowry, R. K., Reed, S. J. B., Nolan, J., Henderson, P., and Long, J. V. P., Lithium tracer-diffusion in an alkali-basalt melt -- an ion-microprobe determination, *Earth Planet. Sci. Lett., 53*, 36-41, 1981.

118. Mackwell, S. J., and Kohlstedt, D. L., Diffusion of hydrogen in olivine: implications for water in the mantle, *J. Geophys. Res., 95*, 5079-5088, 1990.

119. Magaritz, M., and Hofmann, A. W., Diffusion of Sr, Ba, and Na in obsidian, *Geochim. Cosmochim. Acta, 42*, 595-605, 1978a.

120. Magaritz, M., and Hofmann, A. W., Diffusion of Eu and Gd in basalt and obsidian, *Geochim. Cosmochim. Acta, 42*, 847-858, 1978b.

121. Manning, J. R., *Diffusion Kinetics for Atoms in Crystals*, 257 pp., Van Nostrand, Princeton, 1968.

122. Matano, C., On the relation between the diffusion coefficients and concentrations of solid metals, *Japan J. Phys., 8*, 109-113, 1933.

123. McDougall, I., Harrison, T. M., *Geochronology and Thermochronology by the $^{40}Ar/^{39}Ar$ Method*, 212 pp., Oxford Univ. Press, New York, 1988.

124. Meisner, D. J., Cationic diffusion in olivine to 1400°C and 35 kbar, in *Geochemical Transport and Kinetics*, edited by Hofmann, A. W., Giletti, B. J., Yoder, H. S., Jr., and Yund, R. A., pp. 117-129, Carnegie Institution of Washington, Washington, 1974.

125. Morioka, M., Cation diffusion in olivine -- I. Cobalt and magnesium, *Geochim. Cosmochim. Acta, 44*, 759-762, 1980.

126. Morioka, M., Cation diffusion in olivine -- II. Ni-Mg, Mn-Mg, Mg and Ca, *Geochim. Cosmochim. Acta, 45*, 1573-1580, 1981.

127. Morioka, M., and Nagasawa, H., Diffusion in single crystals of melilite: II. Cations, *Geochim. Cosmochim. Acta, 55*, 751-759, 1991a.

128. Morioka, M., and Nagasawa, H., Ionic diffusion in olivine, in *Diffusion, Atomic Ordering, and Mass Transport: Selected Problems in Geochemistry*, edited by Ganguly, J., pp. 176-197, Springer-Verlag, New York, 1991b.

129. Morishita, Y., Giletti, B. J., and Farver, J. R., Strontium and oxygen self-diffusion in titanite (abstract), *Eos Trans. AGU, 71*, 652, 1990.

130. Muehlenbachs, K., and Connolly, C., Oxygen diffusion

in leucite, in *Stable Isotope Geochemistry: A Tribute to Samuel Epstein,* edited by Taylor, H. P., Jr., O'Neil, J. R., and Kaplan, I. R., pp. 27-34, The Geochemical Society, San Antonio, Texas, 1991.

131. Nye, J. F., *Physical Properties of Crystals,* 322 pp., Clarendon Press, Oxford, 1957.

132. Onsager, L., Reciprocal relations in irreversible processes, I, *Physical Rev., 37,* 405-426, 1931a.

133. Onsager, L., Reciprocal relations in irreversible processes II, *Physical Rev., 38,* 2265-2279, 1931b.

134. Onsager, L., Theories and problems of liquid diffusion, *Ann. N.Y. Acad. Sci., 46,* 241-265, 1945.

135. Philibert, Jean, *Atom Movements: Diffusion and Mass Transport in Solids,* 577 pp., Editions de Physique, Les Ulis, France, 1991.

136. Price, G. D., Diffusion in the titanomagnetite solid solution, *Mineral. Mag., 44,* 195-200, 1981.

137. Prigogine, I., *Introduction to Thermodynamics of Irreversible Processes,* 147 pp., Wiley Interscience, New York, 1967.

138. Rapp, R. A., and Watson, E. B., Monazite solubility an dissolution kinetics: implications for the thorium and light rare earth chemistry of felsic magmas., *Contrib. Mineral. Petrol., 94,* 304-316, 1986.

139. Reiss, Howard, *Methods of Thermodynamics,* 217 pp., Blaisdell Publishing Company, Waltham, Massachusetts, 1965.

140. Rovetta, M. R., Holloway, J. R., and Blacic, J. D., Solubility of hydroxl in natural quartz annealed in water at 900°C and 1.5 GPa, *Geophys. Res. Lett.,* 13, 145-148, 1986.

141. Ryerson, F. J., Diffusion measurements: experimental methods, *Methods Exp. Phys., 24A,* 89-130, 1987.

142. Ryerson, F. J., Durham, W. B., Cherniak, D. J., and Lanford, W. A., Oxygen diffusion in olivine: effect of oxygen fugacity and implications for creep, *J. Geophys. Res., 94,* 4105-4118, 1989.

143. Ryerson, F. J., and McKeegan, K. D., Determination of oxygen self-diffusion in åkermanite, anorthite, diopside, and spinel: Implications for oxygen isotopic anomalies and the thermal histories of Ca-Al-rich inclusions, *Geochim. Cosmochim. Acta, 58(17),* in press, 1994.

144. Sautter, V., Jaoul, O., and Abel, F., Aluminum diffusion in diopside using the $^{27}Al(p,\gamma)^{28}Si$ nuclear reaction: preliminary results, *Earth Planet. Sci. Lett., 89,* 109-114, 1988.

145. Shaffer, E. W., Shi-Lan Sang, J., Cooper, A. R., and Heuer, A. H., Diffusion of tritiated water in b-quartz, in *Geochemical Transport and Kinetics,* edited by Hofmann, A. W., Giletti, B. J., Yoder, H. S., Jr., and Yund, R. A., pp. 131-138, Carnegie Institution of Wash., Washington, 1974.

146. Sharp, Z. D., Giletti, B. J., and Yoder, H. S., Jr., Oxygen diffusion rates in quartz exchanged with CO_2, *Earth Planet. Sci. Lett., 107,* 339-348, 1991.

147. Shewmon, Paul G., *Diffusion in Solids,* 246 pp., Minerals, Metals & Materials Society, Warrendale, PA, 1989.

148. Smigelskas, A. D., and Kirkendall, E. O., Zinc diffusion in alpha brass, *Am. Inst. Min. Metal. Engrs. Trans., 171,* 130-142, 1947.

149. Sneeringer, M., Hart, S. R., and Shimizu, N., Strontium and samarium diffusion in diopside, *Geochim. Cosmochim. Acta, 48,* 1589-1608, 1984.

150. Tiernan, R. J., Diffusion of thallium chloride into single crystals and bicrystals of potassium chloride, Ph.D. thesis, MIT, Cambridge, Massachusetts, 1969.

151. Trull, T. W., and Kurz, M. D., Experimental measurement of ^3He and ^4He mobility in olivine and clinopyroxene at magmatic temperatures, *Geochim. Cosmochim. Acta, 57,* 1313-1324, 1993.

152. Watson, E. B., Calcium diffusion in a simple silicate melt to 30 kbar, *Geochim. Cosmochim. Acta, 43,* 313-322, 1979a.

153. Watson, E. B., Diffusion of cesium ions in H_2O-saturated granitic melt, *Science, 205,* 1259-1260, 1979b.

154. Watson, E. B., Diffusion in fluid-bearing and slightly-melted rocks: experimental and numerical approaches illustrated by iron transport in dunite, *Contrib. Mineral. Petrol., 107,* 417-434, 1991a.

155. Watson, E. B., Diffusion of dissolved CO_2 and Cl in hydrous silicic to intermediate magmas, *Geochim. Cosmochim. Acta, 55,* 1897-1902, 1991b.

156. Watson, E. B., and Baker, D. R., Chemical diffusion in magmas: an overview of experimental results and geochemical applications, in *Physical Chemistry of Magmas, Advances in Physical Geochemistry, 9,* edited by Perchuk, L., and Kushiro, I., pp. 120-151, Springer-Verlag, New

York, 1991.

157. Watson, E. B., Harrison, T. M., and Ryerson, F. J., Diffusion of Sm, Ar, and Pb in fluorapatite, *Geochim. Cosmochim. Acta, 49,* 1813-1823, 1985.

158. Watson, E. B., and Lupulescu, A., Aqueous fluid connectivity and chemical transport in clino-pyroxene-rich rocks, *Earth Planet. Sci. Lett., 117,* 279-294, 1993.

159. Watson, E. B., Sneeringer, M. A., and Ross, A., Diffusion of dissolved carbonate in magmas: experimental results and applications, *Earth Planet. Sci. Lett., 61,* 346-358, 1982.

160. Wendlandt, R. F., Oxygen diffusion in basalt and andesite melts: experimental results and discussion of chemical versus tracer diffusion, *Contrib. Mineral. Petrol., 108,* 463-471, 1991.

161. Wendt, R. P., The estimation of diffusion coefficients for ternary systems of strong and weak electrolytes, *J. Phys. Chem., 69,* 1227-1237, 1965.

162. Winchell, P., The compensation law for diffusion in silicates, *High Temp. Sci., 1,* 200-215, 1969.

163. Yund, R. A., Diffusion in feldspars, in *Feldspar Mineralogy,* edited by Ribbe, P. H., pp. 203-222, Mineralogical Society of America, Washington, 1983.

164. Yund, R. A., Quigley, J., and Tullis, J., The effect of dislocations on bulk diffusion in feldspars during metamorphism, *J. Metamorph. Geol., 7,* 337-341, 1989.

165. Yurimoto, H., Morioka, M., and Nagasawa, H., Diffusion in single crystals of melilite: I. oxygen, *Geochim. Cosmochim. Acta, 53,* 2387-2394, 1989.

166. Zhang, Y., Stolper, E. M., and Wasserburg, G. J., Diffusion in rhyolite glasses, *Geochim. Cosmochim. Acta, 55,* 441-456, 1991a.

167. Zhang, Y., Stolper, E. M., and Wasserburg, G. J., Diffusion of a multi-species component and its role in oxygen and water transport in silicates, *Earth Planet. Sci. Lett., 103,* 228-240, 1991b.

Infrared, Raman and Optical Spectroscopy of Earth Materials

Q. Williams

1. INTRODUCTION

There are a number of ways in which light interacts with condensed matter. In most cases, these are associated with the electric field of an incident photon interacting with the vibrational or electronic states of materials. The presence and strength of these interactions are principally dependent on bonding properties, the configurations of electronic states, and local symmetry of ions, molecules or crystals. Here, three separate types of light-mediated spectroscopies of interest to geophysics and mineral sciences are examined: 1) infrared spectroscopy; 2) Raman spectroscopy; and 3) optical absorption spectroscopy.

Each of these techniques provides complementary information on the vibrational and electronic properties of Earth materials. The bonding properties which produce infrared and Raman-active vibrational bands provide not only a useful fingerprinting technique for determining the presence or abundance of different functional groups, such as hydroxyl or carbonate units, but also (in a bulk sense) fundamentally control a variety of thermochemical properties of materials, including their heat capacity. Moreover, the vibrational spectrum provides basic information on the bond strengths present within a material. Electronic transitions may occur in the infrared, visible or ultraviolet region of the spectrum, and commonly produce the colors of many minerals: these are dictated by the local bonding environments and electronic configurations of different elements in crystals or molecules. The role of electronic transitions in controlling the electrical conductivity and in reducing the radiative thermal conductivity of minerals at high temperatures is briefly reviewed, and their relation to spectroscopic observations summarized. Additionally, observations of optical transitions may be used to constrain the energetic effects of different electronic configurations on element partitioning between sites of different symmetries and distortions.

2. VIBRATIONAL SPECTROSCOPY: INFRARED AND RAMAN TECHNIQUES

The primary feature of importance in vibrational spectroscopy is the interaction between the electric field associated with a photon and changes induced by vibrational motions in the electronic charge distribution within a material. The most common unit utilized to describe the frequency of these vibrational motions (as well as, in many cases, electronic transitions) is inverse centimeters, or wavenumbers (cm^{-1}): these may be readily converted to Hertz by multiplying by the speed of light, c (2.998×10^{10} cm/sec), to the wavelength of light (in cm) by taking its inverse, and to energy by multiplying by c and Planck's constant, h (6.626×10^{-34} J-sec).

Here, we focus on infrared and Raman spectroscopic characterization of vibrational states: in this discussion, we explicitly treat only normal Raman scattering and disregard non-vibrational effects such as Raman electronic and magnon scattering and non-linear effects such as the hyper-Raman and resonance Raman effects, none of which have had significant impact to date in the Earth sciences

Q. Williams, University of California, Santa Cruz, Institute of Tectonics, Mineral Physics Laboratory, Santa Cruz, CA 95064

Mineral Physics and Crystallography
A Handbook of Physical Constants
AGU Reference Shelf 2

(see [10] for discussion of these effects).

In the case of infrared spectroscopy, a photon of energy below ~14000 cm^{-1} (and most often below 5000 cm^{-1}) is either absorbed or reflected from a material through interaction with interatomic vibrations. The primary measurement is thus of the intensity of transmitted or reflected infrared light as a function of frequency. With Raman spectroscopy, a photon of light interacts inelastically with an optic vibrational mode and is either red-shifted towards lower frequency (Stokes lines) or blue-shifted to higher frequency (anti-Stokes lines) by an amount corresponding to the energy of the vibrational mode. When such an interaction occurs with an acoustic mode (or sound wave), the effect is called Brillouin spectroscopy (see the section on elastic constants in this volume). Notably, Raman measurements rely on monochromatic light sources (which, since the 1960's, have almost always been lasers) to provide a discrete frequency from which the offset vibrational modes may be measured. Furthermore, the Raman effect is rather weak: only about 1 in 10^5 to 10^6 incident photons are Raman scattered, necessitating relatively high powered monochromatic sources, sensitive detection techniques, or both.

The vibrational modes measured in infrared and Raman spectroscopy may be described as a function of the force constant of the vibration or deformation and the mass of the participating atoms. For the simplest case of a vibrating diatomic species, this relation may be described using Hooke's Law, in which

$$v_{vib} = (1/2\pi)\sqrt{(K/\mu)}, \qquad (1)$$

where K is the force constant associated with stretching of the diatomic bond, and μ is the reduced mass of the vibrating atoms. Clearly, a range of different force constants are required to treat more complex molecules, which have more possible vibrational modes beyond simple stretching motions (see Figure 1): because of the more complex interatomic interactions, different stretching and bending force constants are necessitated. Approximate equations relating these force constants to the frequencies of vibration for different polyatomic molecules have been reviewed and developed by Herzberg [12], while a detailed review of the frequencies of vibrations of different molecules is given by Nakamoto [18]. Figure 1 shows the vibrational modes expected for an isolated silicate tetrahedra, which comprise symmetric and antisymmetric stretching and bending vibrations. That different force constants are required to describe these vibrations is apparent from the atomic displacements illustrated: for

example, the symmetric stretch of the tetrahedron depends predominantly on the strength of the Si-O bond. For comparison, the symmetric bending vibration is controlled principally by the O-Si-O angle bending force constant, the magnitude of which is dictated in turn largely by repulsive interactions between the oxygen anions. Considerable insight in distinguishing vibrations such as those shown in Figure 1 from one another has been derived from isotopic substitution experiments, which effectively alter the masses involved in the vibration without significantly affecting the force constants.

Within crystalline solids, the number of vibrations potentially accessible through either infrared or Raman spectroscopy in 3n-3, where n is the number of atoms in the unit cell of the material. This simply results from the n atoms in the unit cell having 3 dimensional degrees of freedom during displacements, with three degrees of translational freedom of the entire unit cell being associated with acoustic (Brillouin) vibrations. These 3n-3 vibrations represent the maximum number of observable vibrational modes of a crystal, as some vibrations may be symmetrically equivalent (or energetically degenerate) with one another, and others may not be active in either the infrared or Raman spectra. Because the wavelength of even visible light is long relative to the size of crystallographic unit cells, both infrared and Raman spectroscopy primarily sample vibrations which are in-phase between different unit cells (at the center of the Brillouin zone). Characterizing vibrational modes that are out-of-phase to different degrees between neighboring unit cells requires shorter wavelength probes, such as neutron scattering techniques. These shorter wavelength probes are used to characterize the dispersion (or change in energy) of vibrations as the periodicity of vibrations becomes longer than the unit cell length-scale (that is, vibrations occurring away from the center of the Brillouin zone). Such dispersion is characterized by vibrations of adjacent unit cells being out-of-phase with one another [6].

For crystals, a range of bonding interactions (and thus force constants), the crystal symmetry (or atomic locations), and the masses of atoms within the unit cell are generally required to solve for the different vibrational frequencies occurring in the unit cell. Broadly, the calculation of the vibrational frequencies of a lattice entail examining the interchange between the kinetic energies of the atomic displacements and the potential energy induced by the bonding interactions present in the crystal. The primary difficulties with such calculations lie in determining adequate potential interactions, and in the number of different interatomic interactions included in these calculations. Such computations of lattice dynamics

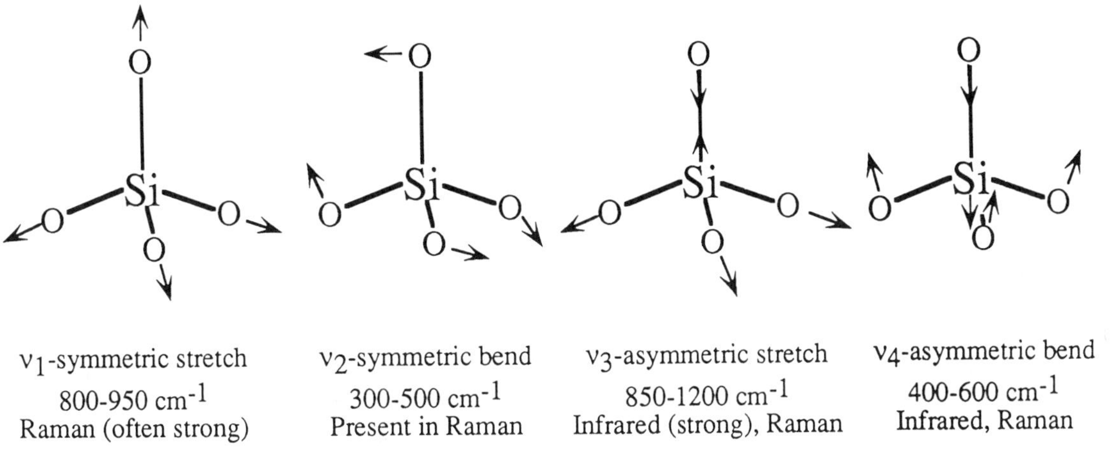

ν_1-symmetric stretch
800-950 cm^{-1}
Raman (often strong)

ν_2-symmetric bend
300-500 cm^{-1}
Present in Raman

ν_3-asymmetric stretch
850-1200 cm^{-1}
Infrared (strong), Raman

ν_4-asymmetric bend
400-600 cm^{-1}
Infrared, Raman

Fig. 1. The normal modes of vibration of an SiO$_4$-tetrahedra. Within isolated tetrahedra, all vibrations are Raman-active, while only the ν_3 and ν_4 vibrations are active in the infrared. However, environments with non-tetrahedral symmetries frequently produce infrared activity of the ν_1 and ν_2 vibrations in silicate minerals [modified from [12] and [18]].

have contributed major insight into the characterization of different types of vibrational modes in complex crystals: the precise equations and means by which such calculations are carried out have been reviewed elsewhere [3, 9].

Whether a given vibrational motion is spectroscopically active under infrared or Raman excitation is governed by selection rules associated with a given crystal structure. These selection rules are primarily governed by the symmetry of the unit cell, and the differing properties of Raman and infrared activity. Infrared activity is produced when a change in the dipole moment of the unit cell occurs during the vibration. It is this change in the electric field associated with the unit cell which interacts with the electric field of the incident photons. Because of this dependence on vibrational changes in the charge distribution, highly symmetric vibrations are often inactive or weak in infrared spectra: for example, within isolated tetrahedra, both the symmetric stretching and symmetric bending vibrations shown in Figure 1 are infrared-inactive.

The physical origin of Raman scattering may be viewed from a simple classical perspective in which the electric field associated with the incident light interacts with the vibrating crystal. In particular, this interaction occurs through the polarizability of the material: that is, its ability to produce an induced dipole in an electric field. It is the oscillations in the polarizability produced by vibrations which cause Raman scattering: if a vibrational

motion produces a change in the polarizability of a material, then the vibration is Raman active. Notably, while the induced dipole is a vector, the polarizability is a second rank tensor, and it is the symmetric properties of this tensor which allow the prediction of which families of modes will be active when observed along different crystallographic directions. Thus, both Raman and infrared activities are governed by interactions of the electric field of the incoming light with vibrationally produced changes in the charge distribution of the molecule or unit cell. Yet, despite Raman active vibrations depending on changes in the polarizability tensor and infrared active vibrations being generated by changes in the dipole moment, the two sets of vibrations are not always mutually exclusive: for non-centrosymmetric molecules (or unit cells), some vibrations can be both Raman and infrared-active. Finally, the intensities of different vibrations in the two different types of spectroscopy may be related to the magnitude of the changes produced in the dipole moment or polarizability tensor in a given vibration.

It is the relation of the change in the polarizability tensor and shifts in the dipole moment to the symmetry of the lattice which enables different vibrations to be associated with different symmetry species, as well as with possible crystallographic directionality for the infrared and Raman modes. The way in which this characterization of vibrational mode symmetries is conducted is through an application of group theory called factor group analysis. This is a means of utilizing the site symmetries of atoms in the primitive unit cell of a crystal (as given, for example, in The International Tables of X-ray Crystallography), coupled with character tables of the site symmetry and crystal symmetry group, and correlation tables between different symmetry groups to determine the number, activity, and symmetry of infrared and Raman vibrations of a crystal. Character tables for different symmetries are given in many texts (for example, 5), and correlation tables are given in [7] and [8]: the latter text provides a number of worked examples of factor group analysis applied to minerals and inorganic crystals.

Infrared spectra may be presented from two different experimental configurations: absorption (or transmission), or reflection. Sample absorbance is dimensionless, and is defined as

$$A = -\log(I/I_o), \qquad (2)$$

where I is the intensity of light transmitted through the sample, and I_o is the intensity of light transmitted through a non-absorbing reference material. Frequently, such data are reported as per cent transmittance, which is simply the ratio I/I_o multiplied by 100 (note that vibrational peaks are negative features in transmittance plots and positive features in absorbance). Because of the intensity of many of the absorptions in rock-forming minerals (greater than ~2 μm thicknesses in the mid-infrared (>~400 cm^{-1}) often render samples effectively opaque), such absorbance or transmission data are often generated from suspensions of powdered minerals or glasses within matrices with indices of refraction which approximate those of minerals. Typical examples of such matrices are pressed salts (typically KBr, KI or CsI) and hydrocarbons (for example, Nujol (paraffin oil) or teflon). While relative absorption intensities and locations can often be accurately determined using these techniques (highly polarizable crystals can present difficulties: see pp. 183-188 of [7]), absolute absorption intensity is compromised and orientation information is lost.

A second means by which infrared spectra are measured is through measurement of reflection spectra. Reflection spectra of crystals sample two components of each vibrational modes: transverse optic and longitudinal optic modes, while absorption spectroscopy ideally samples only the transverse modes. That there are two types of modes in reflection is because each vibrational motion may be described either with atomic displacements oriented parallel (longitudinal modes) and perpendicular (transverse modes) to the wavevector of the incoming light. Those with their atomic displacements oriented

parallel to the wavevector of the incoming light can be subject to an additional electrostatic restoring force produced by the differing displacements of planes of ions relative to one another, an effect not generated by shear. Such transverse and longitudinal modes are thus separated from one another by long-range Coulombic interactions in the crystal [3, 6].

In practice, reflection spectra generally have broad flattened peaks, with the lowest frequency onset of high reflectivity of a peak corresponding approximately to the transverse optic frequency, while the higher frequency drop in reflectivity of a peak is near the frequency of the longitudinal optic vibration. These data are generally inverted for TO and LO frequencies and oscillator strengths (a measure of the intensity of the reflection of the band) using one of two types of analyses: Kramers-Kronig, or an iterative classical dispersion analysis. Such analyses, described in [17] and elsewhere, are designed to model the extreme changes in the optical constants (expressed as the frequency dependent and complex index of refraction or dielectric constant) occurring over the frequency range of an optically active vibration.

For many minerals, the types of vibrational modes may be divided into two loose categories: internal and lattice modes. Internal modes are vibrations which can be associated with those of a molecular unit, shifted (and possibly split) by interaction with the crystalline environment in which the molecular unit is bonded: the vibrations of the silica tetrahedra shown in Figure 1 are typical examples of the types of motions which give rise to internal modes. Such internal modes are typically associated with the most strongly bonded units in a crystal, and thus with the highest frequency vibrations of a given material. We note also that even the simple picture of molecular vibrations is often complicated by the presence of interacting molecular units within a crystal. For example, it is difficult to associate different bands in feldspars with stretching vibrations of distinct AlO_4 or SiO_4 tetrahedra: because of the interlinking tetrahedra, a silica symmetric stretching vibration such as is shown in Figure 1 will involve a stretching motion of the adjoining AlO_4 tetrahedra, and vibrations of these two species must be viewed as coupled within such structures.

Lattice modes comprise both a range of (often comparatively low-frequency) vibrations not readily describable in terms of molecular units, and so-called external modes. External modes are those which involve motions of a molecular unit against its surrounding lattice: for example, displacements of the SiO_4 tetrahedra against their surrounding magnesium polyhedra in forsterite would constitute an external motion of the tetrahedral unit. Typically, such lattice modes are of critical importance in controlling the heat capacity of minerals at moderate temperatures (of order less than ~1000 K). As it is the thermal excitation of vibrational modes which produces the temperature dependence of the lattice heat capacity, considerable effort has been devoted to designing methods to extract thermochemical parameters such as the heat capacity and the Grüneisen parameter (the ratio of thermal pressure to thermal energy per unit volume) from vibrational spectra at ambient and high pressures [13]. A fundamental limitation on such techniques lies in that infrared and Raman spectra access only vibrations which have the periodicity of the unit cell (that is, effects of dispersion are ignored).

Representative Raman spectra of quartz and calcite are shown in Figure 2, along with a powder infrared absorption spectrum of quartz and an unpolarized infrared reflectance spectra of calcite. The factor group analysis of quartz predicts 16 total optically active vibrations, with four being Raman-active (A_1 symmetry type), four being infrared-active (A_2 symmetry), and eight being both Raman and infrared-active (E symmetry: doubly degenerate). For calcite, such an analysis predicts five Raman active vibrations ($1A_{1g}$ and four doubly degenerate E_g symmetry modes) and eight infrared-active vibrations ($3A_{2u}$ and 5 doubly degenerate E_u). Two features are immediately apparent from these spectra: first, the differences between the vibrations of the silicate group and those of the carbonate group are immediately apparent: in the infrared, the asymmetric stretching and out-of-plane bending vibrations of the carbonate group are present near 1410 and 870 cm^{-1}, while the strongest high frequency vibrations in quartz lie near 1100 cm^{-1} (the asymmetric stretches of the silica tetrahedra) and 680-820 cm^{-1} (corresponding to Si-O-Si bending vibrations). Moreover, the LO-TO splitting of the asymmetric stretching vibration of the carbonate group is clearly visible in the reflectance spectrum. Within the Raman spectrum, the most intense vibration of calcite is the symmetric stretching vibration of the carbonate group at ~1080 cm^{-1}, while that within quartz is predominantly a displacement of the bridging oxygens between tetrahedra at 464 cm^{-1}. At lower frequency (below 400 cm^{-1}), the spectrum of the carbonate has Raman bands associated with external vibrations of the carbonate group relative to stationary calcium ions. At lower frequencies than are shown in Figure 2 in the infrared spectrum, both lattice-type vibrations of the calcium ions relative to the carbonate groups and displacements of the carbonate groups and the calcium ions both against and parallel to one another occur.

Figure 3 shows representative vibrational frequencies of different functional, or molecular-like groupings, in mineral spectra. In the case of many minerals with complex structures, normal modes involving motions of the entire lattice, or coupled vibrations of different functional groups may occur, and these are not included in Figure 3. As expected from Equation 1, there is a general decrease in frequency with decreases in bond strength for different isoelectronic and isostructural groupings, such as occurs in the PO_4-SiO_4-AlO_4 sequence of tetrahedral anions.

While both infrared and Raman spectroscopy represent useful techniques for characterizing bonding environments, vibrational mode frequencies and the molecular species present within minerals, infrared spectroscopy has been extensively utilized as a quantitative technique for determining the concentration and speciation of volatile components within samples. Specific examples include the determination of the dissolved molecular water and hydroxyl content of glasses and crystals, the amount of dissolved CO_2 relative to CO_3^{2-} groups, and the amount of impurities within crystals such as diamond and a range of semiconductors. Such determinations depend on an application of the Beer-Lambert Law, in which the amount of absorption is assumed to be proportional to the number of absorbing species present in the sampled region. This may be expressed as

$$c = (M \cdot A)/(\rho \cdot d \cdot \varepsilon), \qquad (3)$$

in which c is the weight fraction of a dissolved species, M is the molecular weight of the dissolved material, A is the absorbance of a diagnostic band of the dissolved species, ρ is the density of the matrix in which the species is suspended, d is the path length through the sample, and ε is the molar absorption coefficient of the suspended species, with units of length^2/mole. It is the determination of this last parameter on which the quantitative application of Equation 3 depends: typically, a non-vibrational means of analysis is used to quantify the amount of dissolved species within a sequence of samples in order to calibrate the value of ε. Several complications exist in the straightforward application of Equation 3: first, the extinction coefficient can be frequency-dependent, in that the amount of absorption can depend on the structural environment in which a given species occurs. Second, for some applications a more appropriate measure of the number of species present is not the amplitude of absorption, but rather the integrated intensity underneath an absorption band (see [21] for a mineralogically-oriented discussion of each of these effects in hydrated species).

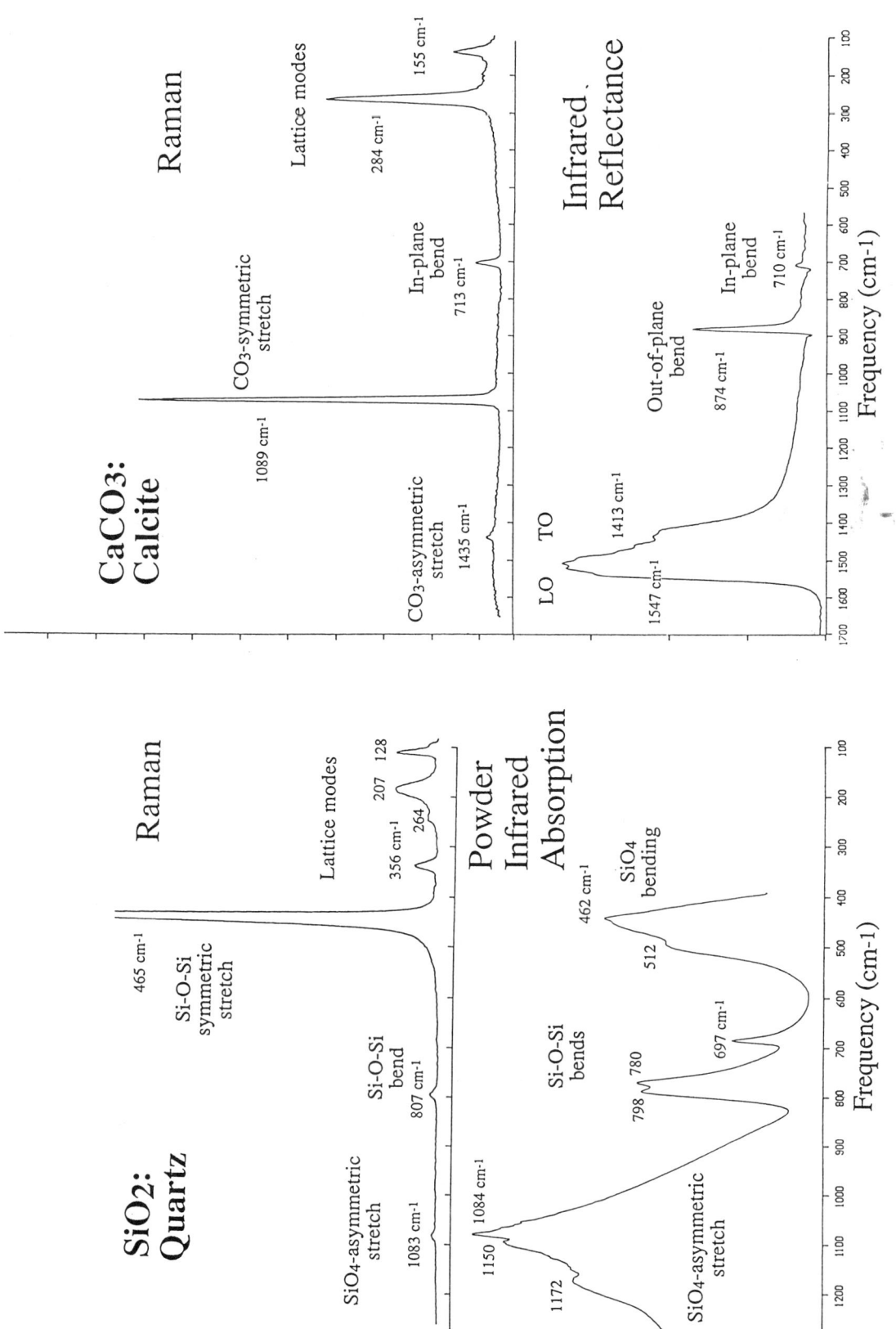

Fig. 2. Infrared and Raman spectra of SiO₂-quartz and CaCO₃-calcite. The infrared spectrum of quartz is an absorption spectrum from powdered material, while that of calcite is an unpolarized reflectance spectrum from a single crystal. Both Raman spectra are from single crystal samples using 514.5 nm excitation.

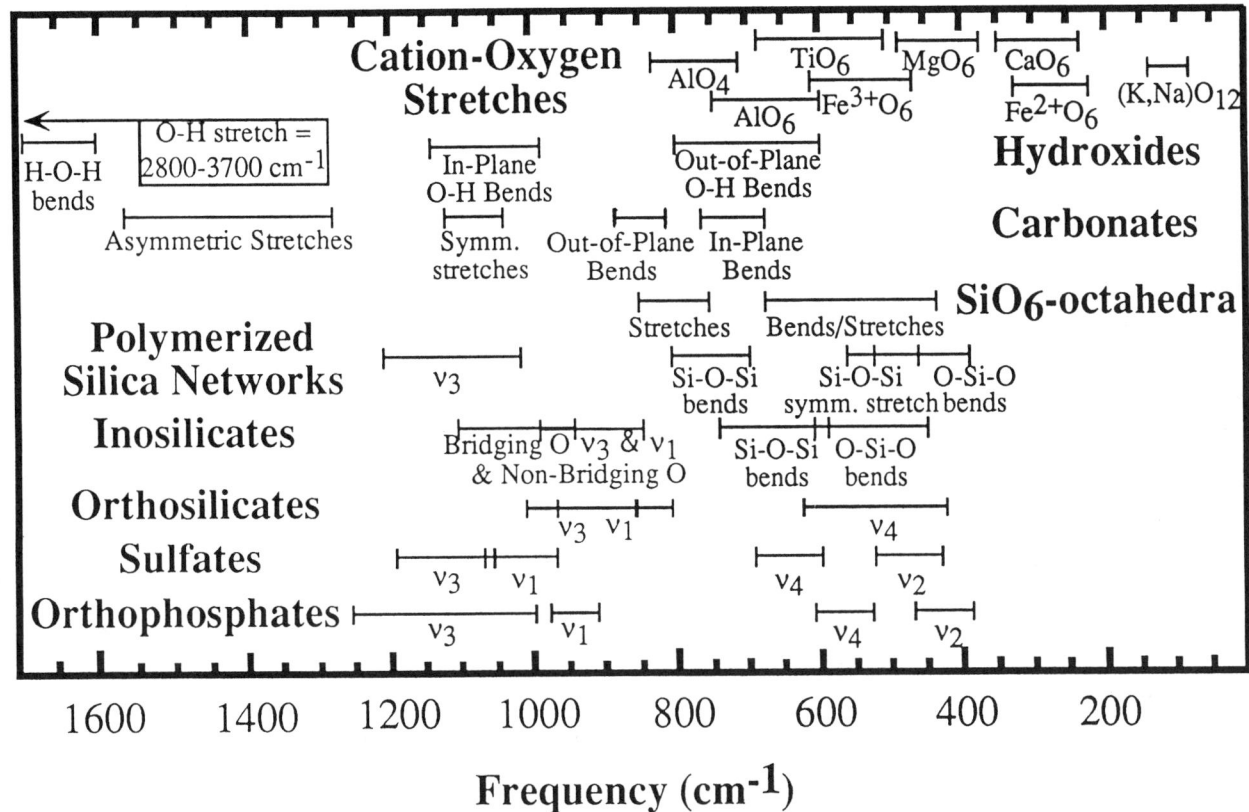

Fig. 3. Approximate frequency range of common vibrations of silicates, oxides and other functional groups within minerals. For tetrahedral species, the mode designations are identical to those shown in Figure 1. Most of the spectral ranges are derived from [3]; that for octahedral silicon is from [24].

The application of Raman spectroscopy to quantitative measurement of molecular species is complicated by practical difficulties in determining the absolute Raman scattering cross-section of different species, and in maintaining a constant volume of Raman excitation from which scattered light is collected. Its utility has thus been principally in determining what species are present in a sample, as opposed to their quantitative abundance. Yet, for analysis of samples whose sizes are of the order of the wavelength of infrared light (such as some fluid inclusions), diffraction of incident infrared light limits the usefulness of infrared techniques. Considerable effort has thus been devoted to characterizing Raman scattering cross-sections of gas-bearing mixtures, with the motivation of establishing Raman spectroscopy as a quantitative, non-destructive compositional probe of fluid inclusion compositions [e.g., 20].

3. OPTICAL SPECTROSCOPY

There are a number of different processes which may generate optical absorption (see [19] for a discussion of different causes of color) in the visible and ultraviolet wavelength range. The three processes of primary importance for minerals are: 1) crystal-field absorptions; 2) intervalence transitions; and 3) charge transfer or absorption into the conduction band (4, 15, and Burns in [2]). The first of these processes involves an ion with a partially filled d- (or in some cases f-) electron shell being incorporated into a crystalline environment whose symmetry produces a difference in energy between the different orbitals. A representative example of this phenomenon is shown in Figure 4: when the five atomic d-orbitals are octahedrally coordinated by anions, the $d_{x^2-y^2}$ and d_{z^2} orbitals, which have lobes of electron density

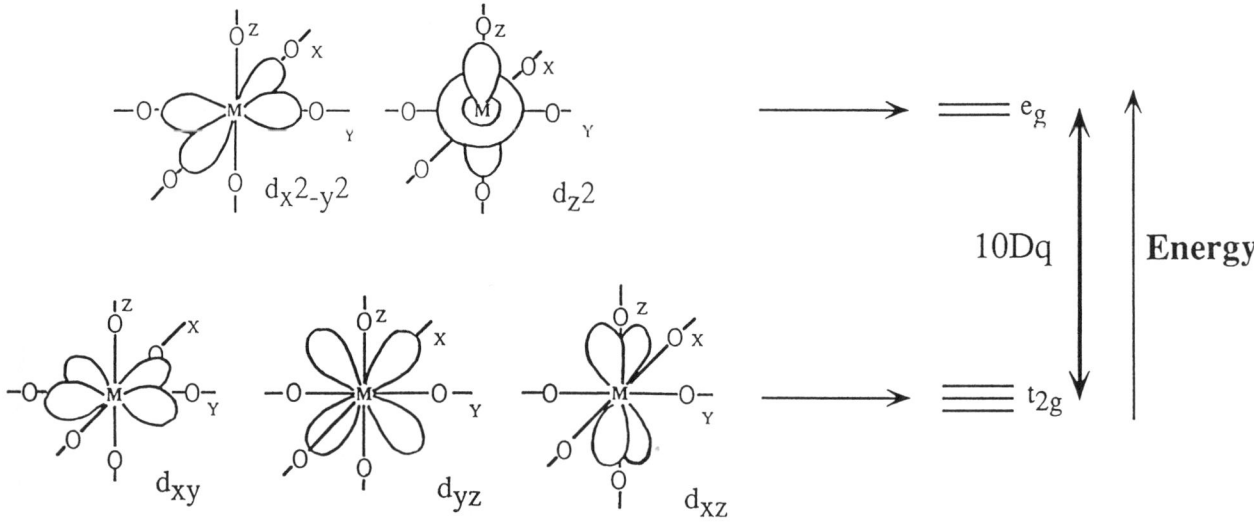

Fig. 4. The effect of octahedral coordination on the different d-orbitals of a transition metal cation (labelled M). The surfaces of the d-orbitals represent contours of constant probability that an electron lies within the boundaries. The repulsive effect produced by close proximity of the negatively charged oxygen cations to the $d_{x^2-y^2}$ and d_{z^2} orbitals separates these in energy from the d_{xy}, d_{yz} and d_{xz} orbitals by an amount referred to as the crystal field splitting energy (also as 10 Dq or Δ_o).

oriented towards the anions, lie at higher energy than the remaining three orbitals because of the larger repulsive interactions between the $d_{x^2-y^2}$ and d_{z^2} orbitals and their neighboring anions. Thus, optical absorption may be produced by excitation of electrons into and between these separated, or crystal-field split, energy levels, the energetic separation between which is often referred to as 10 Dq or Δ_o. Such crystal-field effects are observed within compounds containing such transition metal ions as Fe^{2+}, Fe^{3+}, Mn^{2+}, Ti^{3+}, Cr^{3+}, and is associated with both the colors and fluorescent properties of a range of minerals, gems and laser crystals, including forsteritic olivines (peridot: Mg_2SiO_4 with Fe^{2+}), ruby (Al_2O_3 with Cr^{3+}) and both silicate and aluminate garnets (which may contain a range of transition elements). The preferential occupation of the lower energy levels produces a net crystal field stabilization energy (CFSE) which contributes to the thermochemical energetics of transition metal-bearing crystals. The characteristic magnitude of the CFSE for Fe^{2+} (d^6 configuration) in octahedral coordination in minerals is between about 40 and 60 kJ/mole; that for octahedral Cr^{3+} (d^3) is generally between 200 and 275 kJ/mole. Characteristic octahedral CFSE values for other transition metal cations with unpaired d-electrons generally lie in between these extremes; tetrahedrally coordinated cations typically have CFS energies between ~30 and ~70% that of octahedrally

coordinated cations, thus producing a net preference of many transition metal ions for octahedral sites. For comparison, characteristic cohesive energies of divalent transition metal ions octahedrally coordinated by oxygen are on the order of 3000-4000 kJ/mole [4, 15, 16].

Furthermore, as each orbital can contain two electrons of opposite spins, two possible configurations are possible for octahedrally coordinated cations containing between three and eight d-electrons: high spin configurations, in which the energy required to pair two electrons within an orbital is greater than the crystal field splitting energy, and electrons enter into the higher energy unoccupied $d_{x^2-y^2}$ and d_{z^2} orbitals (e_g states) rather than pairing in the lower energy three-fold t_{2g} states. For comparison, electrons in low spin configurations fully occupy the t_{2g} states before entering the e_g levels. Among major transition elements, iron essentially always occurs in the high spin configuration in minerals: this is simply a consequence of the energy required to pair spins in divalent iron being larger than the crystal field stabilization energy. Estimates of the spin-pairing energy in iron-bearing minerals are poorly constrained, but representative values for this quantity are generally greater than 100 kJ/mole at ambient pressure. Historically, the high spin to low spin transition of iron in crystalline silicates has been frequently invoked as a possible high pressure phenomena. Essentially no experimental evidence exists, however, that indicates that such an electronic transition occurs over the pressure and compositional range of the Earth's mantle.

The values of the crystal-field splitting, discussed in detail elsewhere [4, 15, 16], clearly depend on the valence of the cation (as this controls the amount of occupancy in the d-orbitals, as well as contributing to the amount of cation-anion interaction, through the cation-anion bond strength), the anion present (as the charge and radius of the anion will also control the amount of crystal field splitting which takes place), and the type and symmetry of the site in which the ion sits. Such effects of symmetry are demonstrated by a comparison between octahedral and tetrahedral sites: in tetrahedral sites, the $d_{x^2-y^2}$ and d_{z^2} orbitals lie at lower energy than the three remaining d-orbitals. In more complex environments, such as distorted octahedra or dodecahedral sites, a larger number of crystal field bands may be observed because of further splitting of the d-levels. An example of spectra produced by different polarizations of light incident on Fe_2SiO_4-fayalite, a material with distorted octahedral sites, is shown in Figure 5, combined with an interpretation of the different crystal field transitions present in this material [4]. The lower than octahedral symmetry of the M1 and M2 sites produces more extensive splitting of the energy levels than occurs in sites of ideal octahedral symmetry.

The intensities of crystal-field absorption bands can vary over four orders of magnitude: these are governed both by the abundance of the absorbing cation, in a manner directly analogous to Equation 3, and by a combination of the symmetry of the cation environment and quantum mechanical selection rules. For example, non-centrosymmetric environments (e.g. tetrahedrally coordinated cations) generally produce more intense absorptions. The primary selection rules are related to spin multiplicity (involving conservation of the number of unpaired electrons between the ground and excited state) and to conservation of parity (or, transitions being only allowed between orbitals which differ in the symmetry of their wavefunction: the Laporte selection rule). These selection rules may be relaxed through a range of effects which include orbital interactions, vibrational and magnetic perturbations, and lack of a centrosymmetric ion site, producing both weak spin-forbidden and Laporte-forbidden transitions. Indeed, such crystal field bands are frequently orders of magnitude less intense than the vibrational bands discussed above.

Inter-valence, charge transfer, and valence to conduction band transitions are all terms used to describe absorption mechanisms of similar origins: the transfer of electrons between ions in non-metals via an input of energy in the

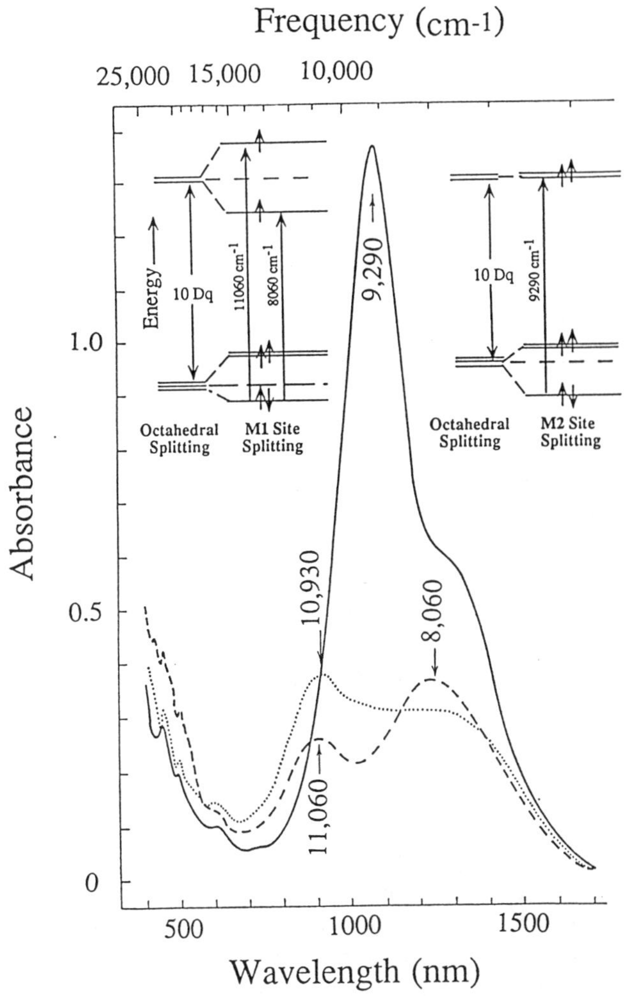

Fig. 5. Polarized crystal field absorption spectra of Fe_2SiO_4-fayalite in three different crystallographic orientations, combined with an interpretation of the observed transitions in terms of the M1 and M2 cation sites of Fe^{2+} in olivine [from 4]. The dotted spectrum is taken parallel to the b-axis of this phase, the dashed line parallel to c, and the solid line parallel to a. As the M1 and M2 sites are each distorted from perfect octahedral symmetry, additional splitting beyond that shown in Figure 4 occurs for the d-orbitals of iron ions within fayalite: this is shown in the insets. Short arrows in insets represent electrons occupying each d-orbital (Fe^{2+} has six d-electrons). The sharp bands above 20,000 cm^{-1} are generated by spin-forbidden transitions.

form of a photon. The term inter-valence transition is generally used to describe processes associated with the transfer of an electron between transition metal ions, especially those which are able to adopt multiple valence states in minerals, such as Fe (+2,+3) and Ti (+3,+4). Characteristically, these bands occur in the visible region of the spectrum, and are most intense in those minerals which have large quantities of transition metals, and relatively short metal-metal distances. As such, these transitions produce the coloration of many iron and iron/titanium-rich oxides and hydroxides.

Charge transfer transitions not only occur between transition metal ions, but also occur in excitation of electrons between cations and anions. Within oxides, such oxygen-to-metal charge transfer absorption bands typically occur at high energies in the ultraviolet, and are extremely intense: often three to four orders of magnitude more intense than crystal field transitions. Effectively, such charge transfer between anions and cations is often associated with delocalization of an electron, or photo-excitation of an electron into the conduction band of a material: the separation in energy between the valence and conduction band of materials is commonly referred to as the band gap. Within many materials, such charge transfer bands appear as an absorption edge, rather than a discrete band: this edge is simply generated because all photons with an energy above that of the edge will produce electron delocalization. This edge may be accompanied by discrete bands at slightly lower energy than the edge: such bands are often associated with a binding, or attractive force between the photo-excited electron and its ion of origin. These bound electron states are referred to as excitons, and are discussed in greater detail in solid state physics texts [e.g., 1]. There is an intimate association between such processes of electron delocalization and the high- and moderate-temperature electrical conductivity of insulators and semiconductors: in materials in which the electrical conductivity is not dominated by defect-related processes, the thermal excitation of electrons into the conduction band represents the primary means by which charge carriers are generated. This relationship is expressed through

$$\sigma = \sigma_0 \exp(-E_g/2kT), \qquad (4)$$

in which σ is the electrical conductivity of the material, E_g is the energy of the band gap, k is Boltzmann's constant (1.381 x 10^{-23} J/K), and T is temperature [e.g., 22]. The pre-exponential factor, σ_0, is generally treated as a constant (although both it and the band gap may be temperature dependent: these dependences are, however,

frequently small), and represents the intercept of a plot of the log of the conductivity and the inverse of temperature.

Absorptions generated by such cation-anion charge delocalization processes are common in ore minerals such as sulfides and arsenides, and produce the opacity or deep red color in the visible of many such minerals. Furthermore, they also induce effective opacity of most

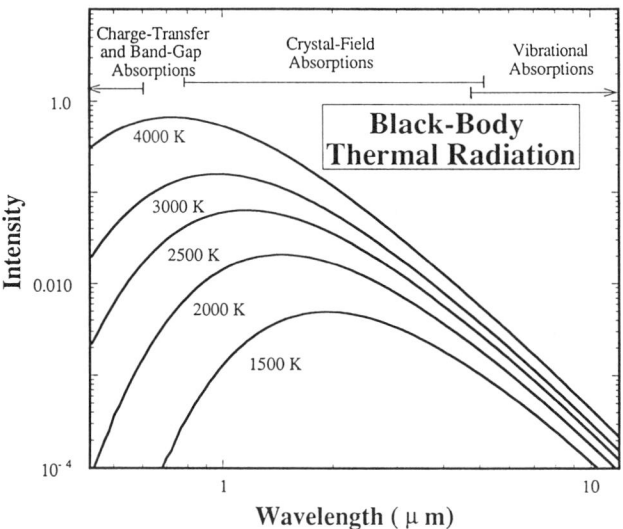

Fig. 6. The dependence of blackbody emission on wavelength at different temperatures, with the wavelength ranges of electronic and vibrational absorptions which inhibit radiative heat transport in mantle minerals illustrated. Vibrational absorptions (both single vibrational transitions and overtones) absorb strongly at wavelengths longer than about 5 μm; if hydroxyl or carbonate units are present, then such vibrational absorptions may extend to 3 μm. Crystal field absorptions of divalent iron in distorted octahedral sites, such as the M2 site in pyroxenes, can occur at wavelengths above 4 μm; similarly, iron in the dodecahedral site of garnets can absorb at longer wavelengths than 2 μm (see Burns in [2]). Among the most intense crystal field absorptions of iron in minerals are those occurring near 1 μm, such as are shown in Figure 5. Such crystal-field absorption bands tend to broaden markedly at high temperatures, increasing the efficiency with which they prevent radiative heat transport. The location of charge-transfer bands is known to be highly sensitive to pressure, with bands in relatively iron-rich phases (such as magnesiowüstite) shifting strongly to longer wavelength with increasing pressure. Such absorptions generate an effective means of impeding radiative conductivity under lower mantle pressure and temperature conditions.

silicates in the far ultraviolet. The role of pressure (in most cases) is to decrease the energy at which such band-gap related absorptions occur (that is, it becomes easier to delocalize electrons as the distance between ions decreases). It is generally believed that among relevant deep Earth constituents only relatively iron-rich materials (such as FeO) will have absorption edges which decrease to zero energy and metallize over the pressure and temperature interval of the Earth's mantle. However, the effect of a lower energy absorption edge in compressed magnesiowüstite ($Mg_xFe_{1-x}O$) may have a significant effect on the electrical conductivity of the deep mantle, in addition to producing an absolute impediment to radiative thermal conductivity at these depths.

This effect is illustrated in Figure 6; the intensity of radiative thermal emission is described by the Planck function,

$$I(v) = 2\pi ehc^2 v^5 (\exp[hv/kT] - 1)^{-1} \qquad (5)$$

in which the emitted intensity, I, is a function of frequency (v), e represents the emissivity of the material (assumed to be unity in the case of a blackbody, and to lie at a wavelength-independent value between zero and one for a greybody), and c, h, k and T are the speed of light, Planck's constant, Boltzmann's constant and temperature, respectively. Figure 6 shows representative values of emitted intensity for a range of temperatures, along with absorption mechanisms of minerals relevant to the Earth's upper and lower mantle. The actual radiative thermal

conductivity of mantle materials naturally incorporates both a frequency-dependent absorption coefficient and index of refraction of the material, as well as the black- or greybody emission of Equation 5 [23]. Figure 6 shows representative values of emitted intensity for a range of temperatures, along with the wavelength range of different absorption mechanisms occurring in minerals relevant to the Earth's upper mantle. Among lower mantle minerals, no optical absorption data have been reported for silicate perovskite; however, the absorption of magnesiowüstite at both ambient and high pressures has been well-characterized [14]. Not shown in Figure 6 are the grain boundary scattering effects which may also act to reduce radiative conductivity over a broad frequency range. Clearly, given the wavelength ranges over which crystal-field absorptions (which broaden in width at high temperatures) and charge-transfer bands occur in transition-metal bearing minerals, radiative conductivity is unlikely to provide a rapid mechanism of heat transport within the deep Earth. Thus, the optical properties of mantle minerals play a passive but seminal role in the thermal and tectonic evolution of the planet: were radiative conductivity an efficient means of heat transport through the deep Earth, thermal convection (and its geochemical and geodynamic consequences) would not be necessitated.

Acknowledgments. I thank the NSF for support, and an anonymous reviewer, T.J. Ahrens and E. Knittle for helpful comments. This is contribution #207 of the Institute of Tectonics at UCSC.

REFERENCES

1. Ashcroft, N.W., and D. Mermin, *Solid-State Physics*, 826 pp., Saunders College Publishing, Philadelphia, 1976.
2. Berry, F.J., and D.J. Vaughan (eds.), *Chemical Bonding and Spectroscopy in Mineral Chemistry*, 325 pp., Chapman and Hall, London, 1985.
3. Born, M. and K. Huang, *The Dynamical Theory of Crystal Lattices*, 420 pp., Oxford Press, New York, 1954.
4. Burns, R.G., *Mineralogical Applica-tions of Crystal Field Theory*, 224 pp., Cambridge University Press, New York, 1970.
5. Cotton, F.A., *Chemical Applications of Group Theory*, 2nd Ed., 386 pp., Wiley-Interscience, New York, 1971.
6. Decius, J.C. and R.M. Hexter, *Molecular Vibrations in Crystals*, 391 pp., McGraw-Hill, New York, 1977.
7. Farmer, V.C. (Ed.), *The Infrared Spectra of Minerals*, 589 pp., Mineralogical Society, London, 1974.
8. Fateley, W.G., F.R. Dollish, N.T. McDevitt and F.F. Bentley, *Infrared and Raman Selection Rules for Molecular and Lattice Vibrations: The Correlation Method*, 222 pp., Wiley-Interscience, New York, 1972.
9. Ghatak, A.K. and L.S. Kothari, *An Introduction to Lattice Dynamics*, 234 pp., Addison Wesley, New York, 1972.
10. Harvey, A.B. (Ed.), *Chemical Applications of Nonlinear Raman Spectroscopy*, 383 pp., Academic Press, San Diego, Calif., 1981.
11. Hawthorne, F.C., (Ed.) Reviews in Mineralogy, Vol. 18, *Spectroscopic methods in*

mineralogy and geology, 698 pp., Min. Soc. Am., Washington, D.C., 1988.
3rd Ed., 1072 pp., Hemisphere, Washington, D.C., 1992.

24. Williams, Q., R. Jeanloz and P. McMillan, Vibrational spectrum of MgSiO$_3$-perovskite: Zero pressure Raman and infrared spectra to 27 GPa, *J. Geophys. Res., 92,* 8116-8128, 1987.

12. Herzberg, G., *Molecular Spectra and Molecular Structure: II. Infrared and Raman Spectra of Polyatomic Molecules,* 632 pp., Van Nostrand Reinhold, New York, 1945.

13. Kieffer, S.W., Heat capacity and entropy: Systematic relations to lattice vibrations, *Rev. Mineral., 14,* 65-126, 1985.

14. Mao, H.K., Charge transfer processes at high pressure, in *The Physics and Chemistry of Minerals and Rocks,* edited by R.G.J. Strens, pp. 573-581, J. Wiley, New York, 1976.

15. Marfunin, A.S., *Spectroscopy, Luminescence and Radiation Centers in Minerals,* 352 pp., Springer-Verlag, New York, 1979.

16. McClure, D.S., Electronic spectra of molecules and ions in crystals. Part II. Spectra of ions in crystals, *Solid State Phys., 9,* 399-525, 1959.

17. McMillan, P.F. and A.M. Hofmeister, Infrared and Raman Spectroscopy, *Rev. Mineral., 18,* 99-159, 1988.

18. Nakamoto, K., *Infrared and Raman Spectra of Inorganic and Coor-dination Compounds,* 4th edition, 484 pp., Wiley-Interscience, New York, 1986.

19. Nassau, K., *The Physics and Chemistry of Color: The Fifteen Causes of Color,* 454 pp., J. Wiley, New York, 1983.

20. Pasteris, J.D., B. Wopenka and J.C. Seitz, Practical aspects of quantitative laser Raman spectroscopy for the study of fluid inclusions, *Geochim. Cosmochim. Acta., 52,* 979-988, 1988.

21. Paterson, M., The determination of hydroxyl by infrared absorption in quartz, silicate glasses and similar materials, *Bull. Minéral., 105,* 20-29, 1982.

22. Solymar, L. and D. Walsh, *Lectures on the Electrical Properties of Materials,* 5th Ed., 482 pp., Oxford Univ. Press, New York, 1992.

23. Siegel, R. and J.R. Howell, *Thermal Radiation Heat Transfer,* 3rd Ed., 1072 pp., Hemisphere, Washington, D.C., 1992.

24. Williams, Q., R. Jeanloz and P. McMillan, Vibrational spectrum of MgSiO$_3$-perovskite: Zero pressure Raman and infrared spectra to 27 GPa, *J. Geophys. Res., 92,* 8116-8128, 1987.

Nuclear Magnetic Resonance Spectroscopy of Silicates and Oxides in Geochemistry and Geophysics

Jonathan F. Stebbins

I. INTRODUCTION

The purpose of this chapter is to very briefly summarize the applications and limitations of NMR spectroscopy in the study of the silicates and oxides that make up the earth's crust and mantle, and to tabulate the most useful data. More extensive reviews have been published recently that give background on the fundamentals, as well as details of applications [2,11,29,33,34,42,61,62,88,116, 141]. Extensive tabulations of NMR data on silicates have also been published [34,43,56,78,118,128,161].

The utility of NMR in understanding the chemistry and physics of materials comes from the small perturbations of nuclear spin energy levels (non-degenerate only in a magnetic field) that are caused by variations in local electron distributions, by the distributions of other neighboring spins (electronic or nuclear), and by the time dependence of these interactions. Any nuclide with non-zero nuclear spin thus can, in principle, be observed by NMR, but the practicality of the experiment varies tremendously. Detection of signals from nuclides with low natural abundance and low resonant frequency is often difficult or impossible, although isotopic enrichment can be useful. The same can be true for even abundant nuclides of heavy elements, which may have extremely wide ranges of frequencies and correspondingly broad NMR spectra. Potentially interesting nuclides are listed in Tables 1 and 2, which contain a simple comparison of the relative ease of observation in a liquid sample, based on resonant frequency, spin quantum number, and abundance. These data serve only for rough comparison, however (especially for solids), because they contain no material-specific information on relaxation rates and line widths.

NMR is an element specific spectroscopy, and spectra are primarily sensitive to short-range effects. Thus, like techniques such as x-ray absorption and Mössbauer spectroscopy, NMR is a good complement to diffraction methods, and is particularly useful in amorphous materials and liquids. NMR can be highly quantitative, with a 1 to 1 correlation between signal intensity and the abundance of a nuclide in a given structural site, regardless of the structure. In practice, of course, experiments must be carried out carefully to be accurately quantitative. A major drawback of NMR is its low sensitivity when compared to spectroscopies involving higher energy transitions (e.g. visible, infrared and Raman). Interpretations of solid-state NMR spectra still rely primarily on empirical correlations, but these are now well understood in a qualitative sense. Another limitation of NMR is a severe one for the study of natural silicate minerals: paramagnetic ions in abundances greater than a few tenths of one percent can broaden NMR spectra to the point of being impossible to observe or interpret [87].

Most of the data reported in the tables has been collected by high resolution magic angles spinning (MAS) techniques, with important contributions from the new technique of dynamic angle spinning (DAS), and from single crystal and static powder spectra. Figure 1 suggests, however, that some of the orientational information that

J. F. Stebbins, Stanford University, Department of Geology, Stanford, CA 94305-2115

Mineral Physics and Crystallography
A Handbook of Physical Constants
AGU Reference Shelf 2

TABLE 1. NMR parameters for some nuclides of interest in geochemistry: spin 1/2 nuclides

Isotope	Natural abundance, %	NMR frequency at 9.4 T, MHz	Receptivity relative to ^{29}Si
^{1}H	99.99	400.0	2700
^{13}C	1.1	100.6	0.48
^{15}N	0.4	40.6	0.010
^{19}F	100	376.4	2252
^{29}Si	4.7	79.5	1
^{31}P	100	161.9	180
^{57}Fe	2.2	12.8	0.002
^{77}Se	7.6	76.3	1.43
^{89}Y	100	19.7	0.32
^{103}Rh	100	12.7	0.09
^{109}Ag	48.2	18.6	0.13
^{113}Cd	12.3	88.7	3.7
^{119}Sn	8.6	149.2	12.2
^{125}Te	7.0	126.2	6.1
^{129}Xe	26.4	111.2	15.4
^{169}Tm	100	33.1	1.5
^{171}Yb	14.3	70.4	2.1
^{183}W	14.4	16.6	0.03
^{195}Pt	33.8	85.6	9.1
^{199}Hg	16.8	71.6	2.7
^{205}Tl	70.5	230.5	377
^{207}Pb	22.6	83.7	5.4

Notes for Tables 1 and 2:
Data are primarily from tabulation in [49]. Noble gasses, lanthanides, less favorable isotopes of elements listed, and unfavorable radioisotopes are excluded. Relative receptivity is calculated for low-viscosity liquid samples. Line width and relaxation behavior in solid samples can make this estimates quite misleading. In particular, nuclides with large quadrupolar moments (Table 2), and heavy nuclides with very large chemical shift ranges may be difficult to observe in solids.

is lost in MAS may prove to be useful in characterizing anisotropic materials.

Quantitative local structural information on minerals and melts is important to many problems in geochemistry and geophysics. NMR has proven to be very useful in this area. A second nearly unique utility of NMR, that is just beginning to be applied in the geosciences, is in studying dynamics at the time scale of seconds to nanoseconds. For many geochemically interesting pro-

cesses, such as diffusion, viscous flow, and displacive phase transitions, the fundamental rate may be within this range.

1.1. Definitions

The chemical shift δ is the perturbation in the resonant (Larmor) frequency ν of a nuclide in a particular chemical environment, caused by screening of the external magnetic field by the surrounding electrons. δ is generally

TABLE 2. NMR parameters for some nuclides of interest in geochemistry: quadrupolar nuclides. See Table 1 for notes.

Isotope	Spin	Natural abundance, %	Quadrupolar moment $\times 10^{28}$, m^{-2}	NMR frequency at 9.4 T, MHz	Receptivity relative to ^{29}Si
^{2}H	1	0.02	0.0028	61.4	0.004
^{6}Li	1	7.4	-0.0008	59.0	1.7
^{7}Li	3/2	92.6	-0.04	155.6	737
^{9}Be	3/2	100	0.05	56.2	37.5
^{10}B	3	19.58	0.085	43.0	10.6
^{11}B	3/2	80.4	0.041	128.4	360
^{14}N	1	99.6	0.01	28.9	2.7
^{17}O	5/2	0.04	-0.026	54.2	0.03
^{23}Na	3/2	100	0.10	105.9	250
^{25}Mg	5/2	10.13	0.22	24.5	0.7
^{27}Al	5/2	100	0.15	104.3	560
^{33}S	3/2	0.8	-0.055	30.7	0.05
^{35}Cl	3/2	75.5	-0.10	39.2	9.6
^{37}Cl	3/2	24.5	-0.08	32.7	1.8
^{39}K	3/2	93.1	0.05	18.7	1.3
^{43}Ca	7/2	0.15	0.2	27.0	0.02
^{45}Sc	7/2	100	-0.22	97.3	819
^{47}Ti	5/2	7.3	0.29	22.5	0.4
^{49}Ti	7/2	5.5	0.24	22.6	0.6
^{51}V	7/2	99.8	-0.05	105.4	1035
^{53}Cr	3/2	9.6	0.03	22.6	0.23
^{55}Mn	5/2	100	0.4	98.8	475
^{59}Co	7/2	100	0.38	94.4	750
^{61}Ni	3/2	1.2	0.16	35.8	0.11
^{63}Cu	3/2	69.1	-0.21	106.1	174
^{65}Cu	3/2	30.9	-0.20	113.7	96
^{67}Zn	5/2	4.1	0.16	25.1	0.32
^{71}Ga	3/2	39.6	0.12	122.2	152
^{73}Ge	9/2	7.8	-0.18	14.0	0.30
^{75}As	3/2	100	0.29	68.7	69
^{81}Br	3/2	49.5	0.31	108.4	133
^{87}Rb	3/2	27.9	0.13	131.3	133
^{87}Sr	9/2	7.0	0.3	17.4	0.51
^{91}Zr	5/2	11.2	-0.21	37.3	2.9
^{93}Nb	9/2	100	-0.22	98.2	1320
^{95}Mo	5/2	15.7	±0.12	26.2	1.4
^{99}Ru	5/2	12.7	0.08	18.4	0.4
^{101}Ru	5/2	17.1	0.44	20.7	0.7

TABLE 2 (continued)

Isotope	Spin	Natural abundance, %	Quadrupolar moment x10^{28}, m^{-2}	NMR frequency at 9.4 T, MHz	Receptivity relative to ^{29}Si
^{105}Pd	5/2	22.2	0.8	18.4	0.7
^{115}In	9/2	95.7	0.83	88.1	910
^{121}Sb	5/2	57.3	-0.28	96.2	251
^{127}I	5/2	100	-0.79	80.5	257
^{133}Cs	7/2	100	-0.003	52.8	130
^{137}Ba	3/2	11.3	0.28	44.7	2.1
^{139}La	7/2	99.9	0.22	56.8	163
^{177}Hf	7/2	18.5	4.5	16.2	0.7
^{181}Ta	7/2	99.99	3	48.0	99
^{187}Re	5/2	62.9	2.2	91.8	238
^{189}Os	3/2	16.1	0.8	31.4	1.0
^{193}Ir	3/2	62.7	1.0	7.6	0.06
^{197}Au	3/2	100	0.59	6.9	0.07
^{201}Hg	3/2	13.2	0.44	26.6	0.5
^{209}Bi	9/2	100	-0.38	64.9	381

expressed in parts per million relative to a standard, with $\delta = 10^6(\nu_{sample} - \nu_{standard})/\nu_{standard}$. The chemical shift is orientation dependent, and is described by the chemical shift anisotropy (CSA) tensor, whose principle components are usually denoted δ_{11}, δ_{22}, and δ_{33} and which has a unique orientation with respect to the local structure (or with respect to crystallographic axes in a crystalline material). The isotropic chemical shift, δ_{iso}, is the average of these three components. For spin 1/2 nuclides, δ_{iso} is observed experimentally in liquids where molecular rotation produces rapid isotropic averaging, and in solids by rapid sample spinning (MAS NMR) at the "magic" angle θ with respect to the external field ($\theta \approx 54.7°$, $1-3\cos^2\theta = 0$).

For nuclides with spin I > 1/2, a total of 2I transitions may be observable. The frequencies of these transitions are controlled by the energy of interactions with the

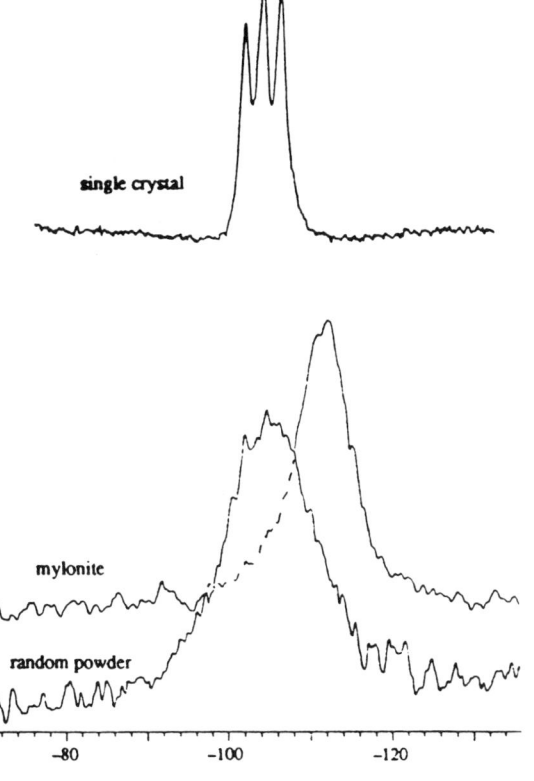

Fig. 1. Static (non-MAS) ^{29}Si spectra for quartz. Top spectrum shows three magnetically inequivalent Si sites; bottom spectrum shows spectrum for randomly oriented powder; center spectrum shows effect of strong preferred orientation in a quartz mylonite [132]. Scales in this and other figures are in ppm.

electric field gradient, often summarized by the nuclear quadrupolar coupling constant $QCC=e^2qQ/h$. Here, eQ is the nuclear quadrupolar moment, and eq is the principle component of the electric field gradient at the site of interest. The full description of the quadrupolar interaction requires the electric field gradient tensor and its orientation relative to the structure. The deviation from cylindrical symmetry of this tensor is given by the asymmetry parameter η, which varies from 0 to 1. In liquids with sufficiently rapid isotropic rotation of molecules, the field gradient and the quadrupolar interaction average to zero and δ_{iso} is observed. In MAS NMR, the central 1/2 to -1/2 transition remains shifted and broadened by a second order quadrupolar interaction; in DAS NMR, δ_{iso} is shifted by the isotropic average of the second order quadrupolar interaction, but is not broadened.

2. APPLICATIONS TO CRYSTALLINE SILICATES AND OXIDES

2.1. ^{29}Si

Isotropic chemical shifts for ^{29}Si and CSA data are listed in Tables 3 through 5. In silicates, the largest effect of structure on chemical shift is that of coordination number. Thus, δ for SiO_6 groups is in the range of about -180 to -220 ppm relative to tetramethyl silane (TMS), and is in nearly all cases between about -65 and -120 ppm for SiO_4 groups. Signals near to -150 ppm are probably from SiO_5 groups [58,166].

The second most important effect is that of the number and identity of first cation neighbors. If bridging oxygens are considered as those shared with tetrahedral Si or Al (or B or P) neighbors, and non-bridging oxygens to be those shared with larger and/or lower charged M cations, an SiO_4 site can be labeled as Q^n, where n designates the number of bridging oxygens (and Q stands for quaternary). For a fixed M cation type, decreasing n by one tends to increase δ by about 10 ppm to less negative, higher frequencies. Each Q^n species thus has a distinct, but somewhat overlapping range of δ. In a similar fashion, the substitution of tetrahedral Al for a tetrahedral Si neighbor tends to increase δ by about 5 ppm. As a result, ^{29}Si MAS NMR spectra of aluminosilicates often have multiple, partially overlapping peaks for Si sites with varying numbers m of Al neighbors [$Q^n(mAl)$]. This effect is particularly obvious and well-exploited in tectosilicates, where it has often provided the key to unraveling quite complex Al/Si ordering patterns. The bond angle between tetrahedra has a related effect, again best calibrated for tectosilicates. Increasing the mean Si-O-T angle systematically decreases δ. A number of semi-empirical correlations

among number of Al neighbors, bond angles, and δ have been developed that allow rather precise estimates of δ for tectosilicate structures [34]. A few correlations have been developed for SiO_4 groups in general that allow estimation of δ for most silicate structures [56,118]. Similar approaches have been taken for SiO_6 groups [47,146].

Structure–δ correlations have been used to derive important constraints on Al/Si ordering in a number of minerals. The greatest efforts have been on synthetic zeolites, because of their tremendous technological importance [34,72]. In geochemistry, the most important examples have been in determining the ordering state in feldspars [64,102,103,120,127,169], other tectosilicates [12,53,95,104,117,148], cordierite [107] and sheet silicates [5,20,52,60,74,112,160]. In several cases, discrepancies in thermodynamic data that had been tentatively ascribed to underestimates of the entropy of disordered synthetic phases have been displaced to other parts of the data base by findings of nearly complete order, such as for sillimanite [59,140] and prehnite [139]. For a few systems, careful combination of NMR spectroscopy with calorimetric observation has provided new insights and details of the energetic control and consequences of Si/Al disorder [17,53,103,106].

Very recently, ^{29}Si MAS NMR has begun to be applied to high pressure mantle phases that contain SiO_6 groups. In $MgSiO_3$ ilmenite and perovskite, and β-Mg_2SiO_4, Mg/Si disorder has not been detected [63,146], but both Mg/Si and Si/Al disorder are significant in majorite and majorite-pyrope solid solutions, as shown in Figure 2 [101,146].

2.2. ^{27}Al

NMR studies of ^{27}Al in minerals have been reviewed recently [66], and data are listed in Tables 6 and 7. Most early studies of this nuclide were of single crystals, with complete determinations of quadrupolar coupling constants, asymmetry parameters, and electric field gradient tensors, but without precise isotropic chemical shifts. The quadrupolar parameters have been shown to be roughly correlated with the extent of distortion of the Al site for both octahedral and tetrahedral geometries [43]. Single-crystal work on $MgAl_2O_4$ spinel allowed quantification of octahedral and tetrahedral Al site occupancies and ordering state [14].

As for ^{29}Si, isotropic shifts for ^{27}Al in oxides (now determined most commonly by high resolution MAS NMR) are most strongly influenced by coordination number. δ values for AlO_4 groups fall roughly in the range of 50 to 90 ppm relative to aqueous $Al(H_2O)_6^{3+}$, and for AlO_6 groups in the range of -10 to 15 ppm.

TABLE 3. ^{29}Si NMR data for Q^0, Q^1, and Q^2 sites in crystalline silicates.

mineral (s)=synthetic	nominal formula	$-\delta_{iso}$[a]	CSA[b]			ref.
			$-\delta_{11}$	$-\delta_{22}$	$-\delta_{33}$	
Q^0 sites:						
chondrodite	Mg$_5$(SiO$_4$)$_2$(OH,F)$_2$	60[c]				[78]
forsterite(s)	Mg$_2$SiO$_4$	61.9	38.8[d]	55.3	95.4	[78,159]
(s)	Li$_2$SiO$_4$	64.9				[78]
monticellite(s)	CaMgSiO$_4$	66	44	60	94	[128]
(s)	NaH$_3$SiO$_4$	66.4				[78]
(s)	Na$_2$H$_2$SiO$_4$·8.5H$_2$O	67.8				[78]
majorite garnet(s)[g]	Mg$_4$Si^{VI}Si$^{IV}_3$O$_{12}$[e]	68 to 90[f]				[101,146]
(s)	Ba$_2$SiO$_4$	70.3				[78]
(s)	α-Ca$_2$SiO$_4$	70.3				[78]
afwillite	Ca$_3$(HSiO$_4$)$_2$·2H$_2$O	71.3,73.3				[34]
larnite(s)	β-Ca$_2$SiO$_4$	71.4				[78]
pyrope(s)	Mg$_3$Al$_2$Si$_3$O$_{12}$[e]	72.0				[101]
Ca-olivine(s)	γ-Ca$_2$SiO$_4$	73.5				[78]
(s)	Ca$_3$SiO$_5$	69 to 75[h]				[78,124]
(s)	CaNaHSiO$_4$	73.5				[78]
phenacite	Be$_2$SiO$_4$	74.2				[78]
rutile	<1% SiO$_2$ in TiO$_2$	77.2				[139]
titanite	CaTiSiO$_5$	79.6				[118]
andalusite	Al$_2$SiO$_5$	79.8	45	78	116	[78,128]
kyanite	Al$_2$SiO$_5$	82.3,83.2	50	79	120	[118,128]
zircon	ZrSiO$_4$	81.6				[78]
piemontite[g]	Ca$_2$(Al,Mn,Fe)$_3$ (Si$_2$O$_7$)(SiO$_4$)OH	81.9				[118]
grossular	Ca$_3$Al$_2$(SiO$_4$)$_3$	83.4				[56]
topaz	Al$_2$SiO$_4$(OH,F)$_2$	85.6				[78]
Q^1 sites:						
(s)	Na$_6$Si$_2$O$_7$	68.4				[78]
(s)	Li$_6$Si$_2$O$_7$	72.4				[34]
gehlenite(s)	Ca$_2$Al$_2$SiO$_7$	72.5	122	74	20	[56]
akermanite(s)	Ca$_2$MgSi$_2$O$_7$	73.7	134	84	1	[56]
rankinite(s)	Ca$_3$Si$_2$O$_7$	74.5,76.0				[78]
hemimorphite	Zn$_4$Si$_2$O$_7$(OH)$_2$·H$_2$O	77.9				[78]
(s)	β-Mg$_2$SiO$_4$	79.0				[146]
lawsonite	CaAl$_2$Si$_2$O$_7$(OH)·H$_2$O	81	123	92	28	[128]
(s)	α-Y$_2$Si$_2$O$_7$	81.6,83.5				[34]
(s)	α-La$_2$Si$_2$O$_7$	83.2				[78]
(s)	Ca$_6$Si$_2$O$_7$(OH)$_6$	82.6	109	109	35	[34,46]

TABLE 3 (continued)

mineral (s)=synthetic	nominal formula	$-\delta_{iso}$[a]	CSA[b]			ref.
			$-\delta_{11}$	$-\delta_{22}$	$-\delta_{33}$	
piemontite[g]	$Ca_2(Al,Mn,Fe)_3$ $(Si_2O_7)(SiO_4)OH$	86.4,90.4				[118]
(s)	$In_2Si_2O_7$	87.7				[78]
(s)	$\beta\text{-}Y_2Si_2O_7$	92.9				[78]
thortveitite	$Sc_2Si_2O_7$	95.3				[78]
zunyite[g]	$Al_{13}Si_5O_{20}(OH)_{14}F_4Cl$[i]	91.2,96.9				[118]
Q^2 sites:						
(s)	Li_2SiO_3	74.5				[78]
(s)	Na_2SiO_3	76.8	18	59	156	[78,136]
(s)	$BaSiO_3$	80	29	71	140	[128]
(s)	$SrSiO_3$	85	30	71	154	[128]
orthoenstatite	$Mg_2Si_2O_6$	82[c]				[78]
clinoenstatite(s)	$Mg_2Si_2O_6$	81.8,84.2	40[j]	73	132	[56,128]
diopside	$CaMgSi_2O_6$	84.8	31	73	148	[118,128]
omphacite	$(Ca,Na)(Mg,Al)Si_2O_6$[i]	85.4				[118]
spodumene	$LiAlSi_2O_6$	91.4	53	81	142	[118]
jadeite	$NaAlSi_2O_6$	91.8				[78]
pyroxene phase(s)[g]	$NaMg_{0.5}Si^{VI}_{0.5}Si^{IV}_2O_6$	92.1,97.6				[146]
alamosite	$Pb_{12}Si_{12}O_{36}$	84.1,86.5, 94.3				[34]
prehnite[g]	$Ca_2Al_2Si_3O_{10}(OH)_2$	84.6[k]				[139]
tchermakitic amphibole(s)[g]	$Ca_2(Mg_4Al)(AlSi_7O_{22})\text{-}(OH)_2$	83.4,87.4				[18]
triple-chain phase(s)[g]	$Na_2Mg_4Si_6O_{16}(OH)_2$	85.3				[18]
Sc-F pargasite (s) (Q^2+Q^3)	$NaCa_2Mg_4ScSi_6Al_2O_{22}\text{-}F_2$	86[m]				[108]
tremolite[g]	$Ca_2Mg_5(Si_4O_{11})_2(OH)_2$	87.2	50	77	137	[18,128]
hillebrandite(s)	$Ca_2SiO_3(OH)_2$	86.3				[78]
pectolite(s)	$Ca_2NaHSi_3O_9$	86.3				[78]
foshagite(s)	$Ca_4Si_3O_9(OH)_2$	84.8,86.4				[78]
xonotlite(s)[g]	$Ca_6Si_6O_{17}(OH)_2$	86.8				[78]
walstromite phase	$Ca_3Si_3O_9$	73.8,78.5, 79.0				[58]
wollastonite	$Ca_3Si_3O_9$	87.6,91.7[l]	24	85	158	[118]
β-wollastonite(s)	$\beta\text{-}Ca_3Si_3O_9$	89.0				[78]

TABLE 3 (continued)

mineral (s)=synthetic	nominal formula	$-\delta_{iso}$[a]	CSA[b]			ref.
			$-\delta_{11}$	$-\delta_{22}$	$-\delta_{33}$	
Q² sites, rings:						
ps-wollastonite(s)	α-$Ca_3Si_3O_9$	83.5				[78]
(s)	$K_4H_4Si_4O_{12}$	87.5	63	63	141	[71,78]
tourmaline	$Na(Mg,Li,Al)_3Al_6Si_6O_{18}$-$(BO_3)_3(OH,F)_4$[i]	88.1				[118]
(s)	$Ba_7Si_7O_{21}\cdot10BaCl_2$	92.5				[34]
benitoite	$BaTiSi_3O_9$	94.2				[78]
wadeite phase(s)[g]	$K_2Si^{VI}Si^{IV}_3O_9$	95.0				[146]
High P phases, uncertain structure:						
"phase E" (s)	$Mg_{2.3}Si_{1.3}H_{2.4}O_6$	75.7[m]				[58]
"phase Y" (s)[g]	$(CaO)_XSi^VSi^{IV}O_4$	80				[146]
(s)	ε-$Na_2Si_2O_5$	80.6,81.8				ee
(s)[g]	$CaSi^{VI}Si^{IV}O_5$	88.9				[146]
(s)[g]	$Na_2Si^{VI}Si^{IV}_2O_7$	94.4				ee
(s)[g]	ζ-$Na_2Si^{VI}Si^{IV}O_5$	97.9				ee
(s)[g]	$Na_2(Si^{VI},Si^{IV})_4O_9$	97.0,107.7, 108.9				ee

Notes for Tables 3, 4, and 5:

†Most intense peak.

Si-O-B bonds are considered as "bridging", placing the Si sites in beryl and datolite in the Q^4 and Q^3 groups, respectively. (a) Chemical shifts are in ppm relative to tetramethylsilane. (b) Unless otherwise noted, principle components of CSA tensor are derived from fitting spinning side bands, and are of relatively low precision; orientation of tensor with respect to crystallographic axes may not be known. In some cases, δ_{iso} and CSA data are from different sources and may therefore appear to be slightly discrepant [δ_{iso} should = $(\delta_{11}+\delta_{22}+\delta_{33})/3$]. Given typical errors in CSA measurements, this is generally not significant. (c) Broad peak (at least 5 ppm width). (d) CSA based on single crystal study. (e) Additional data on solid solution series given in reference. (f) Multiple peaks due to partial disorder among octahedral Si and Mg. Main peak is at -74.3 ppm. (g) Data for another type of Si site listed elsewhere in table or in Table 4 or 5. (h) Nine Si sites, 7 resolved peaks. (i) Approximate formula. (j) CSA data are means for two sites. (k) Two additional peaks indicate a small amount of Si/Al disorder. (l) First peak is for T1 site, second for T2+T3. (m) Broad peak consistent with considerable disorder.(n) Ordering schemes can be complex, but can be characterized for tetrahedral Si and Al as (1) disordered, Al-avoidance violated; (2) ordered according to Al-avoidance only; (3) more ordered than required by Al-avoidance; (4) fully ordered or nearly so; (5) partially disordered. (o) Split peak reported by [4]. (p) CSA components: -43,-59, -147 ppm [128]. (q) CSA components: -49, -83, -129 ppm [128]. (r) CSA components: -107, -107, -59 ppm [128]. (s) small peak due to triple-chain site in partially disordered phase. (t) CSA components: -56, -72, -151 ppm [128]. (u) CSA components: -54, -70, -161 ppm [128]. (v) CSA components from single crystal study: 102.6, 107.0, 109.1 [132]. (w) Approximately 5 peaks resolved for 12 sites. (x) Overlapping peaks due to multiple sites. (y) End member: complex series of peaks for intermediate compositions. (z) Approximately 8 peaks for 15 possible sites. (aa) 8 peaks for 8 sites. (bb) δ_{11}=-173.4 ppm; $\delta_{22} = \delta_{33}$ = -183.1 ppm. (cc) Narrow peak consistent with complete Mg/Si order. (dd) Somewhat broadened peak consistent with some Mg/Si disorder. Reference contains data on solid solution with pyrope. (ee) Kanzaki, Stebbins and Xue, unpublished data.

TABLE 4. ^{29}Si NMR data for Q^3 and Q^4 sites in crystalline silicates. Data for some clay minerals and synthetic silicas, zeolites, as well as silicates with organic ligands, have been excluded for brevity. Data for end-member compositions only. See Table 3 for notes.

mineral (s)=synthetic	nominal formula	$-\delta_{iso}$[a]	ordering state[n]	ref.
Q^3 sites, layer aluminosilicates:				
margarite	$CaAl_2(Al_2Si_2O_{10})(OH)_2$	75.5	4	[78]
phlogopite	$KMg_3AlSi_3O_{10}(F,OH)_2$	84 to 87	5	[60,78]
phlogopite(s)[e]	$KMg_3AlSi_3O_{10}(OH)_2$	83.2,87.0†,90.7	3	[20]
palygorskite	$MgAlSi_4O_{10}(OH)_2{\cdot}4H_2O$	84.9,91.7,96.8		[34]
beidellite	$Na_{0.3}Al_2(Al_{0.3}Si_{2.7}O_{10})(OH)_2$	88,94		[34]
muscovite	$KAl_2SiAl_3O_{10}(OH)_2$	89,85†,81	5	[112]
illite[e]	$KAl_2SiAl_3O_{10}(OH)_2$[i]	91[m]	5	[60]
lepidolite	$KLi_2Al(AlSi_3O_{10})(F,OH)_2$	89[m]	5	[78]
dickite	$Al_4Si_4O_{10}(OH)_8$	90.9		[34]
kaolinite	$Al_4Si_4O_{10}(OH)_8$	91.5[o]		[78]
endellite	$Al_4Si_4O_{10}(OH)_{10}{\cdot}8H_2O$	93.1		[78]
pyrophyllite	$Al_2Si_4O_{10}(OH)_2$	94.0		[112,118]
montmorillonite	$(Al,Mg)_2Si_4O_{10}(OH)_2{\cdot}4H_2O$	93.7		[34]
sepiolite	$Mg_4Si_6O_{15}(OH)_2{\cdot}2H_2O$	92,95,98		[34]
hectorite	$(Mg,Li)_{2.7}Na_{0.3}Si_4O_{10}(OH)_2{\cdot}4H_2O$	95.3		[60]
Q^3 sites, other silicates:				
sapphirine[e]	$(Mg_{3.6}Al_{4.4})(Al_{4.4}Si_{1.6})O_{20}$	73[m]	5	[19]
datolite	$CaBSiO_4(OH)$	83.0[p]		[128]
sillimanite	Al_2SiO_5	86.4[q]	4	[59,118]
tremolite[g]	$Ca_2Mg_5(Si_4O_{11})_2(OH)_2$	91.7[r]		[18]
tremolite(s)[g]	$Ca_2Mg_5(Si_4O_{11})_2(OH)_2$	91.7, 96.9[s]		[18]
tchermakitic amphibole(s)[g]	$Ca_2(Mg_4Al)(AlSi_7O_{22})(OH)_2$	92.1		[18]
triple-chain phase(s)[g]	$Na_2Mg_4Si_6O_{16}(OH)_2$	87.8,91.4		[18]
apophyllite	$KFCa_4Si_8O_{20}{\cdot}H_2O$	92.0		[78]
serpentine	$Mg_3Si_2O_5(OH)_4$	94.0		[78]
talc	$Mg_3Si_4O_{10}(OH)_2$	97.2		[118]
xonotlite(s)[g]	$Ca_6(Si_6O_{17})(OH)_2$	97.6		[78]
(s)	$Li_2Si_2O_5$	92.5[t]		[94]
(s)	$BaSi_2O_5$	93.5		[94]
(s)	α-$Na_2Si_2O_5$	94.5[u]		[94]
(s)	$K_2Si_2O_5$	91.5,93,94.5		[94]
(s)	β-$H_2Si_2O_5$	98.4,101.9,110		[34]
(s)	α-$H_2Si_2O_5$	101.5		[34]

TABLE 4 (continued)

mineral (s)=synthetic	nominal formula	$-\delta_{iso}$[a]	ordering state[n]	ref.
Q^4, silica polymorphs:				
quartz	SiO_2	107.4[v]		[118,126]
coesite	SiO_2	108.1,113.9		[126]
cristobalite	SiO_2	108.5		[118,126]
tridymite	SiO_2	109.3-114.0[w]		[126]
Q^4, feldspars:				
low albite[e]	$NaAlSi_3O_8$	92.3,96.9,104.3	4	[103,127]
high albite[e]	$NaAlSi_3O_8$	91 to 112[x]	5	[169]
microcline[e]	$KAlSi_3O_8$	95.6,97.6,100.6	4	[103,127]
sanidine[e]	$K_{0.6}Na_{0.4}AlSi_3O_8$	97,101[m]	5	[64]
anorthite[e]	$CaAl_2Si_2O_8$	82.7,84.7,89.3[x]	4	[120]
anorthite[e]	$CaAl_2Si_2O_8$ (disordered)	82.5 to 104.5[x]	5	[102]
Q^4, feldspathoids:				
carnegieite(s)	$NaAlSiO_4$	82.2	4	[148]
sodalite	$Na_8Al_6Si_6O_{24}Cl_2$	84.9	4	[118]
nepheline[e]	$Na_3KAl_4Si_4O_{16}$[i]	85.1†,88.4	4	[148]
kalsilite(s)[e]	$KAlSiO_4$	88.8	4	[148]
cancrinite	$Na_8Al_6Si_6O_{24}CO_3$	86.3	4	[118]
scapolite[e]	$Na_4Al_3Si_9O_{24}Cl-$ $Ca_4Al_6Si_6O_{24}CO_3$	92.6,106.2[y]	5	[117]
analcite	$NaAlSi_2O_6 \cdot H_2O$	91.6,96.8†,102.0	3	[95]
leucite	$KAlSi_2O_6$	78.7 to 106.7[z]	3	[95]
Q^4, zeolites:				
thomsonite	$NaCa_2Al_5Si_5O_{20} \cdot 6H_2O$	86.4,89.0,91.7		[118]
scolectite	$CaAl_2Si_3O_{10} \cdot 3H_2O$	86.4,88.8,95.7		[118]
natrolite	$Na_2Al_2Si_3O_{10} \cdot 2H_2O$	87.7†,95.4		[118]
gmelinite	$Na_2Al_2Si_4O_{12} \cdot 6H_2O$	92.0,97.2†,102.5		[72]
chabazite	$CaAl_2Si_4O_{12} \cdot 6H_2O$	94.0,99.4†,104.8		[72]
stilbite	$CaAl_2Si_7O_{18} \cdot 7H_2O$	98,101.5†,108		[72]
harmotone	$BaAl_2Si_6O_{16} \cdot 6H_2O$	95,98.6,102.6,108		[72]
heulandite	$CaAl_2Si_7O_{18} \cdot 6H_2O$	95.0 to 108.0[aa]		[118]
Q^4, others:				
cordierite	$Mg_2Al_4Si_5O_{18}$ (ordered)	79,100†	4	[107]
cordierite	$Mg_2Al_4Si_5O_{18}$ (disordered)	79 to 112[aa]	1	[107]
petalite	$LiAlSi_4O_{10}$	87[c]		[78]

TABLE 4 (continued)

mineral (s)=synthetic	nominal formula	$-\delta_{iso}$[a]	ordering state[n]	ref.
danburite	$CaB_2Si_2O_8$	89		[128]
prehnite	$Ca_2Al_2Si_3O_{10}(OH)_2$	98.6	4	[139]
beryl[e]	$Al_2Be_3Si_6O_{18}$	102.3		[119]
roedderite	$Na_2Mg_5Si_{12}O_{30}$	100.6		[50]
zunyite	$Al_{13}Si_5O_{20}(OH)_{14}F_4Cl$	128.5	4	[118]

TABLE 5. ^{29}Si NMR data for SiO_6 and SiO_5 sites in crystalline silicates. Data for end-member compositions only. See Table 3 for notes.

mineral (s)=synthetic	nominal formula	$-\delta_{iso}$[a]	ref.
SiO_6, known structures:			
thaumasite	$Ca_3Si^{VI}(OH)_6(SO_4)(CO_3)\cdot15H_2O$	179.6[bb]	[48,139]
ilmenite phase(s)	$MgSi^{VI}O_3$	181.0[cc]	[146]
stishovite(s)	$Si^{VI}O_2$	191.3	[146]
perovskite phase(s)	$MgSi^{VI}O_3$	191.7[cc]	[63]
perovskite phase(s)	$CaSi^{VI}O_3$	194.5	[146]
pyroxene phase(s)[g]	$NaMg_{0.5}Si^{VI}_{0.5}Si^{IV}_2O_6$	194.7[cc]	[146]
majorite garnet(s)[g]	$Mg_4Si^{VI}Si^{IV}_3O_{12}$[e]	197.6[dd]	[101,146]
wadeite phase(s)[g]	$K_2Si^{VI}Si^{IV}_3O_9$	203.1	[146]
(s)	$Si^{VI}_5O(PO_4)_6$	214.0, 217.0	[158]
(s)	$Si^{VI}P_2O_7$	220[c]	[158]
SiO_6, uncertain structure:			
(s)[g]	$CaSi^{VI}Si^{IV}O_5$	193.4	[146]
(s)[g]	$\zeta\text{-}Na_2Si^{VI}Si^{IV}O_5$	199.8	ee
(s)[g]	$Na_2Si^{VI}Si^{IV}_2O_7$	200.4	ee
(s)[g]	$Na_2(Si^{VI},Si^{IV})_4O_9$	202.4	ee
"phase X" (s)	$(CaO)_XSi^{VI}O_2$	208.6	[146]
SiO_5, uncertain structure:			
"phase Y"	$(CaO)_XSi^VSi^{IV}O_4$[g]	150.0	[146]

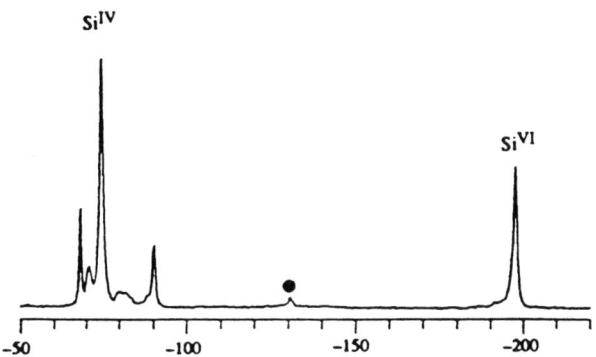

Fig. 2. ^{29}Si MAS spectrum for a high pressure, MgSiO$_3$ garnet. The multiplicity of tetrahedral sites results from partial disorder among six-coordinated Si and Mg neighbors [101]. Black dot marks spinning side band.

TABLE 6. ^{27}Al NMR data AlO$_4$ sites in crystalline silicates and oxides. Data for a number of clay minerals, synthetic zeolites, sheet silicates, and phosphates have been excluded for brevity.

Mineral (s)=synthetic	nominal formula	QCC, MHz	η	$\delta_{iso}^{a,b}$ ppm	ref.
Q^0 sites:					
(s)	Ba$_5$Al$_2$O$_8$	2.3	0.8	80	[92]
Q^0 sites, Al^{VI} neighbors:					
zunyite	Al$^{VI}_{12}$Al^{IV}Si$_5$O$_{20}$(OH)$_{14}$F$_4$Cl			72	[92]
garnet phase(s)	Gd$_3$Al$_2^{VI}$Al$_3^{IV}$O$_{12}$	5.47	0		[43]
garnet phase(s)	Y$_3$Al$_2^{VI}$Al$_3^{IV}$O$_{12}$	6.02	0	76.0	[43,79,129]
spinel (disordered)(s)	(Mg,Al)^{IV}(Al,Mg)$^{VI}_2$O$_4$			72^c	[86,92,163]
(s)	β-Al$_2$O$_3$			64	[92]
(s)	γ-Al$_2$O$_3$			66	[92]
(s)	η-Al$_2$O$_3$			62	[92]
(s)	χ-Al$_2$O$_3$			64	[92]
(s)	BaAl$^{VI}_9$Al^{V}Al$^{IV}_2$O$_{19}$			70	[92]
Q^1 sites:					
(s)	KAlO$_2$·1.5H$_2$O	5.0	0.25	81	[92]
Q^2 sites:					
(s)	KAlO$_2$·H$_2$O	6.5	0.6	83	[92]
(s)	α-BaAl$_2$O$_4$·2H$_2$O	3.4	0.5	81	[92]
	"	5.1	0.9	80	[92]

TABLE 6 (continued)

Mineral (s)=synthetic	nominal formula	QCC, MHz	η	δ_{iso}[a,b] ppm	ref.
Q^3 sites, layer aluminosilicates:[d]					
margarite	$CaAl^{VI}_2(Al^{IV}_2Si_2O_{10})(OH)_2$	4.2		76	[73]
phlogopite[e]	$KMg_3AlSi_3O_{10}(F,OH)_2$			69[c]	[20,60,111]
muscovite	$KAl^{VI}_2SiAl^{IV}_3O_{10}(OH)_2$	2.1		72	[73]
illite[e]	$KAl_2SiAl_3O_{10}(OH)_2$[f]			72.8	[54,60]
hectorite	$(Mg,Li)_3Na_{0.3}Si_4O_{10}(OH)_2 \cdot 4H_2O$			66[c]	[60]
penninite	$(Mg,Al^{VI})_6(Si,Al^{IV})_4O_{10}(OH)_8$[f]	2.8		72	[73]
xanthophyllite	$Ca_2(Mg,Al^{VI})_6(Si,Al^{IV})_4O_{10}(OH)_4$[f]	2.8		76	[73]
Q^3 sites, others:					
(s)	$KAlO_2 \cdot 0.5H_2O$	5.6	0	77	[92]
(s)	$Ca_{12}Al_{14}O_{33}$	11	0.2	85	[92]
sillimanite	$Al^{VI}Al^{IV}SiO_5$	6.77	0.53	64.5	[43,73]
sapphirine	$(Mg_{3.6}Al^{VI}_{4.4})(Al^{IV}_{4.4}Si_{1.6})O_{20}$			75[c]	[19]
Q^4 sites, silica analogs:					
berlinite(s)	$AlPO_4$	4.09	0.37	44.5	[43,93]
tridymite phase(s)	$AlPO_4$	0.75	0.95	39.8	[93]
cristobalite phase(s)	$AlPO_4$	1.2	0.75	42.5	[93]
Q^4 sites, feldspars:					
albite[e]	$NaAlSi_3O_8$	3.29	0.62	63.0	[43,64,73,103,169
microcline[e]	$KAlSi_3O_8$	3.22	0.21	60.9	[43,64,73,103,169
anorthite, 0zi0	$CaAl_2Si_2O_8$	8.5	0.66	62,55[c]	[133][g],[64]
0z00		7.4	0.76		
00i0		6.8	0.65		
m000		6.3	0.88		
m0i0		5.5	0.42		
mz00		4.90	0.75		
0000		4.4	0.53		
mzi0		2.6	0.66		
Q^4 sites, feldspathoids:					
sodalite[e]	$Na_8Al_6Si_6O_{24}Cl_2$	0.94	0.32	62.9	[73,96]
nepheline	$Na_3KAl_4Si_4O_{16}$[f]			61.0,63.5	[53,73]
kalsilite(s)[e]	$KAlSiO_4$			61.7	[53]
scapolite[e]	$Na_4Al_3Si_9O_{24}Cl–Ca_4Al_6Si_6O_{24}CO_3$			58.0[c]	[117]
analcite	$NaAlSi_2O_6 \cdot H_2O$			59.4	[73]
leucite	$KAlSi_2O_6$			61,65,69[c]	[104]

TABLE 6 (continued)

Mineral (s)=synthetic	nominal formula	QCC, MHz	η	δ_{iso}[a,b] ppm	ref.
Q^4 sites, zeolites:					
thomsonite	$NaCa_2Al_5Si_5O_{20} \cdot 6H_2O$			62.7	[73]
scolectite	$CaAl_2Si_3O_{10} \cdot 3H_2O$			62.5,66.4	[73]
natrolite	$Na_2Al_2Si_3O_{10} \cdot 2H_2O$	1.66	0.50	64.0	[43,73]
gmelinite	$Na_2Al_2Si_4O_{12} \cdot 6H_2O$			59.9	[73]
chabazite	$CaAl_2Si_4O_{12} \cdot 6H_2O$			59.4	[73]
mordenite	$(Na_2,K_2,Ca)Al_2Si_{10}O_{24} \cdot 7H_2O$			55.8	[73]
gismondite	$CaAl_2Si_2O_8 \cdot 4H_2O$			56.4	[73]
Q^4 sites, others:					
cordiertite, T_1	$Mg_2Al_4Si_5O_{18} \cdot nH_2O$	10.6	0.38		[43]
T_5		5.6	0.34		[43]
prehnite	$Ca_2Al^{VI}Al^{IV}Si_3O_{10}(OH)_2$	9.0		60	[139,#179]
(s)	β-$LiAlO_2$	1.9	0.56	83.0	[92,125]
(s)	γ-$LiAlO_2$	3.2	0.7	81.3	[92]
(s)	β-$NaAlO_2$	1.4	0.5	80.1	[92]
(s)	$KAlO_2$	1.1	0.7	76.0	[92]
(s)	$BaAl_2O_4$	2.4	0.4	78.0	[92]
(s)	$TlAlO_2$			69	[92]
Ca-aluminates:					
(s)	$Ca_3Al_2O_6$, Al(1)	8.69	0.32	79.5	[122]
"	" , Al(2)	9.30	0.54	78.25	[122]
(s)	$Ca_{12}Al_{14}O_{33}$, Al(1)	9.7	0.40	85.9	[123]
"	" , Al(2)	3.8	0.70	80.2	[123]
(s)	$Ca_4Al_6O_{13}$	2.4	0.95	80.3	[91]
(s)	$Ca_4Al_6O_{13} \cdot 3H_2O$	1.8	0.5	78	[91]
	"	5.4	0.45	79	[91]
(s)	$CaAl_2O_4$[h]	2.5-4.3	0.2-1.0	81.2-86.2	[123]
(s)	$CaAl_4O_7$, Al(1)	6.25	0.88	75.5	[123]
"	" , Al(2)	9.55	0.82	69.5	[123]
(s)	$CaAl^{VI}_9Al^VAl^{IV}_2O_{19}$, Q^0	2.0	≈0	65	[91]

Notes for Tables 6 and 7:
a) Relative to $Al(H_2O)_6^{3+}$. (b) Peak positions corrected for second order quadrupolar shift have been included where possible. Where this correction is not made, MAS peak positions and widths will depend somewhat on the magnetic field used. (c) MAS peak position; δ_{iso} at slightly higher frequency. (d) See [162] for extensive data on clay minerals. (e) Approximate formula. (f) Reference includes data on other solid solution compositions. (g) Reference includes high T study of phase transition. (h) Range of data for six sites.

TABLE 7. ^{27}Al NMR data for AlO$_5$ and AlO$_6$ sites in crystalline silicates and oxides. Data for a variety of synthetic zeolites, sheet silicates, and phosphates have been excluded for brevity. See Table 6 for notes.

Mineral (s)=synthetic	nominal formula	QCC, MHz	η	δ_{iso}[a,b] ppm	ref.
AlO$_5$ sites:					
(s)	Al$^V_2$Si$_2$O$_7$	10.5	0.6	29	[41]
andalusite, Al$_2$	Al^{VI}Al^{V}SiO$_5$	5.9	0.70	36.0	[3,43,73]
augelite	Al^{VI}Al^V(OH)$_3$PO$_4$	5.7	0.85	30.9	[7]
senegalite	Al^{VI}Al^V(OH)$_3$PO$_4\cdot$H$_2$O	≈2.7		36.0	[7]
(s)	Al$_2$Ge$_2$O$_7$	8.8	0.4	36	[80]
(s)	LaAlGe$_2$O$_7$	7.2	0.3	35	[80]
vesuvianite	Ca$_{19}$Al$_{11}$Mg$_2$Si$_{18}$O$_{68}$(OH)$_{10}$[f]			41.1	[100]
AlO$_6$ sites:					
corundum	α-Al$_2$O$_3$	2.39	0	16.0	[43,55]
chrysoberyl, Al$_1$	BeAl$_2$O$_4$	2.85	0.94		[43]
Al$_2$		2.85	0.76		[43]
spinel (ordered)	MgAl$_2$O$_4$	3.68	0		[43]
(disordered)				11[c]	[86,163]
gahnite	ZnAl$_2$O$_4$	3.68	0		[43]
rutile	≈1% Al$_2$O$_3$ in TiO$_2$	2.8	1.0	-6.5	[143]
(s)	Al$_2$TiO$_5$			6[c]	[143]
beryl	Be$_3$Al$_2$Si$_6$O$_{18}$	3.09	0		[43]
euclase	HBeAlSiO$_5$	5.17	0.70		[43]
vesuvianite	Ca$_{19}$Al$_{11}$Mg$_2$Si$_{18}$O$_{68}$(OH)$_{10}$[f]			2.5	[100]
prehnite	Ca$_2$Al^{VI}Al^{IV}Si$_3$O$_{10}$(OH)$_2$	<1		4.5	[66,139]
spodumene	LiAlSi$_2$O$_6$	2.95	0.94		[43]
kyanite, Al$_1$	Al$_2$SiO$_5$	10.04	0.27	15	[43,129]
Al$_2$		3.70	0.89	5.0	[43,73]
Al$_3$		6.53	0.59	7.5	[43,73]
Al$_4$		9.37	0.38	13	[43,129]
sillimanite	Al^{VI}Al^{IV}SiO$_5$	8.93	0.46	4.0	[43,73]
andalusite, Al$_1$	Al^{VI}Al^{V}SiO$_5$	15.6	0.08	12	[23,43]
sapphirine	(Mg$_{3.6}$Al$^{VI}_{4.4}$)(Al$^{IV}_{4.4}$Si$_{1.6}$)O$_{20}$			8[c]	[19]
garnet (s)	Gd$_3$Al$_2^{VI}$Al$_3^{IV}$O$_{12}$	<0.1	0		[43]
garnet "YAG" (s)	Y$_3$Al$_2^{VI}$Al$_3^{IV}$O$_{12}$	0.63	0	0.8	[43,79,129]
grossular	Ca$_3$Al$_2$Si$_3$O$_{12}$	3.61	0		[43]
almandine	(Fe,Mg)$_3$Al$_2$Si$_3$O$_{12}$	1.51	0		[43]
pyrope[f]	Mg$_3$Al$_2$Si$_3$O$_{12}$			2.4[c]	[82,101]

TABLE 7 (continued)

Mineral (s)=synthetic	nominal formula	QCC, MHz	η	δ_{iso}[a,b] ppm	ref.
zoisite, $Al_{1,2}$	$Ca_2Al_3Si_3O_{12}OH$	8.05	0.46		[43]
Al_3		18.5	0.16		[43]
epidote, Al_1	$Ca_2Al_2(Fe,Al)Si_3O_{12}OH$	9.8	0.2		[43]
Al_2		4.6	0.34		[43]
topaz	$Al_2SiO_4F_2$ (AlO_4F_2 site)	1.67	0.38		[43]
margarite	$CaAl^{VI}_2(Al^{IV}_2Si_2O_{10})(OH)_2$	6.3		11	[73]
muscovite	$KAl^{VI}_2SiAl^{IV}_3O_{10}(OH)_2$	2.2		5	[73]
illite[e]	$KAl_2SiAl_3O_{10}(OH)_2$[f]			5.9	[54,60]
penninite	$(Mg,Al^{VI})_6(Si,Al^{IV})_4O_{10}(OH)_8$[f]	1.4		10	[73]
xanthophyllite	$Ca_2(Mg,Al^{VI})_6(Si,Al^{IV})_4O_{10}(OH)_4$[f]	2.0		11	[73]
kaolinite	$Al_4Si_4O_{10}(OH)_8$			4[c]	[60]
pyrophyllite	$Al_2Si_4O_{10}(OH)_2$			4[c]	[60]
smectite	$(Ca,Na)(Al,Mg)_4(Si,Al)_8O_{20}(OH)_4$[f]			4[c]	[60]
gibbsite (s)	$Al(OH)_3$, Al(1)	1.97	0.73	10.4	[123]
	" , Al(2)	4.45	0.44	11.5	[123]
augelite	$Al^{VI}Al^V(OH)_3PO_4$	4.5	1.0	0.3	[7]
senegalite	$Al^{VI}Al^V(OH)_3PO_4 \cdot H_2O$	≈3.8		1.7	[7]
Ca-aluminates:					
(s)	$CaAl_2O_4 \cdot 10H_2O$	2.4		10.2	[123]
(s)	$Ca_3Al_2O_6 \cdot 6H_2O$	0.71	0.09	12.36	[123]
(s)	$Ca_4Al_2O_7 \cdot 13H_2O$	1.8		10.20	[123]
(s)	$CaAl^{VI}_9Al^VAl^{IV}_2O_{19}$	1.5	≈0	9	[91]
	"	<1	≈0	16	[91]
ettringite(s)	$Ca_6Al_2O_9 \cdot 3SO_3 \cdot 32H_2O$	0.36	0.19	13.10	[123]
(s)	$Ca_4Al_2O_7 \cdot SO_3 \cdot 12H_2O$	1.7		11.80	[123]

AlO_5 groups in a few known structures have δ values of about 35 to 40 ppm. The more subtle effects of bond angle and second neighbor identity are similar to those for ^{29}Si [2,66,88].

Many MAS NMR studies of ^{27}Al of solids have reported only peak positions, which are generally shifted to frequencies lower than δ_{iso} by the second order quadrupolar interaction. Quadrupolar shifts can be as large as 20 or more ppm for data collected at relatively low magnetic fields and/or for sites with large field gradients, but can be almost negligible in many minerals at high magnetic fields. Determination of δ_{iso} from MAS spectra can be done in some cases by detailed analysis of spinning sidebands for the satellite transitions [55,79], by collection of spectra at more than one magnetic field [73,103], and, for sites with relatively large QCC (quadrupolar coupling constant) values, by fitting of quadrupolar line shapes [34]. Quadrupolar effects can broaden peaks to the extent of being inadequately narrowed by MAS, or even lost entirely because of instrumental dead time. MAS studies at very high magnetic fields (e.g.14.1 Tesla) can be very helpful. The new techniques of "dynamic angle spinning" (DAS) and "double rotation" (DOR) NMR may prove to be very useful in improving resolution [90,164].

The distinctions among octahedral and tetrahedral Al have in some cases allowed quantification of disorder and Al site occupancy in silicates and oxides. In a number of sheet silicates, for example, ^{27}Al MAS NMR data have been shown to agree well with site assignments based on stoichiometry [5,20,52,60,162]. In $MgAl_2O_4$ spinel [45,86,163], MAS NMR has also been used to estimate the effect of temperature on disorder (Figure 3).

2.3. ^{17}O

Although ^{17}O can be observed at natural abundance in liquids and in highly symmetrical sites in crystalline oxides [6], applications to silicates have generally been limited by the necessity of working on isotopically enriched samples. The data assembled in Table 8 have been obtained at high magnetic fields by the fitting of MAS, DAS spectra, or static spectra. For some materials, the latter can be quite informative, but neglect of chemical shift anisotropy can lead to discrepancies [131].

The most obvious effect on ^{17}O spectra of silicates with tetrahedral Si is the distinction between bridging and non-bridging oxygens.The former generally have much larger QCC values. Indeed, QCC is well correlated with the electronegativity of the neighbor cations and thus with the covalent character of the M-O bonds [115,152,153,154]. Isotropic chemical shifts for non-bridging oxygens vary widely (over 1000 ppm) depending on the nature and number of the coordinating cations. For group II oxides and both bridging and non-bridging oxygens in silicates, there are good correlations between increasing cation size and decreased shielding (higher frequencies or larger chemical shifts) [153,154]. CP MAS NMR has been shown to work well for enhancing signals from oxygens near to protons [157].

DAS and DOR spectra have recently been shown to be remarkably effective for resolving ^{17}O NMR peaks for structurally similar but crystallographically distinct sites in silicates [89].

2.4. Other Nuclides

NMR studies of a variety of other nuclei in solid geological materials have provided important structural information [61]. The large differences between electric field gradients for three- and four-coordinated boron, for example, allow these sites to be easily distinguished by ^{11}B NMR. Many applications of this relationship, as well as the effect of coordination number on δ_{iso}, have been

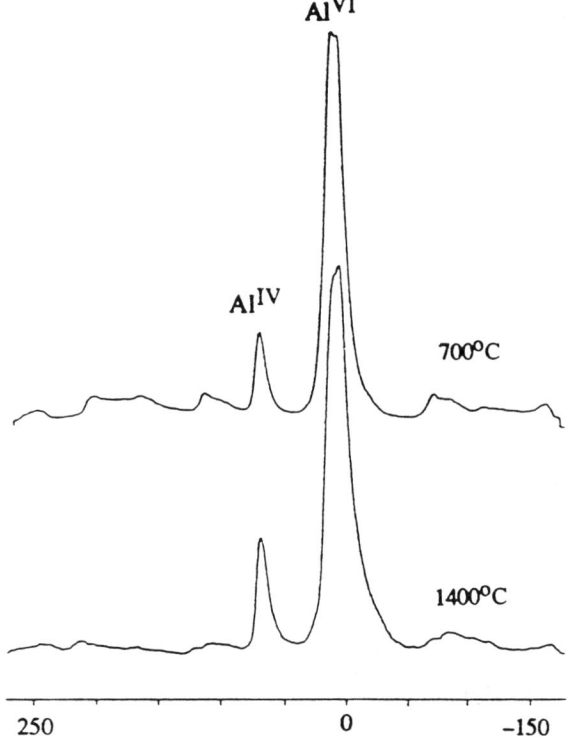

Fig. 3. ^{27}Al MAS spectra for $MgAl_2O_4$ spinel quenched from the two temperatures shown. Features other than the two labeled peaks are spinning side bands. The increase in the intensity of the Al^{IV} peak indicates an increase in disorder with temperature [86].

made to borate and borosilicate glasses [11,155,170]. Recently, similar effects on δ_{iso} have been detected for ^{23}Na in silicates [53,103] as well as aluminofluoride minerals [24]. Studies of other nuclides in silicates and oxides are too numerous to mention in detail, but have recently included ^9Be, ^{13}C, ^{19}F, ^{25}Mg, ^{31}P, ^{35}Cl, ^{39}K, ^{45}Sc, ^{47}Ti, ^{51}V, ^{89}Y, ^{93}Nb, ^{119}Sn, ^{129}Xe, and ^{207}Pb.

2.5. Dynamics in Crystalline Phases

In oxides and silcates, dynamical studies have concentrated on the mechanisms and rates of diffusion of both cations (especially $^7Li^+$ and $^{23}Na^+$) and anions (especially $^{19}F^-$), and on phase transitions in materials with abundant nuclides of high Larmor frequency (e.g. ^7Li, ^{19}F, ^{23}Na, ^{27}Al, ^{93}Nb) [8,109,136,138]. The first of these reports extensive work on perovskite-structured oxides. Most of these studies have been done at relatively low magnetic fields and consist primarily of relaxation time measurements. As such, quantification of results often requires

TABLE 8. ^{17}O NMR data for crystalline oxides and silicates. Some data for synthetic zeolites, as well as oxysalts and oxide superconductors are excluded for brevity. In silicates, "BO" signifies bridging oxygen, "NBO" non-bridging oxygen.

Mineral (s)=synthetic	nominal formula	QCC, MHz	η	δ_{iso}[a], ppm	ref.
miscellaneous oxides:					
(s)	BeO	≈0.02		26[b]	[154]
periclase(s)	MgO	≈0.014		47[b]	[154]
(s)	CaO	<.005		294[b]	[154]
(s)	SrO	<.005		390[b]	[154]
(s)	BaO	<.005		629[b]	[154]
rutile(s)	TiO$_2$	<1.5		590[c]	[6]
anatase(s)	TiO$_2$	<1.1		558[c]	[6]
(s)	Ti$_2$O$_3$	<2.6		503[c]	[6]
baddeleyite	ZrO$_2$	<0.9,1.0		325,402[c]	[6]
(s)	ZrO$_2$ (tetragonal)	<1.4		383[c]	[6]
(s)	87ZrO$_2$·13MgO (cubic)			≈355[d]	[6]
(s)	HfO$_2$	<1.1,0.9		267,335[c]	[6]
(s)	La$_2$O$_3$ (octahedral site)	<1.4		469[c]	[6]
	" (tetrahedral site)	<2.2		590[c]	[6]
(s)	CeO$_2$	≈0		878[c]	[6]
(s)	VO$_2$			≈755,815[d]	[6]
zincite(s)	ZnO	≈0.13		−18[b]	[6,154]
(s)	CdO			≈60	[154]
(s)	HgO(yellow)	7.1		121[e]	[154]
(s)	SnO	<2.3		251[c]	[6]
litharge(s)	PbO	<0.9		294[c]	[6]
cuprite(s)	Cu$_2$O	≈0		−193	[6]
(s)	Ag$_2$O	≈0		−277	[6]
(s)	KMnO$_4$	<0.4		1197	[115]
(s)	K$_2$WO$_4$, site 1[f]			437	[115]
	" , site 2			429	[115]
	" , site 3			422	[115]
aluminum oxides:					
corundum(s)	α-Al$_2$O$_3$, OAl$_4$ site	2.17	0.55	75[c]	[13,156]
(s)	γ-Al$_2$O$_3$, OAl$_4$ site	1.8		73[c]	[156]
(s)	η-Al$_2$O$_3$, OAl$_4$ site	1.6		73[c]	[156]
(s)	δ-Al$_2$O$_3$, OAl$_4$ site	1.6		72[c]	[156]
(s)	θ-Al$_2$O$_3$, OAl$_4$ site	1.2		72[c]	[156]
	" , OAl$_3$ site	4.0	0.6	79[c]	[156]
β-alumina(s)	11Al$_2$O$_3$·Na$_2$O	<2.2		76[c]	[6]

TABLE 8 (continued)

Mineral (s)=synthetic	nominal formula	QCC, MHz	η	δ_{iso}[a], ppm	ref.
hydroxides:					
boehmite(s)	AlO(OH), OAl$_4$ site	1.15	0.13	70.0[c]	[125,156]
"	" , Al$_2$OH site	5.0	0.5	40[c]	[156]
bayerite(s)	Al(OH)$_3$, Al$_2$OH site	6.0	0.3	40[c]	[156]
brucite	Mg(OH)$_2$	6.8	0.0	25[g]	[157]
orthosilicates:					
forsterite(s)	Mg$_2$SiO$_4$, NBO	3.3[h]		72[i]	[89]
"	" , NBO	2.8[h]		64[i]	[89]
"	" , NBO	3.0[h]		49[i]	[89]
larnite(s)	Ca$_2$SiO$_4$, NBO	2.9[h]		134[i]	[89]
"	" , NBO	2.7[h]		128[i]	[89]
"	" , NBO	2.5[h]		122[i]	[89]
"	" , NBO	2.8[h]		122[i]	[89]
chain silicates:					
diopside(s)	CaMgSi$_2$O$_6$, NBO	2.83	0.13	86[i]	[89]
"	" , NBO	2.74	0.00	64[i]	[89]
"	" , BO	4.39	0.36	69[i]	[89]
clinoenstatite(s)	Mg$_2$Si$_2$O$_6$, NBO	2.9[h]		57[i]	[89]
"	" , NBO	3.6[h]		61[i]	[89]
"	" , NBO	3.6[h]		59[i]	[89]
"	" , NBO	4.2[h]		62[i]	[89]
"	" , BO	5.1[h]		70[i]	[89]
"	" , BO	5.2[h]		70[i]	[89]
wollastonite(s)	Ca$_3$Si$_3$O$_9$, NBO	2.3[h]		115[i]	[89]
"	" , NBO	2.6[h]		114[i]	[89]
"	" , NBO	2.2[h]		107[i]	[89]
"	" , NBO	2.0[h]		97[i]	[89]
"	" , NBO	2.9[h]		103[i]	[89]
"	" , NBO	2.6[h]		88[i]	[89]
"	" , BO	4.8[h]		75[i]	[89]
"	" , BO	4.8[h]		75[i]	[89]
"	" , BO	4.7[h]		67[i]	[89]
ps-wollastonite	α-CaSiO$_3$[j], NBO	2.1	0.1	94[c]	[153]
"	" , NBO	2.3	0.1	91[c]	[153]
"	" , BO	3.8	0.2	75[c]	[153]
(s)	α-SrSiO$_3$[j], NBO	2.1	0.1	108[c]	[153]
"	" , NBO	2.2	0.1	105[c]	[153]
"	" , BO	4.1	0.4	80[c]	[153]

TABLE 8 (continued)

Mineral (s)=synthetic	nominal formula	QCC, MHz	η	$\delta_{iso}{}^a$, ppm	ref.
(s)	$BaSiO_3{}^j$, NBO	2.1	0.1	169^c	[153]
"	" , NBO	1.6	0.1	159^c	[153]
"	" , BO	3.7	0.4	87^c	[153]
talc(s)	$Mg_3Si_4O_{10}(OH)_2$, NBO	3.2	0	40^g	[157]
"	" , BO	5.8	0	50^g	[157]
"	" , MgOH site	7.3	0	0^g	[157]
framework silicates:					
cristobalite(s)	SiO_2	5.3	0.13	40^k	[131]
Na-A zeolite(s)	(see reference), Si-O-Al, BO	3.2	0.2	32^c	[152]
Na-Y zeolite(s)	(see reference), Si-O-Al, BO	3.1	0.2	31^c	[152]
"	" , Si-O-Al, BO	4.6	0.1	44^c	[152]
high pressure silicates:					
wadeite phase(s)	$K_2Si^{VI}Si^{IV}{}_3O_9$, Si^{IV}-O-Si^{IV}	4.45	0.35	62.5	[165]
"	" , Si^{IV}-O-Si^{VI}	4.90	0.2	97.0	[165]
stishovite(s)	SiO_2	6.50	0.13	109.5	[165]

Notes for Table 8:
a) Relative to H_2O.   (b) From MAS data, little or no correction for QCC needed.   (c) From simulation of MAS data.   (d) Uncorrected for QCC.   (e) From MAS, corrected using QCC from static spectrum.   (f) For site 1: σ_{11}=564, σ_{22}=530, σ_{33}=217 ppm; for site 2: σ_{11}=567, σ_{22}=518, σ_{33}=202 ppm; for site 3: σ_{11}=561 σ_{22}=497, σ_{33}=208 ppm.   (g) Based on static spectrum, CSA not included.   (h) QCC given is actually the "quadrupolar product". P_Q= QCCx$(1+ \eta^2/3)^{1/2}$; 0.87<QCC<P_Q [89].   (i)Derived from DAS data at two magnetic fields.   (j) More sites are present than are resolved in MAS spectra.   (k) Fit of static spectrum gives σ_{11}=σ_{22}=60, σ_{33}=–10 ppm.

complex models. In a few recent, high resolution studies of silicates, the direct observation of exchange among multiple sites in a crystal offers the possibility of simpler, more direct interpretations. Alkali cation dynamics have been studied in a number of framework silicates, for example [57,144], and can have an important influence even in ortho- and chain-silicates [38]. Structural and dynamical changes occurring during displacive phase transitions in silica polymorphs and their aluminum phosphate analogs have been explored in some detail [99,130,131].

3. APPLICATIONS TO GLASSES, AMORPHOUS SOLIDS, AND MELTS

Techniques such as NMR become particularly important when the absence of long-range structure limits the information obtainable by diffraction methods. Because of the wide and continuous ranges of compositions studied, and the frequent model-dependent nature of the structural conclusions, I have tabulated only a few ^{29}Si MAS data on glasses with the stoichiometry of end-member crystalline phases (Table 9). Other recent reviews do include some tabulations [34], as well as extensive discussion of silicate, oxide, borate, and non-oxide glasses [10,11,29,61,62,145].These contain information on studies of nuclides not included here, especially ^{11}B, ^{19}F, ^{23}Na, and ^{31}P.

3.1. ^{29}Si
The relatively clear relationships between δ_{iso} and Q species in crystalline silicates has led to a number of attempts to quantify their abundance in glasses. Separate

TABLE 9. ^{29}Si NMR data for glasses of simple, crystalline stoichiometry.

Composition	$-\delta_{iso}$, ppm	FWHM, ppm	ref.
SiO_2	111.5	12	[44,98]
$NaAlSi_3O_8$	98.7	16	[98]
$NaAlSi_2O_6$	92.8	18	[94]
$NaAlSiO_4$	86.0	13	[94]
$KAlSi_3O_8$	100.5	15	[94,98]
$CaAl_2Si_2O_8$	87.9	15	[94,98]
$CaAl_2SiO_6$	83.4	11	[94,98]
$Mg_3Al_2Si_3O_{12}$	82.3	17	[70]
$Ca_3Al_2Si_3O_{12}$	80.1	13	[70]
$Na_2Si_4O_9$	105.6,92.2	13,11	[77,166]
$K_2Si_4O_9$	105.1,94.3	11,10	[77,166]
$Li_2Si_2O_5$	102.7*,91.0,81.1*	14*,12,10*	[77]
$Na_2Si_2O_5$	99.5*,88.7,77.7*	12*,10,8*	[77]
$K_2Si_2O_5$	103.3*,90.8,79.5*	11*,11,7*	[77]
Na_2SiO_3	84.8*,76.0,66.8*	10*,7,5*	[77]
$MgSiO_3$	82.3	20	[70]
$CaMgSi_2O_6$	81.3	16	[70]
$CaSiO_3$	80.6	14	[70]

Notes for Table 9:
Uncertainties are generally at least 0.5 ppm in peak positions and widths.
*Partially resolved shoulder

peaks for each Q species are often partially resolved in MAS spectra for alkali silicate glasses (Figure 4). Quantitative interpretations have usually depended on assumptions of Gaussian line shape and curve fitting, but interpretations have converged as data improves [26,28,32,51,134,135]. It is now clear that speciation is more disordered than required by simple stoichiometry, but more ordered than predicted by random mixing of bridging and non-bridging oxygens [94,135]. A number of ther-

modynamic mixing models for these systems have been based on NMR data [9,77]. In a few favorable cases, static spectra can permit quantification of Q species in glasses with fewer curve fitting assumptions than for MAS spectra, because of the contrasts in chemical shift anisotropy [9,32,134].

In binary silicate glasses of alkaline earths, and in aluminosilicate glasses, peaks for Q species are usually unresolved in MAS spectra because of greater overall dis-

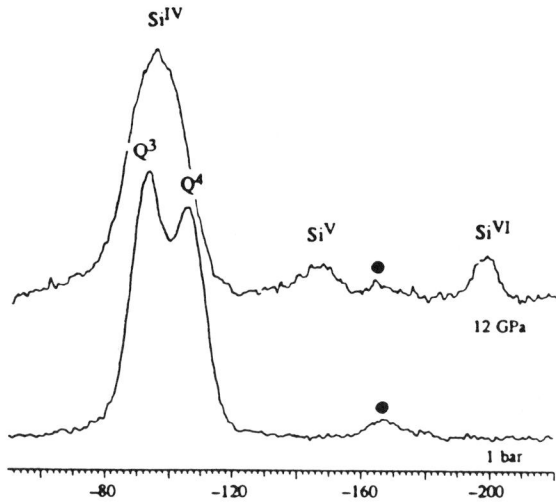

Fig. 4. ^{29}Si MAS spectra for $Na_2Si_4O_9$ glasses quenched from liquids at 12 GPa and at 1 bar pressures. Signals from Si with different coordination numbers, and from different tetrahedral species, are labeled [166]. Black dots mark spinning side bands.

order and variety of first neighbor cation arrangements, making quantitative derivation of speciation highly model-dependent. Studies have included extensive work on alkali and alkaline earth aluminosilicates [35,65,98] and Ca, Mg silicates [70,94]. The narrowing of NMR peaks as Al/Si is increased to 1 suggests that some Si-Al ordering occurs, probably because of at least partial "aluminum avoidance" [94]. For pure SiO_2 glass, bond angle distributions have been derived using empirical correlations between δ and bond angles [44,94,98].

One approach to determining the effect of temperature on liquid structure is to quench liquids at varying rates to produce glasses with different "fictive" or glass transition temperatures (T_g). This approach has been used to show that Q species distributions, as well as overall disorder in bond angles, do indeed become more random with increasing T [9,135].

Studies of alkali silicate glasses quenched from liquids at pressures to 12 GPa have shown the presence of six coordinated Si (Figure 4), and for the first time demonstrated the existence of five coordinated Si [147,166,167]. The latter has now also been detected at low abundance (<0.1%) in glasses formed at 1 bar pressure [137]. High coordinate Si has not been observed in $NaAlSi_3O_8$ or SiO_2 glasses quenched from liquids at high P [150,166]. Six-coordinated Si has also been reported in alkali silicate glasses containing large amounts of P_2O_5 [25,27].

Several NMR studies have suggested that opal-CT, which is characterized by obvious x-ray diffraction peaks and which therefore contains substantial long-range order, is highly disordered with respect to local structure [1,22]. The latter study used the CP MAS technique to study the distribution of silanol groups (Si-OH), which were detectable in most of the samples studied.

3.2. ^{27}Al

As in crystalline materials, interpretation of NMR spectra of quadrupolar nuclides in glasses is complicated by second-order quadrupolar broadening (possibly leading to signal loss from highly distorted sites). It may be impossible to distinguish these effects from those caused by disordered distribution of sites, although again, studies at varying magnetic fields can be useful. Fitting of spinning side bands, when these are observable, can again be useful in determining isotropic chemical shifts [79,129]. Careful quantification of intensities is essential in determining whether all Al has been detected [65].

The most dramatic findings from ^{27}Al MAS NMR in glasses has been the clear detection of five and six coordinated Al in a variety of compositions, including those in the SiO_2-Al_2O_3 binary [105,110,113] and in some phosphorus-rich glasses [27] and boroaluminates [15,97]. In the CaO - Al_2O_3 - SiO_2 ternary, non-tetrahedral Al is abundant only in compositions close to the SiO_2-Al_2O_3 join [114]. In carefully quantified studies of glass compositions with M^{+1}_2O/Al_2O_3 or $M^{+2}O$/Al_2O_3 $\leq$ 1.0, it has been shown that high coordinated Al is undetectable and thus comprises less than a few percent of the total Al [65,98]. However, high-coordinate Al has been detected in only slightly peraluminous glasses near to the $MgAl_2O_4$–SiO_2 join [83].

3.3. ^{17}O

As in crystals, the primary distinction among O sites detected by NMR of glasses is that between bridging and non-bridging oxygens (Figure 5). Static spectra can sometimes be more informative than MAS data. Systematic compositional effects on the chemical shifts for NBO sites have been noted [61], as have pressure effects on O site distribution [165]. Very recently, it has been demonstrated that ^{17}O DAS NMR can accurately quantify bond angles and oxygen site distributions [36].

3.4. Water in Glass

Both static "wideline" and high speed ^{1}H MAS studies have detected and quantified OH$^-$ and molecular H_2O [30,68]. ^{2}H NMR has distinguished among water species in glasses, as well, and has provided some dynamical information [31]. In hydrous $NaAlSi_3O_8$ glass, ^{23}Na, ^{27}Al, and ^{29}Si MAS and CP MAS spectra were interpreted as indicating that OH is bound soley to Na with the possibility of protonated bridging oxygens [67,69], whereas in hydrous binary alkali silicate glasses [85,139] and in SiO_2 [37], ^{29}Si NMR has clearly shown the presence of SiOH groups. ^{11}B, ^{17}O, ^{23}Na, ^{27}Al, and ^{29}Si NMR have been used in several studies of the interaction of silicate glasses with water at ambient to hydrothermal conditions at a range of pH's [16,139,168,169].

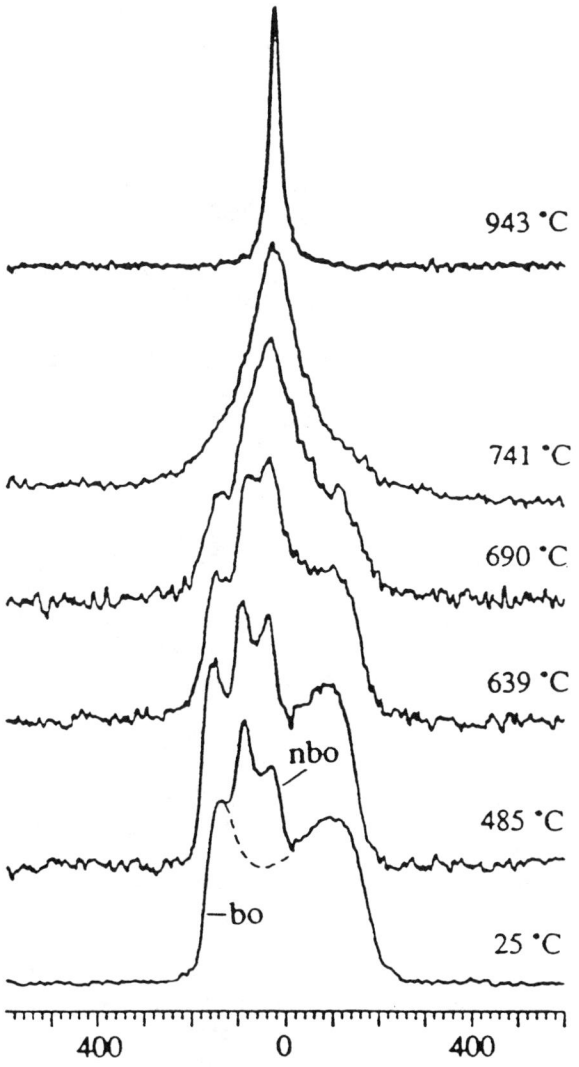

Fig. 5. High temperature ^{17}O spectra for $K_2Si_4O_9$ liquid. "bo" and "nbo" show contributions from bridging and non-bridging oxygens [145]. Note collapse to single, narrow line caused by exchange of species.

3.5. Silicate and Aluminate Liquids

In ionic and partially ionic liquids such as silicates (at least at high silica contents), flow requires the breaking of the strongest bonds in the system (Si-O). As a result, at high temperatures where viscosities are low, silicate species are short lived, with all cations and anions exchanging rapidly among available sites. Under these conditions, therefore, in situ, high temperature NMR reveals only single ^{17}O, ^{23}Na, ^{27}Al, and ^{29}Si peaks, for which orientational effects that lead to broadening in a solid, including chemical shift anisotropy and quadrupolar couplings, are fully averaged [38,81,121,142,149,151]. The average peak position can still give structural information however, since it is a quantitative weighted mean over all species present. This approach has shown, for example, that when aluminum oxides, aluminates, and fluorides melt, the mean Al coordination number decreases [40,81]. A multi-nuclear high T NMR study of alkali aluminosilicate liquids has detected systematic temperature effects on averaged chemical shifts that can be interpreted in terms of structural changes [142]. Studies have now been made to temperatures greater than 2100^0 C, and have included systematic work on the effects of composition on the structure and dynamics in the $CaO-Al_2O_3-SiO_2$ system [21,84].

At lower temperatures, incomplete exchange and averaging occurs, and, in favorable cases, NMR line shapes can be analyzed to measure the rates of exchange. The rate of exchange of Si among various anionic species, and that of O between bridging and non-bridging sites, are fundamentally tied to diffusion and viscous flow. In $K_2Si_4O_9$ liquid, for example (Figure 5), simulations of high temperature ^{17}O and ^{29}Si line shapes gives results that can be used to accurately predict the viscosity, assuming a simple Eyring model for flow [38,145]. 2-D exchange NMR spectroscopy just above the glass transition also suggests that exchange among Q species is of key importance in flow even in the very high viscosity range [39]. A series of relaxation time measurements has shown that Si dynamics are greatly affected by the transition from glass to liquid, and that the interaction of alkali cation with the network can also be detected [38,75,76].

Acknowledgements. I would like to thank my numerous colleagues who sent reprints and preprints and allowed me to include results prior to publication, especially H.J. Jakobsen, B. Phillips, J. Skibsted, and X. Xue. I thank the editor, T. Ahrens, and an anonymous reviewer, and acknowledge the support of the National Science Foundation, grant #EAR9204458.

REFERENCES

1. Adams, S. J., G. E. Hawkes, and E. H. Curzon, A solid state ^{29}Si nuclear magnetic resonance study of opal and other hydrous silicas, *Am. Mineral.*, *76*, 1863-1871, 1991.

2. Akitt, J. W., Multinuclear studies of aluminum compounds, *Prog. NMR Spectros.*, *21*, 1-149, 1989.

3. Alemany, L. B., and G. W. Kirker, First observation of 5-coordinate aluminum by MAS ^{27}Al NMR in well-characterized solids, *J. A. Chem. Soc.*, *108*, 6158-6162, 1986.

4. Barron, P. F., R. L. Frost, J. O. Skjemstad, and A. J. Koppi, Detection of two silicon environments in kaolins by solid-state ^{29}Si NMR, *Nature, 302*, 49-50, 1983.

5. Barron, P. F., P. Slade, and R. L. Frost, Ordering of aluminum in tetrahedral sites in mixed-layer 2:1 phyllosilicates by solid-state high-resolution NMR, *J. Phys. Chem.*, *89*, 3880-3885, 1985.

6. Bastow, T. J., and S. N. Stuart, ^{17}O NMR in simple oxides, *Chem. Phys.*, *143*, 459-467, 1990.

7. Bleam, W. F., S. F. Dec, and J. S. Frye, ^{27}Al solid-state nuclear magnetic resonance study of five-coordinate aluminum in augelite and senegalite, *Phys. Chem. Minerals,* 1990.

8. Borsa, F., and A. Rigmonti, Comparison of NMR and NQR studies of phase transitions in disordered and ordered crystals, in *Structural Phase Transitions-II*, edited by K. A. Muller and H. Thomas, pp. 83-175, Springer-Verlag, New York, 1990.

9. Brandriss, M. E., and J. F. Stebbins, Effects of temperature on the structures of silicate liquids: ^{29}Si NMR results, *Geochim. Cosmochim. Acta, 52*, 2659-2670, 1988.

10. Bray, P. J., J. F. Emerson, D. Lee, S. A. Feller, D. L. Bain, and D. A. Feil, NMR and NQR studies of glass structure, *J. Non-Cryst. Solids, 129*, 240-248, 1991.

11. Bray, P. J., S. J. Gravina, P. E. Stallworth, S. P. Szu, and J. Zhong, NMR studies of the structure of glasses, *Exp. Tech. Phys., 36*, 397-413, 1988.

12. Brown, I. W. M., C. M. Cardile, K. J. D. MacKenzie, M. J. Ryan, and R. H. Meinhold, Natural and synthetic leucites studied by solid state 29-Si and 27-Al NMR and 57-Fe Mossbauer spectroscopy, *Phys. Chem. Minerals, 15*, 78-83, 1987.

13. Brun, E., B. Derighetti, E. E. Hundt, and H. H. Niebuhr, NMR of ^{17}O in ruby with dynamic polarization techniques, *Phys. Lett., 31A*, 416-417, 1970.

14. Brun, E., and S. Hafner, Die elektrische Quadrupolaufspaltung von Al^{27} in Spinell MgAl$_2$O$_4$ und Korund Al$_2$O$_3$, *Zeit. Krist., 117*, 37-62, 1962.

15. Bunker, B. C., R. J. Kirkpatrick, R. K. Brow, G. L. Turner, and C. Nelson, Local structure of alkaline-earth boroaluminate crystals and glasses: II, ^{11}B and ^{27}Al MAS NMR spectroscopy of alkaline-earth boroaluminate glasses, *J. Am. Ceram. Soc., 74*, 1430-1438, 1991.

16. Bunker, B. C., D. R. Tallant, and T. J. Headley, The structure of leached sodium borosilicate glass, *Phys. Chem. Glass., 29*, 106-120, 1988.

17. Carpenter, M. A., Thermochemistry of aluminum/silicon ordering in feldspar minerals, in *Physical Properties and Thermodynamic Behaviour of Minerals*, edited by E. K. H. Salje, pp. 265-324, Reidel, Dordrecht, 1988.

18. Cho, M., and J. F. Stebbins, Structural and cation order/disorder in amphiboles and prehnite: TEM and CPMAS NMR results, *Eos, Trans. Am. Geophys. Union, 71*, 1649, 1990.

19. Christy, A. G., B. L. Phillips, B. K. Güttler, and R. J. Kirkpatrick, A ^{27}Al and ^{29}Si NMR and infrared spectroscopic study of Al-Si ordering in natural and synthetic sapphirines, *Am. Mineral., 77*, 8-18, 1992.

20. Circone, S., A. Navrotsky, R. J. Kirkpatrick, and C. M. Graham, Substitution of $^{[6,4]}$Al in phlogopite: mica characterization, unit-cell variation, ^{27}Al and ^{29}Si MAS-NMR spectroscopy, and Al-Si distribution in the tetrahedral sheet, *Am. Mineral., 76*, 1485-1501, 1991.

21. Coté, B., D. Massiot, F. Taulelle, and J. Coutures, ^{27}Al NMR spectroscopy of aluminosilicate melts and glasses, *Chem. Geol., 96*, 367-370, 1992.

22. de Jong, B. H. W. S., J. van Hoek, W. S. Veeman, and D. V. Manson, X-ray diffraction and ^{29}Si magic-angle-spinning NMR of opals: incoherant long- and short-range order in opal CT, *Am. Mineral., 72*, 1195-1203, 1987.

23. Dec, S. F., J. J. Fitzgerald, J. S. Frye, M. P. Shatlock, and G. E. Maciel, Observation of six-coordinate aluminum in andalusite by solid-state ^{27}Al MAS NMR, *J. Mag. Reson., 93*, 403-406, 1991.

24. Dirken, P. J., J. B. H. Jansen, and R. D. Schuiling, Influence of Octahedral Polymerization on Sodium-23 and Aluminum-27 MAS-NMR in Alkali Fluoroaluminates, *Am. Mineral., 77*, 718-724, 1992.

25. Dupree, R., D. Holland, and M. G. Mortuza, Six-coordinated silicon in glasses, *Nature, 328*, 416-417, 1987.

26. Dupree, R., D. Holland, and M. G. Mortuza, A MAS-NMR investigation of lithium silicate glasses and glass ceramics, *J. Non-Cryst. Solids, 116*, 148-160, 1990.

27. Dupree, R., D. Holland, M. G. Mortuza, J. A. Collins, and M. W. G. Lockyer, Magic angle spinning NMR of alkali phospho-aluminosilicate glasses, *J. Non-Cryst. Solids, 112*, 111-119, 1989.

28. Dupree, R., D. Holland, and D. S. Williams, The structure of binary alkali silicate glasses, *J. Non-Cryst. Solids, 81*, 185-200, 1986.

29. Eckert, H., Structural concepts for

disordered inorganic solids. Modern NMR approaches and strategies, *Ber. Bunsenges Phys. Chem.,* 94, 1062-1085, 1990.

30. Eckert, H., J. P. Yesinowski, L. A. Silver, and E. M. Stolper, Water in silicate glasses: quantitation and structural studies by ^{1}H solid echo and MAS-NMR methods, *J. Phys. Chem.,* 92, 2055-2064, 1988.

31. Eckert, H., J. P. Yesinowski, E. M. Stolper, T. R. Stanton, and J. Holloway, The state of water in rhyolitic glasses, a deuterium NMR study, *J. Non-Cryst. Solids,* 93, 93-114, 1987.

32. Emerson, J. F., P. E. Stallworth, and P. J. Bray, High-field ^{29}Si NMR studies of alkali silicate glasses, *J. Non-Cryst. Solids, 113,* 253-259, 1989.

33. Engelhardt, G., and H. Koller, ^{29}Si NMR of Inorganic Solids, in *Solid-State NMR II: Inorganic Matter,* edited by B. Blümich, pp. 1-30, Springer-Verlag, Berlin, 1994.

34. Engelhardt, G., and D. Michel, *High-Resolution Solid-State NMR of Silicates and Zeolites,* 485 pp., Wiley, New York, 1987.

35. Engelhardt, G., M. Nofz, K. Forkel, F. G. Wihsmann, M. Magi, A. Samosen, and E. Lippmaa, Structural studies of calcium aluminosilicate glasses by high resolution solid state ^{29}Si and ^{27}Al magic angle spinning nuclear magnetic resonance, *Phys. Chem. Glasses,* 26, 157-165, 1985.

36. Farnan, I., P. J. Grandinetti, J. H. Baltisberger, J. F. Stebbins, U. Werner, M. Eastman, and A. Pines, Quantification of the disorder in network modified silicate glasses, *Nature, 358,* 31-35, 1992.

37. Farnan, I., S. C. Kohn, and R. Dupree, A study of the structural role of water in hydrous silica glass using cross-polarization magic angle spinning NMR, *Geochim. Cosmochim. Acta, 51,* 2869-2873, 1987.

38. Farnan, I., and J. F. Stebbins, A high temperature ^{29}Si NMR investigation of solid and molten silicates, *J. Am. Chem. Soc., 112,* 32-

39., 1990.

39. Farnan, I., and J. F. Stebbins, Observation of slow atomic motions close to the glass transition using 2-D ^{29}Si NMR, *J. Non-Cryst. Solids, 124,* 207-215, 1990.

40. Farnan, I., J. F. Stebbins, N. R. Dando, and S. Y. Tzeng, Structure and dynamics of liquids and crystals in the NaF-AlF$_3$-Al$_2$O$_3$ system: high temperature NMR results, *Eos, Trans. Am. Geophys. Union, 72,* 572, 1991.

41. Fitzgerald, J. J., S. F. Dec, and A. I. Hamza, Observation of five-coordinated Al in pyrophyllite dehydroxylate by solid-state ^{27}Al NMR spectroscopy at 14 T, *Am. Mineral., 74,* 1405-1408, 1989.

42. Fyfe, C. A., *Solid State NMR for Chemists,* pp., CFC Press, Guelph, 1983.

43. Ghose, S., and T. Tsang, Structural dependence of quadrupole coupling constant e^{2}qQ/h for ^{27}Al and crystal field parameter D for Fe^{3+} in aluminosilicates, *Am. Mineral., 58,* 748-755, 1973.

44. Gladden, L. F., T. A. Carpenter, and S. R. Elliot, ^{29}Si MAS NMR studies of the spin-lattice relaxation time and bond-angle distribution in vitreous silica, *Phil. Mag. B, 53,* L81-L87, 1986.

45. Gobbi, G. C., R. Christofferson, M. T. Otten, B. Miner, P. Buseck, G. J. Kennedy, and C. A. Fyfe, Direct determination of cation disorder in MgAl$_2$O$_4$ spinel by high-resolution ^{27}Al magic-angle-spinning NMR spectroscopy, *Chem. Lett., 6,* 771-774, 1985.

46. Grimmer, A. R., R. Peter, E. Fechner, and G. Molgedey, High-resolution ^{29}Si NMR in solid silicates. Correlation between shielding tensor and Si-O bond length, *Chem. Phys. Lett., 77,* 331-335, 1981.

47. Grimmer, A. R., F. von Lampe, and M. Mägi, Solid-state high-resolution ^{29}Si MAS NMR of silicates with sixfold coordination, *Chem. Phys. Lett., 132,* 549-553, 1986.

48. Grimmer, A. R., W. Wieker, F. von Lampe, E. Fechner, R. Peter, and G. Molgedey, Hochauflösende ^{29}Si-

NMR an festen Silicaten: anisotropie der chemischen Verschiebung im Thaumasit, *Zeit. Chem., 20,* 453, 1980.

49. Harris, R. K., *Nuclear magnetic resonance spectroscopy,* 250 pp., Pitman, London, 1983.

50. Hartmann, J. S., and R. L. Millard, Gel synthesis of magnesium silicates: a ^{29}Si magic angle spinning NMR study, *Phys. Chem. Minerals, 17,* 1-8, 1990.

51. Hater, W., W. Müller-Warmuth, M. Meier, and G. H. Frischat, High-resolution solid-state NMR studies of mixed-alkali silicate glasses, *J. Non-Cryst. Solids, 113,* 210-212, 1989.

52. Herraro, C. P., M. Gregorkiewitz, J. Sanz, and J. M. Serratosa, ^{29}Si MAS-NMR spectroscopy of mica-type silicates: observed and predicted distributions of tetrahedral Al-Si, *Phys. Chem. Minerals, 15,* 84-90, 1987.

53. Hovis, G. L., D. R. Spearing, J. F. Stebbins, J. Roux, and A. Clare, X-ray powder diffraction and ^{23}Na - ^{27}Al - ^{29}Si MAS-NMR investigation of nepheline-kalsilite crystalline solutions, *Am. Mineral., 77,* 19-29, 1992.

54. Jakobsen, H. J., H. Jacobsen, and H. Lindgreen, Solid state ^{27}Al and ^{29}Si MAS n.m.r. studies on diagenesis of mixed layer silicates in oil source rocks, *Fuel, 67,* 727-730, 1988.

55. Jakobsen, H. J., J. Skibsted, H. Bildsøe, and N. C. Nielsen, Magic-angle spinning NMR spectra of satellite transitions for quadrupolar nuclei in solids, *J. Magn. Reson., 85,* 173-180, 1989.

56. Janes, N., and E. Oldfield, Prediction of silicon-29 nuclear magnetic resonance chemical shifts using a group electronegative approach: applications to silicate and aluminosilicate structures, *J. Am. Chem. Soc., 107,* 6769-6775, 1985.

57. Janssen, R., G. A. H. Tijink, W. S. Veeman, T. L. Maesen, and J. F. Van Lent, High-temperature NMR study of zeolite Na-A: detection of a phase transition, *J. Phys. Chem.,*

93, 899-904, 1989.

58. Kanzaki, M., J. F. Stebbins, and X. Xue, Characterization of crystalline and amorphous silicates quenched from high pressure by ^{29}Si MAS NMR spectroscopy, in *High Pressure Research: Applications to Earth and Planetary Sciences*, edited by Y. Syono and M. H. Manghnani, pp. 89-100, Terra Scientific Publishing Co., Tokyo, 1992.

59. Kerrick, D. M., *The Al$_2$SiO$_5$ Polymorphs*, 406 pp., Mineralogical Society of America, Washington, D.C., 1990.

60. Kinsey, R. A., R. J. Kirkpatrick, J. Hower, K. A. Smith, and E. Oldfield, High resolution aluminum-27 and silicon-29 nuclear magnetic resonance studies of layer silicates, including clay minerals, *Am. Mineral.*, *70*, 537-548, 1985.

61. Kirkpatrick, R. J., MAS NMR spectroscopy of minerals and glasses, in *Spectroscopic Methods in Mineralogy and Geology*, edited by F. C. Hawthorne, pp. 341-403, Mineralogical Society of America, Washington D.C., 1988.

62. Kirkpatrick, R. J., T. Dunn, S. Schramm, K. A. Smith, R. Oestrike, and G. Turner, Magic-angle sample-spinning nuclear magnetic resonance spectroscopy of silicate glasses: a review, in *Structure and Bonding in Noncrystalline Solids*, edited by G. E. Walrafen and A. G. Revesz, pp. 302-327, Plenum Press, New York, 1986.

63. Kirkpatrick, R. J., D. Howell, B. L. Phillips, X. D. Cong, E. Ito, and A. Navrotsky, MAS NMR spectroscopic study of Mg^{29}SiO$_3$ with perovskite structure, *Am. Mineral.*, *76*, 673-676, 1991.

64. Kirkpatrick, R. J., R. A. Kinsey, K. A. Smith, D. M. Henderson, and E. Oldfield, High resolution solid-state sodium-23, aluminum-27, and silicon-29 nuclear magnetic resonance spectroscopic reconnaissance of alkali and plagioclase feldspars, *Am. Mineral.*, *70*, 106-123, 1985.

65. Kirkpatrick, R. J., R. Oestrike, J. C.A. Weiss, K. A. Smith, and E. Oldfield, High-resolution ^{27}Al and ^{29}Si NMR spectroscopy of glasses and crystals along the join CaMgSi$_2$O$_6$ - CaAl$_2$SiO$_6$, *Am. Mineral.*, *71*, 705-711, 1986.

66. Kirkpatrick, R. J., and B. L. Phillips, ^{27}Al NMR spectroscopy of minerals and related materials, *Applied Mag. Reson., in press*, 1992.

67. Kohn, S. C., R. Dupree, and M. G. Mortuza, The interaction between water and aluminosilicate magmas, *Chem. Geol.*, *96*, 399-409, 1992.

68. Kohn, S. C., R. Dupree, and M. E. Smith, ^{1}H NMR studies of hydrous silicate glasses, *Nature, 337*, 539-541, 1989.

69. Kohn, S. C., R. Dupree, and M. E. Smith, A multinuclear magnetic resonance study of the structure of hydrous albite glasses, *Geochim. Cosmochim. Acta, 53*, 2925-2935, 1989.

70. Libourel, G., C. Geiger, L. Merwin, and A. Sebald, High-resolution solid-state ^{29}Si and ^{27}Al MAS NMR spectroscopy of glasses in the system CaSiO$_3$-MgSiO$_3$-Al$_2$O$_3$, *Chem. Geol.*, *96*, 387-397, 1991.

71. Lippmaa, E., M. Mägi, A. Samosen, and G. Engelhardt, Structural studies of silicates by solid-state high-resolution ^{29}Si NMR spectroscopy, *J. Amer. Chem. Soc.*, *103*, 4992-4996, 1980.

72. Lippmaa, E., M. Mägi, A. Samosen, M. Tarmak, and G. Engelhardt, Investigation of the structure of zeolites by solid-state high-resolution ^{29}Si NMR spectroscopy, *J. Am. Chem. Soc.*, *103*, 4992-4996, 1981.

73. Lippmaa, E., A. Samosen, and M. Mägi, High resolution ^{27}Al NMR of aluminosilicates, *J. Am. Chem. Soc.*, *108*, 1730-1735, 1986.

74. Lipsicas, M., R. H. Raythatha, T. J. Pinnavaia, I. D. Johnson, R. F. Giese Jr., P. M. Costanzo, and J. L. Robert, Silicon and aluminum site distributions in 2:1 layered silicate clays, *Nature, 309*, 604-606, 1984.

75. Liu, S. B., A. Pines, M. Brandriss, and J. F. Stebbins, Relaxation mechanisms and effects of motion in albite (NaAlSi$_3$O$_8$) liquid and glass: a high temperature NMR study, *Phys. Chem. Minerals, 15*, 155-162, 1987.

76. Liu, S. B., J. F. Stebbins, E. Schneider, and A. Pines, Diffusive motion in alkali silicate melts: an NMR study at high temperature, *Geochim. Cosmochim. Acta, 52*, 527-538, 1988.

77. Maekawa, H., T. Maekawa, K. Kawamura, and T. Yokokawa, The structural groups of alkali silicate glasses determined from ^{29}Si MAS-NMR, *J. Non-Cryst. Solids, 127*, 53-64, 1991.

78. Mägi, M., E. Lippmaa, A. Samosen, G. Engelhardt, and A. R. Grimmer, Solid-state high-resolution silicon-29 chemical shifts in silicates, *J. Phys. Chem.*, *88*, 1518-1522, 1984.

79. Massiot, D., C. Bessada, J. P. Coutures, and F. Taulelle, A quantitative study of ^{27}Al MAS NMR in crystalline YAG, *J. Magn. Res.*, *90*, 231-242, 1990.

80. Massiot, D., A. Kahn-Harari, D. Michel, D. Müller, and F. Taulelle, Aluminum-27 MAS NMR of Al$_2$Ge$_2$O$_7$ and LaAlGe$_2$O$_7$: two pentacoordinated aluminum environments, *Mag. Res. Chem.*, *28*, S82-88, 1990.

81. Massiot, D., F. Taulelle, and J. P. Coutures, Structural diagnostic of high temperature liquid phases by ^{27}Al NMR, *Colloq. Phys.*, *51-C5*, 425-431, 1990.

82. McMillan, P., M. Akaogi, E. Ohtani, Q. Williams, R. Nieman, and R. Sato, Cation disorder in garnets along the Mg$_3$Al$_2$Si$_3$O$_{12}$-Mg$_4$Si$_4$O$_{12}$ join: an infrared, Raman, and NMR study, *Phys. Chem. Minerals, 16*, 428-435, 1989.

83. McMillan, P. F., and R. J. Kirkpatrick, Al coordination in magnesium aluminosilicate glasses, *Am. Mineral.*, *77*, 898-900, 1992.

84. McMillan, P. F., B. T. Poe, B. Coté, D. Massiot, and J. P. Coutures, ^{27}Al NMR linewidths for SiO$_2$-Al$_2$O$_3$ liquids: structural re-

laxation and a mechanism for viscous flow in high silica aluminosilicate liquids, *Eos, Trans. Am. Geophys. Union*, 72, 572, 1991.

85. Merwin, L., H. Keppler, and A. Sebald, ^{29}Si MAS NMR evidence for the depolymerization of silicate melts by water, *Eos, Trans. Am. Geophys. Union*, 72, 573, 1991.

86. Millard, R. L., R. C. Peterson, and B. K. Hunter, Temperature dependence of cation disorder in $MgAl_2O_4$ spinel using ^{27}Al and ^{17}O magic-angle spinning NMR, *Am. Mineral.*, 77, 44-52, 1992.

87. Morris, H. D., S. Banks, and P. D. Ellis, ^{27}Al NMR spectroscopy of iron-bearing montmorillonite clays, *J. Phys. Chem.*, 94, 3121-3129, 1990.

88. Mueller, D., D. Hoebbel, and W. Gessner, ^{27}Al studies of aluminosilicate solutions. Influences of the second coordination sphere on the shielding of aluminum, *Chem. Phys. Lett.*, 84, 25-29, 1981.

89. Mueller, K. T., J. H. Baltisberger, E. W. Wooten, and A. Pines, Isotropic chemical shifts and quadrupolar parameters for oxygen-17 using dynamic-angle spinning NMR, *J. Phys. Chem.*, 96, 7001-7004, 1992.

90. Mueller, K. T., Y. Wu, B. F. Chmelka, J. Stebbins, and A. Pines, High-resolution oxygen-17 NMR of solid silicates, *J. Am. Chem. Soc.*, 113, 32-38, 1990.

91. Müller, D., W. Gessner, A. Samosen, and E. Lippmaa, Solid state ^{27}Al NMR studies on polycrystalline aluminates of the system $CaO-Al_2O_3$, *Polyhedron*, 5, 779-785, 1986.

92. Müller, D., W. Gessner, A. Samosen, E. Lippmaa, and G. Scheler, Solid-state aluminum-27 nuclear magnetic resonance chemical shift and quadrupole coupling data for condensed AlO_4 tetrahedra, *J. Chem. Soc. Dalton Trans.*, 1986, 1277-1281, 1986.

93. Müller, D., E. Jahn, G. Ladwig, and U. Haubenreissee, High-resolution solid-state ^{27}Al and ^{31}P NMR: correlation between chemical shift and mean Al-O-P angle in $AlPO_4$ polymorphs, *Chem. Phys. Lett.*, 109, 332-336, 1984.

94. Murdoch, J. B., J. F. Stebbins, and I. S. E. Carmichael 70, 332-343., High-resolution ^{29}Si NMR study of silicate and aluminosilicate glasses: the effect of network-modifying cations, *Am. Mineral.*, 70, 332-343, 1985.

95. Murdoch, J. B., J. F. Stebbins, I. S. E. Carmichael, and A. Pines, A silicon-29 nuclear magnetic resonance study of silicon-aluminum ordering in leucite and analcite, *Phys. Chem. Minerals*, 15, 370-382, 1988.

96. Nielsen, N. C., H. Bildsoe, H. J. Jakobsen, and P. Norby, 7Li, 23Na, and 27Al quadrupolar interactions in some aluminosilicate sodalites from MAS n.m.r. spectra of satellite transitions, *Zeolites*, 11, 622-631, 1991.

97. Oestrike, R., A. Navrotsky, G. L. Turner, B. Montez, and R. J. Kirkpatrick, Structural environment of Al dissolved in $2PbO \cdot B_2O_3$ glass used for solution calorimetry: an ^{27}Al NMR study, *Am. Mineral.*, 72, 788-791, 1987.

98. Oestrike, R., W. H. Yang, R. Kirkpatrick, R. L. Hervig, A. Navrotsky, and B. Montez, High-resolution ^{23}Na, ^{27}Al, and ^{29}Si NMR spectroscopy of framework aluminosilicate glasses, *Geochim. Cosmochim. Acta*, 51, 2199-2209, 1987.

99. Phillips, B., R. J. Kirkpatrick, and J. G. Thompson, ^{27}Al and ^{31}P NMR spectroscopy of the α-β transition in $AlPO_4$ cristobalite: evidence for a dynamical order-disorder transition, *Eos, Trans. Am. Geophys. Union*, 71, 1671, 1990.

100. Phillips, B. L., F. M. Allen, and R. J. Kirkpatrick, High-resolution solid-state ^{27}Al NMR spectroscopy of Mg-rich vesuvianite, *Am. Mineral.*, 72, 1190-1194, 1987.

101. Phillips, B. L., D. A. Howell, R. J. Kirkpatrick, and T. Gasparik, Investigation of cation order in $MgSiO_3$-rich garnet using ^{29}Si and ^{27}Al MAS NMR spectroscopy, *Am. Mineral.*, 77, 704-712, 1992.

102. Phillips, B. L., R. J. Kirkpatrick, and M. A. Carpenter, Investigation of short-range Al,Si order in synthetic anorthite by ^{29}Si MAS NMR spectroscopy, *Am. Mineral.*, 77, 484-495, 1992.

103. Phillips, B. L., R. J. Kirkpatrick, and G. L. Hovis, ^{27}Al, ^{29}Si, and ^{23}Na MAS NMR study of an Al,Si ordered alkali feldspar solid solution series, *Phys. Chem. Minerals*, 16, 262-275, 1988.

104. Phillips, B. L., R. J. Kirkpatrick, and A. Putnis, Si,Al ordering in leucite by high-resolution ^{27}Al MAS NMR spectroscopy, *Phys. Chem. Minerals*, 16, 591-598, 1989.

105. Poe, B. T., P. F. McMillan, C. A. Angell, and R. K. Sato, Al and Si coordination in $SiO_2-Al_2O_3$ liquids and glasses: A study by NMR and IR spectroscopy and MD simulations, *Chem. Geol.*, 96, 241-266, 1992.

106. Putnis, A., Solid state NMR spectroscopy and phase transitions in minerals, in *Physical Properties and Thermodynamic Behaviour of Minerals*, edited by E. K. H. Salje, pp. 325-358, Reidel, Dordrecht, 1988.

107. Putnis, A., C. A. Fyfe, and G. C. Gobbi, Al,Si ordering in cordierite using "magic angle spinning" NMR, *Phys. Chem. Minerals*, 12, 211-216, 1985.

108. Raudsepp, M., A. C. Turnock, F. C. Hawthorne, B. L. Sherriff, and J. S. Hartman, Characterization of synthetic pargasitic amphiboles $(NaCa_2Mg_4M^{3+}Si_6Al_2O_{22}(OH,F)_2$; $M^{3+} = Al$, Cr, Ga, Sc, In) by infrared spectroscopy, Rietveld structure refinement and ^{27}Al, ^{29}Si and ^{19}F MAS NMR spectroscopy, *Am. Mineral.*, 72, 580-593, 1987.

109. Rigamonti, A., NMR-NQR studies of structural phase transitions, *Adv. Phys.*, 33, 115-191, 1984.

110. Risbud, S. H., R. J. Kirkpatrick, A. P. Taglialavore, and B. Montez, Solid-state NMR evidence of 4-, 5-, and 6-fold aluminum sites in roller-

quenched SiO_2-Al_2O_3 glasses, *J. Am. Ceram. Soc.*, *70*, C10-C12, 1987.

111. Sanz, J., and J. L. Robert, Influence of structural factors on ^{29}Si and ^{27}Al NMR chemical shifts of 2:1 phyllosilicates, *Phys. Chem. Minerals, 19*, 39-45, 1991.

112. Sanz, J., and J. M. Serratosa, ^{29}Si and ^{27}Al high-resolution MAS-NMR spectra of phyllosilicates, *J. Am. Chem. Soc., 106*, 4790-4793, 1984.

113. Sato, R. K., P. F. McMillan, P. Dennison, and R. Dupree, High resolution ^{27}Al and ^{29}Si MAS NMR investigation of SiO_2-Al_2O_3 glasses, *J. Phys. Chem, 95*, 4484, 1991.

114. Sato, R. K., P. F. McMillan, P. Dennison, and R. Dupree, A structural investigation of high alumina content glasses in the CaO-Al_2O_3-SiO_2 system via Raman and MAS NMR, *Phys. Chem. Glasses, 32*, 149-154, 1991.

115. Schramm, S., and E. Oldfield, High-resolution oxygen-17 NMR of solids, *J. Am. Chem. Soc., 106*, 2502-2506, 1984.

116. Sebald, A., MAS and CP/MAS NMR of Less Common Spin-1/2 Nuclei, in *Solid-State NMR II: Inorganic Matter*, edited by B. Blümich, pp. 91-132, Springer-Verlag, Berlin, 1994.

117. Sherriff, B. L., H. D. Grundy, and J. S. Hartman, Occupancy of T sites in the scapolite series: a multinuclear NMR study using magic-angle spinning, *Can. Mineral., 25*, 717-730, 1987.

118. Sherriff, B. L., H. D. Grundy, and J. S. Hartman, The relationship between ^{29}Si MAS NMR chemical shift and silicate mineral structure, *Eur. J. Mineral., 3*, 751-768, 1991.

119. Sherriff, B. L., H. D. Grundy, J. S. Hartman, F. C. Hawthorne, and P. Cerny, The incorporation of alkalis in beryl: multi-nuclear MAS NMR and crystal-structure study, *Can. Mineral., 29*, 271-285, 1991.

120. Sherriff, B. L., and J. S. Hartman, Solid-state high-resolution ^{29}Si NMR of feldspars: Al-Si disorder and the effects of paramagnetic centres, *Can. Mineral., 23*, 205-212, 1985.

121. Shimokawa, S., H. Maekawa, E. Yamada, T. Maekawa, Y. Nakamura, and T. Yokokawa, A high temperature ($1200°C$) probe for NMR experiments and its application to silicate melts, *Chem. Lett., 1990*, 617-620, 1990.

122. Skibsted, J., H. Bildoe, and H. J. Jakobsen, High-speed spinning versus high magnetic field in MAS NMR of quadrupolar nuclei. ^{27}Al MAS NMR of 3CaO·Al_2O_3, *J. Mag. Reson., 92*, 669-676, 1991.

123. Skibsted, J., E. Henderson, and H. J. Jakobsen, Characterization of calcium aluminate phases in cements by high-performance ^{27}Al MAS NMR spectroscopy, *Inorg. Chem., in press*, 1992.

124. Skibsted, J., J. Hjorth, and H. J. Jakobsen, Correlation between ^{29}Si NMR chemical shifts and mean Si-O bond lengths for calcium silicates, *Chem. Phys. Let., 172*, 279-283, 1990.

125. Skibsted, J., N. C. Nielsen, H. Bilsoe, and H. J. Jakobsen, Satellite transitions in MAS NMR spectra of quadrupolar nuclei, *J. Mag. Reson., 95*, 88-117, 1991.

126. Smith, J. V., and C. S. Blackwell, Nuclear magnetic resonance of silica polymorphs, *Nature, 303*, 223-225, 1983.

127. Smith, J. V., C. S. Blackwell, and G. L. Hovis, NMR of albite-microcline series, *Nature, 309*, 140-143, 1984.

128. Smith, K. A., R. J. Kirkpatrick, E. Oldfield, and D. M. Henderson, High-resolution silicon-29 nuclear magnetic resonance spectroscopic study of rock-forming silicates, *Am. Mineral., 68*, 1206-1215, 1983.

129. Smith, M. E., F. Taulelle, and D. Massiot, Getting more out of MAS NMR studies of quadrupolar nuclei, *Bruker Rep., 1990-2*, 16-18, 1990.

130. Spearing, D. R., I. Farnan, and J. F. Stebbins, NMR lineshape and T_1 relaxation study of the $\alpha-\beta$ phase transitions in quartz and cristo-

balite, *Eos, Trans. Am. Geophys. Union, 71*, 1671, 1990.

131. Spearing, D. R., I. Farnan, and J. F. Stebbins, Dynamics of the $\alpha-\beta$ phase transitions in quartz and cristobalite as observed by in-situ high temperature ^{29}Si and ^{17}O NMR, *Phys. Chem. Min., 19*, 307-321, 1992.

132. Spearing, D. R., and J. F. Stebbins, The ^{29}Si NMR shielding tensor in low quartz, *Am. Mineral., 74*, 956-959, 1989.

133. Staehli, J. L., and D. Brinkmann, Assignment and structural dependence of electric field gradients in anorthite and simple field gradient calculations in some aluminosilicates, *Zeit. Krist., 140*, 374-392, 1974.

134. Stebbins, J. F., Identification of multiple structural species in silicate glasses by ^{29}Si NMR, *Nature, 330*, 465-467, 1987.

135. Stebbins, J. F., Effects of temperature and composition on silicate glass structure and dynamics: Si-29 NMR results, *J. Non-Cryst. Solids, 106*, 359-369, 1988.

136. Stebbins, J. F., NMR spectroscopy and dynamic processes in mineralogy and geochemistry, in *Spectroscopic Methods in Mineralogy and Geology*, edited by F. C. Hawthorne, pp. 405-430, Mineralogical Society of America, Washington D.C., 1988.

137. Stebbins, J. F., Experimental confirmation of five-coordinated silicon in a silicate glass at 1 atmosphere pressure, *Nature, 351*, 638-639, 1991.

138. Stebbins, J. F., Nuclear magnetic resonance at high temperature, *Chem. Rev., 91*, 1353-1373, 1991.

139. Stebbins, J. F., Nuclear magnetic resonance spectroscopy of geological materials, *Mat. Res. Bull., 17*, 45-52, 1992.

140. Stebbins, J. F., C. W. Burnham, and D. L. Bish, Tetrahedral disorder in fibrolitic sillimanite: constraints from ^{29}Si NMR spectroscopy, *Am. Mineral., 78*, 461-464, 1993.

141. Stebbins, J. F., and I. Farnan, NMR spectroscopy in the earth sciences: structure and dynamics, *Science, 245*, 257-262, 1989.

142. Stebbins, J. F., and I. Farnan, The effects of temperature on silicate liquid structure: a multi-nuclear, high temperature NMR study, *Science, 255*, 586-589, 1992.

143. Stebbins, J. F., I. Farnan, and I. Klabunde, Aluminum in rutile (TiO_2): Characterization by single crystal and MAS NMR, *J. Am. Ceram. Soc., 11*, 2198-2200, 1989.

144. Stebbins, J. F., I. Farnan, E. H. Williams, and J. Roux, Magic angle spinning NMR observation of sodium site exchange in nepheline at 500° C, *Phys. Chem. Minerals, 16*, 763-766, 1989.

145. Stebbins, J. F., I. Farnan, and X. Xue, The structure and dynamics of alkali silicate liquids: one view from NMR spectroscopy, *Chem. Geol., 96*, 371-386, 1992.

146. Stebbins, J. F., and M. Kanzaki, Local structure and chemical shifts for six-coordinated silicon in high pressure mantle phases, *Science, 251*, 294-298, 1990.

147. Stebbins, J. F., and P. McMillan, Five- and six- coordinated Si in $K_2Si_4O_9$ glass quenched from 1.9 GPa and 1200° C, *Am. Mineral., 74*, 965-968, 1989.

148. Stebbins, J. F., J. B. Murdoch, I. S. E. Carmichael, and A. Pines, Defects and short range order in nepheline group minerals: a silicon-29 nuclear magnetic resonance study, *Phys. Chem. Mineral., 13*, 371-381, 1986.

149. Stebbins, J. F., J. B. Murdoch, E. Schneider, I. S. E. Carmichael, and A. Pines, A high temperature nuclear magnetic resonance study of ^{27}Al, ^{23}Na, and ^{29}Si in molten silicates, *Nature, 314*, 250-252, 1985.

150. Stebbins, J. F., and D. Sykes, The structure of $NaAlSi_3O_8$ liquids at high pressure: new constraints from NMR spectroscopy, *Am. Mineral., 75*, 943-946, 1990.

151. Taulelle, F., J. P. Coutures, D. Massiot, and J. P. Rifflet, High and very high temperature NMR, *Bull. Magn. Reson., 11*, 318-320, 1989.

152. Timken, H. K. C., N. Janes, G. L. Turner, S. L. Lambert, L. B. Welsh, and E. Oldfield, Solid-state oxygen-17 nuclear magnetic resonance spectroscopic studies of zeolites and related systems, *J. Am. Chem. Soc., 108*, 7236-7241, 1986.

153. Timken, H. K. C., S. E. Schramm, R. J. Kirkpatrick, and E. Oldfield, Solid-state oxygen-17 nuclear magnetic resonance spectroscopic studies of alkaline earth metasilicates, *J. Phys. Chem., 91*, 1054-1058, 1987.

154. Turner, G. L., S. E. Chung, and E. Oldfield, Solid-state oxygen-17 nuclear magnetic resonance spectroscopic study of group II oxides, *J. Mag. Res., 64*, 316-324, 1985.

155. Turner, G. L., K. A. Smith, R. J. Kirkpatrick, and E. Oldfield, Boron-11 nuclear magnetic resonance spectroscopic study of borate and borosilicate minerals and a borosilicate glass, *J. Mag. Reson., 67*, 544-550, 1986.

156. Walter, T. H., and R. Oldfield, Magic angle spinning oxygen-17 NMR of aluminum oxides and hydroxides, *J. Phys. Chem., 93*, 6744-6751, 1989.

157. Walter, T. H., G. L. Turner, and E. Oldfield, Oxygen-17 cross polarization NMR spectroscopy of inorganic solids, *J. Magn. Reson., 76*, 106-120, 1988.

158. Weeding, T. L., B. H. W. S. de-Jong, W. S. Veeman, and B. G. Aitken, Silicon coordination changes from 4-fold to 6-fold on devitrification of silicon phosphate glass, *Nature, 318*, 352-353, 1985.

159. Weiden, N., and H. Rager, The chemical shift of the ^{29}Si nuclear magnetic resonance in a synthetic single crystal of Mg_2SiO_4, *Z. Naturforsch., 40A*, 126-130, 1985.

160. Weiss, C. A., Jr., R. J. Kirkpatrick, and S. P. Altaner, Variations in interlayer cation sites of clay minerals as studied by (133)Cs MAS nuclear magnetic resonance spectroscopy, *Am. Mineral., 75*, 970-982, 1990.

161. Wilson, M. A., *NMR Techniques and Applications in Geochemistry and Soil Chemistry*, 353 pp., Pergamon, Oxford, 1987.

162. Woessner, D. E., Characterization of clay minerals by ^{27}Al nuclear magnetic resonance spectroscopy, *Am. Mineral., 74*, 203-215, 1989.

163. Wood, B. J., R. J. Kirkpatrick, and B. Montez, Order-disorder phenomena in $MgAl_2O_4$ spinel, *Am. Mineral., 71*, 999-1006, 1986.

164. Wu, Y., B. Q. Sun, A. Pines, A. Samosen, and E. Lippmaa, NMR experiments with a new double rotor., *J. Magn. Reson., 89*, 297-309, 1990.

165. Xue, X., J. F. Stebbins, and M. Kanzaki, Oxygen-17 NMR of quenched high-pressure crystals and glasses, *Eos, Trans. Am. Geophys. Union, 72*, 572, 1991.

166. Xue, X., J. F. Stebbins, M. Kanzaki, P. F. McMillan, and B. Poe, Pressure-induced silicon coordination and tetrahedral structural changes in alkali silicate melts up to 12 GPa: NMR, Raman, and infrared spectroscopy, *Am. Mineral., 76*, 8-26, 1991.

167. Xue, X., J. F. Stebbins, M. Kanzaki, and R. G. Tronnes, Silicon coordination and speciation changes in a silicate liquid at high pressures, *Science, 245*, 962-964, 1989.

168. Yang, W. A., and R. J. Kirkpatrick, Hydrothermal reaction of a rhyolitic-composition glass: a solid-state NMR study, *Am. Mineral., 75*, 1009-1019, 1990.

169. Yang, W. H., R. J. Kirkpatrick, and D. M. Henderson, High-resolution ^{29}Si and ^{27}Al NMR spectroscopic study of Al-Si disordering in annealed albite and oligoclase, *Am. Mineral., 71*, 712-726, 1986.

170. Zhong, J., and P. J. Bray, Change in boron coordination in alkali borate glasses, and mixed alkali effects, as elucidated by NMR, *J. Non-Cryst. Solids, 111*, 67-76, 1989.

Mössbauer Spectroscopy of Minerals

Catherine McCammon

1. INTRODUCTION

Since the discovery of the Mössbauer effect in 1958, numerous applications in a wide variety of scientific disciplines have been described. Of the more than 30,000 papers published as of 1993, at least 2000 contain results of studies on minerals (as estimated from data provided by the Mössbauer Effect Data Center, USA). This chapter provides a reference to Mössbauer data for 108 minerals containing ^{57}Fe and 18 containing ^{119}Sn, accompanied by reference material on Mössbauer spectroscopy.

2. THEORY

The Mössbauer effect is the recoilless absorption and emission of γ-rays by specific nuclei in a solid [81, 82], and provides a means of studying the local atomic enviroment around the nuclei.

The interactions between the nucleus and the atomic electrons depend strongly on the electronic, chemical and magnetic state of the atom. Information from these hyperfine interactions is provided by the hyperfine parameters, which can be determined experimentally from the line positions in a Mössbauer spectrum (Figure 1). A typical experimental spectrum is illustrated in Figure 2. Table 1 describes the hyperfine parameters as well as other observables. Formulae relating the Mössbauer line positions and the hyperfine parameters are given in Table 2. Suggested references for further information are listed in Table 3.

3. EXPERIMENT

A transmission Mössbauer spectrometer is very simple, and typically consists of a γ-ray source, the absorber (sample) and a detector. The source is moved relative to the absorber, shifting the energy spectrum due to the Doppler effect. Spectra are commonly plotted as percent transmission versus source velocity (energy). Selected references to important experimental considerations are given in Table 4, while Table 5 lists some common applications of Mössbauer spectroscopy to mineral studies. This chapter only includes references to transmission studies; however the technique can also be performed in a scattering geometry to study surface properties (e.g., [105, 121, 127]).

4. MINERAL DATA

Over 100 different Mössbauer transitions have been observed, although unfavourable nuclear properties limit the number of commonly used nuclei. ^{57}Fe is by far the most popular isotope, followed by ^{119}Sn. Both the 14.4 keV transition in ^{57}Fe and the 23.88 keV transition in ^{119}Sn involve a spin change of $3/2 \to 1/2$, and therefore have similar hyperfine properties. ^{57}Fe Mössbauer data of selected minerals are listed in Tables 6 through 10, while ^{119}Sn data are listed in Table 11. The data were chosen from the literature as being typical for each mineral; however since hyperfine parameters often depend on chemical composition, particle size, thermal history and degree of crystallinity, the data should be considered representative

C. McCammon, Bayerisches Geoinst., Postfach 10 12 51, D-8580 Bayreuth, Germany

Mineral Physics and Crystallography
A Handbook of Physical Constants
AGU Reference Shelf 2

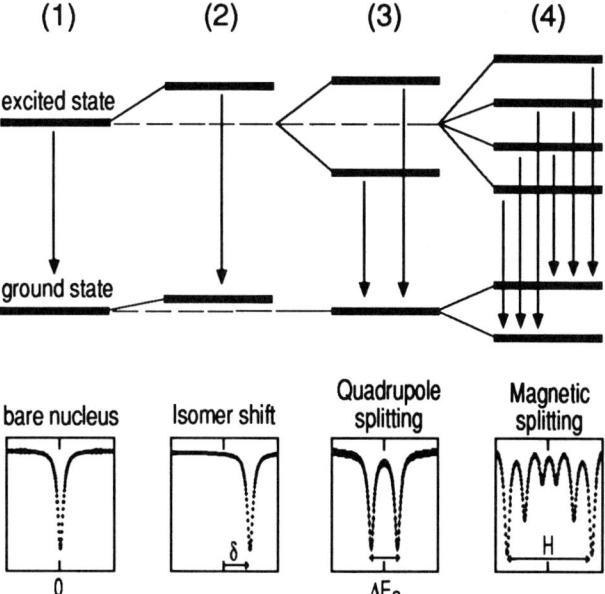

Fig. 1. Illustration of hyperfine interactions for ^{57}Fe nuclei, showing the nuclear energy level diagram for (1) a bare nucleus, (2) electric monopole interaction (isomer shift), (3) electric quadrupole interaction (quadrupole splitting), and (4) magnetic dipole interaction (hyperfine magnetic splitting). Each interaction is shown individually, accompanied by the resulting Mössbauer spectrum.

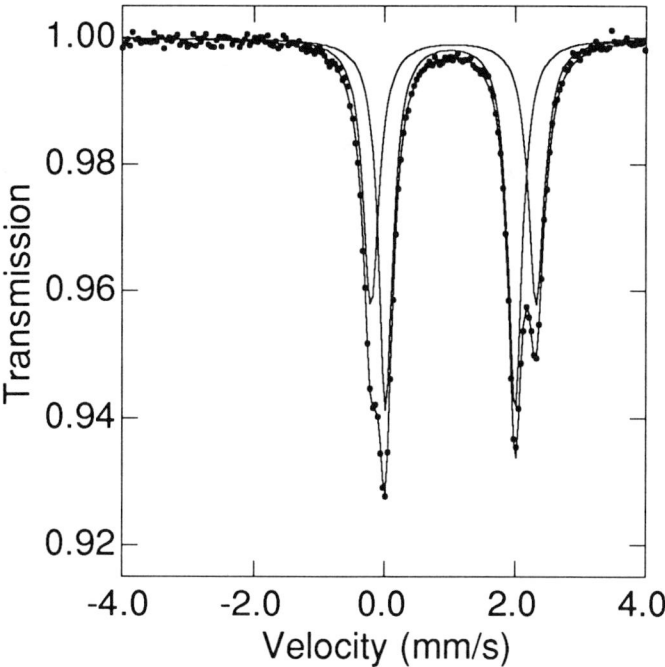

Fig. 2. Mössbauer spectrum of orthopyroxene with composition $Fe_{0.8}Mg_{0.2}SiO_3$ showing two quadrupole doublets, one corresponding to Fe^{2+} in the M1 site (45% of total area) and one corresponding to Fe^{2+} in the M2 site (55% of total area).

only. For more complete information on specific minerals, one should consult resources such as the Minerals Handbook published by the Mössbauer Effect Data Center (see Table 3). Minerals are listed by name except when part of a larger structure group, e.g. $Fe_3Al_2Si_3O_{12}$ is listed under garnet, not almandine. Chemical compositions are given exactly as reported by the authors (even if the resulting compositions are not electrostatically neutral). Data for differing compositions are provided for the major mineral groups to illustrate the dependence of hyperfine parameters on composition. The relative areas of subspectra can be used as a rough approximation to relative abundance, e.g. [97], but note that site proportions often vary between different samples of the same mineral. For example, the amount of Fe^{3+} may depend strongly on fO_2 conditions, and the distribution of iron cations between different crystallographic sites may be a function of thermal history. Most spectra were fitted to Lorentzian lineshapes; the few exceptions are noted in the tables.

TABLE 1. Description of Mössbauer parameters

Name	Unit	Description
Isomer shift (δ)	mm s^{-1}	Energy difference between source and absorber nuclei resulting from effects including differences in valence state, spin state and coordination of absorber atoms. Experimentally one observes a single line shifted from a reference zero point by the isomer shift plus the second-order Doppler shift (SOD), a small thermal shift due to atomic vibrations.
Centre shift (CS)	mm s^{-1}	The experimental shift of the centroid of a Mössbauer spectrum from a zero reference point. The contribution from the SOD is similar in most standard materials, so for purposes of comparison the isomer shift is often taken to be equal to the centre shift.
Quadrupole splitting (ΔE_Q)	mm s^{-1}	Splitting of the energy levels caused by interaction between the nuclear quadrupolar moment and an electric field gradient at the nucleus, and depends on the valence and spin state of the absorber atoms, as well as the coordination and degree of distortion of the crystallographic site. Experimentally one observes a doublet in ^{57}Fe and ^{119}Sn spectra with components of equal intensity and linewidth in the ideal random absorber case. The quadrupole splitting is given by the energy separation between components.
Hyperfine magnetic field (H)	Tesla	Interaction of the dipole moment of the nucleus and a hyperfine magnetic field causes a splitting of the nuclear energy levels, resulting in six peaks for ^{57}Fe spectra in the simplest case. For an ideal random absorber with no quadrupole interaction the linewidths of the peaks are equal with intensity ratio 3:2:1:1:2:3. The separation of peaks 1 and 6 is proportional to the magnitude of the hyperfine magnetic field.
Line width (Γ)	mm s^{-1}	Full width at half maximum of the peak height. Peaks can be broadened beyond the natural line width by effects due to equipment (vibrational, geometrical, thermal, and electronic problems), the source (self-absorption resulting from decay), and the sample (thickness broadening, next-nearest-neighbour effects, and dynamic processes such as relaxation).
Relative area (I)	—	Relative proportion of subspectrum area to the total area. Each site normally contributes a subspectrum (e.g. a quadrupole doublet) whose area is approximately related to the relative abundance of that particular site within the absorber.

TABLE 2. Determination of line positions for ^{57}Fe 14.4 keV transition

Hyperfine interactions present	Line positions
– electric monopole	$L_1 = CS$
– electric monopole + quadrupole	$L_1 = CS + \frac{1}{2}\,\Delta E_Q$ $L_2 = CS - \frac{1}{2}\,\Delta E_Q$
– electric monopole + magnetic dipole ($\Delta E_Q = 0$) – electric monopole + quadrupole + magnetic dipole (special case of axially symmetric electric field gradient and $|\mu_N g_I H| \gg |\Delta E_Q|$) $\mu_N g_{1/2} = 0.11882$ mm s^{-1} T^{-1} $\mu_N g_{3/2} = 0.067899$ mm s^{-1} T^{-1}	$L_1 = \frac{1}{2}\,\mu_N H\ (\ 3\,g_{3/2} - g_{1/2}\ ) + CS + \frac{1}{2}\,\Delta E_Q$ $L_2 = \frac{1}{2}\,\mu_N H\ (\ \ g_{3/2} - g_{1/2}\ ) + CS - \frac{1}{2}\,\Delta E_Q$ $L_3 = \frac{1}{2}\,\mu_N H\ (\ \ g_{3/2} - g_{1/2}\ ) + CS - \frac{1}{2}\,\Delta E_Q$ $L_4 = \frac{1}{2}\,\mu_N H\ (\ \ g_{3/2} + g_{1/2}\ ) + CS - \frac{1}{2}\,\Delta E_Q$ $L_5 = \frac{1}{2}\,\mu_N H\ (\ -g_{3/2} + g_{1/2}\ ) + CS - \frac{1}{2}\,\Delta E_Q$ $L_6 = \frac{1}{2}\,\mu_N H\ (\ -3g_{3/2} + g_{1/2}\ ) + CS + \frac{1}{2}\,\Delta E_Q$
– electric monopole + quadrupole + magnetic dipole (general case)	Requires calculation of the complete interaction Hamiltonian (e.g. [71]). There are eight lines involving the following hyperfine parameters: isomer shift (δ), hyperfine magnetic field (H), quadrupole splitting (ΔE_Q), the polar (θ) and azimuthal (φ) angles relating the direction of H to the electric field gradient (EFG), and the asymmetry parameter of the EFG (η).

TABLE 3. Suggested references for Mössbauer spectroscopy

Type	Reference
Book	Bancroft, G.M. *Mössbauer Spectroscopy. An Introduction for Inorganic Chemists and Geochemists.* McGraw Hill, New York, 1973. Cranshaw, T.E., Dale, B.W., Longworth, G.O. and Johnson, C.E. *Mössbauer Spectroscopy and its Applications*, Cambridge University Press, Cambridge, 1986. Dickson, D.P. and Berry, F.J. (eds.) *Mössbauer Spectroscopy*, Cambridge University Press, Cambridge, 1986. Gibb, T.C. *Principles of Mössbauer Spectroscopy*, Chapman and Hall, London, 1977. Gonser, U. (ed.) *Mössbauer Spectroscopy*, Topics in Applied Physics, Vol. 5, Springer-Verlag, Berlin, 1975. Greenwood, N.N. and Gibb, T.D. *Mössbauer Spectroscopy*, Chapman and Hall, London, 1971. Gütlich, P., Link, R. and Trautwein, A., *Mössbauer Spectroscopy and Transition Metal Chemistry*, Springer-Verlag, Berlin, 1978. Hawthorne, F.C. (ed.) *Spectroscopic Methods in Mineralogy and Geology*, Rev. Mineral. Vol. 18, Mineralogical Society of America, 1988. See Chapter on Mössbauer Spectroscopy, F.C. Hawthorne, pp. 255-340. Mitra, S. *Applied Mössbauer Spectroscopy, Theory and Practice for Geochemists and Archeologists*, Pergamon Press, Oxford, 1992. Robinson, J.W. (ed.) *Handbook of Spectroscopy, Vol. 3*, CRC Press, Inc., Boca Raton, USA, 1981. See Chapter on Mössbauer Spectroscopy, J.G. Stevens (ed.), pp. 403-528.

TABLE 3. (continued)

Type	Reference
Journal	Analytical Chemistry (American Chemical Society, Washington DC) contains biennial reviews (starting in 1966) of Mössbauer spectroscopy, see for example Vol. 62, pp. 125R-139R, 1990.
	Hyperfine Interactions (J.C. Baltzer AG, Basel) publishes proceedings from various Mössbauer conferences, see for example Vol. 68-71, 1992.
	Mössbauer Effect Reference and Data Journal (Mössbauer Effect Data Center, Asheville, NC) contains references and Mössbauer data for nearly all Mössbauer papers published.
Data Resource	Stevens, J.G., Pollack, H., Zhe, L., Stevens, V.E., White, R.M. and Gibson, J.L. (eds.) *Mineral: Data* and *Mineral: References*, Mössbauer Handbook Series, Mössbauer Effect Data Center, University of North Carolina, Asheville, North Carolina, USA, 1983.
	Mössbauer Micro Databases (Mössbauer Effect Data Center, Asheville, NC) cover many topics including Minerals. Databases are set up to run on IBM-compatible microcomputers and can be searched using various options.
	Mössbauer Effect Data Center Mössbauer Information System (maintained by the Mössbauer Effect Data Center, Asheville, NC) contains extensive bibliographic and Mössbauer data entries compiled from the Mössbauer literature. Searches of the database are possible; contact the Mössbauer Effect Data Center for details.

TABLE 4. Methodology References

Experimental aspect	Reference
Absorber thickness	[74, 99]
Geometric effects	[16, 28]
Absorber homogeneity	[18, 50]
Preferred orientation of absorber	[95, 96]
Saturation effects	[97, 99, 120]
Isomer shift reference scales	[116]
Goodness of fit criteria	[31, 37, 58, 103]
Conventions for reporting Mössbauer data	[117]

TABLE 5. Applications in mineralogy

Application	Reference
Oxidation state, including intervalence charge transfer	[10, 15, 20]
Site occupancies, including $Fe^{3+}/\Sigma Fe$	[14, 32, 97, 114]
Site coordination	[15, 22]
Semi-quantitative phase analysis	[13, 85]
Phase transitions	[66, 108]
Magnetic structure	[23, 25]

TABLE 6. ^{57}Fe Mössbauer data for selected silicate minerals

Absorber	T	CS(Fe) mm s^{-1}	ΔE_Q mm s^{-1}	H Tesla	I	site	Ref
Amphibole structure							
$Mg_{5.7}Fe_{1.3}Si_8O_{22}(OH)_2$	RT	1.16(1)	2.76(1)		0.07	$^{VI}Fe^{2+}$	[107]
		1.13(1)	1.81(1)		0.93	$^{VI}Fe^{2+}$	
$Fe_{6.2}Mg_{0.8}Si_8O_{22}(OH)_2$	RT	1.16(1)	2.79(1)		0.69	$^{VI}Fe^{2+}$	[53]
		1.07(1)	1.55(1)		0.31	$^{VI}Fe^{2+}$	

TABLE 6. (continued)

Absorber	T	CS(Fe) mm s⁻¹	ΔE_Q mm s⁻¹	H Tesla	I	site	Ref
$Ca_2XSi_8O_{22}(OH)_2$[a]	77 K	1.27(1)	3.17(1)		0.59	$^{VI}Fe^{2+}$	[113]
$X=Mg_{4.7}Fe_{0.3}$		1.30(1)	2.39(1)		0.31	$^{VI}Fe^{2+}$	
		1.27(1)	1.86(1)		0.10	$^{VI}Fe^{2+}$	
$Na_{1.8}Ca_{0.1}XSi_8O_{22}(OH)_2$	RT	1.14(1)	2.87(1)		0.36	$^{VI}Fe^{2+}$	[40]
$X=Fe_{4.6}Mg_{0.3}$		1.12(1)	2.36(1)		0.19	$^{VI}Fe^{2+}$	
		0.40(1)	0.44(1)		0.45	$^{VI}Fe^{3+}$	
Andalusite							
$(Al_{0.96}Fe_{0.03}Mn_{0.01})_2SiO_5$	RT	0.34(1)	1.76(1)			$^{VI}Fe^{3+}$	[1]
Babingtonite							
$Ca_2Fe_{1.7}Mn_{0.3}Si_5O_{14}(OH)$	RT	1.20(1)	2.44(1)		0.40	$^{VI}Fe^{2+}$	[21]
		0.41(1)	0.86(1)		0.60	$^{VI}Fe^{3+}$	
Chlorite							
$XSi_{2.9}Al_{2.4}(OH)_{7.9}O_{10}$[b]	RT	1.14(3)	2.67(5)		0.70	$^{VI}Fe^{2+}$	[33]
$X=Mg_{2.2}Fe_{2.3}Mn_{0.1}$		1.16(3)	2.38(5)		0.21	$^{VI}Fe^{2+}$	
		0.23(5)	0.70(3)		0.09	Fe^{3+}	
Chloritoid							
$Fe_{1.7}Mg_{0.3}Al_4Si_2O_{10}(OH)_4$	RT	1.15(1)	2.41(1)		0.98	$^{VI}Fe^{2+}$	[57]
		0.29(1)	0.98(1)		0.02	$^{VI}Fe^{3+}$	
Clay minerals[c]							
Cordierite							
$Al_3Mg_{1.9}Fe_{0.2}AlSi_5O_{18}$	RT	1.22(1)	2.31(1)		0.94	$^{VI}Fe^{2+}$	[47]
		1.21(1)	1.60(1)		0.06	channel Fe^{2+}	
Epidote structure							
$Ca_2XSi_3O_{12}(OH)$	RT	0.36(1)	2.06(1)		0.92	$^{VI}Fe^{3+}$	[36]
$X=Al_{2.2}Fe_{0.8}$		0.30(2)	1.54(3)		0.08	$^{VI}Fe^{3+}$	
$YAl_{1.7}Fe_{1.2}Si_3O_{12}(OH)$	RT	1.08(1)	1.67(1)		0.58	$^{VI}Fe^{2+}$	[36]
$Y=Ca_{1.2}Ce_{0.5}La_{0.2}$		1.20(4)	1.90(8)		0.09	Fe^{2+}	
		0.35(1)	1.94(1)		0.33	$^{VI}Fe^{3+}$	
Garnet structure							
$Fe_3Al_2Si_3O_{12}$	RT	1.29(1)	3.51(1)			$^{VIII}Fe^{2+}$	[89]
$Fe^{2+}_3Fe^{3+}_2Si_3O_{12}$	RT	1.31(1)	3.46(1)		0.54	$^{VIII}Fe^{2+}$	[131]
quenched from 9.7 GPa,1100°C		0.36(1)	0.24(1)		0.46	$^{VI}Fe^{3+}$	
$Mg_2Ca_{0.5}Fe_{0.5}XSi_3O_{12}$	RT	1.28(1)	3.56(1)		0.84	$^{VIII}Fe^{2+}$	[7]
$X=Al_{1.5}Cr_{0.5}$		0.36(1)	0.33(1)		0.16	$^{VI}Fe^{3+}$	
$Ca_3Fe_2Si_3O_{12}$	RT	0.41(1)	0.55(1)			$^{VI}Fe^{3+}$	[7]
$Ca_{2.8}Fe_{0.7}Al_{1.3}Si_3O_{12}$	RT	1.26(1)	3.49(1)		0.17	$^{VIII}Fe^{2+}$	[7]
		0.39(1)	0.58(1)		0.83	$^{VI}Fe^{3+}$	
$Mg_{0.9}Fe_{0.1}SiO_3$	RT	1.26(1)	3.60(1)		0.80	$^{VIII}Fe^{2+}$	[92]
quenched from 18 GPa,1800°C		1.11(1)	1.39(1)		0.10	$^{VI}Fe^{2+}$	
		0.31(5)	0.48(5)		0.10	Fe^{3+}	
Grandidierite							
$Mg_{0.9}Fe_{0.1}Al_3BSiO_9$	RT	1.11(1)	1.73(1)		0.94	$^{V}Fe^{2+}$	[109]
		0.33(1)	1.20(1)		0.06	Fe^{3+}	
Ilvaite							
$CaFe_3Si_2O_8(OH)$[d]	RT	1.03(2)	2.48(2)		0.27	$^{IV}Fe^{2+}$	[73]
		1.06(2)	2.01(2)		0.35	$^{VIII}Fe^{2+}$	
		0.48(2)	1.32(2)		0.38	$^{VIII}Fe^{3+}$	

TABLE 6. (continued)

Absorber	T	CS(Fe) mm s^{-1}	ΔE_Q mm s^{-1}	H Tesla	I	site	Ref
Kyanite							
$(Al_{0.98}Fe_{0.02})_2SiO_5$	RT	0.38(2)	0.99(2)			$^{VI}Fe^{3+}$	[94]
Mica group[b]							
$K_{0.9}Na_{0.1}XAlSi_3O_{10}(OH)_2$	RT	1.21(1)	2.99(1)		0.08	$^{VI}Fe^{2+}$	[41]
$X=Al_{1.7}Fe_{0.2}Mg_{0.1}$		1.14(1)	2.12(1)		0.05	$^{VI}Fe^{2+}$	
		0.36(1)	0.86(1)		0.87	$^{VI}Fe^{3+}$	
$KXAl_{0.8}Si_3O_{10}(OH)_2$	RT	1.12(1)	2.63(1)		0.38	$^{VI}Fe^{2+}$	[39]
$X=Mg_{2.6}Fe_{0.6}$		0.19(1)	0.56(1)		0.62	$^{IV}Fe^{3+}$	
$KXAlSi_3O_{10}(OH)_2$	RT	1.02(1)	2.52(1)		0.59	$^{VI}Fe^{2+}$	[76]
$X=Mg_{1.6}Fe_{1.2}Mn_{0.1}Ti_{0.1}$		1.06(1)	2.08(1)		0.33	$^{VI}Fe^{2+}$	
		0.31(1)	0.80(1)		0.08	$^{VI}Fe^{3+}$	
$CaXAl_{2.7}Si_{1.2}O_{10}(OH)_2$	RT	1.06(1)	2.34(1)		0.30	$^{VI}Fe^{2+}$	[67]
$X=Mg_{2.3}Al_{0.7}Fe_{0.1}$		0.28(1)	0.66(1)		0.70	$^{IV}Fe^{3+}$	
Olivine							
Fe_2SiO_4	310 K	0.89(2)	1.91(2)		0.48	$^{VI}Fe^{2+}$	[111]
		0.95(2)	2.39(2)		0.52	$^{VI}Fe^{2+}$	
$CaFeSiO_4$	400 K	0.84(2)	1.33(2)		0.70	$^{VI}Fe^{2+}$	[111]
		0.19(4)	1.23(4)		0.30	$^{VI}Fe^{3+}$	
$Mg_{0.83}Fe_{0.17}SiO_4$	310 K	0.94(2)	1.98(2)		0.51	$^{VI}Fe^{2+}$	[111]
		0.99(2)	2.36(2)		0.41	$^{VI}Fe^{2+}$	
		0.23(4)	0.70(4)		0.08	$^{VI}Fe^{3+}$	
$Fe^{2+}_{0.6}Fe^{3+}SiO_4$	290 K	1.13(1)	2.75(2)		0.41	$^{VI}Fe^{2+}$	[68]
		0.39(1)	0.91(2)		0.59	$^{VI}Fe^{3+}$	
Orthoclase							
$KAl_{0.95}Fe_{0.05}Si_3O_8$	RT	0.46(1)	0.68(1)			$^{IV}Fe^{3+}$	[19]
Osumilite							
$XMg_{1.4}Fe_{0.9}Al_{4.4}Si_{10.3}O_{30}$	RT	1.20(1)	2.35(1)		0.68	$^{VI}Fe^{2+}$	[46]
$X=K_{0.9}Na_{0.1}$		1.14(1)	1.86(1)		0.32	channel Fe^{2+}	
Perovskite structure							
$Mg_{0.95}Fe_{0.05}SiO_3$	RT	1.12(1)	1.58(1)		0.92	$^{XII}Fe^{2+}$	[80]
quenched from 25 GPa,1650°C		0.44(5)	0.98(5)		0.08	Fe^{3+}	
Pyrophyllite							
$Fe_2Mg_{0.1}Al_{0.1}Si_4O_{10}(OH)_2$	RT	0.36(1)	0.18(1)		0.85	$^{VI}Fe^{3+}$	[26]
		0.43(4)	1.22(8)		0.07	$^{VI}Fe^{3+}$	
		0.14(4)	0.59(8)		0.08	$^{IV}Fe^{3+}$	
Pyroxene structure							
$FeSiO_3$	RT	1.18(1)	2.49(1)		0.54	$^{VI}Fe^{2+}$	[38]
		1.13(1)	1.91(1)		0.46	$^{VI}Fe^{2+}$	
	77 K	1.30(1)	3.13(1)		0.50	$^{VI}Fe^{2+}$	
		1.26(2)	2.00(1)		0.50	$^{VI}Fe^{2+}$	
$Mg_{0.85}Fe_{0.15}SiO_3^e$	77 K	1.29(1)	3.06(1)		0.20	$^{VI}Fe^{2+}$	[12]
		1.28(1)	2.16(1)		0.80	$^{VI}Fe^{2+}$	
$CaFeSi_2O_6$	RT	1.19(1)	2.22(1)			$^{VI}Fe^{2+}$	[38]
$CaMg_{0.9}Fe_{0.2}Si_{1.9}O_6$	RT	0.42(1)	1.07(1)		0.50	$^{VI}Fe^{3+}$	[54]
		0.14(1)	1.62(1)		0.50	$^{IV}Fe^{3+}$	
$NaFeSi_2O_6$	RT	0.39(1)	0.30(1)			$^{VI}Fe^{3+}$	[10]

TABLE 6. (continued)

Absorber	T	CS(Fe) mm s^{-1}	ΔE_Q mm s^{-1}	H Tesla	I	site	Ref
CaFeAlSiO$_6$	RT	0.22(1)	1.58(2)		0.11	$^{IV}Fe^{3+}$	[3]
		0.35(1)	0.99(2)		0.87[f]	$^{VI}Fe^{3+}$	
Serpentine							
(Mg$_{0.99}$Fe$_{0.01}$)$_3$Si$_2$O$_5$(OH)$_4$	RT	1.12(1)	2.70(1)		0.68	$^{VI}Fe^{2+}$	[102]
antigorite		0.36(4)	0.70(5)		0.32	$^{VI}Fe^{3+}$	
(Mg$_{0.93}$Fe$_{0.07}$)$_3$Si$_2$O$_5$(OH)$_4$	RT	1.14(1)	2.74(2)		0.39	$^{VI}Fe^{2+}$	[102]
chrysotile		0.38(3)	1.08(1)		0.29	$^{VI}Fe^{3+}$	
		0.27(4)	0.30(3)		0.32	$^{VI}Fe^{3+}$	
(Mg$_{0.13}$Fe$_{0.87}$)$_3$Si$_2$O$_5$(OH)$_4$	RT	1.15(2)	2.79(1)		0.30	$^{VI}Fe^{2+}$	[102]
lizardite		1.16(1)	2.21(2)		0.52	$^{VI}Fe^{2+}$	
		0.36(4)	0.70(5)		0.18	$^{VI}Fe^{3+}$	
Sillimanite							
(Al$_{0.98}$Fe$_{0.02}$)$_2$SiO$_5$	RT	0.38(2)	1.11(3)		0.79	$^{VI}Fe^{3+}$	[101]
		0.16(50)	0.5(10)		0.21	$^{IV}Fe^{3+}$	
Smectite minerals							
Ca$_{0.2}$XSi$_{3.6}$Al$_{0.4}$O$_{10}$(OH)$_2$	RT	0.37(1)	0.23(1)		0.65	$^{VI}Fe^{3+}$	[110]
X=Fe$_{1.9}$Mg$_{0.1}$		0.37(1)	0.65(1)		0.35	$^{VI}Fe^{3+}$	
Ca$_{0.2}$XSi$_{3.5}$Al$_{0.3}$O$_{10}$(OH)$_2$	RT	0.24(1)	0.54(1)		0.09	$^{IV}Fe^{3+}$	[110]
X=Fe$_{1.4}$Mg$_{1.2}$		0.35(1)	0.81(1)		0.55	$^{VI}Fe^{3+}$	
		0.37(1)	1.35(1)		0.30	$^{VI}Fe^{3+}$	
		1.13(1)	2.65(1)		0.06	$^{VI}Fe^{2+}$	
Spinel structure							
γ-Fe$_2$SiO$_4$	RT	1.09(1)	2.62(1)		0.93	$^{VI}Fe^{2+}$	[92]
quenched from 8 GPa,1000°C		0.18(5)	0.37(5)		0.07	Fe^{3+}	
γ-Mg$_{0.85}$Fe$_{0.15}$SiO$_4$	RT	1.05(1)	2.78(1)		0.94	$^{VI}Fe^{2+}$	[92]
quenched from 18 GPa,1700°C		0.27(5)			0.06	Fe^{3+}	
Staurolite							
XAl$_9$Si$_4$O$_{20}$(OH)$_2$	RT	0.96(1)	2.50(1)		0.23	$^{IV}Fe^{2+}$	[5]
X=Fe$_{1.1}$Mg$_{0.5}$Zn$_{0.3}$Ti$_{0.1}$		0.98(1)	2.13(1)		0.40	$^{IV}Fe^{2+}$	
		0.92(1)	1.17(1)		0.31	$^{IV}Fe^{2+}$	
		0.60(1)	0.83(1)		0.06	$^{VI}Fe^{3+}$	
Talc							
(Mg$_{0.9}$Fe$_{0.1}$)$_3$Si$_4$O$_{10}$(OH)$_2$	RT	1.15(1)	2.63(1)			$^{VI}Fe^{2+}$	[90]
Titanite							
CaTi$_{0.9}$Fe$_{0.1}$SiO$_5$	RT	0.21(1)	1.25(1)		0.14	$^{IV}Fe^{3+}$	[61]
		0.35(1)	0.96(1)		0.55	$^{VI}Fe^{3+}$	
		0.48(1)	0.81(1)		0.31	$^{VI}Fe^{3+}$	
Wadsleyite							
β-(Mg$_{0.84}$Fe$_{0.16}$)$_2$SiO$_4$	RT	1.06(1)	2.76(1)		0.19	$^{VI}Fe^{2+}$	[92]
quenched from 15.5 GPa,1800°C		1.09(3)	2.29(3)		0.77	$^{VI}Fe^{2+}$	
		0.27(5)	0.37(5)		0.04	Fe^{3+}	
Yoderite (Mg$_2$Al$_{3.6}$Fe$_{0.3}$Mn$_{0.1}$)$_6$Al$_2$Si$_4$O$_{18}$(OH)$_2$							
	RT	0.36(1)	1.00(1)			$^{V}Fe^{3+}/^{VI}Fe^{3+}$	[2]

[a] see [45] for a detailed discussion of calcic amphibole data

[b] spectra are more realistically described with hyperfine parameter distributions, see [98]

[c] see [59] for a compilation of data

[d] spectral data were fitted using a relaxation model

[e] site distribution depends strongly on thermal history, see e.g. [112]

[f] small amount of additional component present

TABLE 7. ^{57}Fe Mössbauer data for selected oxide and hydroxide minerals

Absorber	T	CS(Fe) mm s^{-1}	ΔE_Q mm s^{-1}	H Tesla	I	site	Ref
Akaganéite							
β FeOOH	RT	0.39(1)	0.95(1)		0.39	$^{VI}Fe^{3+}$	[84]
		0.38(1)	0.55(1)		0.61	$^{VI}Fe^{3+}$	
Feroxyhite							
δ'-FeOOH	RT	0.4(1)	- 0.1(1)	44.8(5)	0.60	$^{VI}Fe^{3+}$	[35]
		0.4(1)	+1.1(1)	39.3(5)	0.40	$^{IV}Fe^{3+}$	
Ferrihydrite							
$Fe_5HO_8 \cdot 4H_2O$ [a]	RT	0.35(1)	0.62(1)			$^{VI}Fe^{3+}$	[86]
Goethite							
α-FeOOH[b]	RT	0.35(1)	- 0.3(1)	38.4(5)		$^{VI}Fe^{3+}$	[42]
Haematite							
α-Fe_2O_3[c]	RT	0.38(5)	- 0.21(5)	52.1(5)		$^{VI}Fe^{3+}$	[122]
Ilmenite							
$FeTiO_3$	RT	1.07(1)	0.70(1)			$^{VI}Fe^{2+}$	[44]
Lepidochrocite							
γ-FeOOH	RT	0.30(1)	0.55(1)			$^{VI}Fe^{3+}$	[65]
Magnesiowüstite	RT	1.06(1)	0.53(1)			$^{VI}Fe^{2+}$	[72]
$Mg_{0.8}Fe_{0.2}O$							
Maghemite							
γ-Fe_2O_3	RT	0.22(5)	+0.08(5)	50.2(1)	0.33	$^{IV}Fe^{3+}$	[11]
		0.37(5)	+0.02(4)	50.5(1)	0.67	$^{VI}Fe^{3+}$	
Perovskite							
$Ca_{1.1}Ti_{0.8}Fe_{0.1}O_3$	RT	0.35(5)	0.34(5)			$^{VI}Fe^{3+}$	[83]
Pseudobrookite							
Fe_2TiO_5	RT	0.37(1)	0.52(1)		0.54	$^{VI}Fe^{3+}$	[29]
		0.37(1)	0.90(1)		0.46	$^{VI}Fe^{3+}$	
Spinel structure							
Fe_3O_4	310 K	0.63(1)	0.05(10)	45.7(1)	0.46	$^{VI}Fe^{2.5+}$	[56]
		0.63(1)	0.05(10)	44.6(1)	0.15	$^{VI}Fe^{2.5+}$	
		0.27(1)		48.9(1)	0.39	$^{IV}Fe^{3+}$	
$FeCr_2O_4$	RT	0.90(1)				$^{IV}Fe^{2+}$	[93]
$FeAl_2O_4$	RT	0.91(1)	1.57(1)			$^{IV}Fe^{2+}$	[93]
$ZnFe_2O_4$	RT	0.33(1)	0.41(1)			$^{VI}Fe^{3+}$	[78]
$MgFe_2O_4$	RT	0.37(1)		51.0(2)	0.36	$^{IV}Fe^{3+}$	[91]
quenched from 1000°C		0.48(1)		52.6(2)	0.64	$^{VI}Fe^{3+}$	
$Zn_{0.7}Mg_{0.15}Fe_{0.15}Al_2O_4$	RT	0.29(2)	0.78(2)		0.11	$^{VI}Fe^{3+}$	[128]
		0.92(2)	0.23(2)		0.76	$^{IV}Fe^{2+}$	
		0.89(2)	0.81(2)		0.13	$^{IV}Fe^{2+}$	
Fe_2TiO_4[d]	RT	0.83(1)	1.91(8)			Fe^{2+}	[77]
Tapiolite							
$FeTa_2O_6$	RT	1.11(2)	3.15(5)			$^{VI}Fe^{2+}$	[106]
Wüstite							
$Fe_{0.95}O$ [e]	RT	1.00(1)	0.22(1)		0.43	$^{VI}Fe^{2+}$	[79]
		0.93(1)	0.42(1)		0.48	$^{VI}Fe^{2+}$	
		0.60(5)			0.09	$^{IV}Fe^{3+}$	

[a] spectra data were fitted with a distribution model

[b] see [87] for a discussion of the effect of Al substitution and varying crystal size

[c] see [88] for a discussion of the effect of Al substitution and varying crystal size

[d] octahedral and tetrahedral sites in Fe_2TiO_4 have been distinguished using external magnetic fields [123]

[e] there is considerable controversy over fitting models, see [75] for a review

TABLE 8. ^{57}Fe Mössbauer data for selected sulphide, selenide and telluride minerals

Absorber	T	CS(Fe) mm s^{-1}	ΔE_Q mm s^{-1}	H Tesla	I	site	Ref
Arsenopyrite							
FeAsS	RT	0.26(3)	1.15(3)			$^{VI}Fe^{2+}$	[64]
Berthierite							
FeSb$_2$S$_4$	RT	0.83(2)	2.69(2)			$^{VI}Fe^{2+}$	[17]
Bornite							
Cu$_5$FeS$_4$	RT	0.39(1)	0.22(1)			$^{IV}Fe^{3+}$	[27]
Chalcopyrite							
CuFeS$_2$	RT	0.25(3)		35.7(5)		$^{IV}Fe^{3+}$	[63]
Cobaltite							
(Co,Fe)AsS	RT	0.26(1)	0.45(1)			$^{VI}Fe^{2+}$	[124]
Cubanite							
CuFe$_2$S$_3$ (ortho)	RT	0.43(1)	1.2	33.1(5)		$^{IV}Fe^{2.5+}$	[63]
CuFe$_2$S$_3$ (cubic)	RT	0.72(1)	0.20(1)		0.46	$^{IV}Fe^{2+}$	[49]
		0.22(1)			0.54	$^{IV}Fe^{3+}$	
Löllingite structure							
FeAs$_2$	RT	0.30(1)	1.65(1)			$^{VI}Fe^{2+}$	[64]
FeSb$_2$	RT	0.45(1)	1.28(2)			$^{VI}Fe^{2+}$	[119]
Marcasite structure							
FeS$_2$	RT	0.27(1)	0.51(1)			$^{VI}Fe^{2+}$	[119]
FeSe$_2$	RT	0.39(1)	0.58(1)			$^{VI}Fe^{2+}$	[119]
FeTe$_2$	RT	0.47(1)	0.50(1)			$^{VI}Fe^{2+}$	[119]
Pentlandite							
Fe$_{4.2}$Co$_{0.1}$Ni$_{4.7}$S$_8$	RT	0.36(1)	0.32(1)		0.82	$^{IV}Fe^{2.5+}$?	[69]
		0.65(1)			0.18	$^{VI}Fe^{2+}$	
Pyrite							
FeS$_2$	RT	0.31(1)	0.61(1)			$^{VI}Fe^{2+}$	[119]
Pyrrhotite							
Fe$_{0.89}$S	285 K	0.69(1)	- 0.48	30.2(5)	0.41	$^{VI}Fe^{2+}$	[70]
		0.68(1)	- 0.59	25.7(5)	0.36	$^{VI}Fe^{2+}$	
		0.67(1)	- 0.45	23.1(5)	0.23	$^{VI}Fe^{2+}$	
Sphalerite							
Zn$_{0.95}$Fe$_{0.05}$S	RT	0.67(3)			0.54	$^{IV}Fe^{2+}$	[43]
		0.67(3)	0.60(10)		0.46	$^{IV}Fe^{2+}$	
Stannite							
Cu$_2$FeSnS$_4$	RT	0.57(1)	2.90(1)			$^{IV}Fe^{2+}$	[49]
Sternbergite							
AgFe$_2$S$_3$	RT	0.39(2)	1.07(2)	27.8(2)		$^{IV}Fe^{2.5+}$	[129]
Tetrahedrite							
Cu$_{8.9}$Ag$_2$Fe$_{1.1}$Sb$_4$S$_{12.8}$	RT	0.58(1)	2.28(1)		0.60	$^{IV}Fe^{2+}$	[24]
		0.37(1)	0.33(1)		0.40	$^{III}Fe^{3+}$	
Thiospinel minerals							
FeNi$_2$S$_4$	RT	0.29(1)	0.54(1)			$^{VI}Fe^{2+}$	[125]
FeCr$_2$S$_4$	RT	0.59(1)	0.00(2)		0.93	$^{IV}Fe^{2+}$	[100]
		0.58(1)	0.70(1)		0.06^a	$^{IV}Fe^{2+}$	
Fe$_3$S$_4$	RT	0.55(1)		31.0(5)	0.66	$^{VI}Fe^{2.5+}$?	[115]
		0.26(1)		31.1(5)	0.34	$^{IV}Fe^{3+}$	

TABLE 8. (continued)

Absorber	T	CS(Fe) mm s^{-1}	ΔE_Q mm s^{-1}	H Tesla	I	site	Ref
FeIn$_2$S$_4$	RT	0.88(1)	3.27(1)			^{VI}Fe^{2+}	[49]
Co$_{2.9}$Fe$_{0.1}$S$_4$	RT	0.25(1)	0.25(1)		0.45	^{VI}Fe^{3+}	[130]
		0.23(1)			0.55	^{IV}Fe^{3+}	
Troilite							
FeS	RT	0.76(4)	- 0.88	31.0(5)		^{VI}Fe^{2+}	[55]
Wurtzite							
Zn$_{0.95}$Fe$_{0.05}$S	RT	0.69(3)			0.54	^{IV}Fe^{2+}	[43]
		0.69(3)	0.56(10)		0.46	^{IV}Fe^{2+}	

[a] small amount of additional component present

TABLE 9. ^{57}Fe Mössbauer data for selected carbonate, phosphate, sulphate and tungstate minerals

Absorber	T	CS(Fe) mm s^{-1}	ΔE_Q mm s^{-1}	H Tesla	I	site	Ref
Siderite							
FeCO$_3$	RT	1.24(1)	1.80(1)			^{VI}Fe^{2+}	[48]
Ankerite							
Ca$_{1.1}$X(CO$_3$)$_2$							
X=Mg$_{0.5}$Fe$_{0.3}$Mn$_{0.1}$	RT	1.25(1)	1.48(3)			^{VI}Fe^{2+}	[34]
Ferberite							
FeWO$_4$	RT	1.11(2)	1.49(3)			^{VI}Fe^{2+}	[51]
Jarosite							
KFe$_3$(SO$_4$)$_2$(OH)$_6$	RT	0.40(5)	1.15(5)			^{VI}Fe^{3+}	[62]
Wolframite							
Fe$_{0.5}$Mn$_{0.5}$WO$_4$	RT	1.13(2)	1.53(3)			^{VI}Fe^{2+}	[51]
Vivianite							
Fe$_3$(PO$_4$)$_2$·8H$_2$O	RT	1.21(1)	2.98(1)		0.22	^{VI}Fe^{2+}	[10]
		1.18(1)	2.45(1)		0.21	^{VI}Fe^{2+}	
		0.38(1)	1.06(1)		0.38	^{VI}Fe^{3+}	
		0.40(1)	0.61(1)		0.19	^{VI}Fe^{3+}	

TABLE 10. ^{57}Fe Mössbauer data for other minerals

Absorber	T	CS(Fe) mm s^{-1}	ΔE_Q mm s^{-1}	H Tesla	I	site	Ref
Iron							
α-Fe	298 K	0.00	+0.001(2)	33.04(3)		Fe^0	[126]
Kamacite							
~Fe$_{0.95}$Ni$_{0.05}$	RT	0.02(1)		33.8(7)		Fe^0	[30]
Taenite							
Fe$_{1-x}$Ni$_x$ $x < 0.3$	RT	- 0.08(1)	0.40(2)			Fe^0	[4]
FeNi	RT	0.02(1)		28.9(2)		Fe^0	[4]

TABLE 11. ^{119}Sn Mössbauer data for selected minerals

Absorber	T	CS(SnO$_2$) mm s^{-1}	ΔE_Q mm s^{-1}	I	site	Ref
Berndtite, SnS$_2$	RT	1.03(5)			^{VI}Sn^{4+}	[8]
Cassiterite, SnO$_2$	RT	0.00	0.40(5)		^{VI}Sn^{4+}	[132]
Garnet structure						
Ca$_3$Fe$_{1.8}$Al$_{0.1}$Sn$_{0.1}$Si$_3$O$_{12}$	RT	- 0.14(5)	0.42(5)		^{VI}Sn^{4+}	[9]
YCa$_2$Sn$_2$Fe$_3$O$_{12}$	RT	0.07(5)	0.42(5)		^{VI}Sn^{4+}	[9]
Herzenbergite, SnS	RT	3.23(3)	0.85(5)		^{VI}Sn^{2+}	[8]
Incaite, Pb$_{3.5}$FeSn$_4$Sb$_2$S$_{13.5}$	RT	1.13(4)		0.66	^{VI}Sn^{4+}	[6]
		3.29(5)	0.98(5)	0.34	Sn^{2+}	
Malayite, CaSnSiO$_5$	RT	- 0.07(2)	1.32(4)		^{VI}Sn^{4+}	[104]
Mawsonite, Cu$_6$Fe$_2$SnS$_8$	RT	1.46(5)	0.00(5)		^{IV}Sn^{4+}	[132]
Ottemannite, Sn$_2$S$_3$	RT	3.48(5)	0.95(5)	0.29	Sn^{2+}	[8]
		1.10(5)		0.71	^{VI}Sn^{4+}	
Romarchite, SnO	RT	2.64(2)	1.31(1)		^{VI}Sn^{2+}	[60]
Spinel structure						
Co$_2$SnO$_4$	RT	0.30(4)	0.80(8)		^{VI}Sn^{4+}	[52]
Mn$_2$SnO$_4$	RT	0.25(4)	0.75(8)		^{VI}Sn^{4+}	[52]
Zn$_2$SnO$_4$	RT	0.24(4)	0.75(8)		^{VI}Sn^{4+}	[52]
Mg$_2$SnO$_4$	RT	0.12(4)	1.20(8)		^{VI}Sn^{4+}	[52]
Stannite, Cu$_2$Fe$_{0.9}$Zn$_{0.1}$SnS$_4$	RT	1.45(5)	0.00(5)		^{IV}Sn^{4+}	[132]
Stannoidite, Cu$_8$(Fe$_{0.8}$Zn$_{0.2}$)$_3$Sn$_2$S$_{12}$	RT	1.48(5)	0.00(5)		^{IV}Sn^{4+}	[132]
Tin						
α-Sn	300 K	2.02(2)			Sn^0	[118]
β-Sn	300 K	2.55(1)			Sn^0	[118]

Acknowledgments. I am grateful to G. Amthauer, H. Annersten, J. Cashion, E. Murad, G. Rossman and F. Seifert for valuable comments on the manuscript.

REFERENCES

1. Abs-Wurmbach, I., Langer, K., Seifert, F. and Tillmanns, E., The crystal chemistry of (Mn^{3+}, Fe^{3+})-substituted andalusites (viridines and kannonaite), (Al$_{1-x-y}$Mn$^{3+}_x$Fe$^{3+}_y$)$_2$(O|SiO$_4$): crystal structure refinements, Mössbauer, and polarized optical absorption spectra. *Z. Krist., 155*, 81-113, 1981.

2. Abu-Eid, R. M., Langer, K. and Seifert, F., Optical absorption and Mössbauer spectra of purple and green yoderite, a kyanite-related mineral. *Phys. Chem. Minerals, 3*, 271-289, 1978.

3. Akasaka, M., ^{57}Fe Mössbauer study of clinopyroxenes in the join CaFe^{3+}AlSiO$_6$-CaTiAl$_2$O$_6$. *Phys. Chem. Minerals, 9*, 205-211, 1983.

4. Albertsen, J. F., Aydin, M. and Knudsen, J. M., Mössbauer effect studies of taenite lamellae of an iron meteorite Cape York (III.A). *Phys. Scripta, 17*, 467-472, 1978.

5. Alexander, V. D., Iron distribution in staurolite at room and low temperatures. *Amer. Mineral., 74*, 610-619, 1989.

6. Amthauer, G., Crystal chemistry and valencies of iron, antimony, and tin in franckeites. *Neues Jahr. Mineral., Abhand., 153*, 272-278, 1986.

7. Amthauer, G., Annersten, H. and Hafner, S. S., The Mössbauer spectrum of ^{57}Fe in silicate garnets. *Z. Krist., 143*, 14-55, 1976.

8. Amthauer, G., Fenner, J., Hafner, S., Holzapfel, W. B. and Keller, R., Effect of pressure on resistivity and Mössbauer spectra of the mixed valence compound Sn$_2$S$_3$. *J. Chem. Phys., 70*, 4837-4842, 1979.

9. Amthauer, G., McIver, J. R. and

Viljoen, E. A., ^{57}Fe and ^{119}Sn Mössbauer studies of natural tin-bearing garnets. *Phys. Chem. Minerals, 4,* 235-244, 1979.

10. Amthauer, G. and Rossman, G. R., Mixed valence of iron in minerals with cation clusters. *Phys. Chem. Minerals, 11,* 37-51, 1984.

11. Annersten, H. and Hafner, S. S., Vacancy distribution in synthetic spinels of the series Fe_3O_4-γ-Fe_2O_3. *Z. Krist., 137,* 321-340, 1973.

12. Annersten, H., Olesch, M. and Seifert, F., Ferric iron in orthopyroxene: A Mössbauer study. *Lithos, 11,* 301-310, 1978.

13. Bancroft G. M., *Mössbauer Spectroscopy. An Introduction for Inorganic Chemists and Geochemists.* McGraw Hill, New York, 1973.

14. Bancroft, G. M., Quantitative site population in silicate minerals by the Mössbauer effect. *Chem. Geol., 5,* 255-258, 1969.

15. Bancroft, G. M., Maddock, A. G. and Burns, R. G., Applications of the Mössbauer effect to silicate mineralogy – I. Iron silicates of known crystal structure. *Geochim. Cosmochim. Acta, 31,* 2219-2246, 1967.

16. Bara, J. J. and Bogacz, B. F., Geometric effects in Mössbauer transmission experiments. *Möss. Effect Ref. Data J., 3,* 154-163, 1980.

17. Bonville, P., Garcin, C., Gerard, A., Imbert, P. and Wintenberger, M., ^{57}Fe Mössbauer absorption study in berthierite ($FeSb_2S_4$). *Hyper. Inter., 52,* 279-290, 1989.

18. Bowman, J.D., Kankelheit, E., Kaufmann, E.N. and Persson, B., Granular Mössbauer absorbers. *Nucl. Instr. Meth., 50,* 13-21, 1967.

19. Brown, F. F. and Pritchard, A. M., The Mössbauer spectrum of iron orthoclase. *Earth Planet. Sci. Lett., 5,* 259-260, 1969.

20. Burns, R. G., Intervalence transitions in mixed-valence minerals of iron and titanium. *Ann. Rev. Earth Planet. Sci., 9,* 345-383, 1981.

21. Burns, R. G. and Dyar, M. D., Crystal chemistry and Mössbauer spectra of babingtonite. *Amer. Mineral., 76,* 892-899, 1991.

22. Burns, R. G. and Solberg, T. C., ^{57}Fe-bearing oxide, silicate, and aluminosilicate minerals. In *Spectroscopic Characterization of Minerals and their Surfaces,* L.M. Coyne, S.W.S. McKeever, D.F. Blake (eds.), ACS Symposium Series Vol. 415, American Chemical Society, Washington DC, pp. 262-283, 1990.

23. Campbell, S. J., Introduction to Mössbauer studies of magnetic materials. *Austr. J. Phys., 37,* 429-447, 1984.

24. Charnock, J. M., Garner, C. D., Pattrick, R. A. D. and Vaughn, D. J., EXAFS and Mössbauer spectroscopic study of Fe-bearing tetrahedrites. *Min. Mag., 53,* 193-199, 1989.

25. Coey, J. M. D., Mössbauer spectroscopy of silicate minerals. In *Chemical Applications of Mössbauer Spectroscopy, Vol. 1,* G.L. Long (ed.). Plenum Press, New York, pp. 443-509, 1984.

26. Coey, J. M. D., Chukhrov, F. K. and Zvyagin, B. B., Cation distribution, Mössbauer spectra and magnetic properties of ferripyrophyllite. *Clays & Clay Miner., 32,* 198-204, 1984.

27. Collins, M. F., Longworth, G. and Townsend, M. G., Magnetic structure of bornite, Cu_5FeS_4. *Can. J. Phys., 59,* 535-539, 1981.

28. Crespo, D. and Parellada, J., Geometrical effects on line shape and background in experimental Mössbauer spectra. *Hyper. Inter., 29,* 1539-1542, 1986.

29. Cruz, J. M. R., Morais, P. C. and Skeff Neto, K., On the spin-glass transition in pseudobrookite. *Phys. Lett. A, 116,* 45-47, 1986.

30. Danon, J., Scorzelli, R. B., Sousa Azevedo, I. and Christophe-Michel-Lévy, M., Iron-nickel superstructure in metal particles of chondrites. *Nature, 281,* 469-471, 1979.

31. Daniels, J.M., A note on the criteria for fitting functions to measured points. *Can. J. Phys., 59,* 182-184, 1981.

32. DeGrave, E. and Van Alboom, A., Evaluation of ferrous and ferric Mössbauer fractions. *Phys. Chem. Minerals, 18,* 337-342, 1991.

33. De Grave, E., Vandenbruwaene, J. and Van Bockstael, M., ^{57}Fe Mössbauer spectroscopic analysis of chlorite. *Phys. Chem. Minerals, 15,* 173-180, 1987.

34. De Grave, E. and Vochten, R., An ^{57}Fe Mössbauer study of ankerite. *Phys. Chem. Minerals, 12,* 108-113, 1985.

35. Dészi, I., Keszthelyi, L., Kulgawczuk, D., Molnár, B. and Eissa, N. A., Mössbauer study of β- and δ-FeOOH and their disintegration products. *Phys. Stat. Sol., 22,* 617-629, 1967.

36. Dollase, W. A., Mössbauer spectra and iron-distribution in the epidote-group minerals. *Z. Krist., 138,* 41-63, 1973.

37. Dollase, W. A., Statistical limitations of Mössbauer spectral fitting. *Amer. Mineral., 60,* 257-264, 1975.

38. Dowty, E. and Lindsley, D. H., Mössbauer spectra of synthetic hedenbergite-ferrosilite pyroxenes. *Amer. Mineral., 58,* 850-868, 1973.

39. Dyar, M. D. and Burns, R. G., Mössbauer spectral study of ferruginous one-layer trioctahedral micas. *Amer. Mineral., 71,* 955-965, 1986.

40. Ernst, W. G. and Wai, C. M., Mössbauer, infrared, x-ray and optical study of cation ordering and dehydrogenation in natural and heat-treated sodic amphiboles. *Amer. Mineral., 55,* 1226-1258, 1970.

41. Finch, J., Gainsford, A. R. and Tennant, W. C., Polarized optical absorption and ^{57}Fe Mössbauer study of pegmatitic muscovite. *Amer. Mineral., 67,* 59-68, 1982.

42. Forsyth, J. B., Hedley, I. G. and Johnson, C. E., The magnetic structure and hyperfine field of

goethite (α-FeOOH). *J. Phys. C.*, *Ser. 2*, *1*, 179-188, 1968.

43. Gerard, A., Imbert, P., Prange, H., Varret, F. and Wintenberger, M., Fe^{2+} impurities, isolated and in pairs, in ZnS and CdS studied by the Mössbauer effect. *J. Phys. Chem. Solids*, *32*, 2091-2100, 1971.

44. Gibb, T. C., Greenwood, N. N. and Twist, W., The Mössbauer spectra of natural ilmenites. *J. Inorg. Nucl. Chem.*, *31*, 947-954, 1969.

45. Goldman, D. S., A reevaluation of the Mössbauer spectroscopy of calcic amphiboles. *Amer. Mineral.*, *64*, 109-118, 1979.

46. Goldman, D. S. and Rossman, G. R., The site distribution of iron and anomalous biaxiality in osumilite. *Amer. Mineral.*, *63*, 490-498, 1978.

47. Goldman, D. S., Rossman, G. R. and Dollase, W. A., Channel constituents in cordierite. *Amer. Mineral.*, *62*, 1144-1157, 1977.

48. Grant, R. W., Wiedersich, H., Muir, J. A. H., Gonser, U. and Delgass, W. N., Sign of the nuclear quadrupole coupling constants in some ionic ferrous compounds. *J. Chem. Phys.*, *45*, 1015-1019, 1966.

49. Greenwood, N. N. and Whitfield, H. J., Mössbauer effect studies of cubanite ($CuFe_2S_3$) and related iron sulphides. *J. Chem. Soc. A*, 1697-1699, 1968.

50. Guettinger, T.W. and Williamson, D.L., Quantitative Mössbauer spectroscopy of nonuniform absorbers: Basic concepts. *Nucl. Instr. Meth. Phys. Res.*, *B42*, 268-276, 1989.

51. Guillen, R., Regnard, J. R. and Amossé, J., Mössbauer study of natural wolframites. *Phys. Chem. Minerals*, *8*, 83-86, 1982.

52. Gupta, M. P. and Mathur, H. B., Mössbauer spectra of oxidic spinels containing Sn^{4+} ion. *J. Phys. Chem. Solids*, *29*, 1479-1481, 1968.

53. Hafner, S. S. and Ghose, S., Iron and magnesium distribution in cummingtonites $(Fe,Mg)_7Si_8O_{22}(OH)_2$. *Z. Krist.*, *133*, 301-326, 1971.

54. Hafner, S. S. and Huckenholz, H. G., Mössbauer spectrum of synthetic ferridiopside. *Nature*, *233*, 9-11, 1971.

55. Hafner, S. S. and Kalvius, G. M., The Mössbauer resonance of ^{57}Fe in troilite (FeS) and pyrrhotite ($Fe_{0.88}S$). *Z. Krist.*, *123*, 443-458, 1966.

56. Häggström, L., Annersten, H., Ericsson, T., Wäppling, R., Karner, W. and Bjarman, S., Magnetic dipolar and electric quadrupolar effects on the Mössbauer spectra of magnetite above the Verwey transition. *Hyper. Inter.*, *5*, 201-214, 1978.

57. Hålenius, U., Annersten, H. and Langer, K., Spectroscopic studies on natural chloritoids. *Phys. Chem. Minerals*, *7*, 117-123, 1981.

58. Hawthorne, F. C. and Waychunas, G. A., Spectrum-fitting methods. In *Spectroscopic Methods in Mineralogy and Geology*, F.C. Hawthorne (ed.), Rev. Mineral. Vol. 18, Mineralogical Society of America, pp. 63-98, 1988.

59. Heller-Kallai, L. and Rozenson, I., The use of Mössbauer spectroscopy of iron in clay mineralogy. *Phys. Chem. Minerals*, *7*, 223-238, 1981.

60. Herber, R. H., Mössbauer lattice temperature of tetragonal (P4/nmm) SnO. *Phys. Rev. B*, *27*, 4013-4017, 1983.

61. Holényi, K. and Annersten, H., Iron in titanite: A Mössbauer spectroscopic study. *Can. Mineral.*, *25*, 429-433, 1987.

62. Hrynkiewicz, A. S., Kubiz, J. and Kulgawczuk, D. S., Quadrupole splitting of the 14.4 keV gamma line of ^{57}Fe in iron sulphates of the jarosite group. *J. Inorg. Nucl. Chem.*, *27*, 2513-2517, 1965.

63. Imbert, P. and Wintenberger, M., Étude des propriétés magnétiques et des spectres d'absorption par effet Mössbauer de la cubanite et de la sternbergite. *Bull. Soc. Fr. Mineral. Cristallogr.*, *90*, 299-303, 1967.

64. Ioffe, P. A., Tsemekhman, Parshukova, L. N. and Bobkovskii, A. G., The chemical state of the iron atoms in FeS_2, FeAsS, and $FeAs_2$. *Russ. J. Inorg. Chem.*, *30*, 1566-1568, 1985.

65. Johnson, C. E., Antiferromagnetism of γ-FeOOH: A Mössbauer effect study. *J. Phys. C, Ser. 2*, *2*, 1996-2002, 1969.

66. Johnson, C. E., The Mössbauer effect and magnetic phase transitions. *Hyper. Inter.*, *49*, 19-42, 1989.

67. Joswig, W., Amthauer, G. and Takéuchi, Y., Neutron-diffraction and Mössbauer spectroscopic study of clintonite (xanthophyllite). *Amer. Mineral.*, *71*, 1194-1197, 1986.

68. Kan, X. and Coey, J. M. D., Mössbauer spectra, magnetic and electrical properties of laihunite, a mixed valence iron olivine mineral. *Amer. Mineral.*, *70*, 576-580, 1985.

69. Knop, O., Huang, C., Reid, K. I. G., Carlow, J. S. and Woodhams, F. W. D., Chalcogenides of the transition elements. X. X-ray, neutron, Mössbauer, and magnetic studies of pentlandite and the π phases (Fe,Co,Ni,S), Co_8MS_8, and $Fe_4Ni_4MS_8$ (M=Ru,Rh,Pd). *J. Solid State Chem.*, *16*, 97-116, 1976.

70. Kruse, O., Mössbauer and X-ray study of the effects of vacancy concentration in synthetic hexagonal pyrrhotites. *Amer. Mineral.*, *75*, 755-763, 1990.

71. Kündig, W., Evaluation of Mössbauer spectra for ^{57}Fe. *Nucl. Instr. Meth.*, *48*, 219-228, 1967.

72. Kurash, V. V., Goldanskii, V. I., Malysheva, T. V., Urusov, V. S., Kuznetsov, L. M. and Moskovkina, L. A., Mössbauer effect study of the solid solutions MgO-$Fe_{1-x}O$. *Inorg. Mater.*, *8*, 1395-1400, 1972.

73. Litterst, F. J. and Amthauer, G., Electron delocalization in ilvaite, a reinterpretation of its ^{57}Fe

Mössbauer spectrum. *Phys. Chem. Minerals, 10*, 250-255, 1984.

74. Long, G. L., Cranshaw, T. E. and Longworth, G., The ideal Mössbauer effect absorber thickness. *Möss. Effect Ref. Data J., 6*, 42-49, 1983.

75. Long, G. L. and Grandjean, F., Mössbauer effect, magnetic and structural studies of wüstite, $Fe_{1-x}O$. *Adv. Solid State Chem., 2*, 187-221, 1991.

76. Longworth, G., Townsend, M. G., Provencher, R. and Kodama, H., Magnetic interaction in biotite and oxidised biotites. *Phys. Chem. Minerals, 15*, 71-77, 1987.

77. Malysheva, T. V., Polyakova, N. P. and Mishin, N. E., Mössbauer spectroscopy study of lunar soil sampled by Luna 24 space probe. *Geokhimiya*, 835-841, 1978.

78. Marshall, C. P. and Dollase, W. A., Cation arrangement in iron-zinc-chromium spinel oxides. *Amer. Mineral., 69*, 928-936, 1984.

79. McCammon, C. A. and Price, D. C., Mössbauer spectra of Fe_xO ($x >$ 0.95). *Phys. Chem. Minerals, 11*, 250-254, 1985.

80. McCammon, C. A., Rubie, D. C., Ross II, C. R., Seifert, F. and O'Neill, H. St. C., Mössbauer spectra of $^{57}Fe_{0.05}Mg_{0.95}SiO_3$ perovskite at 80 K and 298 K. *Amer. Mineral., 77*, 894-897, 1992.

81. Mössbauer, R. L., Kernresonanzfluoresent von Gammastrahlung in Ir^{191}. *Z. Phys., 151*, 124-143, 1958.

82. Mössbauer, R. L., Kernresonanzfluoresent von Gammastrahlung in Ir^{191}. *Naturwiss., 45*, 538-539, 1958.

83. Muir, I. J., Metson, J. B. and Bancroft, G. M., ^{57}Fe Mössbauer spectra of perovskite and titanite. *Can. Mineral., 22*, 689-694, 1984.

84. Murad, E., Mössbauer and X-ray data on β-FeOOH (akaganéite). *Clay Minerals, 14*, 273-283, 1979.

85. Murad, E., Application of ^{57}Fe Mössbauer spectroscopy to problems in clay mineralogy and soil science: Possibilities and limitations. In *Advances in Soil Science, Vol. 12*, B.A. Stewart (ed.), Springer-Verlag, New York, pp. 125-157, 1990.

86. Murad, E., Bowen, L. H., Long, G. L. and Quin, T. G., The influence of crystallinity on magnetic ordering in natural ferrihydrites. *Clay Minerals, 23*, 161-173, 1988.

87. Murad, E. and Schwertmann, U., The influence of aluminium substitution and crystallinity on the Mössbauer spectra of goethite. *Clay Miner., 18*, 301-312, 1983.

88. Murad, E. and Schwertmann, U., Influence of Al substitution and crystal size on the room-temperature Mössbauer spectrum of hematite. *Clays & Clay Miner., 34*, 1-6, 1986.

89. Murad, E. and Wagner, F. E., The Mössbauer spectrum of almandine. *Phys. Chem. Minerals, 14*, 264-269, 1987.

90. Noack, Y., DeCarreau, A. and Manceau, A., Spectroscopic and oxygen isotopic evidence for low and high temperature origin of talc. *Bull. Minéral., 109*, 253-263, 1986.

91. O'Neill, H. St. C., Annersten, H. and Virgo, D., The temperature dependence of the cation distribution in magnesioferrite ($MgFe_2O_4$) from powder XRD structural refinements and Mössbauer spectroscopy. *Amer. Mineral., 77*, 725-740, 1992.

92. O'Neill, H. St. C., McCammon, C. A., Canil, D., Rubie, D. C., Ross II, C. R. and Seifert, F., Mössbauer spectroscopy of mantle transition zone phases and determination of minimum Fe^{3+} content. *Amer. Mineral., 78*, 462-466, 1993.

93. Osborne, M. D., Fleet, M. E. and Bancroft, G. M., Fe^{2+}-Fe^{3+} ordering in chromite and Cr-bearing spinels. *Contrib. Mineral. Petrol., 77*, 251-255,

1981.

94. Parkin, K. M., Loeffler, B. M. and Burns, R. G., Mössbauer spectra of kyanite, aquamarine, and cordierite showing intervalence charge transfer. *Phys. Chem. Minerals, 1*, 301-311, 1977.

95. Pfannes, H.D. and Fischer, H., The texture problem in Mössbauer spectroscopy. *Appl. Phys., 13*, 317-325, 1977.

96. Pfannes, H. D. and Gonser, U., Goldanskii-Karyagin effect versus preferred orientations (texture). *Appl. Phys., 1*, 93-102, 1973.

97. Rancourt, D. G., Accurate site populations from Mössbauer spectroscopy. *Nucl. Instr. Meth. Phys. Res., B44*, 199-210, 1989.

98. Rancourt, D. G., Dang, M. Z. and Lalonde, A. E., Mössbauer spectroscopy of tetrahedral Fe^{3+} in tri-octahedral micas. *Amer. Mineral., 77*, 34-43, 1992.

99. Rancourt, D.G., McDonald, A.M., Lalonde, A.E. and Ping, J.Y., Mössbauer absorber thicknesses for accurate site populations in Fe-bearing minerals. *Amer. Mineral., 78*, 1-7, 1993.

100. Riedel, E. and Karl, R., Mössbauer studies of thiospinels. IV. The system $FeCr_2S_4$-Fe_3S_4. *J. Solid State Chem., 38*, 48-54, 1981.

101. Rossman, G. R., Grew, E. S. and Dollase, W. A., The colors of sillimanite. *Amer. Mineral., 67*, 749-761, 1982.

102. Rozenson, I., Bauminger, E. R. and Heller-Kallai, L., Mössbauer spectra of iron in 1:1 phyllosilicates. *Amer. Mineral., 64*, 893-901, 1979.

103. Ruby, S.L. Why MISFIT when you have χ^2? In *Mössbauer Effect Methodology, Vol. 8*, I.J. Gruverman and C.W. Seidel (eds.), Plenum Press, New York, pp. 263-276.

104. Sanghani, D. V., Abrams, G. R. and Smith, P. J., A structural investigation of some tin-based coloured ceramic pigments. *Trans. J. Br. Ceram. Soc., 80*, 210-214, 1981.

105. Sawicki, J. A. and Sawicka, B. D.,

Experimental techniques for conversion electron Mössbauer spectroscopy. *Hyper. Inter.*, *13*, 199-219, 1983.

106. Schmidbauer, E. and Lebküchner-Neugebauer, J., ^{57}Fe Mössbauer study on compositions of the series $Fe^{3+}TaO_4$-$Fe^{2+}Ta_2O_6$. *Phys. Chem. Minerals*, *15*, 196-200, 1987.

107. Seifert, F., Compositional dependence of the hyperfine interaction of ^{57}Fe in anthophyllite. *Phys. Chem. Minerals*, *1*, 43-52, 1977.

108. Seifert, F., Phase transformation in minerals studied by ^{57}Fe Mössbauer spectroscopy. In *Absorption Spectroscopy in Mineralogy*, A. Mottana and F. Burragato (eds.), Elsevier, Amsterdam, pp. 145-170, 1990.

109. Seifert, F. and Olesch, M., Mössbauer spectroscopy of grandidierite, $(Mg,Fe)Al_3BSiO_9$. *Amer. Mineral.*, *62*, 547-553, 1977.

110. Sherman, D. S. and Vergo, N., Optical (diffuse reflectance) and Mössbauer spectroscopic study of nontronite and related Fe-bearing smectites. *Amer. Mineral.*, *73*, 1346-1354, 1988.

111. Shinno, I., A Mössbauer study of ferric iron in olivine. *Phys. Chem. Minerals*, *7*, 91-95, 1981.

112. Skogby, H., Order-disorder kinetics in orthopyroxenes of ophiolite origin. *Contrib. Mineral. Petrol.*, *109*, 471-478, 1992.

113. Skogby, H. and Annersten, H., Temperature dependent Mg-Fe-cation distribution in actinolite-tremolite. *Neues Jahr. Mineral.*, *Monatsh.*, *5*, 193-203, 1985.

114. Skogby, H., Annersten, H., Domeneghetti, M. C., Molin, G. M. and Tazzoli, V., Iron distribution in orthopyroxene: A comparison of Mössbauer spectroscopy and x-ray refinement results. *Eur. J. Mineral.*, *4*, 441-452, 1992.

115. Spender, M. R., Coey, J. M. D. and Morrish, A. H., The magnetic properties and Mössbauer spectra of synthetic samples of Fe_3S_4. *Can. J. Phys.*, *50*, 2313-2326, 1972.

116. Stevens, J. G., Isomer Shift Reference Scales. *Hyper. Inter.*, *13*, 221-236, 1983.

117. Stevens, J. G., Nomenclature and conventions for reporting Mössbauer spectroscopic data. In *CRC Handbook of Spectroscopy, Vol. 3*, J.W. Robinson (ed.), CRC Press, Inc., Boca Raton, USA, pp. 520-522, 1981.

118. Stevens, J. G. and Gettys, W. L., Mössbauer isomer shift reference scales. In *Mössbauer Isomer Shifts*, G.K. Shenoy and F.E. Wagner (eds.), North-Holland Publ. Co., Amsterdam, pp. 901-906, 1978.

119. Temperley, A. A. and Lefevre, H. W., The Mössbauer effect in marcasite structure iron compounds. *J. Phys. Chem. Solids*, *27*, 85-92, 1966.

120. Trooster, J. M. and Viegers, M. P. A., Effect of sample thickness on the linewidth, intensity, and absorption area. In *CRC Handbook of Spectroscopy, Vol. 3*, J.W. Robinson (ed.), CRC Press, Inc., Boca Raton, USA, pp. 465-476, 1981.

121. Ujihira, Y., Analytical applications of conversion electron Mössbauer spectrometry (CEMS). *Rev. Anal. Chem.*, *8*, 125-177, 1985.

122. Van der Woude, F., Mössbauer effect in α-Fe_2O_3. *Phys. Stat. Sol.*, *17*, 417-432, 1966.

123. Vanleerberghe, R. and Vandenberghe, R. E., Determination of the quadrupole splitting distributions of the A- and B-site ferrous ions in Fe_2TiO_4. *Hyper. Inter.*, *23*, 75-87, 1985.

124. Vaughan D. J., Craig J. R., *Mineral Chemistry of Metal Sulphides*. Cambridge University Press, Cambridge, 1978.

125. Vaughan, D. J. and Craig, J. R., The crystal chemistry of iron-nickel thiospinels. *Amer. Mineral.*, *70*, 1036-1043, 1985.

126. Violet, C. E. and Pipkorn, D. N., Mössbauer line positions and hyperfine interactions in α iron. *J. Appl. Phys.*, *42*, 4339-4342, 1971.

127. Wagner, F. E., Applications of Mössbauer scattering techniques. *J. Phys.*, *Colloq.*, *37*, 673-689, 1976.

128. Warenborgh, J. C., Annersten, H., Ericsson, T., Figueiredo, M. O. and Cabral, J. M. P., A Mössbauer study of natural gahnite spinels showing strongly temperature-dependent quadrupole splitting distributions. *Eur. J. Mineral.*, *2*, 267-271, 1990.

129. Wintenberger, M., André, G., Perrin, M., Garcin, C. and Imbert, P., Magnetic structure and Mössbauer data of sternbergite $AgFe_2S_3$, an intermediate valency Fe compound. *J. Magn. Magn. Mater.*, *87*, 123-129, 1990.

130. Woodhams, F. W. D. and Knop, O., Chalkogenides of the transition elements. XI. Mössbauer ^{57}Fe spectrum of the spinel $Co_{2.94}Fe_{0.06}S_4$ between 10 K and room temperature. *Can. J. Chem.*, *55*, 91-98, 1977.

131. Woodland, A. and O'Neill, H. St. C., Synthesis and stability of $Fe_2^{2+}Fe_2^{3+}Si_3O_{12}$ ("skiagite") garnet and phase relations with almandine-"skiagite" solid solutions. *Amer. Mineral.*, in press, 1993.

132. Yamanaka, T. and Kato, A., Mössbauer effect study of ^{57}Fe and ^{119}Sn in stannite, stannoidite, and mawsonite. *Amer. Mineral.*, *61*, 260-265, 1976.

Subject Index

349